# SHALE
**Techniques and Technology for Improving Solids Control Management**

# SHAKERS
## and DRILLING FLUID SYSTEMS

Gulf Publishing Company
Houston, Texas

# SHALE
### Techniques and Technology for Improving Solids Control Management
# SHAKERS
### and DRILLING FLUID SYSTEMS

American Association of Drilling Engineers

**Shale Shakers and Drilling Fluid Systems**

Copyright © 1999 by Gulf Publishing Company, Houston, Texas. All rights reserved. This book, or parts thereof, may not be reproduced in any form without express written permission of the publisher.

Gulf Publishing Company
Book Division
P.O. Box 2608 ☐ Houston, Texas 77252-2608

10 9 8 7 6 5 4 3 2 1

**Library of Congress Cataloging-in-Publication Data**

Shale shakers and drilling fluid systems.
    p.   cm.
   Includes bibliographical references and index.
   **ISBN 0-88415-948-5** (acid-free)
  1. Shale shakers.  2. Drilling muds.  I. Gulf Publishing Company.
TN871.27.S53  1999
622'.3381—dc21                                       99-13067
                                                                              CIP

Printed in the United States of America.

Printed on acid-free paper (∞).

# Shale Shakers
# and
# Drilling Fluid Systems

The following members of the Joint Industry Shaker Technology Committee, individually and collectively, have made contributions to this book:

| | | |
|---|---|---|
| James Andrews | Robert Lee | Robert Murphy |
| C. S. Adkins, Jr | Albert Lew | Carter Ness |
| Jason Bradley | Walter Liljestrand | John Oeffner |
| Bill Cagle | Bob Line | Bob Palmer |
| Tom Carter | Bill Love | Nace Peard |
| Roger DeSpain | Charles Marshall | Elvis Rich |
| Bob DeWolfe | Steve Matlock | Leon Robinson |
| Robert Dugal | Bob McKenzie | Tim Sneider |
| Matt Frankl | James Merrill | Ken Seyffert |
| Charles Grichar | Mark Morgan | Brad Smolen |
| Jerry Haston | Keith Morton | Wiley Steen |
| Ben Hiltl | Mike Montgomery | Grant Young |
| Michael Kargl | Ron Morrison | |
| Gordon Lawson | Bernard Murphy | |

# DISCLAIMER

For the purposes of this disclaimer, "committee" is defined as the committee of industrial experts sponsored by the American Association of Drilling Engineers who have individually and jointly written and edited the text material. The text material contained herein is defined, for the purposes of this disclaimer, as the "work."

The committee makes no warranties, express, implied, or statutory, with respect to the work, including, without limitation, any warranties of merchantability, or fitness for a particular purpose; and/or any warranties of the safety or results of the recommendations contained herein.

The committee does not guarantee results or safety. All interpretations used to create the work, and all recommendations based upon such interpretations, are opinions of a group of experts assembled to discuss the inferences from measurements and empirical relationships, and on assumptions, which inferences and assumptions are neither infallible nor necessarily the opinion of all of the individual members of the committee, and competent specialist may differ. In addition, such interpretations, recommendations, and descriptions may involve the opinion and judgment of the user of this technology. Anyone using this information has full responsibility for all actions, interpretations, recommendations, and descriptions based upon this work. The committee cannot and does not warrant the accuracy, correctness, or completeness of any action, interpretation, recommendation, or description. Under NO circumstance should any action, interpretation, recommendation, or description be relied upon as the basis for any drilling, completion, well-site activity, production, or any financial decision, or any procedure involving any risk to the safety of any drilling venture, drilling rig, or drilling crews, or any other individual. The user of this technology has full, and complete responsibility for all decisions concerning any procedure or information resulting from application of technology described in the work. Any person, company, or other entity using the technology contained in the work agrees that the committee shall have no liability to the user of this technology or to any third party for any ordinary, special, or consequential damages or losses which may arise directly, or indirectly, by reason of using the information contained in the work. Users of this technology shall protect, indemnify, hold harmless, and defend committee from any loss, cost, damage, or expense, including attorneysí fees, arising from any claim asserted against committee that is in any way associated with the matters set forth in this disclaimer.

In summary, the technology described in this work is the consensus of a group of experts, but the application of this technology must be done in a professional manner that does not risk safety of personnel or equipment. Suggestions made in this work do not relieve the user from the ultimate responsibility of applying this information in a safe manner.

# Contents

*Acknowledgments, xi*

*Preface, xiii*

## PART ONE
## *Historical Perspective*

The Evolution of Solids Separation Devices 6 •
Index to Archival *Composite Catalog* Pages 11

## PART TWO
## *Shale Shakers*

### CHAPTER ONE
### *Introduction*
### 87

Why Control Drilled Solids? 85 • Filter Cakes 87 • Plastic Viscosity 88 • Wear 88
Carrying Capacity 88 • Drilling Fluid Costs 89 • Waste Management 90

### CHAPTER TWO
### *The Role of Shale Shakers*
### 91

How a Shale Shaker Screens Fluid 92 • Shaker Description 93
Shale Shaker Limits 94 • Shaker Development Summary 96

### CHAPTER THREE
## *Shale Shaker Design*
## *97*

Shapes of Motion 97 • Vibrating Systems 102 • Deck Design 103
"G"-Factor 103 • Power Systems 105

### CHAPTER FOUR
## *Shaker Applications*
## *109*

Selection of Shale Shakers 109 • Selection of Shaker Screens 109
Cascade Systems 111 • Dryer Shakers 113 •
Non-Oilfield Drilling Uses of Shale Shakers 114

### CHAPTER FIVE
## *Shaker User's Guide*
## *115*

Installation 115 • Operation 116 • Maintenance 116
General Guidelines 117 • Comparison and Analysis of
Shale Shaker Performance 117 • Conclusions 119

### CHAPTER SIX
## *Shale Shaker Screens*
## *120*

Common Screen Cloth Weaves 120

### PART THREE
# *Solids Control Management*

### CHAPTER SEVEN
## *Solids Control Equipment*
## *139*

Suction and Testing Section 139 • Additions Section 140 • Removal Section 141
Piping and Equipment Arrangement 141 • Surface Tanks 142 • Gumbo
Removal 143 • Sand Traps 143 • Degassers 143 • Hydrocyclones 143
Mud Cleaners 154 • Centrifuges 155 • Dewatering Unit 162
Removal Section Arrangement 162

**CHAPTER EIGHT**

*Dilution*

*164*

Introduction 164 • Effect of Drilled Solids Reaching the Surface 165 • Optimum Removal Efficiencies 167 Determining Removal Efficiencies 167

**CHAPTER NINE**

*Cut Points*

*170*

**CHAPTER TEN**

*Calculating Drilled Solids Concentrations*

*173*

Procedure for Determining Accurate, Low-Gravity Solids 173 Calculation of Barite Discard 176 • Calculating Solids Discard as Whole Drilling Fluid 177

**CHAPTER ELEVEN**

*Centrifugal Pumps*

*178*

Commonly Used Oil Field Pumps 178 • The Centrifugal Pump 179 Centrifugal Pump Performance Curves 184 • Head Losses Through Pipe 187 • Pumping Viscous Liquids 194 • Anatomy of Centrifugal Pumps 196 • Vibration 197 • Summary 198

**CHAPTER TWELVE**

*Electric Motors*

*201*

Voltage 201 • Temperature Considerations 202 • Motor Installation and Troubleshooting 202 • Electric Motor Standards 204 Enclosure and Frame Desginations 204 • Hazardous Locations 210 Specific Motor Applications 213

**CHAPTER THIRTEEN**

*Solids Dewatering*

*216*

Introduction 216 • Procedure 216 • Polymer Technology 218 Coagulation 219 • Flocculation 220 • Emulsion Polymers 220 Coagulation Concepts and Mechanisms 223

**APPENDIX A**
..........
*Derrickman's Pages*
*224*

**APPENDIX B**
..........
*Equipment Guidelines*
*226*

**APPENDIX C**
..........
*Pre-Well Checklist*
*235*

**APPENDIX D**
..........
*Troubleshooting Guide*
*237*

**APPENDIX E**
..........
*Shaker Manufacturers*
*240*

........
*Glossary*
*279*

.....
*Index*
*333*

# Acknowledgments

This handbook has been written by a group of dedicated engineers, scientists, and drilling operations personnel to explain many of the complex functions of drilled solids management. These people toiled without pay and dedicated many hours of their own time to this endeavor.

Our committee has been fortunate to have several members who made significant contributions and are no longer active in the solids control industry. In recognizing these individuals we, in our small way, seek to preserve their names for future generations:

**George Stonewall Ormsby** is famous for contributions to understanding all aspects of drilled solids control and management. His name is well known to early students of this technology because he became an evangelical prophet for this technology long before others actually recognized and understood the importance of good solids control. He was a major contributor and technically competent editor of the *IADC Mud Equipment Manual.* George has retired and is enjoying life. We miss his wisdom, controversies, guidance, outspoken comments, and understanding. We would like to name George as one of the technical giants of this technology.

**Walter Liljestrand**, although retired and well past his 80th birthday, actively assisted in this book. As our text developed, he contributed sound engineering logic in very simple terms. Before retiring, Walter invented and developed the atmospheric degasser. He, also, was a member of the IADC Mud Equipment Manual Committee, author of the *Centrifugal Pump Handbook,* and co-author of the *Degasser Handbook*. His impact on this committee, however, reached far beyond these achievements. Walter was a natural teacher and possessed the ability to explain very complex engineering technology clearly and concisely.

**Gordon Lawson** joined our IADC Mud Equipment Committee shortly after it was formed and remained a faithful contributor until his death several years ago. His expertise was to take technology from textbooks and apply that technology and concepts at drilling rigs. His career started with a major solids control service company and he then became a contract drilling supervisor. With an understanding of the fundamental principals of drilling, he brought practical experience to the committee and to the editing process.

**Grant Bingham** made an indelible impact on drilling and solids control. After serving in the United States Marine Corps in the South Pacific during World War II, Grant returned to Princeton to receive his master's degree in engineering. Grant began working for Shell as roughneck and eventually wrote several extensive papers on drilling theory while working for Shell Development. Grant's career eventually led him to solids control. While working for Milchem, an offshoot of the Milwhite Company, his research converted the Mobil rotary mud separator (RMS) from a research tool, with an infinite number of variable changes, into a field-usable centrifuge. Under Grant's leadership, a hydraulic test shaker was constructed that permitted testing of different types of motion and deck angles. As a result of these tests, one of the oil field's first linear motion shakers was developed. Grant's interest in solids control was exhibited through years of service in organizations such as the IADC Mud Equipment Committee until his final retirement.

# Preface

In 1974 the Mud Equipment Manual Committee (MEMC) was formed as a subcommittee of the IADC Drilling Technology Committee. This group wrote the IADC Mud Equipment Manual. It comprised many interested and informed industry participants. The task required 10 years to complete. All but two of the original committee members remained for the entire time as more experts joined the effort.

The charge of that committee was to address the factors relating to mechanical solids control and their relationship to overall drilling fluid system performance, and to publish findings in an "easy-to-read" format for dissemination to drilling engineers and drilling crew. The culmination of this work was the publication of eleven handbooks. The books included an *Introduction to Drilling Fluids and Solids Control Treatment, Drilling Fluid System Arrangements,* a handbook on each piece of mechanical solids control equipment and accessory equipment such as pumps, valves, and disposal systems. The last handbook was published in the early 1980s.

Soon afterwards, the Drilling Technology Committee of the IADC formed another subcommittee named the Rig Instrumentation and Measurements (RIM) Committee. Most of the original members of the MEMC became members of the RIM Committee. The RIM Committee formed six task groups: Safety, Rig Floor Instrumentation, Data Telemetry, Solids Control, Well Control, and Measurement While Drilling (MWD). The Safety Task Group published its documentation in one of the IADC/SPE conference preprints; the Data Telemetry Task Group developed a standard for data transmission that became an API Recommended Practice; the MWD Task Group became an international society sponsored by the Society of Professional Well Log Analysts. The Solids Control Task Group decided that the innovations in shale shakers and screen designs require a rewrite of *Handbook 3: Shale Shakers.*

A situation arose that caused the Solids Control Task Group to seek another sponsor. The American Association of Drilling Engineers (AADE) and the American Filtration Society (AFS) sponsored a Second Shale Shaker Technology Conference held in Houston, Texas on Feb. 6–7, 1991. The AADE steadfastly supported the rewrite effort for the new Shale Shaker Handbook and was responsible for expanding the handbook into this textbook on Solids Control Equipment.

All manufacturers of shakers and screens were invited to join the committee and an active campaign was conducted to solicit all viewpoints and their assistance in developing this book. It is greatly expanded from the initial book that precipitated the endeavor. This book is written for a broad audience: derrickmen and drilling foremen will find practical help; drilling engineers will find design and technical data to assist in understanding drilled solids management. Most members of the committee not only edited the text but made written contributions.

This book is dedicated to improving the understanding of drilled solids management. A shale shaker is the first in a series of devices that remove undesirable drilled solids from a drilling fluid. The other equipment, gumbo busters, desanders, desilters, mud cleaners, and centrifuges are described as well as the tank arrangements necessary for correct performance. An understanding of technology is frequently revealed by the history of developments. The first section describes events and equipment created as solids removal equipment and drilling fluid technology matured. This history is traced through advertising material printed in the *World Oil's Composite Catalogs.* As the book developed, the committee felt that it should also explain many other facets of drilled solids management on a drilling rig. For this reason, the book also contains information about:

- Dilution
- Calculating drilled solids concentration in a discard stream (or in the drilling fluid)
- Dewatering

- Centrifugal pumps
- Electric motors
- Clear, specific directions for rig personnel to ensure the proper operation of the shale shaker

In this book bold-faced words are further defined in the Glossary at the end of the book. Such words may appear more than once in the text if it seems appropriate to assure the reader that a definition is available.

This book required more than nine years to write and edit. Patience, dedication, many long hours of writing and editing, and evaluation of the latest technology has been contributed by members of this committee. The total value of the professional time required to create this book is enormous; some half jokingly say it rivals the budget of some states. Many members of this committee have been volunteers for more than 20 years—the industry owes them abundant accolades, which they will accept in lieu of royalties from the sales of this book. The committee would also like to express its appreciation to Gulf Publishing Company for publishing this book in a dismal oil-patch economic environment.

Leon Robinson, Ph.D.
Committee chairman

# SHALE
Techniques and Technology for Improving
Solids Control Management
# SHAKERS
and DRILLING FLUID SYSTEMS

# Historical Perspective

Drilling fluid was used in the mid-1800s in cable tool (percussion) drilling to suspend the cuttings until bailed from the drilled hole.[1] With the advent of rotary drilling in the water well-drilling industry, drilling fluid was well understood to cool the drill bit and to suspend drilled cuttings for removal from the wellbore. Clays were being added to the drilling fluid by the 1890s and by the time Spindletop was discovered in 1901, it was considered necessary to have suspended solids (clays) in the drilling fluid to support the walls of the borehole. These solids (clays) resulted from the **disaggregation**[2] of formations penetrated by the drill bit.[2] If the penetrated formations failed to yield sufficient clay in the drilling process, clay was mined on the surface from a nearby source and added to the drilling fluid. These were native muds created either by "mud making formations" or, as mentioned, by adding specific materials from a surface source.

Drilling fluid was recirculated and water was added to maintain the best weight and viscosity for specific drilling conditions. Cuttings, or pieces of formation (small rocks) that were not dissolved by water, required removal from the drilling fluid to continue the drilling operation. Under the sole discretion of the driller or tool pusher, a system of pits and ditches was dug on-site to separate cuttings from the drilling fluid by gravity settling (gravity forced the cuttings to deposit in the pits and ditches). This system included a ditch from the well, or possibly a bell nipple, settling pits, and a suction pit from which the "clean" mud was picked up by the mud pump and recirculated.

Mud was circulated through these pits, and sometimes a partition was placed in the settling pits to accelerate removal of unwanted sand and cuttings. This partition extended to within a foot or two of the bottom of the pit, thereby forcing all the mud to move downward under the partition and up again to flow into the ditch to the suction pit. Much of the heavier material settled, by gravity, in the bottom of the pit. With time, the pits filled with cuttings and the fluid became too thick to pump because of the finely ground cuttings being carried along in the drilling fluid. To remedy this problem, jets were placed in the settling pits to move the unusable mud to a reserve pit. Then, water was added to thin the mud and drilling resumed.

In the late 1920s, drillers started looking to see how other industries resolved similar problems. It was discovered that ore dressing plants and coal tipples were using:

1. Fixed bar screens placed on an incline
2. Revolving drum screens
3. Vibrating screens

The latter two methods were adopted for removing cuttings from drilling fluids.

The revolving drum, or barrel-type screens (called trommel screens), were widely used with the early low-height substructures. These units could be placed in the ditch or incorporated into the flow line from the wellbore. The mud flowing into the machine turned a paddle wheel that rotated the drum screen, through which the drilling fluid passed. The screen used at this time was very coarse, or 4 to 12 mesh. These units were quite popular because no electricity was required and the settling pits did not fill up as quickly. Today, revolving drum units have just about disappeared.

The vibrating screen, or shaker, became the first line of defense in the solids removal chain and for many years was the only machine used. Early shakers were generally used in dry sizing applications

---

[1] For a discussion on cable tool drilling, read *History of Oil Well Drilling* by J. E. Brantley.

[2] Bold-faced words are defined in the Glossary, pages 276–329.

**Earthen Pit Design.** The earthen pit system was designed as a settling pit that overflowed through a ditch into the suction pit. A duplex mud pump suction is shown at the far end of the suction pit. Drilled solids levels were poorly controlled. Mud treatment consisted of dilution to decrease the drilled solids concentration and viscosity. Fortunately, the well was normally pressured and was easy to drill.

**Earthern Mud Pits.** A drilling rig in East Texas drilling a 2,800-foot well using earthen mud pits. Cuttings, for geological examination, were collected on a piece of hardware cloth at the end of the flow line. Solids control equipment was not used on this well.

and went through several modifications before arriving at a basic type and size for drilling. The first modification reduced the size and weight of the unit for transporting between locations. The name "shale shaker" was adopted to distinguish the difference between shakers (classifiers) used in mining and shale shakers used in oil well drilling since both were obtained from the same suppliers.

Other modifications included a 4' × 5' hook strip screen that tensioned from the sides with tension bolts. Motion was elliptical, which made a downslope necessary to move cuttings off the screen. Screen mesh was limited to 20- to 30-square mesh (838 to 541 microns). This unit was the workhorse of the industry until the late 1960s. Even though superseded by circular motion and linear motion shakers, standard shale shakers are still in demand and being manufactured today.

In the late 1920s and early 1930s, larger oil companies organized research laboratories and began exploring oil well drilling problems. They began to understand that the smaller cuttings, or particles, left in drilling fluid were also detrimental to the drilling process and another ore dressing machine was introduced from the mining industry—the cone classifier. This machine, combined with the concept of a centrifugal separator taken from the dairy industry, became the hydrocyclone desander. The basic principle behind separating heavier and coarser materials from the drilling fluid is the centrifugal action of rotating the volume of the sand-laden mud to the outer limit, or periphery, of the cone. The heavier particles exit the bottom of the cone, and the cleaner drilling fluid rises to the top and exits as the effluent. The desander, ranging in size from 6 to 12 inches in diameter, removes most solids larger than 30 to 60 microns. Desanders have been considerably refined through the use of more abrasion-resistant materials and more accurately defined body geometry and are an integral part of most solids separation systems today.

After the development of the oil-field desander, it became apparent that side wall sticking of the drill string on the borehole wall was generally associated with soft, thick filter cakes. Using the already existing desander design, a 4-inch desander was introduced in the early 1960s. Results were better than anticipated and included longer bit life, reduced pump repair costs, increased penetration rates, and lower mud costs. These smaller hydrocyclones became known as "desilters" since they remove a much smaller particle, called silt (15 to 30 microns), which is smaller than "API sand."

As barite and other compounds were developed to improve drilling, drilling fluid became very complex. Also, the liquid phase for carrying solids was being reduced by the addition of barite and other compounds. The shaker removed the larger cuttings (larger than 541 microns, or 30 mesh), and the desanders and desilters removed the smaller particles (60 to 15 microns). However, the intermediate-size particles (from 541 to 60 microns) were still left in the drilling fluid.

Intermediate-size particle removal led to the development of circular motion, or tandem, shakers. Development was slow for these fine-screen, high-speed shakers for two reasons. First, screen technology was insufficiently developed for screen strength, so screen life was short. There was insufficient mass in the screen wires to properly secure the screens without tearing. Second, the screening basket required greater development expertise than required for earlier modifications in solids removal equipment.

During this time, major oil company research recognized the problems associated with ultra-fines (colloidals) in sizes of 10 microns or less. These ultra-fines "tied up" large amounts of liquid and created viscosity problems that could only be solved by adding water (dilution). Centrifuges had been used in many industries for years and were adapted to drilling operations in the early 1950s. They were first used on weighted muds to remove and discard colloidals—fine particles smaller than 2 to 4 microns—and to save larger particle-size barite (weighting material) and some drilled solids.

In recent years, a centrifuge was applied to unweighted drilling fluids to reduce and discharge fine solids in the active mud system. This application saves the more expensive liquid phase of the mud for reuse. Dilution is minimized, thereby reducing mud cost; however, these machines are quite expensive and require a great deal of care.

In the early 1970s, the mud cleaner was developed as an addition to the desander and desilter for reducing loss in the expensive liquid phase. Hydrocyclones discard a slurry, including the liquid phase, which can be expensive over time. The mud cleaner takes the underflow from a bank of hydrocyclones and introduces the slurry to a very fine, pretensioned vibrating screen. The expensive liquid phase and most of the barite pass through the screen and back into the system while the larger solids are discarded. This was the first successful application of a screen, bonded to a rigid frame, using very fine screens. Many mud cleaners have screen cleaners, or sliders, which are circular plastic pieces that vibrate against the bottom of the screen to prevent screen blinding. In weighted muds, screens of 200 mesh (74 microns) can be used, which is the upper size for commerical barite. For unweighted muds, the smallest practical size is 250 mesh (58 microns) for economical operation.

**A Mud Cleaner in the First Field Application.** The mud cleaner consisted of a bank of 20, 4-inch Pioneer hydrocyclones mounted above a stainless steel, double-deck, round Sweco shaker. The mud cleaner was used during 40 days of drilling from 11,000 to 16,000 feet, with 11 ppg mud, through gas sands with formation pressures equivalent to 2.2 ppg mud weight. Stuck pipe and lost circulation experienced on earlier wells was eliminated. The mud cleaner was deactivated for drilling from 16,000 to 16,200 feet to "find" a caseing seat. Torque and drag on the drill string increased significantly, and wiper trips with the drill string were required between logging runs. The mud cleaner was re-activated and "cleaned" the drilling fluid for one circulation through a 150 x 150 mesh screen, followed by one circulation using a 200 x 200 mesh screen before running $9\frac{5}{8}$-inch casing in a $12\frac{1}{4}$-inch hole. No problems were encountered whie running the casing or during the cementing operation.

A more recent development, introduced in the 1980s, is the linear shaker. Developments in screen technology have made it possible for pretensioned screens to be layered for obtaining very precise cuts while still maintaining an economical screen life. Linear motion is the best conveying motion to move the solids off the screen, and it is possible to convey cuttings "uphill." Screens can be elevated to retain the cuttings longer to obtain a dryer-reduced liquid content discharge. Also, finer screens, with smaller openings, can be used on the linear motion shaker. One application of linear shakers is to screen the underflow from desanders and desilters rather than using a mud cleaner. This device is called a "mud conditioner."

Present technology includes liquid salvage—dewatering or solids flocculation—that strips the liquid phase from solids and returns an almost clear stream of water into the mud system. This process includes a decanting centrifuge with pre-mixed polymers injected into the feed line of the centrifuge causing flocculation. The solids are coalesced inside the centrifuge resulting in separation of solids from the liquid, and the solids are then discarded.

A recent innovation for environmental purposes and more liquid retention, is the dryer. The discharge from linear shakers, desanders, and desilters is flowed across another linear shaker with even finer screens (down to 450 mesh, or 32 microns) and usually a larger screening surface. Any liquid that escapes can be retained in the sump. The sump pump returns the liquid to the active system, usually to the centrifuge feed tank.

These systems, or combinations of the various items discussed above, meet most environmental requirements and conserve expensive liquid phases. The desirable effect is to close the loop on liquid discharges, leaving a damp, semi-dry solid mass to remove for disposal.

**A Mud Cleaner in the First Commercial Application.** This installation, in Pecan Island Field in Louisiana, occurred within one year after the patent was submitted. The mud cleaner consisted of a manifold arranged for eight, 4-inch hydrocyclones mounted above a single-deck, round Sweco shaker. This was the first well in this field that casing could be reciprocated while cementing.

**The First Mud Cleaner Installation.** In 1971, Power Rig #10 was drilled M. J. Foster in South Louisiana. The mud cleaner concept had previously been tested in a 1,900-foot well near Houston, Texas. This unit consisted of a Sweco shale shaker, five feet in diameter, receiving solids from the underflow of two, 12-inch desanders and twenty, 4-inch desilters while circulating a potassium chloride drilling fluid. The rig shown has a degasser mounted on the mud tank in front of the mud cleaner skid and a centrifuge mounted on top of the tank after the mud cleaner. The mud cleaner skid was covered by a tin roof with additional lighting to facilitate sampling.

**The Second Field Model.** The mud cleaner included 2 open top, single-deck Sweco shale shakers processing underflow from sixteen, 4-inch Magcobar desilters.

Disposal, in some locations, consists of a cuttings injection system that blends cuttings into a slurry inside a specially modified centrifugal pump (shear pump). This slurry is pumped down the well annulus between two strings of casing, or the casing and wellbore.

An innovation recently made available on the Gulf coast is the "gumbo chain," or gumbo screen belt. It is used to discard gumbo and large, pliable cuttings typical of coastal and offshore drilling. The gumbo chain is a special conveyor, in a channel or trough, which drags gumbo and large, pliable cuttings out of the drilling fluid. The gumbo screen belt is an endless belt of 5–10 mesh synthetic cloth that moves gumbo up out of a liquid pool. This operation reduces the severe screen loading problems caused by gumbo in typical screening operations. The devices can be mounted in the flow line or as an accessory unit in the solids control system.

In summary, specific requirements for a given drilling fluid system and prevailing economic factors, will dictate the need for using specific items of solids removal equipment. The solids removal system may be simple using one shale shaker, or very complex using a multiple of one or more different items discussed above for complete separation of solids from liquid. Factors controlling the items used in a drilling fluid system include economics and the ever-increasing requirements of the environmental sector.

## THE EVOLUTION OF SOLIDS SEPARATION DEVICES

As mentioned previously, during the 1920s, the oil well drilling industry recognized the need to remove cuttings from the drilling fluid. The mining industry used stationary, vibrating screens for classifying larger pieces of ore from smaller pieces. These screens were adapted to separate cuttings from the drilling fluid in an evolutionary process, which has continued through time to yield the sophisticated solids control systems used today. The early machines were called "screens" not "shakers," as reflected in the 1930 *Oil Weekly* article reprinted on pages 13–14. The oilwell industry adapted mining equipment by redesigning it to handle viscous liquids, reducing its weight, and mounting it on skids to facilitate rig moves. The history of this evolution has been documented in the *World Oil's Composite Catalog*.

The following is a general chronology of solids separation devices used in the oilwell drilling industry. Unfortunately, some early versions are not included because the scope was limited to the *Composite Catalog* from the first edition (1930) through the 1978–1979 edition. The following descriptions are brief but represent the best general information available for each device. Additional information for each unit is presented in the following reproductions of original advertising pages taken from archival copies of the *World Oil's Composite Catalog*.

**1932 Baroid Sales Company**—The Baroid "Lemco mud screen" was one of the first screens developed in the late 1920s, with screen cloth approximately 30 mesh and an extremely high angle with an unbalanced elliptical motion.

**1934 Shaffer Tool Works**—The "Shaffer vibrating mud screen" is another early machine using somewhat larger dimensions—48" × 55"—while still maintaining a 4 × 5 relationship and using a high angle with elliptical motion.

**1935 Allis-Chalmers Mfg. Co.**—This early device, as with all of these units was used in the mining industry as a classifier. As it was built to be stationary, it was rather bulky and had to be "skidded" and then reduced in weight for installation on a drilling rig and to make it easily transportable.

**1936 Link-Belt Company**—Probably the best known of the early shale shakers, this unit was adapted from the mining industry to handle rotary drilling applications in drilling mud, as well as the first designed for installation on a drilling rig.

**1937 Gulf Engineers, Inc.**—The "Keil vibrating screen" was another adaptation designed for the oil industry. This unit, again, used elliptical motion and was "skidded" for moving.

**1938 The Jeffery Manufacturing Co.**—The Jeffery "Blue Streak Shale Shaker," primarily used in the midwest, was the first to advertise using the words "shale shaker."

**1938 Lucey Export Corporation**—The "Hudson-Boucher automatic shale separator" was the first barrel-type separator to be offered that became well known along with the "Linda K" and "Thompson." These units were commonly used with a very low substructure because they could be set on the ground and the drilling fluid flowed through them without requiring a great deal of floor height.

**1938 Vernon Tool Company, Ltd.**—With the "McNeely," designs began to resemble the oilfield units of today. It was well skidded, well supported, and boasted a real vibrator mounted centrally to the deck.

**1939 International Nickel Co., Inc.**—International Nickel introduced stainless screens—in this case, a monel screen—for use in high $H_2S$, salt water, and sour crude. Many of the early shale shaker screens were not stainless, which did not become standard until the late 1930s.

**1939 Link-Belt Company**—When examining this unit, it is interesting to see how a standard piece of equipment used in the mining industry was modified for use in the drilling industry and how the variations in the link belt unit evolved over time.

**1939 W-K-M Company, Inc.**—The W-K-M mud screen was a short-lived adaptation of the barrel-type mud separator. It had a screen exterior with a conveyor that carried the solids through the screen, discharging liquid back into earthen pits.

**1940 Chain Belt Company of Milwaukee**—Rex mud conditioners were another adaptation from the mining industry that were offered in competition with the Link Belt.

**1940 Gulf Engineers, Inc.**—The "Jitterbug" unit was introduced and was a name that stuck with shale shakers for many years. This unit emphasizes how much shale shakers changed in a relatively short period of time.

**1940 Hutchinson Engineering Works**—Hutchinson Engineering became the first to mass produce shakers specifically designed for drilling fluid use. Their "Rumba" became the standard for the oil industry until the mid-1980s.

**1943–1944 Hutchinson Engineering Works**—Although many changes were being made in the style of shakers, they continued to use the same elliptical motion and required gravity to remove cuttings from the screens.

**1943–1944 Overstrom & Sons**—The first Overstrom unit used a roll of wire mounted on the side of the shaker with two long clamps. The screen was pulled tight across the shaker and clamped down. When a tear occurred in the screen, a knife was used to cut the screen and the damaged portion was discarded. A new portion of screen was unrolled, pulled over the shaker frame, and clamped down. There were no hook strips or method of tightening the screen down to the shaker other than the clamps on either side of the frame.

**1946–1947 Link-Belt Company**—It is interesting to examine the suspension of the link belt in operation. Resembling a truck suspension, it worked quite well in its day and time.

**1946–1947 Sunshine Iron Works**—This unit was another of the early barrel-type shakers that worked quite well and would remove extremely large cuttings. It was rare to find a screen size finer than 10 mesh on this type of separator.

**1946–1949 Thompson Tool Co.**—The Thompson tool shale separator became the standard in the industry for the barrel-type shaker. There were many of these machines built and were still being used into the early 1970s in many old, relatively shallow, fields using earthen pits.

**1950 Link-Belt Company**—When comparing this 1950 version of the link belt to the earlier units, although somewhat more sophisticated, they look much the same. Pages are included to show the many parts comprising a unit. There were a number of screen meshes that were used and, interestingly, some rectangular-type meshes or openings, or combinations used that later became known as the "b-type" screens.

**1951 Thompson Tool Co.**—Opening the big round cover, where the drilling fluid entered the unit, exposed large paddles that turned in much the same manner as a mill wheel for grinding grain. These paddles used the fluid flow through the pipe to rotate themselves. They were connected to a shaft that turned the spiral drum so that the unit was self-powered and ran strictly off the gravity flow of mud through the unit.

**1952–1953 Link-Belt Company**—In the early 1950s, innovative mud boxes were developed, which were commonly called possum bellies, or back tanks, for more equal distribution of mud flow across the mud shaker screens. Conveyors were also introduced to move the solids away from the shakers to an area where they could be more conveniently handled, which was necessary particularly in offshore operations.

**1952–1953 Vernon Tool Co., Ltd.**—As today's mud systems are rather sophisticated, it is interesting to look at the early introduction of centrifugal separators, hydrocyclones, desanders and desilters. This early unit may possible originate from the grain or feed industry.

**1954–1955 Hutchison Manufacturing Company**—The Hutchinson, which later became known as Hutchison-Hayes "Rumba" shakers, illustrates one of the first schematics showing how separation occurred and how "conveyance off the screen" actually worked. A cartoon depicts how the device handled the discards.

**1954–1955 Thompson Tool Co.**—Thompson was one of the first to introduce galvanized parts. Soon, other manufacturers were using galvanized parts, which continued quite extensively from 1950 through the 1970s.

**1955–1956 C. F. Hickman Company**—The "Linda K," first introduced by C. F. Hickman, was a barrel-type separator used quite heavily in the south and southwest. It had the same basic application as the Thompson, using mud flow to power the barrel. In most cases, the barrel diameter was larger than the Thompson but the operation was pretty much the same. This unit is still manufactured today by Funston Supply in Wichita Falls, Texas.

**1955–1956 Merco Centrifugal Co.**—The Merco concentrator was a very early unit that controlled the specific gravity, was a relatively high rpm unit, and used centrifugal force in its separation.

**1955–1956 Medearis Oil Well Supply Corp.**—Medearis is one of the early fabricators that built mud tanks for surface tankage rather than using earthen pits. These tanks had provisions for mounting solids separation equipment on top of the mud tanks.

**1957 Hutchison Manufacturing Company**—This illustration provides an excellent view of how the motion of the shaker handled the separation of materials.

**1957 Thompson Tool Co.**—Known for their barrel-type separators, this was Thompson Tool's introduction into the vibrating screen-type separators. This shaker was introduced as a galvanized unit.

**1958–1959 Medearis Oil Well Supply Corp.**—Medearis introduced their first shakers as accessories to their tank fabrication. They used elliptical-style shakers and some, in time, became extremely high angle and used relatively high rpm motors.

**1960–1961 Hutchison Manufacturing Company**—These "Rumba" units could be obtained in two different configurations: either as an overslung or an underslung unit. The underslungs became much more popular because they lasted longer and provided better use of the screen by directing the flow across the entire surface. However, the overslung screen handled gumbo much better and allowed for easier removal of large pieces.

# HISTORICAL PERSPECTIVE

A circular motion Baroid double-deck shale shaker.

**1960–1961 Thompson Tool Co.**—Quite fearless in their early endeavors, Thompson Tool was not adverse to experimenting with new methods. They introduced an extremely efficient desander and a packless centrifugal pump to feed these units, which was quite innovative for the time.

**1966–1967 Baroid Division National Lead Company**—Baroid introduced the double-deck shaker, as well as circular motion, to the oil field shale shaker. In the beginning, it was extremely difficult to maintain the integrity of the screens, but as these units became more popular and personnel became more familiar with them, they found they could run much finer screens—up to 80 mesh. These units could also remove considerably more solids than the current shakers that were only using 30-square mesh screens.

**1966–1967 Medearis Oil Well Supply Corp.**—Medearis introduced a cuttings washing system that transported the cuttings from the shaker directly into a washing system that adequately removed the drilling mud from the cuttings. This enabled the cuttings to be discharged overboard and the fluid returned back to the mud system or held as makeup. This was an important early device for offshore drilling, particularly on the west coast.

**1968–1969 S. N. Marep**—Although most of the oil field equipment was built in the United States, the S. N. Marep shaker was built in France and several other shakers were manufactured in Romania and Russia.

**1968–1969 Thompson Tool Co.**—This unit has an adjustable orifice for controlling the underflow. The advertisement depicts a good cross-section showing the desander hydocyclone in operation.

**1970–1971 Centrifuge, Inc.**—Centrifuge, Inc. introduced the "Derrick Mfg. Co." machines to the oil field. These were some of the first "Derrick" industrial units that were modified for oil field operation and possessed both an extremely high angle and high rpm. This is where the term "high speed" originated from for fine-screen shakers. Although the centrifuge had been used in the oil field for some time, during the early 1970s they became an integral part of the solids control system to remove colloidals from drilling fluids.

**1974–1975 Dahlory, Inc.**—Dahlory began sloping its screen downhill at extremely high angles, that were being used at this time. The units still used elliptical motion but in a more efficient manner.

**1974–1975 Swaco/Dresser**—The Swaco division of Dresser introduced their vibrating screens using elliptical motion. The vibrator was positioned so that they could run finer screens, and it competed very efficiently with Baroid's double-deck screen. One of the advantages of this machine was the "openness" of its design—one could actually see the separation occurring. This also enabled broken screens to be seen and replaced much more quickly and easily.

**1974–1975 Hutchison-Hayes International, Inc.**—Hutchison-Hayes, formerly Hutchinson Manufacturing, introduced a cascade-type unit where elliptical motion was used on both the top and bottom shakers; however, the bottom shakers had the vibrator moved out very close to the discharge end of the shaker to attempt to increase conveyance. The process included scalping off the shale (the large cuttings or

Centrifuge Inc. unbalanced elliptical motion shale shaker.

IMCO triple-deck shale shaker. Designed by Louis Brandt prior to starting The Brandt Company where he designed the Brandt double-deck shale shaker.

gumbo on the top shaker) and then passing the underflow across the second set of shakers to produce a finer cut. The Hutchison-Hayes "Rumba SCS" was the first cascade-type unit advertised in the *Composite Catalog;* however, the concept was originally introduced by The Brandt Company.

**1974–1975 Imco Services**—Imco introduced a circular motion shaker that was the first triple deck—probably the only triple deck used to any extent in the oil field. It was used as a rental shaker to compete with the Baroid "Double Deck" and Swaco "Super-Screen" units and was designed and manufactured by The Brandt Company.

**1974–1975 Milchem, Inc.**—The Milchem shaker was another adaptation of a mining unit that was redesigned for oil field use and was an outgrowth of the Payne-Harris shaker. The rotary mud separator was the first centrifuge introduced that was different than those currently being used and was originally developed by Mobil Oil Company.

**1976–1977 The Brandt Company**—This was the first advertisment for Brandt, which began in the early 1970s with tandem shakers that used circular motion. These became the standard for the industry for many years and were manufactured in junior, singles, duals, triples, and quads. Another unit advertised was the cascade-type that used elliptical motion on the top deck to remove coarse cuttings and gumbo and then directed the flow across the bottom circular motion to produce a finer cut. In later years, the circular motion machines were used as the scalpers over linear units. Also depicted is one of the first introductions of the mud cleaner, originally referred to as the silt separator.

**1976–1976 Dreco**—The Dreco unit and its operation was basically the same as the Milchem unit.

**1976–1977 Swaco/Dresser**—Desilter cones mounted over the back of the Swaco super-screen shaker, produced a mud cleaner. This is an early rendition of what is now known as the mud conditioner. Also shown is the super clone

Milchem, Inc. elliptical motion shale shaker.

An elliptical motion Magcobar "super-screen" shale shaker, a refinement of the original units.

centrifuge, which was the "poor man's" centrifuge where water was readily available, but maintenance was probably extremely difficult.

**1976–1977 G&C Enterprises, Inc.**—The "Shimmy Shaker" was an air-powered shaker that required an extremely large volume of air to operate and, in most cases, an additional compressor was added to rig equipment when it was in operation. This unit worked quite well but was inefficient in maintaining operations.

**1976–1977 Medearis Oil Well Supply Corp.**— This advertisement takes another look at the Medearis unit using extremely high angle shakers and still using elliptical motion.

**1978–1979 The Brandt Company**—Brandt developed the use of a canted blade impeller for agitation of mud in mud tanks during oil field drilling. They also developed the use of a chart to determine the turnover rate (TOR) which was copied by a number of users throughout the industry. One of the early versions of a cuttings cleaner is also depicted.

**1978–1979 Picenco International, Inc.**—Pioneer did a tremendous amount of research and testing with hydrocyclones separation in its desanders and desilters. Additionally, they were leaders in the use of internal coatings. Their units consisted of cast iron bodies that were later replaced by polymers using various elastomers for the bodies. This reduced erosion, increased the life, and reduced the cost so that parts could be replaced more efficiently.

**1978–1979 Sweco, Inc.**—The Sweco Sand Separator became known as the "mud cleaner" and was the first introduced into the oil field in 1971.

## INDEX TO ARCHIVAL *COMPOSITE CATALOG* PAGES

| *Composite Catalog* Year | (Page No.) | Manufacturers | See Page |
|---|---|---|---|
| 1932 | (46) | Baroid Sales Company | 15 |
| 1934 | (468) | Shaffer Tool Works | 16 |
| 1935 | (74) | Allis-Chalmers Mfg. Co. | 17 |
| 1936 | (817) | Link-Belt Company | 18 |
| 1937 | (708) | Gulf Engineers, Inc. | 19 |
| 1937 | (709) | Gulf Engineers, Inc. | 20 |
| 1938 | (1150) | The Jeffery Manufacturing Co. | 21 |
| 1938 | (1262) | Lucey Export Corporation | 22 |
| 1938 | (2049) | Vernon Tool Company, Ltd. | 23 |
| 1939 | (1142) | The International Nickel Co. Inc. | 24 |
| 1939 | (1346) | Link-Belt Company | 25 |
| 1939 | (2226) | W-K-M Company, Inc. | 26 |
| 1940 | (613) | Chain Belt Company Of Milwaukee | 27 |
| 1940 | (885) | Gulf Engineers, Inc. | 28 |
| 1940 | (1117) | Hutchison Engineering Works | 29 |
| 1943–1944 | (1342) | Hutchison Engineering Works | 30 |
| 1943–1944 | (2297) | Overstrom & Sons | 31 |
| 1946–1947 | (2119) | Link-Belt Company | 32 |
| 1946–1947 | (2122) | Link-Belt Company | 33 |
| 1946–1947 | (3638) | Sunshine Iron Works | 34 |
| 1946–1947 | (3639) | Sunshine Iron Works | 35 |
| 1946–1947 | (3736) | Thompson Tool Co. | 36 |
| 1946–1947 | (3738) | Thompson Tool Co. | 37 |
| 1948–1949 | (4124) | Thompson Tool Co. | 38 |
| 1948–1949 | (4125) | Thompson Tool Co. | 39 |

## INDEX TO ARCHIVAL COMPOSITE CATALOG PAGES (continued)

| *Composite Catalog* Year | (Page No.) | Manufacturers | See Page |
|---|---|---|---|
| 1950 | (2902) | Link-Belt Company | 40 |
| 1950 | (2903) | Link-Belt Company | 41 |
| 1950 | (2904) | Link-Belt Company | 42 |
| 1950 | (2905) | Link-Belt Company | 43 |
| 1950 | (2906) | Link-Belt Company | 44 |
| 1951 | (4874) | Thompson Tool Co. | 45 |
| 1952–1953 | (3067) | Link-Belt Company | 46 |
| 1952–1953 | (5056) | Vernon Tool Co., Ltd. | 47 |
| 1954–1955 | (2329) | Hutchison Manufacturing Company | 48 |
| 1954–1955 | (2330) | Hutchison Manufacturing Company | 49 |
| 1954–1955 | (4885) | Thompson Tool Co. | 50 |
| 1955–1956 | (2280) | C. F. Hickman Company | 51 |
| 1955–1956 | (2281) | C. F. Hickman Company | 52 |
| 1955–1956 | (3313) | Merco Centrifugal Co. | 53 |
| 1955–1956 | (3316) | Medearis Oil Well Supply Corp. | 54 |
| 1957 | (5092) | Thompson Tool Co. | 55 |
| 1958–1959 | (3391) | Medearis Oil Well Supply Corp. | 56 |
| 1958–1959 | (3393) | Medearis Oil Well Supply Corp. | 57 |
| 1960–1961 | (2770) | Hutchison Manufacturing Company | 58 |
| 1960–1961 | (5304) | Thompson Tool Co. | 59 |
| 1966–1967 | (618) | Baroid Division National Lead Company | 60 |
| 1966–1967 | (619) | Baroid Division National Lead Company | 61 |
| 1966–1967 | (3238) | Medearis Oil Well Supply Corp. | 62 |
| 1968–1969 | (3059) | S. N. Marep | 63 |
| 1968–1969 | (4782) | Thompson Tool Co. | 64 |
| 1970–1971 | (1005) | Centrifuges Inc. | 65 |
| 1970–1971 | (1009) | Centrifuges Inc. | 66 |
| 1974–1975 | (1349) | Dahlory, Inc. | 67 |
| 1974–1975 | (1742) | Swaco/Dresser | 68 |
| 1974–1975 | (1743) | Swaco/Dresser | 69 |
| 1974–1975 | (1744) | Swaco/Dresser | 70 |
| 1974–1975 | (2714) | Hutchison-Hayes International, Inc. | 71 |
| 1974–1975 | (2986) | Imco Services | 72 |
| 1974–1975 | (3606) | Milchem, Inc. | 73 |
| 1974–1975 | (3607) | Milchem, Inc. | 74 |
| 1976–1977 | (710) | The Brandt Company | 75 |
| 1976–1977 | (711) | The Brandt Company | 76 |
| 1976–1977 | (712) | The Brandt Company | 77 |
| 1976–1977 | (1851) | Dreco | 78 |
| 1976–1977 | (2121) | Swaco/Dresser | 79 |
| 1976–1977 | (2478) | G&C Enterprises, Inc. | 80 |
| 1976–1977 | (4205) | Medearis Oil Well Supply Corp. | 81 |
| 1978–1979 | (861) | The Brandt Company | 82 |
| 1978–1979 | (862) | The Brandt Company | 83 |
| 1978–1979 | (5412) | Picenco International, Inc. | 84 |
| 1978–1979 | (5414) | Picenco International, Inc. | 85 |
| 1978–1979 | (5970) | Sweco, Inc. | 86 |

Reprint from *The Oil Weekly*, October 17, 1930.

*Vibrating screen in use at General Petroleum Corporation's Moco 218B in the Maricopa Flat district*

# Vibrating Screen Cleans Mud— Makes Big Saving

### By BRAD MILLS
### *Staff Representative*

THE use of a special type of vibrating screen to remove sand, shales and foreign matter from rotary drilling mud has proven a very satisfactory innovation at General Petroleum Corporation's Moco 218-B in the Maricopa district, where one of the first installations of its kind has functioned without interruption for two and one-half months. Application to the petroleum industry of a practice long used in other industries for removing undesirable portions of certain mixtures is considered one of the most important engineering developments of the year.

While General Petroleum Corporation's installation at Moco 218-B in the Maricopa field is only one of several now in use in California, the careful record kept on its operation and the success of its action make it perhaps the most interesting of a dozen screens now in use in at least four districts. Installed in the nature of an experiment, the layout has been operated from the beginning with few changes in the original design.

The growing tendency toward reclaiming rotary drilling mud has been apparent for more than a year and many operators have kept careful records on the cost of this phase of drilling. Thousands of tons of mud has been discarded in the Maricopa district over a comparatively short period, the cost of replacing that lost having been a good percentage of the total drilling cost. Shales and sand at certain depths have rendered the mud useless in very short periods. Early in the development campaign in the district new mud was mixed as needed and most operators thought that nothing could be done about the high cost of circulation. The installation of a vibrating mud screen at Moco 218-B followed a decision to reduce mud costs and maintenance of pump parts. It was obvious that the expense involved with experiments in a new type of installation was small when compared with the possibility of arriving at an improved operating technique.

### The Apparatus

The installation consists chiefly of a rectangular wire screen cloth mounted on a steel frame and a small electric motor. The screen is 30-mesh and four by five feet. The screen frame is mounted on four snail house type (coil) springs. On one side of the frame and attached to it is a short shaft and pulley to handle the belt connected to the motor.

The screen is placed in the regular mud circuit so that the entire volume passes over or through it. The electric motor is mounted a short distance above the upper end of the screen and on one side of it. The screen is tilted at an easy angle to facilitate the drop of the sand and shales, but the motion of the frame and screen while in normal action does not make a very delicate adjustment necessary. A short slide is placed immediately below the screen to give the coarse particles momentum. A one-inch perforated pipe crosses the slide immediately above it, the play of water from the pipe washing the sand and cuttings to a distant point by means of a galvanized iron trough. The ditch or trough leading to the screen from the well is the same width as the screen, with diamond-shaped wooden blocks placed just ahead of the short drop onto the screen to insure even distribution.

The screen is naturally placed immediately above the ditch carrying the mud back to the pits for recirculation, the reclaimed fluid and that suitable for further use dropping through the screen and flowing by gravity to be picked up by the pumps. The terrain is ideal at Moco 218-B for the installation but similar layouts could be used in almost any location.

The motor and screen are kept in operation as long as circulation is maintained but are stopped when the shutdown involves any appreciable period. The screen gets its action from the belt-driven motor but its efficiency is attributed to the snail house springs. The vibration is so intense when the motor has attained its full speed that it approaches a musical hum. The eye cannot detect its action but a touch with the foot or hand gives an idea of the vibration. The heart of the cleaning action lies in the snail house springs supporting the screen frame.

The engineer who designed the screen for oil field use worked for some time to obtain an elliptical motion. This was found to be the most desirable for removing the coarse particles as well as causing the mud to drop through without hesitation. The four by five screen seems to be the ideal size for an individual well.

The action of the screen is familiar to industrial engineers who have used them in other industries, but the average petroleum engineer has been given a chance to study the installation only during the past year. By mounting the screen frame to the center of the snail house springs, it is obvious that a motor driving a pulley solidly attached to the frame would bring about the vibrating action, which might be termed a takeoff from the age-old sifter action.

Steam turbines have been used on other installations and may even be used on this particular layout, but the crew is optimistic over the uniform action caused by the electric motor.

A flat endless belt is used, but future installations will carry "V" belts. The motor is mounted on slide rails and adjustment of the belt is a simple matter. When the motor is started or stopped, the screen and frame pass through critical periods of speed, vibrating vigorously for a few seconds. This is caused by the action of the snail house springs. The action is very smooth at regular running speeds.

The vibrating screen frame should not touch the mud ditch, the constant action causing leaks under ordinary circumstances. The same applies to the screen frame supports. It is a simple matter to support the screen above the ditch.

When the installation was first made, the sand and cutting were allowed to fall near the bottom of the screen. This necessitated almost constant removal of these cuttings and added to labor costs. The galvanized iron trough was installed and enough water played on the drop at the lower end of the screen to carry the coarse materials to a safe distance. This eliminated the extra labor cost. The well has the natural advantage of a hillside location, however, besides being far removed from an inhabited district. It might be possible to design the proper decline at any well, the flow of water partially governing the efficiency of the removal. When certain formations were drilled through, the sand and shale cuttings accumulated at a rapid rate, giving some idea of the reasons for former pump repairs and addition of fresh mud.

When this article was prepared, the installation had been in use 920½ hours. During the time, the screen cloth had been replaced four times. There was no further cost of replacements. The operating cost, which included chiefly power and water, was about $2 per day. The same mud had been used since July 23, the date of installation, without interruption or substantial additions. No pump parts have been replaced during the period, while the pump had not lost suction during drilling operations.

The elliptical action of the screen causes a portion of the materials not wanted in the mud to form into small balls. These balls often carry a percentage of very fine sand which might otherwise find its way through the screen.

A large by-pass around the vibrating screen has been installed for emergenices. It empties into the ditch below the screen.

The average cost of changing rotary drilling mud in the Maricopa district runs between 40 and 50 cents per barrel; so it is evident that constant reclaiming of the original mixture is a very profitable practice.

*Side view of installation, showing mud ditch leading to sump, galvanized trough for carrying away cuttings and sand, and by-pass*

# BAROID SALES COMPANY
### 837 JACKSON STREET          LOS ANGELES, CALIF.
*"A Complete Drilling Mud Service"*

DISTRIBUTORS

| California | Gulf Coast | Mid-Continent | Export |
|---|---|---|---|
| Carl Ingalls, Inc. | Beaumont Cement Sales Co. | Bridgeport Machine Co. | Oil Well Supply Co. |
| Oil Well Supply Co. | Lucey Products Corp. | National Supply Co. | National Supply Co. |
|  | Peden Company |  | Oil Well Engineering Co., Ltd. |

## PRODUCTS:
- BAROID—Extra-Heavy Colloidal Drilling Mud
- AQUAGEL—Trouble-Proof Colloidal Drilling Mud
- LEMCO MUD SCREEN—For Removing Cuttings and Abrasive Material From Drilling Mud
- STABILITE—A Chemical Mud Thinner.
- VISCO—Petroleum Emulsion Breaker.

### BAROID
### Extra Heavy Colloidal Drilling Mud

Baroid is a specially processed pulverized Barytes, manufactured and sold under U. S. Patent No. 1,575,945. It is a uniform, water-ground product of approximately 4.2 Specific Gravity, almost twice that of ordinary clay, and is used for controlling oil, gas and water pressures encountered during drilling.

Baroid is compounded and processed so that it has excellent suspending and wall-building characteristics. Baroid will not settle from the mud, even though circulation is suspended for months. The fine particle size of Baroid, together with the suspensoid incorporated during manufacture, causes the building of a mud wall in the uncased hole which prevents loss of fluid from the hole, prevents caving and prevents penetration of mud into the formation. This thin wall is quickly washed out when the well is ready to be put on production. Baroid possesses excellent lubricating qualities and will greatly lengthen the life of equipment used in the circulating system.

Where a drilling mud weighing between 75 lbs. and 90 lbs. per cu. ft. (Sp. G. 1.20 to 1.50) is required, Baroid in sufficient quantities can be added directly to the mud in the circulating system. For preparing drilling muds heavier than 90 lbs. per cu. ft. it is recommended that Baroid and water alone be used. Pumpable muds weighing as much as 145 lbs. per cu. ft. (19.4 lbs. per gallon) have been prepared from Baroid.

### AQUAGEL
### Trouble-Proof Colloidal Drilling Mud

Aquagel is a specially selected and prepared clay which contains a high concentration of gel type colloids. This type of earth is found in small quantities in all clays and is responsible for most of the desirable qualities of drilling muds. It is the gel-type colloids which give clays their wall-building properties in the hole.

Aquagel and water alone make an excellent drilling mud which weighs 64 lbs. per cu. ft. (Sp. G. 1.03) and has excellent sealing and wall-building characteristics. One hundred forty to 180 barrels of such mud can be prepared from each ton of Aquagel, this effecting a great saving in initial cost of mud.

In cases where a mud heavier than the above is required, the ordinary native mud made in the course of drilling usually will provide adequate weight and when this is compounded with Aquagel, in quantities of from 2% to 3% by weight, an excellent drilling mud will result.

Aquagel, when properly used in drilling muds provides cheap insurance against costly fishing jobs. Aquagel is being used daily in practically all oil fields of the world to:
- Prevent Loss of Circulation.
- Prevent Caving.
- Prevent Stuck Pipe and Casing.
- Reduce Abrasion, with consequent saving in replacement of pump parts, etc.
- Suspend Drilling Mud Solids.
- Seal Off Fissures and Caves.
- Prevent settling of mud solids and cuttings in hole during suspension of circulation.
- Prevent building up thick filter cake on wall of hole.
- Seal off minor flows of gas and water.
- Seal off flows of Artesian water encountered during drilling.
- Insure landing of long strings of casing.
- Insure positive water-tight shutoffs in an admixture with cement (U. S. Patent Re-issue No. 17,207).

### LEMCO MUD SCREEN
### For Removing Cuttings and Abrasive Material from Drilling Muds

Ground Space Required 4'0" x 6'0".
Head Room 2'6" (slope 24° to 28°).
POWER UNITS SUPPLIED:
  Motor: 1 H.P. 220/440 volts, 50/60 cycle;
  or Steam Turbine: 1 H.P. at 220 lbs. pressure, Pyle-National.
CAPACITY WITH:
  30B mesh screen cloth assembly—750 gal. per minute
  30 mesh screen cloth assembly—650 gal. per minute
  40 mesh screen cloth assembly—550 gal. per minute
SUPPORTS—2" x 12" timbers
  Shipping weight—Motor drive—800 lbs., Turbine drive—900 lbs.

The Lemco mud screen has been used for over three years and under many kinds of drilling conditions as a successful means of eliminating sand and cuttings from the drilling mud.

In the motor-driven unit the vibrating mechanism is directly connected to the motor; in the turbine-driven unit the vibrators are driven through a reduction gear from the turbine.

Screen cloth assemblies are furnished in 30B, 30A, 30 and 40 mesh, as drilling conditions demand. These assemblies are fabricated in our shops and the method of mounting is such as to insure a considerably greater life for the screen cloth than can be obtained in any other manner.

The foregoing screening capacities are for "average conditions," that is, for a mud of 80 lbs. per cu. ft. and of 10 centipoise viscosity. As the weight is decreased or the viscosity increased, this capacity will be reduced proportionally.

The removal of sand and cuttings from the mud will result in lower pump equipment maintenance costs, which will pay for the screen installation many times over. Where weight-material is being used in the mud, the Lemco screen will show even greater economy due to prolonging the efficiency of the mud by removal of both cuttings and gas.

Nobody that has used a "Lemco" screen has ever been satisfied to drill another well without one.

### STABILITE
### Chemical Mud Thinner

Stabilite is a liquid chemical compound, developed after several years research and field tests, for reducing the viscosity of drilling muds without destroying their wall-building properties. By reducing the viscosity of the drilling mud, the weight can be built up several pounds per cubic foot by adding ordinary clay without resulting in too great a viscosity. Stabilite is valuable whenever you want to thin your mud and is particularly valuable when mud becomes gas-cut. It can be quickly and easily added to the mud and is very economical.

### VISCO
### Petroleum Emulsion Breaker

An improved chemical compound for the treatment of emulsified petroleum. There are a number of different formulas for the efficient treatment of the many types of cut oil and our field men will be glad to cooperate with you in any of your treating problems.

Visco Emulsion Breaker is used in the field in the same manner as other chemical emulsion breakers, but works on an entirely different principle. Its use will materially reduce your treating costs. Fully protected by U. S. Letters Patent.

Write us for booklets descriptive of these products, or write us in detail with reference to your specific mud problems. With service men in all important fields, we will be glad to give you the benefit of our experience free of charge.

# SHAFFER TOOL WORKS — BREA, CALIFORNIA

## SHAFFER DRIFT INDICATOR

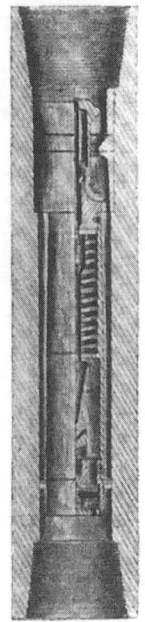

The Shaffer Drift Indicator is a patented tool which is run on the drill pipe above the bit or drill collar. It consists of a sub made of heat treated alloy steel in which is mounted a valve operated by a plumb bob. This bob is free to swing in all directions. Below the plumb bob is a cup shaped pin. The force of the circulation tends to close the valve. But if the hole is within required limits of straightness the plumb bob hits the pin thereby holding the valve off the seat so the circulation is continuous. The diameter of the cup shaped pin may be changed according to the limits of straightness desired, which may be anything up to 5 degrees. When the hole is more than the desired degree off vertical the plumb bob will miss the pin and allow valve to close, thus shutting off circulation.

To find out if the well is off vertical it is only necessary to stop the tools and circulation for a very brief time and then start circulation again. If the hole is within the required limits of straightness circulation will go on as usual and drilling be resumed. But when deflection from vertical, greater than the amount for which the machine is set, takes place the valve is closed thereby shutting off the pump. Before being able to proceed it is necessary to lift the drill pipe to a point above the crooked spot. The drill is then held up and hole is reamed straight. In this way a continuous check is made and the hole is straightened without having to disconnect the swivel and kelly or waiting until the tools are withdrawn from the hole.

## SHAFFER VIBRATING MUD SCREEN

The Shaffer Vibrating Mud Screen provides an effective and economical method of reclaiming and cleaning rotary mud at the well. Even gassing mud can be treated with this device. The Shaffer Screen is a simple mechanism and has been thoroughly proven under all kinds of drilling con-

## SHAFFER-BOLES FORMATION TESTER

The Shaffer-Boles Formation Tester often saves its cost many times over by positively proving whether the well is capable of flowing gas or oil or whether further work will be needed. Not only might you save the cost of the casing itself, but also the cement job, the drilling out, and the possibilities of necessary re-cementing. Moreover, it eliminates the necessity of reducing the size of the hole when the tester proves the absence of commercial production and further drilling is necessary. This device gives reliable samples and flow tests. The Shaffer-Boles Formation Tester in conjunction with Shaffer-Phipps Packer enables you to secure a positive test of water shut off. Simplicity of construction and operation make the Shaffer-Boles Formation Tester outstanding. It has few working parts — and no springs or trips. A hard faced valve which is at all times under positive and easy control from the surface, may be opened and closed as many times as required during testing operations. Three revolutions of the pipe to the left open this valve fully — three turns to the right close it tight. No fractional turns or other precise mechanical operations have to be employed. Each valve is tested under 3000 pounds pressure to insure it against leakage. A by-pass consisting of a left-hand joint opens the tester below the main valve so that the fluid pressure can be equalized on both sides of the packer; and this joint can be unscrewed only after the tester valve has been positively closed.

Type C is used for open hole testing where no shoulder is provided. The packer is set by lowering the weight of the drill pipe which shears a pin. The length of tail piece may be varied to control the depth at which the packer is desired to be set.

ditions. There are no complicated parts. It is noted for its long life and ease of adjustments. Any work on it can be done with ordinary tools at the derrick.

The screen, which is 48 inches wide by 55 inches long, should be set on a 27 degree inclination and is placed in the mud system so that as the mud emerges from under the derrick floor it flows over the screen. The rapidity of vibration causes all the mud to flow through the screen while particles larger than the mesh are retained and fall into the rejection pit. Special adjustments on the screen assembly together with the peculiar method of vibrating make the machine unique and exceptionally efficient. The vibrating mechanism and the method of mounting the screen (rubber insulated from metal parts) is such as to give the screen long life. A 2" rotating cam shaft, equipped with two 1/8" offset eccentric bearings, produce a 1/4" vibration with each revolution. Two connecting rods run to a cross bar which is attached to the screen assembly. A steam driven turbine equipped with a governor permits regulation of vibrations to the most efficient speed of operation.

# ALLIS-CHALMERS MANUFACTURING CO. MILWAUKEE, WIS.

*"Rannett" Centrifugal Pipe Line Pumping Unit.*

can be installed outdoors if desired. These units are suitable for crude oil or gasoline pipe line service either main line or booster and similar high pressure applications. Three sizes are available, a 4" size with 200-hp. motor, a 6" size with 400 hp. motor and an 8" size with 700 hp. motor which will take care of a pumping range of from 7000 to 38,000 barrels per day against 800 pounds pressure. These units are known as the "Rannett" type after the inventors Messrs. Moran and Bennett, and were developed and built by Allis-Chalmers. This highly advanced, new type of pipe line pumping unit is now made available to the oil industry for pipe line pumping service. **Additional information will be furnished on request.**

## RECIPROCATING PIPE-LINE PUMPS

**Forged or Cast Steel Pumps** are built for capacities of 5000 to 50,000 bbls. per day and pressures up to 1000 lbs.

*Forged Steel Oil Pipe Line Pumping Unit.*

per sq. in. Can be furnished complete with motors or Diesel oil engines coupled through gears or Texrope drives and also with direct-connected steam engines. *Bulletin 1646.*

## ROTARY COMPRESSORS

**Rotary Compressors and Dry Vacuum Pumps** of the multi-cellular, sliding-vane type, in both water-cooled and air-cooled designs, are built for a range of volumes from 50 to 2000 cfm, at pressures up to 150 lbs., and vacuums up to 29.85" mercury. The air-cooled units are suitable for pressures up to 10 lbs. G., and vacuums up to 18" mercury, and the water-cooled units for higher pressures and vacuums. These Rotary Compressors are especially suitable for low-pressure air and gas lift and repressuring, for gas booster service and for air agitation work in refineries.

*Rotary Compressor for Gas Lift. Units may be direct-connected motor driven or driven by engines through Texrope drives or flat belts.*

The smooth operation, small dimensions, continuous delivery and absence of maintenance requirements are a few of the advantages presented. The units operate at standard motor speeds and the design is extremely simple, involving no valves or complicated motions. All working parts are totally enclosed, but easily accessible. *Leaflet 2159A.*

## VIBRATING SCREENS

The "Aero-Vibe" Vibrating Screen "floats in the air" suspended on springs. It has an adjustable, vibrating motion produced by counter-weighted wheels on the drive shaft.

*"Aero-Vibe" Vibrating Screen for drilling mud*

Bearings are anti-friction, Alemite lubricated. Power is through Texrope V-belt from a steam turbine or electric motor. Feed and discharge boxes for drilling mud and discharge spout for cuttings can be provided. The unit is suited for rough service.

## TRACTORS

*Model "L" Track-Type Tractor 76 Drawbar Horsepower*

**Oil Field Tractors** are built in a variety of sizes to suit the job. **Model "L"** has a drawbar pull of 76 horsepower and weighs approximately 22,000 pounds. This tractor has six speeds forward and two reverse. It will fit into the heaviest hauling conditions. **Model "K"** has a drawbar pull of 45 horsepower and weighs approximately 11,000 pounds. It is equipped with 3 forward speeds, one reverse. **Model "M"** has a drawbar rating of 29 horsepower, weighs approximately 6,200 pounds and is equipped with four forward speeds—one reverse.

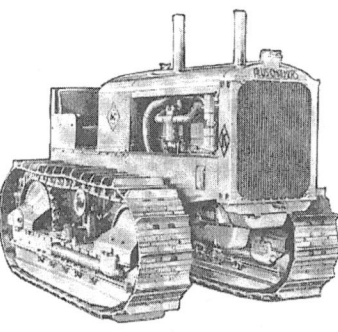

*Model "K" Track-Type Tractor 45 Drawbar Horsepower*

All Allis-Chalmers track-type tractors are equipped with renewable sleeve engines of the latest type—all are of unit construction and special A-C tracks of the long-wearing ground-gripping type. **Track-type Wagons** with logging or oilfield bunks which may be used in tandem for extra long pipe are available in capacities up to 30 tons.

**Oil Field Winches, Wheel-Type Winches and Winch Tractors** are handled exclusively by <u>Fred E. Cooper, Tulsa, Oklahoma</u>, whose advertisement appears on **next page.**

*Model "M" Track-Type Tractor 29 Drawbar Horsepower*

# LINK-BELT COMPANY

5465

**Sales Offices and Warehouse Stocks at**

| PHILADELPHIA | DALLAS | LOS ANGELES | INDIANAPOLIS |
|---|---|---|---|
| 2045 W. Hunting Park Ave. | 413 Second Ave. | 361 So. Anderson St. | 220 S. Belmont Ave. |

**Other Sales Offices**

HOUSTON, Fred Wallace, 4317 Montrose Blvd., P. O. Box 1082     OKLAHOMA CITY, W. H. Abele, Box 305, Route 4
EXPORT SALES: New York, N. Y. 2680 Woolworth Bldg. Cable and Radio Address: "LINKBELT-NEW YORK."
SOLD BY MOST SUPPLY HOUSES

## THE LINK-BELT VIBRATING SCREEN FOR ROTARY MUD

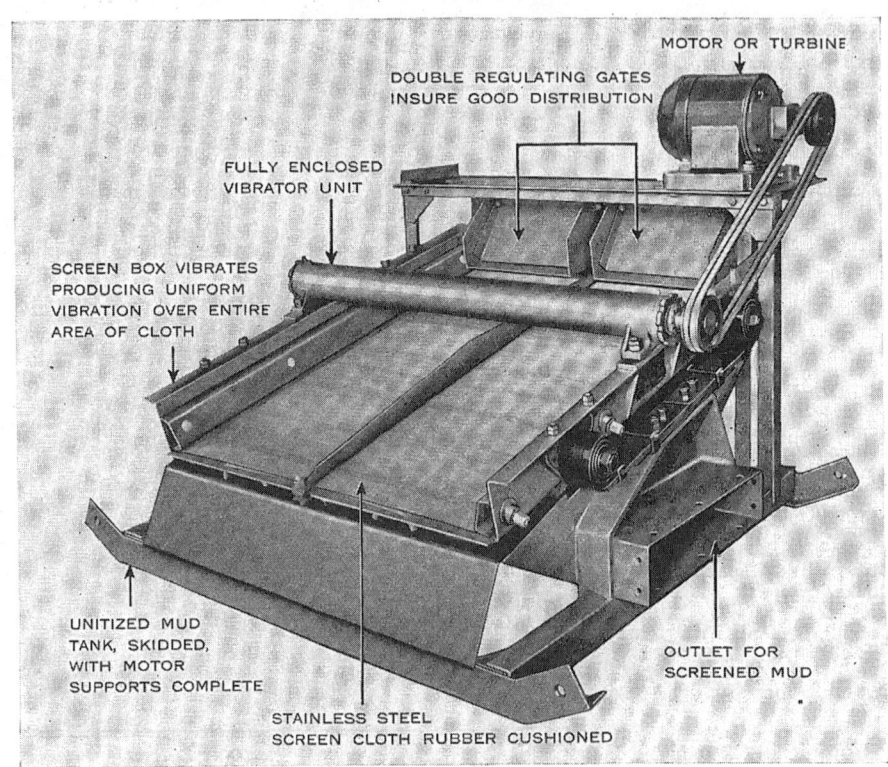

Clean mud of uniform consistency and weight can be maintained automatically with the Link-Belt Vibrating Screen. This efficient unit not only rejects large, heavy cuttings from the mud as it flows over the screen surface on its way to the slush pit, but eliminates fine sand and shale as well. The action of the screen also thoroughly degasses and reconditions the mud for re-use without loss of good mud or a reduction in its tenacity. It further simplifies the entire mud system by the reduction in size of mud storage and settling pits.

**The Link-Belt Unitized Vibrating Screen for Rotary Mud includes Screen Box, fitted with single piece, rubber cushioned stainless steel cloth with tension members; Belt Drive; Mud Tank with two-way outlet, made self contained for supporting screen box and electric motor or steam turbine; Distributing Chute with control gates and necessary tools.**

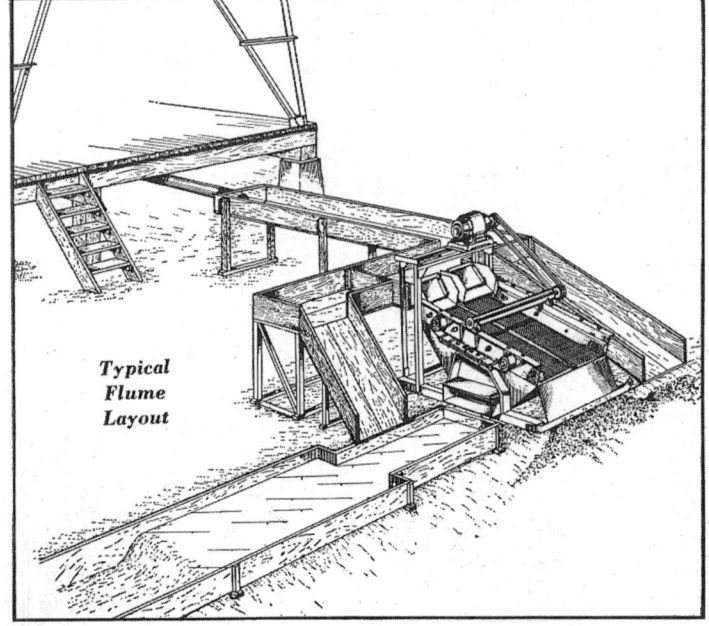

*Typical Flume Layout*

Consider these advantages of mud automatically cleaned and maintained in good condition by the Link-Belt Vibrating Screen:

—Reduces the hazards of gas blowouts.

—Makes a better and smoother well.

—Saves drilling time by reducing the possibility of drill pipe sticking and of having to drill through settled cuttings when changing bits.

—Lengthens the life of pump parts, swivels, drill pipes, rotary hose and stand pipe connections, as well as bits and reamers.

The Link-Belt screen has many exclusive advantages such as—larger capacity, because of larger screening area—fine mesh screen cloth, which produces a cleaner mud—longer life of both screen cloth and entire unit—ease of transportation and minimum head room required. Send for Booklet 1572.

# GULF ENGINEERS, Inc.
## HOUSTON, TEXAS

PHONE FAIRFAX 1348            P. O. Box 182

Oklahoma and Kansas Distributor KEIL VIBRATING SCREENS: CARSON MACHINE & SUPPLY CO., OKLAHOMA CITY, OKLA.

### PRODUCTS:

- Keil Vibrating Screens
- Turbodrive Steam Turbines
- Chiksan All-Steel Rotary Hose
- Chiksan Aluminum Dock Loading Hose
- Chiksan Ball-bearing Swivel Joints
- Chiksan 3-Way Disc Bits; Chiksan 3-Way Reamers
- Bettis Drill Pipe and Casing Protectors
- Twin-Seat Slush Pump Valves
- "Varco" Hardened and Ground Slush Pump Liners
- "Varco" Hardened and Ground Piston Rods
- "Lamb" Solid Slush Pump Pistons
- Volz Steam Specialties
- Volz Steam Pump Valve Motion
- Grove Universal Regulator
- Harbison-Fisher "Tuff Temper" Balls and Seats
- Oil-Bath Stuffing Boxes, and Wire Line Oil-Savers

### EXCLUSIVE FEATURES OF THE KEIL VIBRATING SCREEN:

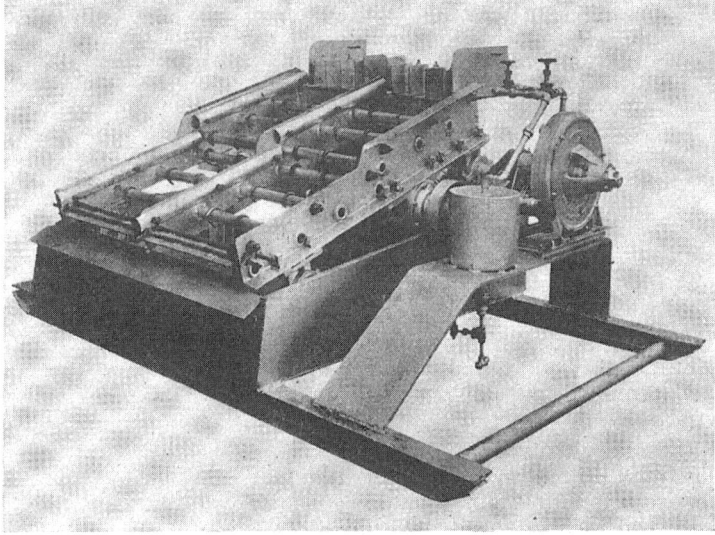

1. STEP DECK—This permits the mud laden rejects to turn over as they cascade to the lower screen sections. This frees the recoverable mud and permits it to pass through the screen cloth. This results in larger screen capacity and less loss of usable mud.

2. FOUR SMALLER SCREEN CLOTHS—This reduces the cloth upkeep to the replacement of a small section at a time, rather than the entire screen cloth when any one section should fail. A smaller size screen cloth is also easier to install, can be kept in more uniform tension and cannot be distorted like larger screen cloths when applied.

3. The combination of step deck and sectional cloths also permits the use of two different mesh screen cloths on the same unit, a decided advantage.

4. There are no springs, cams and loose moving parts on the Keil Mud Screen, all resilient parts being made of rubber. This reduces shut-down possibilities because of breakage and wear. All other parts are of steel.

5. A rugged construction throughout permits the Keil Screen to be operated with a suitable steam turbine, if electric motive power is not available or is more expensive.

6. Lubrication is provided for by two Alemite fittings, one on each bearing in the rotor shaft, preventing entry of sand and injurious particles into the bearings.

7. SCREEN OBTAINABLE ON UNITIZED MOUNTING—We strongly recommend the unitized assembly as shown above. The slight increase in cost of this assembly over the plain screen and collecting tank is rapidly offset by the saving in installation cost and the longer life and better general operation of the equipment, because the danger of uneven assembly and misalignment of parts is eliminated.

# GULF ENGINEERS, Inc.      HOUSTON, TEXAS

**SPECIFICATIONS AND PARTS FOR KEIL VIBRATING SCREEN**

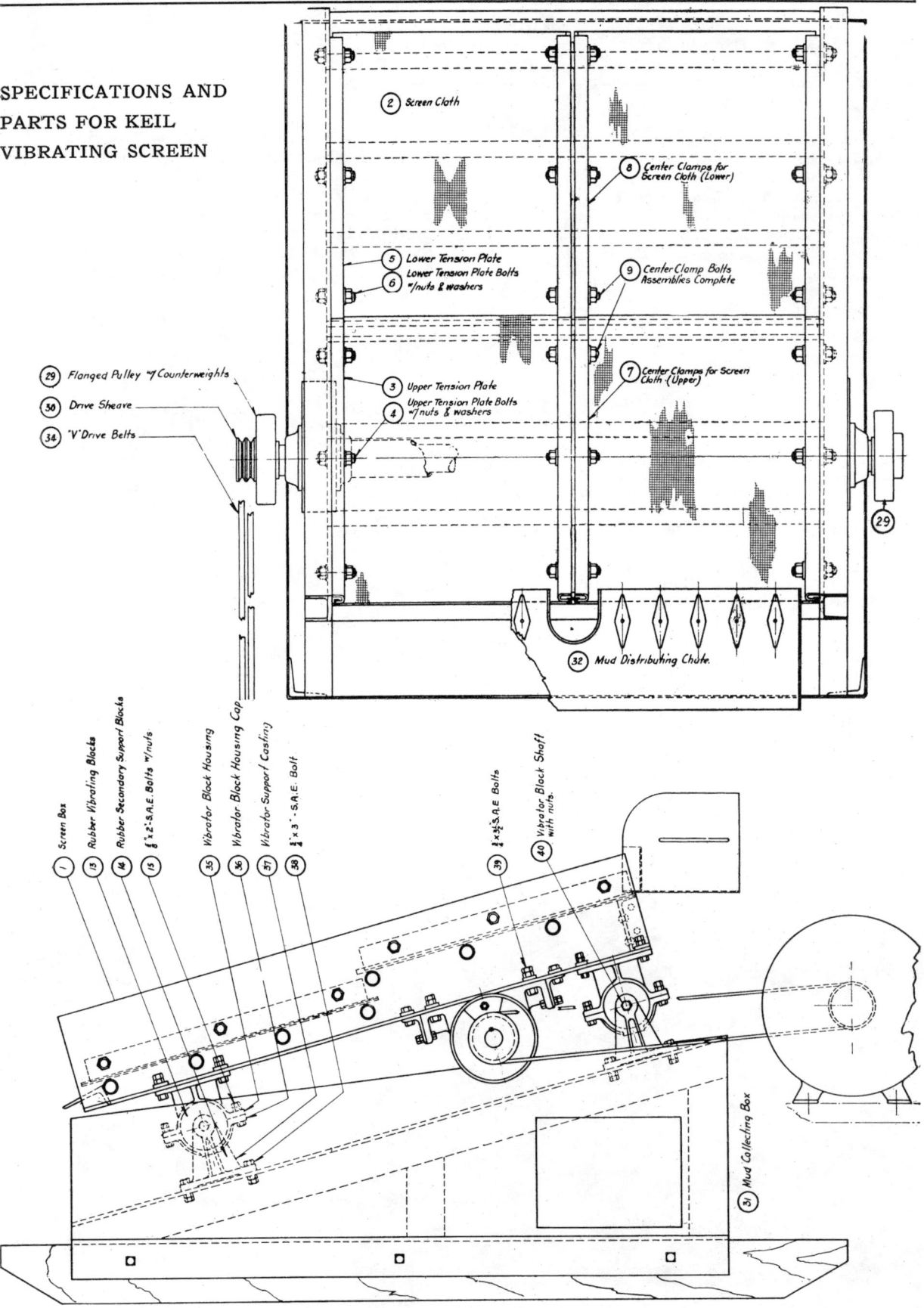

HISTORICAL PERSPECTIVE 21

# THE JEFFREY MANUFACTURING COMPANY
### 876-99 North Fourth Street, Columbus, Ohio
*Sales Office: 3632 Purdue Street, Dallas, Texas*

## OIL WELL CHAINS

Whether drilling shallow or deep, in hard or easy going, you will find it money-saving practice ALWAYS to use a better class of oil well chains ... such as the Jeffrey 103 RC or the 124 RC in the **Rotor Chief** series.

Both of these Jeffrey chains are built to the Standards of the American Petroleum Institute. In them you will find plus values derived by discriminating selection of alloy steels, balanced construction without excess weight, and precision manufacture. Broken in for you at the factory ... by fine finishing of the bearing surfaces to accurate measurements ... they will not stretch or elongate during the first several thousand feet of drilling. And with the initial set taken out, these chains naturally suffer less subsequent wear.

*A.P.I. No. 4—Jeffrey 124 RC*

Your guarantee of "Better" in Jeffrey oil well chains is underwritten by new machine tools in our chain plant ... by a new heat-treating plant equipped with latest control instruments ... and by experience accumulated through 60 years.

Try a better chain on your next job ... ask for **Rotor Chief**.

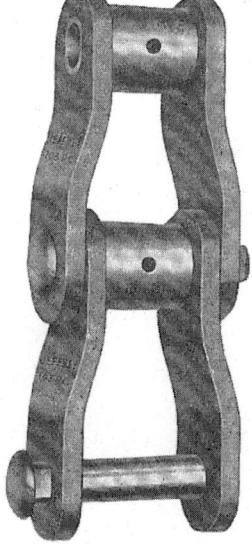

*A.P.I. No. 3—Jeffrey 103 RC*

## SHALE SHAKERS

The Jeffrey Blue Streak Shale Shaker for conditioning rotary mud has been proved by actual tests in the oil fields. It will speed up your rotary drilling operations and, more important, it will save you money.

By eliminating ALL harmful sand, shale and other abrasive material, it adds much to the life of pumps,

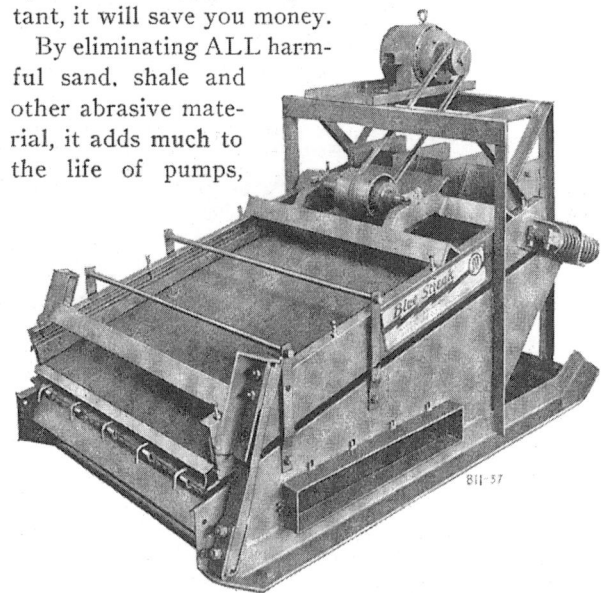

valves, swivels, rotary hose, standpipe connections, etc. The Jeffrey Blue Streak has high capacity and requires little upkeep attention.

The lively business action at every point on the large (4'x6') screen area is made all the more effective by taut stretching crosswise and lengthwise. The Blue Streak is self-contained and of rigid construction, making installation and moving easy. It has a simple vibrator with an unbalanced pulley on a stationary shaft, running on ball bearings in an oil bath ... a provision for adjusting the vibrating stroke while screen is running.

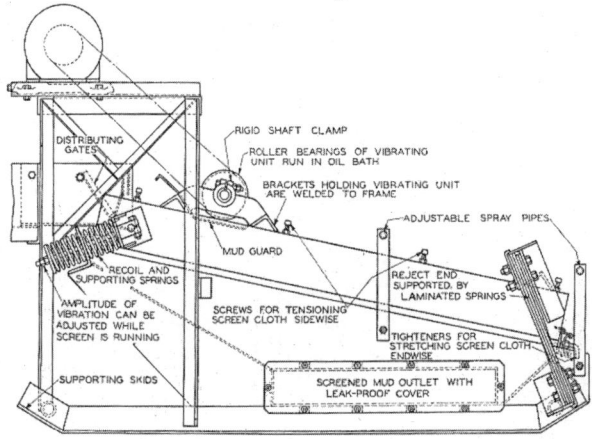

Send for Bulletin No. 669, which goes into detail.

## LUCEY EXPORT CORPORATION
*Exclusive Export Distributor*
### LUCEY PRODUCTS CORPORATION
Tulsa, Oklahoma

### HUDSON-BOUCHER AUTOMATIC SHALE SEPARATOR

The Hudson-Boucher Automatic Shale Separator is of unique design, and entirely different from any other machine which has been introduced for the separation of cuttings and other foreign matter from drilling fluid.

Operated by the gravity flow of the drilling fluid against turbine blades, the machine requires no other motive power. Unlike the conventional "shaker" type of separator, the Hudson-Boucher machine effects complete separation by the rotation of the screen unit as the fluid passes through. Clean mud passes through the screen, down into the return trough, while cuttings are discharged out the open end of the screen cylinder.

The turbine shaft is mounted on roller bearings which are fully sealed off against the entrance of fluid.

The screen proper is 6' 0" long and may be adjusted to proper tension by manipulation of the adjusting bars which run the length of the screen. At normal adjustment it rotates at 50 r.p.m.

On top of the housing a manifold is provided for the introduction of steam and water to thoroughly clean both the screen and cuttings. This feature provides for the salvage of much useful mud that ordinarily clings to the cuttings and is lost forever.

The extreme simplicity of design and construction are outstanding features of the Hudson-Boucher Automatic Shale Separator. With but one moving part, it seldom requires adjustment, repairs or new parts. It may truly be said that the "first cost is last cost."

#### BOXED FOR EXPORT

| | |
|---|---|
| Length Overall | 11' -6" |
| Width Overall | 3' -6" |
| Height Overall | 4' -8" |
| Net Weight | 1150 lb. |
| Gross Weight | 1750 lb. |

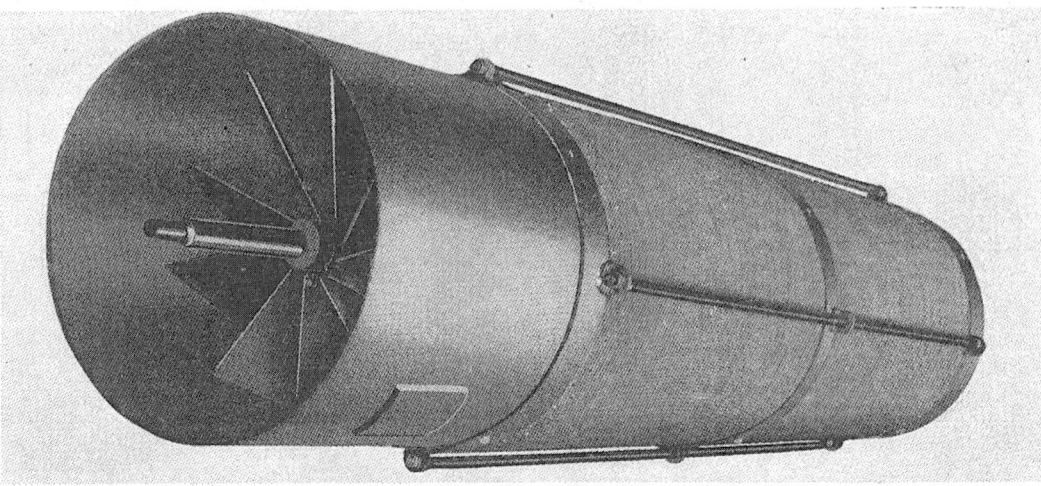

Complete specifications and prices upon request.

# VERNON TOOL COMPANY, LTD.

**2740 East 37th Street**      **Los Angeles, California, U. S. A.**

*Gulf Coast and Mid-Continent Representative*
**McNEELY MATERIALS COMPANY**
2935 Jensen Drive, Houston, Texas

*Kansas and Oklahoma*
**THE BRIDGEPORT MACHINE COMPANY**
*Export:*
**THE NATIONAL SUPPLY CORP.**
**OIL WELL SUPPLY CO.**

### PRODUCTS:

| | | | |
|---|---|---|---|
| McNeely Vibrating Rotary Mud Screens | Suter Heavy-Duty Engines | Steel Engine Foundations | Pitmans |
| Tool Joints | Pumping Units | A.P.I. Gages | Hoist Drum Brake Rims |
| Straight Line Pumping Adjusters | Drill Collars | Fish Tail Bits | Production Crown Blocks |
| | Kellys | Rig Irons | Forged Steel Fittings |
| | Subs | Tubing Heads | Gasoline Storage Tanks |
| | | Christmas Trees | |

## McNEELY VIBRATING MUD SCREEN—MODEL E

Combining large capacity, low maintenance cost and long life, the new, improved Model E McNeely Vibrating Mud Screen fully meets the requirements of the oil industry for an economical and reliable shaker screen to remove sand and shale from drilling mud. Capacity has been increased 15%, screen cloth life doubled, and the vibrating frame double reinforced to withstand severe overspeeding.

The McNeely Screen, equipped with the customary 30 mesh wire cloth, will continuously handle the entire output of the largest mud pumps in most instances. Capacity varies with the character of the formation being drilled, but averages in excess of 650 gallons per minute. This high volume is largely achieved because of the curved screening surface—an exclusive McNeely feature.

Screen cloth is unusually long lived. Operators all over the world report from four to six months service from a single set of cloth. The cloth, made from heavy, corrosive-proof Monel wire, is divided into two sections. Each half is independently tensioned and only one half need be replaced at one time.

The screen is mounted on heavy skids, and steel approach and discharge ditches are incorporated in the unit. This makes installation and removal easy and economical.

A high frequency vibration is imparted equally to all points on the screening surface through rotation of an unbalanced shaft mounted between two oversize, self-aligning ball bearings. The entire vibrating frame is in turn mounted on four pure gum rubber blocks.

The double reinforced, all-welded vibrating frame and sturdy base are built to give years of satisfactory, economical performance in the most severe oil field service.

The McNeely Vibrating Mud Screen may be driven by a 3 H.P. electric motor, gasoline or gas engine, or steam turbine. The alternating current electric motor is recommended whenever power is available. Write for Bulletin No. 101.

### MODEL E BRIEF SPECIFICATIONS

| | |
|---|---|
| Average Capacity, gallons per minute | 650 |
| Horsepower Required | 3 |
| Recommended Speed Range, R.P.M. | 3000-3400 |
| Headroom Required, inches | 25½ |
| Screen Shaft Pulley, 4.0" P.D. | 2VB |
| Net Weight, pounds | 1625 |
| Domestic Boxed Weight, pounds | 1900 |
| Export Boxed Weight, pounds | 2075 |
| Boxed Cubic Displacement, cubic feet | 185 |
| Overall Length, inches | 93 |
| Overall Height, inches | 40 |
| Overall Width, inches | 72 |

# THE INTERNATIONAL NICKEL COMPANY, Inc.

*Producers of Monel, Nickel, and Nickel Alloys*
67 Wall Street, New York
*Distributors throughout the United States*

## PRODUCTS

MONEL AND NICKEL strip, sheet, plate, rods, flats, angles, special shapes, tubing and pipe, forgings, castings, wire, hardware and fittings. Monel-jacketed polished rods.

## SPECIAL PRODUCTS

Inconel, "Z" Nickel (heat-treatable) "K" Monel (heat-treatable). "R" Monel (free machining), "H" and "S" Monel castings having high hardness, 175-275 and 275-325 Brinell, respectively.

## MONEL IN PRODUCTION EQUIPMENT

Monel is a technically controlled two-thirds nickel, one-third copper alloy, highly corrosion resistant, tougher and stronger than mild steel. These properties make it highly desirable for production equipment such as that described below.

## MONEL-JACKETED POLISHED RODS

*Cross section of 1⅛" diameter Monel jacketed polished rod with core of S.A.E. 4615 steel. Note tight bond between Monel and the steel core.*

Polished rods are now available from stock in 11 and 15 foot lengths, in diameters of 1⅛", 1¼" and 1½". Other sizes are available on order. The jacket is Monel .049" in thickness, and the core is S.A.E. 4615 steel. Production men have long been aware that Monel polished rods outlast plain carbon steel from ten to twenty times. Monel-jacketed polished rods were tested for a period of four years before being placed on the market. The price of the new rod is less than half that of solid Monel, yet it provides the surface smoothness and corrosion resistance of Monel with the high strength of an alloy steel core. Further information on request.

## GAS LIFT DEVICES

Monel springs and needles of gas lift devices resist corrosion by hydrogen sulfide, brine, sour oil, the abrasion of entrained sand, and the cutting action of high pressure gas. "K" Monel is also used in some makes of these units.

## MONEL MUD SCREENS

Vibrating Mud screens of Monel resist the effect of hydrogen sulfide, salt water, and sour crudes, which are so corrosive to ordinary metals. Its high mechanical strength, resistance to abrasion and fatigue make it ideal for this service. Rectangular mesh is usually used.

Pure nickel well strainers are being used to guard against loss of efficiency due to corrosion, and are coming into greater prominence as old wells are being reworked.

*Rectangular Monel wire screen as used on various makes of vibrating mud screens.*

## OTHER USES OF MONEL

Meter parts, wire rope and screen, hardware and fittings, valve parts, valve trim, "Durabla" pump valve discs, springs, etc.

## "K" MONEL

"K" Monel is a hardenable alloy possessing all the corrosion resistance of regular Monel. Hardnesses over 300 B.H.N. and strengths over 165,000 psi. are induced by heat treatment after fabrication. It is non-magnetic, a property which makes it extremely valuable for well sounding devices.

## DROPS AND SEATS AND BALLS AND SEATS

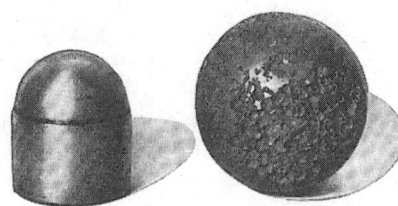

*Left: "K" Monel drop valve after 14 months' service in oil well, pumping corrosive crudes, salt water, and hydrogen sulfide. Spherical ball to its right shows the effect of the same crude on a steel ball after 5 days. Maximum life in this field for materials other than "K" Monel is approximately 45 days.*

"K" Monel drops and seats, and balls and seats, are sold under the following manufacturers' trade names: KA Alloy, Lumonel, Kamonex, Tuff-Temper. The hardness of "K" Monel, its high strength and corrosion resistance, and its toughness make it the longest lived—and therefore the most economical—material for drop and seat and ball and seat valves. Send for further details.

## GAS LIFT DEVICES

Monel springs and needles of gas lift devices resist corrosion by hydrogen sulfide, brine and sour oil. Since it is a very tough and wear-resisting metal it also withstands better than most materials the abrasion of entrained sand and the cutting action of high pressure gas. "K" Monel is also used in some makes of these units.

*Left: Gas lift device with Monel working parts. Monel valves and springs resist corrosion and abrasion better than any other material used.*

## FREE LITERATURE

"Nickel Bearing Alloys in the Production and Drilling of Petroleum."

C-2 Monel and Nickel in Oil Refining

H-1 Monel Accessories

H-3 Monel Wire Screen and Filter Cloth

T-5 Engineering Properties of Monel

T-9 Engineering Properties of "K" Monel

List "B"—Listings of available publications on Monel, nickel, and Inconel.

# LINK-BELT COMPANY

SALES OFFICES AND WAREHOUSE STOCKS AT

| PHILADELPHIA | DALLAS | LOS ANGELES | INDIANAPOLIS |
| --- | --- | --- | --- |
| 2045 W. Hunting Park Ave. | 413 Second Ave. | 361 S. Anderson St. | 220 S. Belmont Ave. |

OTHER SALES OFFICES: HOUSTON, 517 Southern Standard Life Bldg. TULSA, 1527 S. Atlanta Place
EXPORT SALES: New York, N. Y., 2680 Woolworth Bldg. Cable and Radio Address: "LINKBELT"

SOLD BY MOST SUPPLY HOUSES

## LINK-BELT VIBRATING ROTARY MUD SCREEN
### Reconditions—Degasses—Cleans—Maintains Uniform Weight of Mud

**Made in Two Sizes, 48″ x 60″ and 24″ x 48″, for Shallow, Medium and Deep Well Drilling.**

Mud cleaned with the Link-Belt vibrating screen—

Makes a better and smoother wall, and lessens the likelihood of cave-ins; and will help to drill a straight hole.

Lengthens the life of pump parts, swivels, drill pipe, rotary hose and stand pipe connections. Also bits and reamers.

Permits better control of gas pressures. Should a blowout develop, the clean mud always available in the pit can be pumped back into the hole, and drilling operations continued with minimum delay.

Saves drilling time. After coming out of the hole to change bits, it is much easier to get back on bottom through properly screened mud. In the case of unscreened mud, heavy shale particles settle out during the course of the round trip when circulation cannot be maintained, making it necessary to sometimes drill as much as several hundred feet of settled shale before new hole can be resumed.

### MANY UNIQUE FEATURES

Uniform vibration of the entire screening area. No dead spots.

All rotating parts are carried on anti-friction self-aligning roller bearings enclosed in a single dirt and weather-proof housing.

Screen cloth rests on hollow gum rubber tubes. An important, exclusive, patented and proved feature which assures maximum service, and assists in preventing loss of good mud in the rejects. This feature permits the use of finer mesh screen cloths than ordinarily could be used.

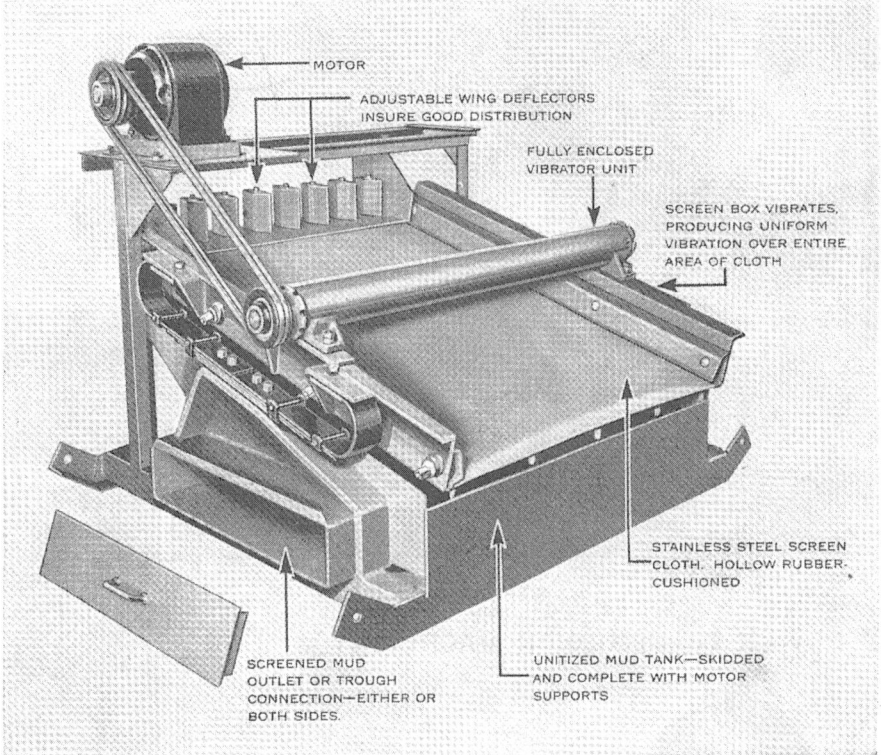

*The Link-Belt Unitized Vibrating Screen for Rotary Mud includes Screen Box, fitted with single piece, hollow-rubber cushioned stainless steel cloth with tension members; Belt Drive; Mud Tank with two-way outlet, made self-contained for supporting screen box and motor (we recommend the use of 2 horsepower, 1800 r.p.m., splash proof electric motor but a steam turbine may be used when required); Distributing Chute with control gates and necssary tools.*

A small, compact, self-contained unit requiring minimum head room, and readily transportable from one location to another.

Change of screen cloth can usually be made in not more than fifteen minutes.

The screen cloth is of fine mesh, enabling it not only to reject large, heavy cuttings, but also sand, thus thoroughly cleaning the mud.

Large screening area results in maximum capacity. The Link-Belt screen will handle more mud for a given size screen cloth opening.

The unusually long life obtained from the screen cloth is one of the reasons for the low operating cost of the Link-Belt unit.

Comparatively low speed at which the Link-Belt screen operates, results in longer life of all parts.

The feed chute is an integral part of the Link-Belt unit. This eliminates all guesswork in setting, to get the proper flow of fluid to the screen surface.

### USED THROUGHOUT THE WORLD

Link-Belt Vibrating Mud Screens have been and are being used successfully in practically every oil field in the world employing the rotary system of drilling: United States, Mexico, Canada, Argentine, Colombia, Peru, Venezuela, Trinidad, Austria, Germany, Italy, Jugoslavia, Rumania, Iran, Irak, India, Japan, Persia, etc.

*Where an abnormal shale problem justifies, as the job pictured here, a Link-Belt double screen hook-up is employed.*

# W-K-M COMPANY, Inc.   W-K-M   HOUSTON, TEXAS

## W-K-M ROTARY SLIPS
### Type 35—Friction Slips

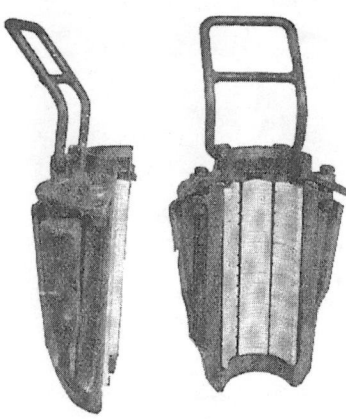

The latest advancement in rotary slips comes from the pioneer slip manufacturer in this improved design, thoroughly field tested and now protecting the longest pipe strings. Allows more gripping area, greater holding capacity, and a soft action. Takes hold or releases instantly, and will not slip, cut or bottleneck your pipe. Extreme light weight permits ease of handling. Highly flexible. Should wear occur on segments, they're readily replaced on the derrick floor, or redressed, at a nominal cost, making slips same as new. Long life assured by highest quality one-piece, drop-forged and heat-treated segments.

## W-K-M TUBING SLIPS

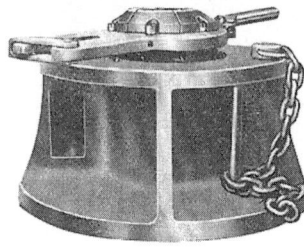

*Series 12,000*

The original flexible, one-man tubing slip is furnished in all sizes to 4½-inch inclusive. Slips are forged steel throughout; and spider is top-grade steel casting. Segments may be redressed any number of times. Made in Regular type, for depths to 3000 feet, Deepwell type, for depths to 6000 feet, and the newest and latest Series 12,000, for depths to 12,000 feet.

*Write for Pamphlet No. 320.*

*Deepwell Type*     *Regular Type*

## W-K-M MUD SCREEN — SERIES 1100

The W-K-M Mud Screen employs a very simple means of operation; the flow of returns from the well under a pelton wheel rotates in turn the screening drum. This is not a whipping action, and consequently screen cloths last indefinitely.

Operating expense is exactly nothing . . . maintenance cost is negligible due principally to the total absence of turbines, motors or generators. Nor is there any destructive vibration to cause frequent shut-downs for repairs or screen cloth replacements.

Good clean samples of cuttings may be taken at any time, as they are thoroughly cleaned of mud before being discharged.

For extremely heavy going, a mud motor, connected to your mud pump, provides that extra power or accelerated rotation to discharge unusually heavy loads of shale, or to quickly clean out the screen.

All parts of the W-K-M Mud Screen are designed for ease of operation and long life. Frictionless bearings carry

all moving shafts and are fitted for Alemite lubrication. Driving gears run in a bath of oil.

All bearings are thoroughly sealed off against the entrance of mud or other foreign matter.

Final drive for drum is by a rigidly mounted, cut steel sprocket, meshing with a special high-grade roller chain, which lasts indefinitely.

*Write for fully descriptive literature.*

## W-K-M KING PIN BRAKE LINING

Designed particularly to dissipate the extreme heat generated in the brake flanges, and to avoid scoring or grooving of brake flanges, naturally increasing the life of both lining and flanges. The pins spaced throughout the lining not only conduct the heat away from the flanges, but provide a thin film over the braking surface, preventing scoring, and maintaining a constant friction. Pins are forced into the lining by special, patented machines, under terrific hydraulic pressure, where they are expanded after the manner of a rivet, and become an integral part of the lining. Prices of King Pin Lining are no higher than for ordinary lining.

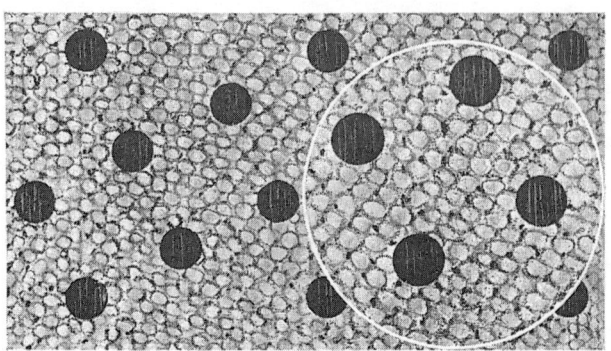

# CHAIN BELT COMPANY *of* MILWAUKEE

## REX OIL WELL DRILLING CHAINS

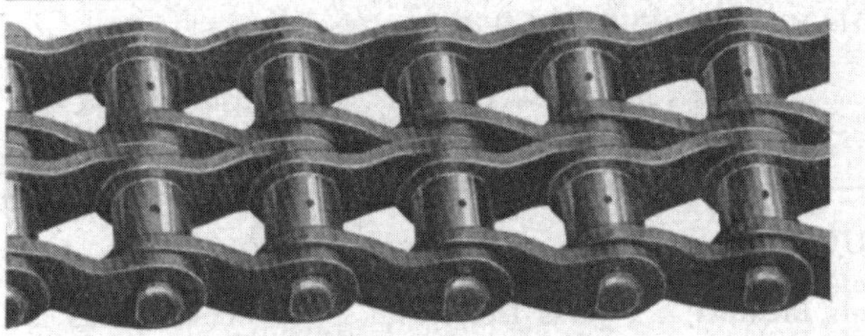

*Rex Universal Double 3125*

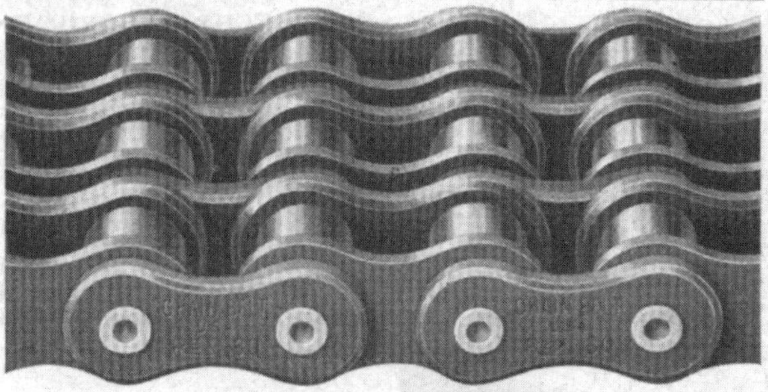

*Rex Triple Strand Roller Chain*

### REX UNIVERSAL

Rex Universal Chabelco is furnished both as a single and double strand chain for desired strength. It is made along the same precision lines as Rex Deepwell and Rex Champion with alloy steel parts, ground pins and bushings, off-set side bars, and Rex Unit Link construction. Rex Universal chains are new standard 3⅛-inch pitch chains for rotary drilling. These chains have the pitch substantially shortened so that much higher drilling speeds are possible. Rex Universal Chains are not interchangeable with A.P.I. 3 or A.P.I. 4 chains and should be ordered for new draw works or complete replacement of both chain and sprockets on existing rigs.

### REX ROLLER CHAINS

Rex Roller Chains are built with the same high standards of design and manufacture that distinguish Rex rotary drilling chains. They are the product of skilled engineers, experienced in chain requirements. All parts are carefully selected for the part they must perform and are highly treated and finished. The result is that these chains are durable in service, smooth in operation, quiet at high speeds and uniformly high in quality. They are used in the oil field for various purposes, as drilling rig drives, as well as for a wide variety of pumps, slush pumps and similar auxiliary applications.

### REX ROLLER CHAINS—Manufacturers' Standard Series

| REX CHAIN No. | Dimensions, Inches | | | Average Ultimate Strength—Lbs. | | | |
|---|---|---|---|---|---|---|---|
| | Pitch | Roller Width | Roller Diameter | Single Strand | Double Strand | Triple Strand | Quadruple Strand |
| 35...... | 3/8 | 3/16 | 0.200 | 2,100 | 4,200 | 6,300 | 8,400 |
| 40...... | 1/2 | 5/16 | 5/16 | 3,700 | 7,400 | 11,100 | 14,800 |
| 41...... | 1/2 | 1/4 | .306 | 2,000 | ...... | ...... | ...... |
| 50...... | 5/8 | 3/8 | .400 | 6,100 | 12,200 | 18,300 | 24,400 |
| 60...... | 3/4 | 1/2 | 15/32 | 8,500 | 17,000 | 25,500 | 34,000 |
| 80...... | 1 | 5/8 | 5/8 | 14,500 | 29,000 | 43,500 | 58,000 |
| 100...... | 1¼ | 3/4 | 3/4 | 24,000 | 48,000 | 72,000 | 96,000 |
| 120...... | 1½ | 1 | 7/8 | 34,000 | 68,000 | 102,000 | 136,000 |
| 140...... | 1¾ | 1 | 1 | 46,000 | 92,000 | 138,000 | 184,000 |
| 160...... | 2 | 1¼ | 1⅛ | 58,000 | 116,000 | 174,000 | 232,000 |
| 200...... | 2½ | 1½ | 1 9/16 | 95,000 | 190,000 | 285,000 | 380,000 |

In Chain Numbers—Prefix "D" denotes double strand; "E" denotes triple strand; "F" denotes quadruple strand. Data on Special Chains on request.

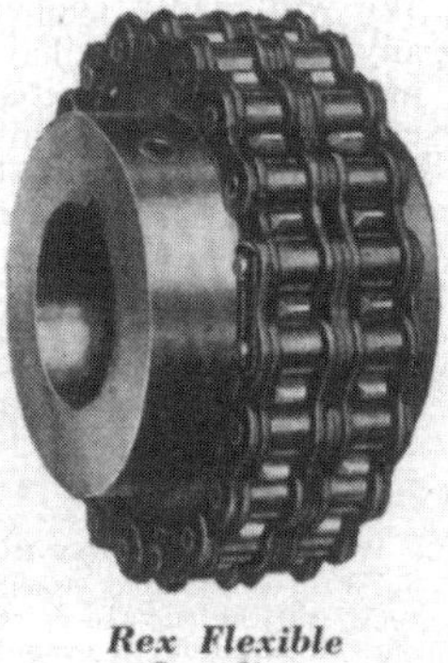

*Rex Flexible Coupling*

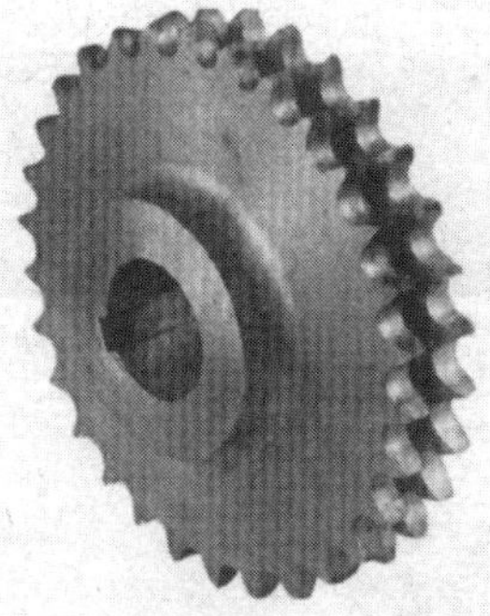

*Rex Double Cut-Tooth Sprocket*

### REX MUD CONDITIONERS

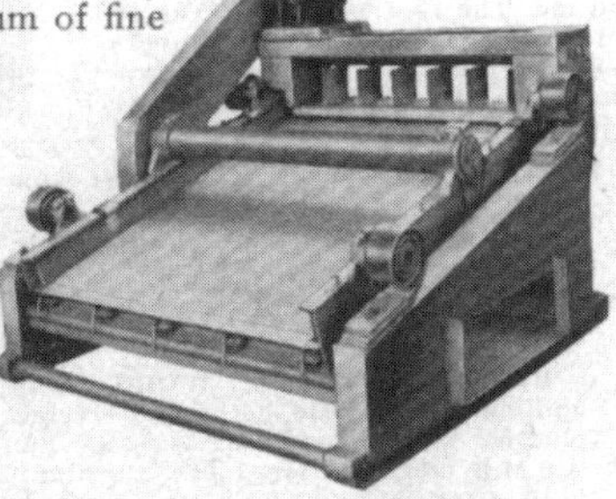

Rex Mud Conditioners produce clean mud of uniform consistency and weight, free from shale and clay cuttings and with a minimum of fine sand. Clean mud saves drilling time, lengthens the life of drilling tools and equipment, and reduces the hazards of gas blowouts. Rex Mud Conditioners simplify the entire mud system by eliminating large settling pits and by reducing the quantity of mud in circulation.

### SPECIFICATIONS: REX MUD CONDITIONERS

Screen Size: 5′0″ x 4′0″
Clear Screen Area: 20 square feet.
Net Weight: 1550 Lbs. (Approx.)
Shipping Weight: 1700 lbs. (Domestic) 2000 lbs. (Export) Approx.
H. P. Required: 3
Capacity: 700 g.p.m. (Approx.)

### REX FLEXIBLE COUPLINGS

Rex Flexible Couplings are of the new double chain type which permits boring for larger shafts and affords greater horsepower capacity for a given outside diameter. They are simple in design and are of rugged all-steel construction with no intricate or delicate parts to get out of order or to break in service. They are quickly connected or disconnected.

### REX CUT TOOTH SPROCKETS

Rex Cut Tooth Sprockets are cut with a tooth form designed for maximum efficiency, quiet operation and long life for both chain and sprockets. They are made in single and multiple widths for all standard roller chains. Sprocket accuracy is essential to satisfactory roller chain performance. The pitch must be exact, the tooth profile must be held within close limits. Rex Cut Tooth Sprockets fulfill these requirements.

**Other Rex Products:** Rex Conveyors for handling fuller's earth and petroleum coke; Rex barrel and package elevators; Rex Speed Prime Water Pumps, Rex Concrete Mixers, Rex Moto-Mixers, Rex Pumpcretes.

**CHAIN BELT COMPANY, 1614 W. Bruce St., Milwaukee, Wisconsin**

Cable Address: "Beltchain"

**REX OIL WELL CHAINS CARRIED IN STOCK BY**

**MID-CONTINENT AND GULF COAST DISTRIBUTORS**

Frick-Reid Supply Corporation
Lucey Products Corporation
Norvell-Wilder Supply Company
Rex Supply Company
Industrial Supply Company, Inc.
Bethlehem International Supply Company
Murray-Brooks Hdwe. Company, Ltd.
Coastal Supply Company
Dunigan Tool and Supply Co.

**DISTRIBUTED IN CALIFORNIA BY LEADING SUPPLY HOUSES**

**Export:** CHAIN BELT COMPANY, News Building, New York; LUCEY EXPORT CORPORATION, Woolworth Building, New York; WILLIAM E. KNIGHT, 14-15 Lancaster Place, Strand, London W. C. 2, England

# GULF ENGINEERS, Inc.        HOUSTON, TEXAS

## "JITTERBUG" SHALE SHAKER

### For Standard Duty Use

This shaker embodies all of the features of the KEIL Heavy Duty Shale Shaker, including four piece screen cloth, step deck construction and complete rubber mounting.

A lighter rotor assembly is furnished on this shaker for use where the heavier type would not be necessary. The "Jitterbug" is sold completely equipped with screen cloth and portable steel skid base.

## GULF ENGINEERS BUG BLOWER

### SPECIFICATIONS:

**Propeller:**
56" 3 Blade Cast Aluminum.

**Guard:**
1½" Safety Mesh Expanded Metal, welded to frame.

**Turbine:**
Gulf Engineers Turbodrive, 7½ H.P. at 1200 r.p.m.

Maximum Steam Pressure 350 lbs.
Shipping Wt. 500 lbs.

# HUTCHISON ENGINEERING WORKS   Chicago, Ill., Houston, Tex.

## STEAM TURBINES

*For driving Mud Screens and Belted Generators*

| Size of Turbine in Inches | Catalog Number of Turbine | Turbine Speed Ranges | | Steam Inlet Pipe Size | Exhaust Connec. Pipe Size | Shipping Wt., Lbs. | Price |
|---|---|---|---|---|---|---|---|
| | | Min. | Max. | | | | |
| 9 | 9 DS | 1600 | 4000 | ¾" | 2" | 225 | $305.00 |
| 12 | 12 DSST | 1600 | 4000 | 1" | 3" | 350 | 465.00 |

The fact that FLOLIGHT turbines can be adjusted for speed while running under load makes them a most desirable source of power for this service. Our engineering department will be glad to have you call on them regarding speed of turbine and pulley sizes pertaining to installations of every kind.

A steam strainer is furnished with all turbines. The horse power of turbines is dependent upon steam pressure and speed. For example: (Slower speeds less horse power.)

At 250 lbs. steam pressure and 3600 RPM turbine speed, the 9 inch turbine will deliver 9 horse power or pull a 5 KW generator and the 12 inch turbine will deliver 17 horse power or pull a 12 KW generator. Please give speed and steam pressures.

## "FLOAIR" FORGE BLOWERS

No. 4 "Floair" Blower

No. 925 "Floair" Blower

**These blowers operate at a speed of 3400 RPM**

The No. 4 is capable of supplying enough air at sufficiently high pressures to heat the largest bits, having a maximum capacity of 650 cubic feet of air per minute at zero; free opening and 460 cubic feet at 2.6 inches static pressure (average operating condition). It is powered with a ½ HP heavy duty motor and furnished with a fitting tapped for standard 4 inch pipe for connecting up the blower to the forge.

The No. 925 is for service on smaller tools, although large tools may be heated, as it is equipped with a narrow highly efficient type of fan, but it will, of course, take more time than with the larger blower. It has a maximum capacity of 325 cubic feet per minute at zero; free opening and 250 cubic feet at 2.46 inches static pressure; powered with a ¼ HP heavy duty type of motor. The fitting on the blower is tapped for 2½ inch standard pipe to facilitate connecting it up to the forge.

| Floair Blower Number | CATALOG NUMBERS | | | | List Price F.O.B. Factory | Domestic Shipping Weights | |
|---|---|---|---|---|---|---|---|
| | DC 110-V | Single Phase 60 Cycle AC 110-V | Single Phase 60 Cycle AC 220-V | Three Phase 60 Cycle AC 110-V | | Net Lbs. | Ship. Lbs. |
| Type G-4 with Grinding Wheel | 40-D | 41-IR | 42-IR | 44-TF | $105.00 | 127 | 150 |
| Type 4 less Grinding Wheel | 45-D | 46-IR | 47-IR | 49-TF | 90.00 | 121 | 144 |
| Type 925 with Grinding Wheel | 9251-D | 9252-IR | 9253-IR | 9254-TF | 85.00 | 96 | 121 |
| Type 925 less Grinding Wheel | 9255-D | 9256-IR | 9257-IR | 9258-TF | 75.00 | 90 | 115 |

## "RUMBA" No. 1 MUD SCREEN (Shale Shaker)
(Patent Applied For)

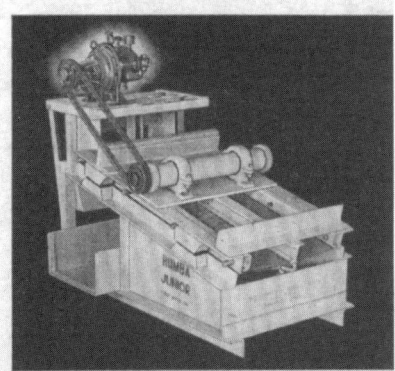

Equipped with underslung top and side driven, quick change, long life screen cloth; a rubber cushion mounted vibrator to give maximum life to the screen frame, which is also fully rubber floated. A mud bypass is provided and the entire unit completely unitized with a mud tank to equalize flow distribution—ready for the field, on 10-foot skids. Screen size 48" x 60".

No. 4860—LESS DRIVING UNIT
  Shipping weight 2500 pounds.
  LIST PRICE F.O.B. Houston, Texas.................$1045.00
No. 4860-T—WITH STEAM TURBINE
  Shipping weight 2725 pounds.
  LIST PRICE, F.O.B. Houston, Texas................$1345.00
No. 4860-M—WITH ELECTRIC MOTOR
  Shipping weight 2700 pounds.
  LIST PRICE, F.O.B. Houston, Texas................$1250.00
No. 4860-E—WITH GASOLINE ENGINE
  Shipping weight 2950 pounds.
  LIST PRICE, F.O.B. Houston, Texas................$1340.00

Export shipping weight approximately 3500 pounds. Export dimensions, length 10' 7", width 8' 3", height 4' 1". Plus $80.00 for export boxing.

The above prices include all belts and pulleys. Motor starting box is included with Cat. No. 4860-M.

27 inches of head room required.

## "RUMBA" JUNIOR MUD SCREEN
(Shale Shaker)
(Patent Applied For)

This smaller Junior type of vibrating screen is offered for service where the conditions will permit the use of a smaller screen. The general construction is the same as the Number One Shaker, except that it is narrower.

We retain the length of the larger shaker so as to be able to catch high velocity, or larger volumes of mud on the longer screen cloth. Cloth size 27" x 60".

Prices are F.O.B. Houston, Texas, and without power.
Price Vibrating Screen only...........................$575.00
Price Unitized with 9' 6" Skids and built in mud
  tank to reduce velocity flow......................$695.00
Price Unitized with Mud By-passes..................$765.00

## MOTOR AND STARTING BOX

Weatherproof Flolight 2 H.P. Shale Shaker Motor with 10-foot all-rubber 3-conductor cable, attachment plug and pulley. Shipping weight 175 pounds. Price..................$127.00
Motor Belt Tightening Slide Rails. Shipping Weight 20 pounds. Price $6.00
Weatherproof Flolight Motor Starting Box with Circuit Breaker Receptacle completely wired. Shipping weight 40 pounds. Price..................$37.50

## HUTCHISON ENGINEERING WORKS — Chicago, Ill., Houston, Tex.

### MUD SCREENS—(Shale Shakers)

**PATENT ALLOWED**

### THIS RUBBER MOUNTED VIBRATOR IS THE FEATURE THAT MAKES "RUMBA" MUD SCREENS (SHALE SHAKERS) LAST

As the frame that carries the screen cloth is driven by a vibrator that is fully CUSHIONED IN 5 RUBBER LINED PILLOW BLOCKS which are bolted to 5 steel lateral struts, under which the screen cloth is stretched; it is plain that the steel frame will not fatigue and break; and that the screen cloth is positively driven through these 5 lateral frame members that are 5 feet long; they drive the screen cloth through two side hooks and center struts or a total of 25 feet of friction hold; with this hold on the screen cloth it cannot whip and break as it must move with the frame. When screen cloths on "RUMBA" shakers need replacement they are really worn out by the abrasive materials going over them, they are not whipped out.

A further advantage of this design is that it only takes three rubber cushion tubes for fenders between the screen cloth and the steel frame.

### LARGE AND SMALL SHAKER SPECIFICATIONS

**Screen Cloth** for No. 4860-B has 19 square feet of screening area. Price $1195.00.

**Screen Cloth** for No. 2760-B has 10 square feet of screening area. Price $895.00.

**Skids:** Are 9' 6" long, just the right length for truckmen to handle.

**Main Frame:** Is made of 3/8" structural steel. It will take years to rust out.

**Mud Boxes:** Are made of 12 gauge steel, red leaded with a coat of machinery enamel over red lead.

**Screen Top Supports:** Top vibrates on top of 4 heavy rubber blocks that toe into stirrups. The full height of the rubber blocks is used. There are no bolts holding the top to these blocks. Top is full floating.

**Mud By-pass:** Is 8 5/8" OD with 8 V threads. The 45-degree bends are welded, and by-pass turns on the threads that are screwed into a collar that is welded into the side of the mud box.

**Bearings:** Heavy roller bearings support the shafts on both sizes. When greased, the new grease forces the old grease out of the bearings to an over-flow.

*This rear and side view of both large and small model shakers shows the collar welded in the back of the mud tank to which the 8 5/8" O.D. 8 V-thread flow line from the well is to be connected. This design saves building mud troughs and a lot of rigging-up time. The mud by-pass is shown in the up position to run the mud over the screen cloth.*

**FRONT VIEW OF MOTOR DRIVEN SHAKER LARGE MODEL**
*Showing the underslung screen cloth and the mud by-pass down in position to by-pass the returns. (Motor is not included.) In Catalog No. 4860-B.*

**FRONT VIEW OF THE JUNIOR MODEL WITH A 4-CYLINDER ENGINE DRIVE.**

(As this is a 24-hour service we recommend that a 4-cylinder slow speed engine be used.)

This small shaker was designed for small power rigs or for two of them to be used on large rigs. (Engine is not included.) In the Catalog No. 2760-B.

# OVERSTROM & SONS

**2217 WEST MISSION ROAD** — ENGINEERS AND MANUFACTURERS — **ALHAMBRA, CALIFORNIA**

Overstrom Products May Be Ordered Direct from the Manufacturer or Through Your Regular Supply House

**REPRESENTATIVES**

**MEDEARIS OILWELL SUPPLY CO.**, *Exclusive California Distributors*, 8638 Otis Street, South Gate, California
**PETROLEUM MACHINERY CORPORATION**, *Export Representatives*, 30 Rockefeller Plaza, New York City, N. Y.

## OVERSTROM *VIBRATING* ROTARY MUD SCREENS

U. S. Patents No. 1,621,949; 1,713,143; 1,995,435; 2,062,760; 2,136,950; 2,199,596; 2,204,379. Other patents pending.

Overstrom Vibrating Rotary Mud Screens thoroughly remove sand and cuttings from the mud and ensure that only clean fluid will be returned to the pumps.

The vibrating screen box operates in perfect balance at all times. This balance is obtained by mounting the springs with their centers on a horizontal line passing through the center of gravity of the screen box. The vibrator element is in alignment with the center of gravity of the screen box. The vibrator means is an unbalanced pulley shaft mounted on two self-aligning S.K.F. spherical roller bearings, shaft and bearings being completely enclosed and running in a bath of oil. This mechanism is simple and sturdy and the vibrator element can be reversed end for end to permit driving from either side. Oversize springs transmit vibration to the screen box most effectively and without waste of power. An exclusive feature, the **flexible closure seal**, interposed between the discharge end of the mud screen and the mud tank, prevents any possible leakage of the screenings back into the cleaned mud.

Stainless steel screen cloth is stretched in two directions over curved longitudinal supports which are provided with rubber inserts secured in dovetail slots. The tautness and adequate cushioned support of the screen cloth greatly prolongs its service life. As metal binders are required only on the transverse edges, replacement screen cloth of 20, 30, 40, or 60 mesh is shipped rolled in small containers.

Overstrom Vibrating Rotary Mud Screens embody thoroughly proved and practical features which have been developed through experience gained over a period of more than 20 years devoted to designing and manufacturing vibrating screen equipment.

Write for bulletin MS-20A.

*Model S.M. 1 (4½' x 5') Overstrom Vibrating Rotary Mud Screen. Note the rigid, unitized construction. The Model M.S.-34 (3' x 4') machine is smaller, but similar in design and construction.*

### SPECIFICATIONS
Model S.M.-1 (4½'x5'0") Rotary Mud Screen
POWER—Electric Motor ..........2 H.P.
  Steam turbine or gasoline
  engine ................3 H.P.
Note: Prime mover is extra, and not furnished with mud screen unless specified. When ordering, give complete specifications of prime mover desired.
SCREEN CLOTH, stainless steel, 20 or 30 mesh, regularly furnished with machine. (Stainless steel cloth of 40 or 60 mesh will be furnished on order at extra cost).
R.P.M. of Vibrator Pulley.....1750 to 1800
V-Belt, one required..............No. 97-B
(Belt and Belt Guard furnished with machine)
Vibrator Shaft Pulley (furnished with machine) 7.4 P.D. Single "B" Groove, 1¼" bore, ¼" keyway.
Headroom required for
  installation ....................27 inches
Weight, uncrated ............1550 Pounds
Weight, boxed for
  export shipment ...........2550 Pounds
Volume, boxed for
  export shipment .........164½ Cu. Ft.

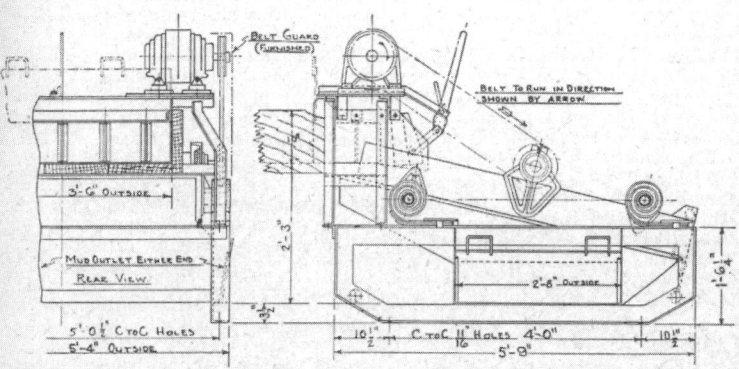

Dimensional Data of Model S.M. 1 (4½'x5') shown in Diagram.

## LINK-BELT COMPANY

### LINK-BELT MUD SCREENS FOR SHALLOW, MEDIUM AND DEEP OIL DRILLING

The Link-Belt Mud Screen will give everything a driller could ask for in design, performance, efficiency and dependability. Units are regularly equipped with screen box, fitted with a single-piece stainless steel screen cloth cushioned on Neoprene strips; cloth tensioning members; "V" belt drive (single for 24" x 48" size and double for 48" x 60" size), mud tank with two-way outlet, made self-contained for supporting screen box and motor; distributing chute with control gates; and the necessary tools.

**Requires Little Power.** We recommend the use of an 1800 r.p.m. splashproof electric motor (1½ h.p. for 24" x 48" and 2 h.p. for 48" x 60" screen), but a steam turbine, gas or Diesel engine may be used.

**Efficient Separation and Great Wear Resistance.** Swinging inlet gates deflect mud downward against a perforated bottom plate, assuring good distribution and control of flow—less wear on screen cloth and effective utilization of entire screening area. When not in service, gates can be closed tight to halt the flow of mud.

**Great Capacity** with minimum loss of good mud. Screen cloth and box vibrate as a unit, with uniform intensity—no dead areas—quick discharge of cuttings.

**Clean Mud.** Fine-mesh stainless steel screen cloth rejects sand and shale as well as larger cuttings, thoroughly cleaning the mud. Cloth is mounted on acid- oil- and heat-resisting Neoprene tubes for maximum life. Cloth, when it does wear out, is easy to replace. Cuttings are prevented from entering the cleaned mud trough by an effective sealing means.

**Easily Transportable.** Unit is small, compact, and self-contained . . . requires minimum head room.

**Simple Installation.** Outlets from the mud tank are provided at either or both sides for easy connection with flumes.

**Rotating Parts Have Long Life Expectancy.** All rotating parts are carried on anti-friction self-aligning roller bearings, enclosed in a single, dirt- and weather-proof housing.

# HISTORICAL PERSPECTIVE 33

## LINK-BELT COMPANY

### LINK-BELT MUD SCREEN—TYPE NRM-124—MODEL NO. 45—24"x48"

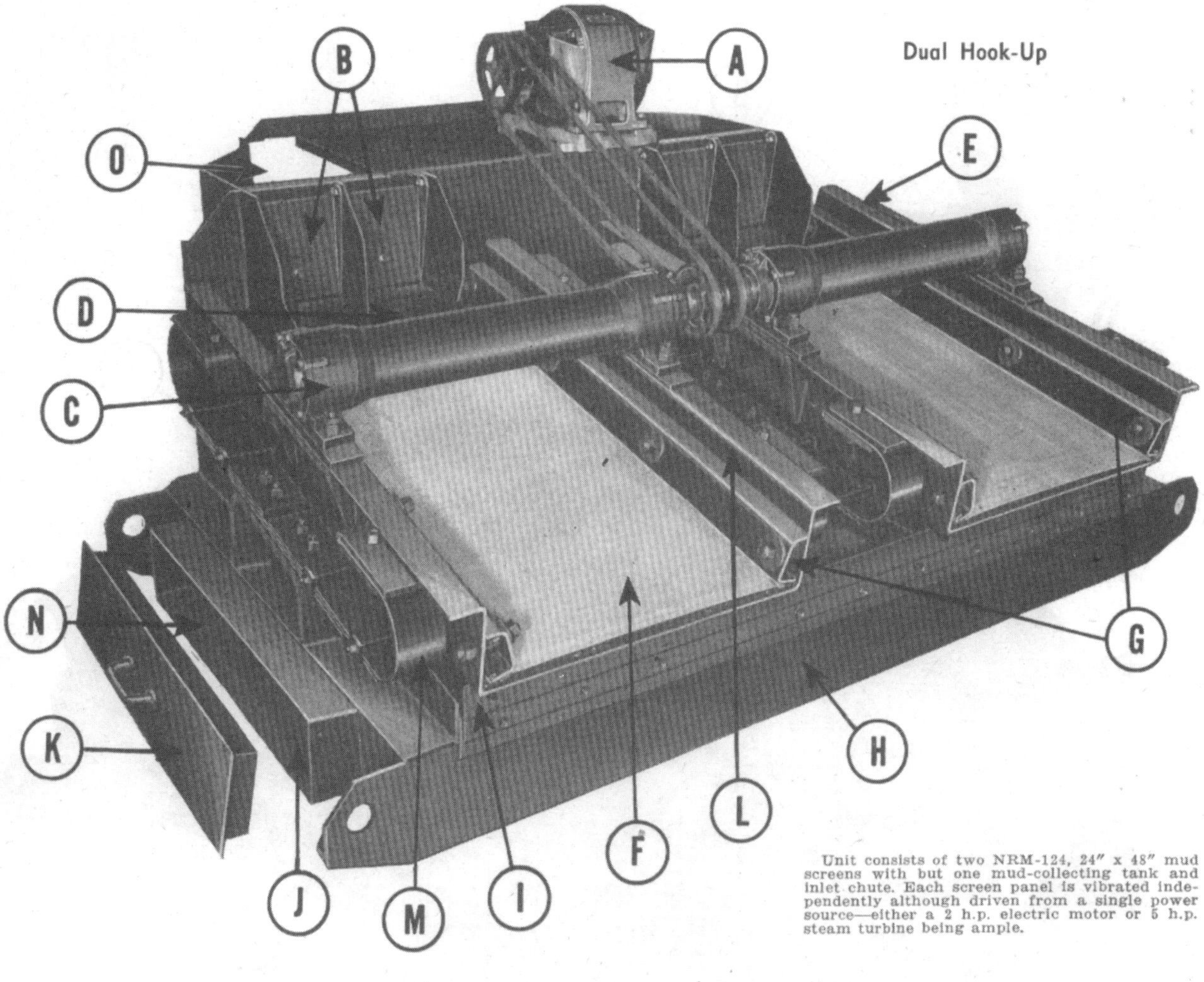

Dual Hook-Up

Unit consists of two NRM-124, 24" x 48" mud screens with but one mud-collecting tank and inlet chute. Each screen panel is vibrated independently although driven from a single power source—either a 2 h.p. electric motor or 5 h.p. steam turbine being ample.

A—One motor or turbine drives both screens.
B—Adjustable swinging deflecting gates distribute and control flow of mud; gates of either screen can be closed to shut off flow.
C—Fully enclosed roller bearing vibrator unit
D—Perforated bottom plate reduces wear on screen cloth and permits utilizing full area
E—Screen box vibrates, producing uniform vibration over entire area of cloth
F—Stainless steel screen cloth cushioned on flexible tubes gives maximum life
G—Tension members permit quick replacement of cloth
H—Heavy skids permit easy portability
I—Rubber seal to prevent rejects falling back into screened mud
J—Screened mud outlet or trough connection—either or both sides
K—Mud outlet plug
L—Two NRM-124 screens, each 2 feet wide by 4 feet long
M—Patented spring assemblies give sharpest reactions and permit maximum movement without breakage
N—One collecting tank for both screens
O—One mud inlet chute for both screens

The Type NRM-124 unitized mud screen includes: screen boxes fitted with single-piece rubber-cushioned stainless steel cloth with tension members; V-belt drives consisting of sheave for rotor, single X-Belt and motor sheave; mud tank with two-way outlet made self-contained for supporting motor or steam turbine and screen boxes; distributing chute; adjustable gates; necessary tools. Prices are f.o.b. or f.a.s. points designated.

### PRICE OF COMPLETE SCREEN—LESS MOTOR

| | | |
|---|---|---|
| F.O.B. | Los Angeles, California | $1,182.50 |
| | Houston, Texas | 1,182.50 |
| | Dallas, Texas | 1,199.00 |
| F.A.S. | New York (Packed for Export) | 1,160.50 |

Guard for V-Belt drive: $38.50 additional.
Extra screen cloth: Price on application.

### APPROXIMATE WEIGHTS

Complete screen, net, 2,100 lbs. Packed for domestic shipment, 2,480 lbs. Packed for export, New York, 2,850 lbs. (1,300 kilos). Measurements: 175 cu. ft.

| SUNSHINE IRON WORKS | | ODESSA, TEXAS |
|---|---|---|
| 601 WEST MURPHY ST. |  | PHONE 1336 |

**Manufacturers of Oil Field Specialties**

## S-I-W ROTATING MUD SCREEN

The S-I-W Mud Screen has been designed by practical men who know and appreciate the importance of removing shale from drilling mud efficiently and thoroughly. From their many years experience they have developed numerous features to improve the screening of mud and the operation of the screen which are incorporated in the S-I-W MUD SCREEN. It is the most complete unit of its kind on the market.

### SIMPLICITY OF DESIGN

This is a mud driven screen, so finely balanced that the power developed by the weight of mud falling from a one inch pipe will rotate the screen. This feature makes it possible to use the S-I-W rotating screen on jobs where ordinarily nothing but a motor driven screen could be used. Especially is this true in deep drilling where small drill pipe is used, and the returns are too small to operate other rotating screens. Its sturdy construction and capacity is also sufficient to handle the flow from an eight inch pipe. The S-I-W screen has two pillow block ball bearings, mounted away from mud assuring protection against abrasives.

The power wheel is the large end of the screen. Along the inside of its outer rim 16 fins, 4½" by 9¾" are welded. The mud flow, entering the top side of the drum, is directed sideways onto the power wheel fins, causing the drum to revolve and rotate the screen.

### CONE SHAPED SCREEN

This screen is set in the machine so that the lower side is about level, causing a sharp taper on the top side. In this position the mud easily flows through the bottom and into the pit, while shale clings to the screen and is carried upward until falling to the bottom. Each fall brings the shale closer to the screen outlet, thus feeding it out of the screen without the aid of a screw type feed within the screen drum. This is especially advantageous in sticky formations which clog the drive and require constant washing to prevent the screen from choking up, frequently requiring so much water that the mud becomes too thin.

### RIGIDLY CONSTRUCTED

The construction of the S-I-W Mud Screen is such that it has been known to run at high speed for more than an hour in gas blow-outs without damage. The drum can be removed in a few minutes to repair the screen or set in an exchange.

A special valve is provided which can be used in slowing the speed of the drum, where the flow of mud is large with a small amount of shale to screen out. This helps prolong the life of the screen and bearings. There are approximately 36 square feet of standard mesh stainless Rek-tang screen in the Model "B" S-I-W Mud Screen. The usual procedure of connecting the Mud Screen is to butt the inlet against the flow pipe and join them with a flexible coupling.

### MUD BY-PASS

This feature permits the mud flow to be cut past the screen, to save hulls, jell flakes, etc., by diverting the flow directly into the pit. This by-pass is effected by a lever operated, quick opening valve on the mud inlet. This feature will save hours of valuable drilling time when circulation is lost and it becomes necessary to remove the screen to prevent the loss of mud thickening materials.

### SAMPLE CATCHER

All S-I-W Mud Screens have a conveniently located sample catcher with which adequate mud samples may be taken as frequently as desired.

### CHEMICAL MIXER

This accessory is available for Models B and J. It is a steel box mounted above the inlet in front of the drum. The mixing paddle is operated by a cam on the end of the mud screen shaft. Simple in design, with few parts to wear or give trouble, it thoroughly mixes such chemicals as salt and soda into the mud. Users report the chemical mixer a valuable adjunct to the S-I-W Mud Screen. When ordering specify whether or not it is to be included with the screen.

### AVAILABLE IN THREE SIZES

S-I-W Screens are available in three sizes for quick delivery. Specifications are shown below; prices will be furnished upon request.

## SPECIFICATIONS

| MODEL NO. | Inlet Size, Inches | Length | Width | Height | Height, Ground to Bottom of Inlet | Screen Area | Shaft Bearing Size, Inches | Weight, Lbs. | Recommended Service |
|---|---|---|---|---|---|---|---|---|---|
| E | | | | | 19" | | | 550 | Slim Hole |
| B | 8½" O.D. | 10' 6" | 40" | 48" | 29" | 36 Sq. Ft. | 1¼" | 1,550 | Standard |
| J | 10¾" O.D. | 10' 6" | | | | | 1½" | 1,800 | Extreme |

# SUNSHINE IRON WORKS  ODESSA, TEXAS

## S-I-W ROTATING MUD SCREEN PARTS LIST

Mud screen parts should be ordered by name as shown below. To insure shipment of correct parts, Model number of the screen for which they are intended should be given.

Part listed as "Upper Screen Section" is for large end of screen; "Lower Section" is for small end. All bolts necessary for installation are furnished with each section. Coarse or fine mesh should be specified as desired. Prices furnished upon request.

Fig. 1

### MUD SCREEN PARTS LIST

Upper Screens
Lower Screens
Screen complete with shaft and power wheel
Screen Shaft, 1¼"
Screen Shaft, 1-7/16"
Bearings
By-Pass flap
Chute
Plug
Chute latch
Step
Cover for screen
  (6) Sample catcher bucket
  (7) Sample catcher trough
  (8) Sample catcher brackets
  (9) Bearing housing
  (10) Belt protector for housing

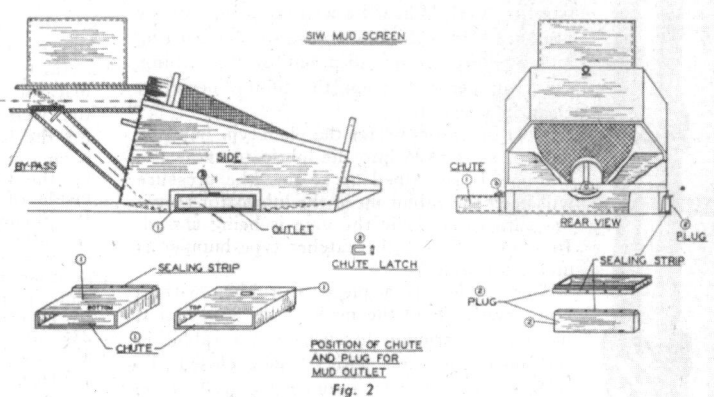

Fig. 2

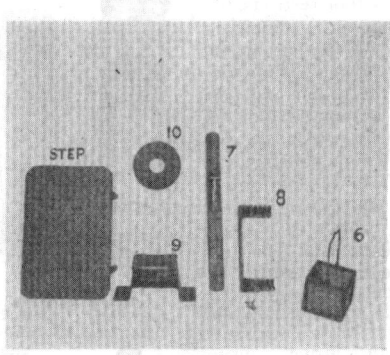

Fig. 4

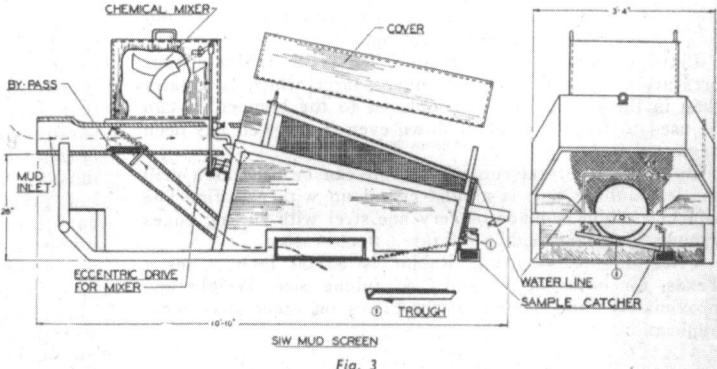

Fig. 3

## CHEMICAL MIXER PARTS

Chemical mixers are available for Models B and J. Prices are identical for identical parts in either model. Prices furnished upon request.

### CHEMICAL MIXER PARTS LIST

(1) Cam
    Cam Bearing
(2) Paddle
(3) Pitman rod
    Pull spring
(4) Cross beam less bearing
(5) Knuckle joint
    Small bronze bushings, per set of three
    Push spring

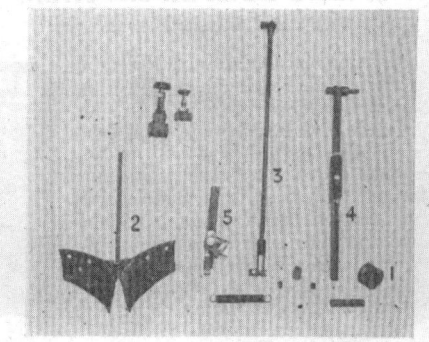

Fig. 5

(Continued on Next Page)

# THOMPSON TOOL CO.

**Sold Through Supply Stores Everywhere**

  Phone 3521         Iowa Park, Texas

### THE NEW AND IMPROVED *Thompson* SHALE SEPARATOR AND SAMPLE MACHINE

### MODEL "DW" For Deep Wells

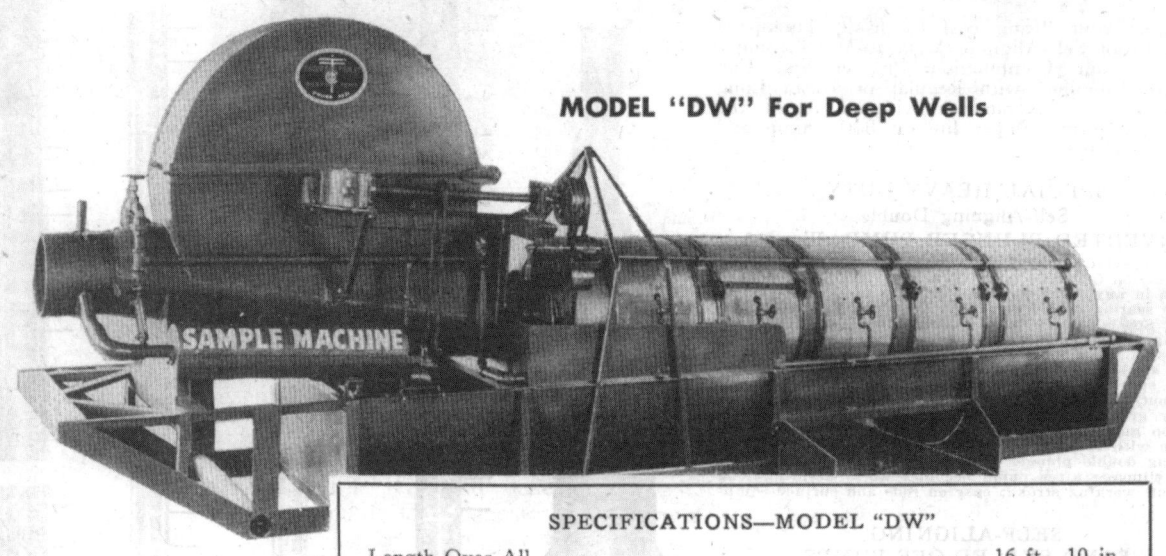

| SPECIFICATIONS—MODEL "DW" | |
|---|---|
| Length Over All | 16 ft. 10 in. |
| Height Over All | 6 ft. |
| Width | 3 ft. 5 in. |
| Flow Line Connection Above Base | 2 ft. 7 in. |
| Size of Flow Line Connection | 10 in. |
| Weight With Sample Machine Attached | 2,550 pounds |

The new Thompson "DW" Model has been improved and increased in size to the extent it is capable of handling the flow of drilling mud from the largest mud pumps now in operation. The Power Wheel has been redesigned and increased in size for greater efficiency. The number of water jets has been increased and relocated on the side of the drum where their function will be unhampered by any wind.

Like all Thompson Shale Separators, this new model is the simplest in construction and operation of any similar machine and definitely the most economical method of removing shale and abrasives from mud while drilling. A large power wheel operated from the flow of mud, furnishes the necessary motive power. No motors or other auxiliary units to maintain, repair, or replace. Through the use of V-Belts and gears ... no chains ... the power is transmitted to the rotary screen drum. Here the mud is strained through a fine meshed screen ... reconditioned mud flows back into the mud pit ... the shale and abrasives are ejected through the rear of the machine. By thoroughly cleaning and reconditioning drilling mud, wear and tear on costly drilling equipment is reduced. Clean mud also means greater efficiency and economies in drilling.

The Thompson Shale Separator is engineered so that all moving parts are readily accessible ... easily removed ... replaceable at a low cost. Fabricated of sturdy materials for long life. All joints and seams are welded. Entire machine is mounted on a skid frame.

NOW AVAILABLE ARE THREE MODELS ... each designed to tailor-fit your requirements from the smallest to the largest well. Because of its greater efficiency and economy of operation, many major oil companies and independent drilling operators have adopted the Thompson Shale Separator as standard equipment. You too, will find that this machine will not only increase your drilling efficiency, but will pay for itself in the savings it effects. Immediate delivery can be made through your supply store or direct from the factory.

## THOMPSON TOOL CO.　　　　　　　　　　　　　　　IOWA PARK, TEXAS

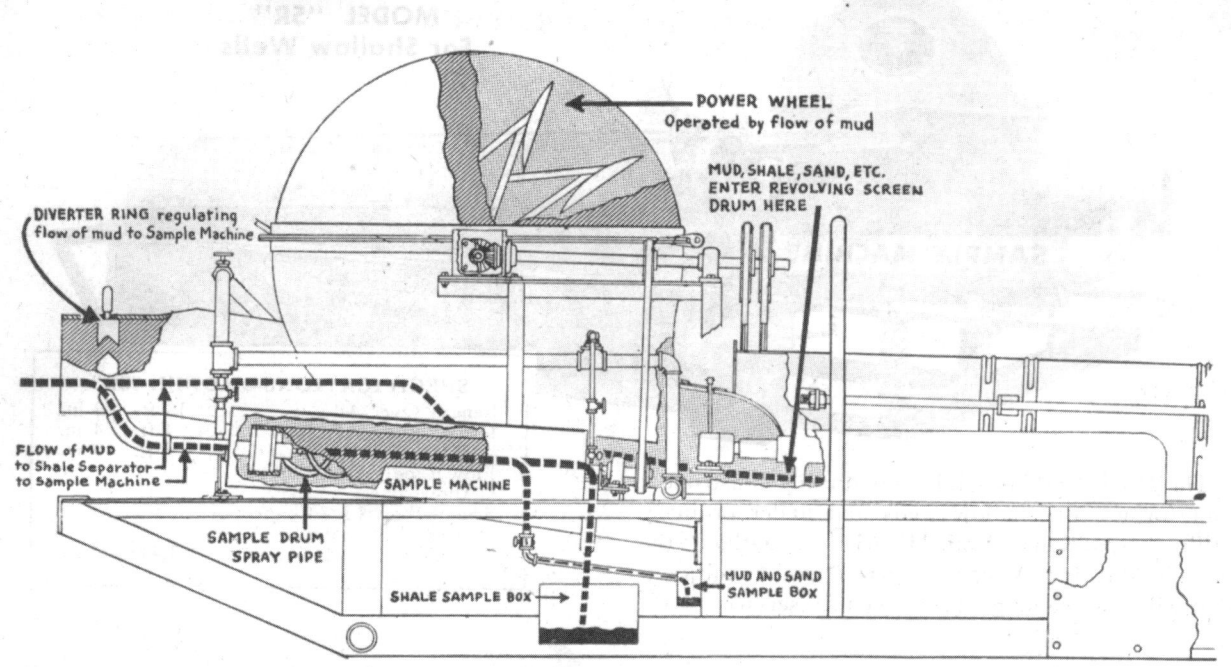

### POWER WHEEL

The Power Wheel, which is operated by the flow of mud from the well, is the main motive power for the Thompson Shale Separator and Sample Machine. The newly redesigned Power Wheel is now lighter, but stronger... the slightest pressure of drilling mud places the machine in operation. Fabricated of steel and welded into one piece. Mounted on self-aligning, pillow block, free-rolling ball bearings that are fully enclosed... easier operation... longer life.

### SAMPLE MACHINE

The easiest, quickest, and most economical method of obtaining accurate foot by foot samples of cuttings and mud. Just push a lever and part of the flow of drilling mud is diverted into the Sample Machine. Here the mud is separated into... shale and abrasives... drilling mud... and deposited into two, easily accessible, catch basins. These samples are invaluable in geological research. Sample Machine is standard equipment on all Thompson Shale Separators, unless otherwise specified.

# THOMPSON TOOL CO.      IOWA PARK, TEXAS

### PROTECTIVE COVER FOR ALL MODELS

Additional protection for water jets and lines on side of THOMPSON Separator can be obtained with streamlined steel cover, pictured at right. Shipping weight is 150 pounds for "SRF" and "LRF" models and 250 pounds for "DWF" model.

### THOMPSON POWER WHEEL

The new, lighter and stronger Power Wheel (shown at left) is the main motive power for the THOMPSON Shale Separator and Sample Machine. Fabricated of steel and welded into one piece, mounted on self-aligning pillow blocks with fully enclosed ball bearings, the Power Wheel sets the machine in operation with the slightest pressure.

### THOMPSON SAMPLE MACHINE

The easiest, quickest and most dependable method for obtaining samples of drilling mud is the THOMPSON Sample Machine, standard equipment on all models of THOMPSON Separators unless specified. It separates mud from abrasives, deposits samples of each into two accessible catch basins. Oil field men say this is the most accurate Sample Machine on the market.

### IMPROVED TYPE SCREEN DRUM

The secret of the efficient operation of all three models of THOMPSON Shale Separator is the spiral metal screen drum. Rigid and durable, the drum has adjustable stretcher rods to keep screen tight and smooth. The screen, attached by steel clamp bands, can be replaced easily and quickly. Spiral design conveys shale and abrasives to end of drum where it is ejected automatically.

## THOMPSON TOOL CO.  IOWA PARK, TEXAS

### THE THOMPSON SHALE SEPARATOR ON THE JOB . . .

*Above is an on-the-job photograph of the THOMPSON SHALE SEPARATOR AND SAMPLE MACHINE. Note the full flow of reconditioned mud pouring back into the pit. The THOMPSON Separator is considered standard equipment by many leading oil companies.*

In the field, under actual drilling conditions, The THOMPSON Shale Separator and Sample Machine has shown its great value. Used continuously for several years by some of the world's largest oil companies, The THOMPSON has met every test. There is a THOMPSON model to fit your drilling needs. In removal of damaging shale and abrasives, in thoroughly reconditioning drilling mud, giving accurate samples of cuttings, in generally saving the driller time and money—the THOMPSON Shale Separator and Sample Machine is worth many times its low cost. Write to THOMPSON for full details.

**THOMPSON TOOL COMPANY**
Iowa Park, Texas

# Link-Belt Company

## NRM-124-S AND NRM-145-S MODEL NO. 49 SHALE SHAKERS

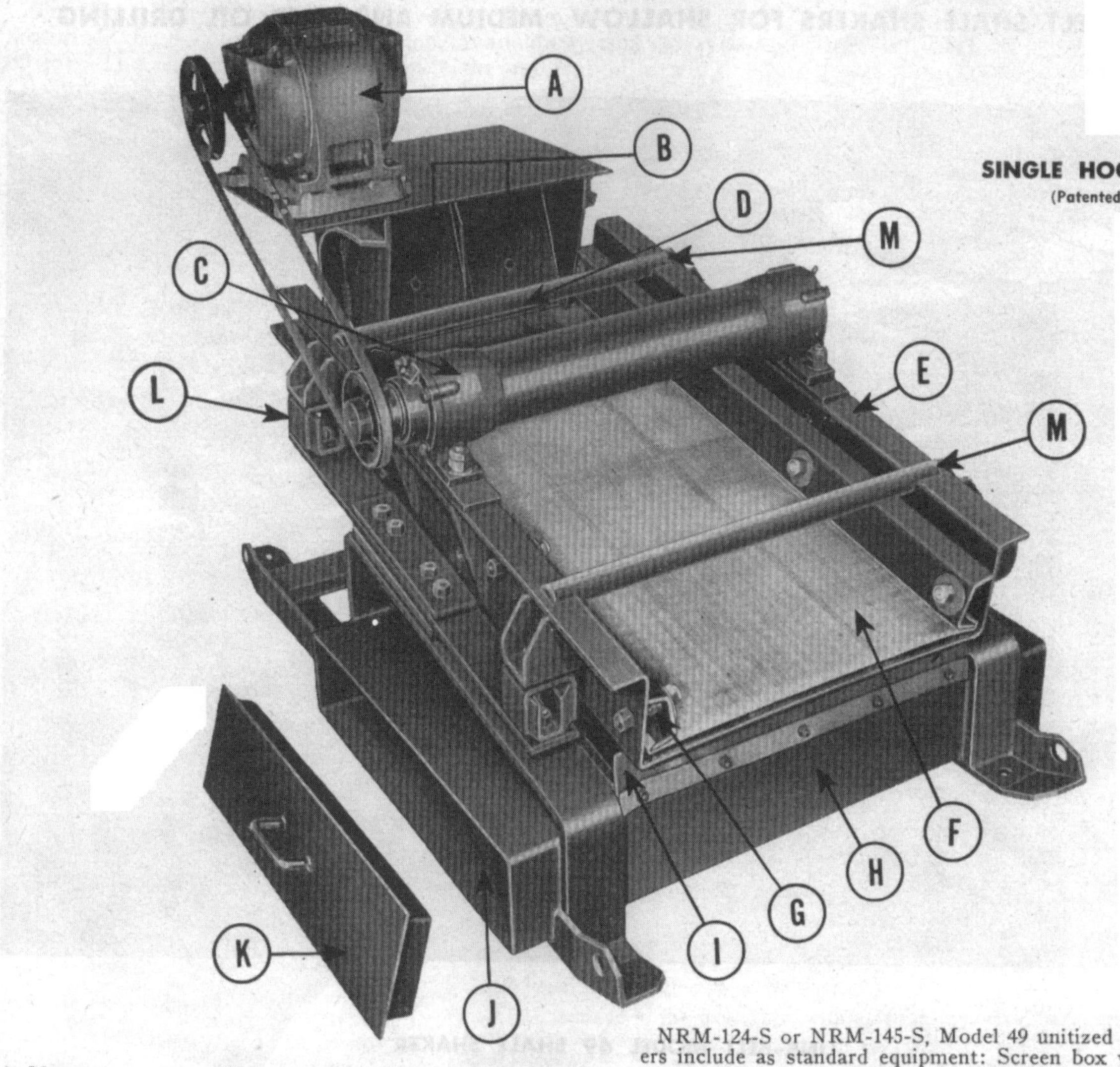

**SINGLE HOOK-UP**
(Patented)

A. Motor or steam turbine may be used (1 HP motor on NRM-124-S, 2 HP motor on NRM-145-S, 5 HP turbine or equivalent internal combustion engine).
B. Adjustable swinging gates assure good distribution, control flow and provide means for shutting off flow completely.
C. Fully enclosed, roller bearing vibrator imparts full vibration to screen; can be quickly detached by removing only four bolts.
D. Perforated bottom plate protects screen cloth against abrasive wear and permits utilizing entire screening area.
E. Screen box vibrates, producing uniform vibration over entire area of cloth.
F. Stainless steel screen cloth cushioned on elastic strips gives maximum life.
G. Tension members permit quick changes of cloth.
H. Unitized Mud tank skidded and complete with motor supports and inlet chute.
I. Rubber seal to prevent rejects falling back into screened mud.
J. Screened mud outlet or trough connection either or both sides.
K. Mud outlet plug.
L. Combination rubber in shear and steel leaf spring assembly maintains high energy vibration of deck assuring movement of shale.
M. Sturdy pipe cross bracing stiffens deck.

NRM-124-S or NRM-145-S, Model 49 unitized shale shakers include as standard equipment: Screen box with 20 x 20 mesh stainless steel cloth and tension members; V-belt with vibrator sheave; self contained, skid mounted, mud collecting tank with two-way outlet and flow distributing flume with gates; and support for electric motor, steam turbine or engine.

Domestic and Canadian prices are F.O.B. designated points.
Export prices are F.A.S. designated points, packed for export.

### PRICE AND SHIPPING POINT OF COMPLETE SHAKER—LESS MOTOR

| | | |
|---|---|---|
| F.O.B. | Los Angeles, California. Houston, Texas. Casper, Wyoming. | See Current Price List |
| F.A.S. | New York, N. Y. (Packed for Export). Houston, Texas (Packed for Export). | |

Guard for V-Belt Drive: Price on application.
Extra Screen Cloth: Price on application.
Additional price for galvanizing on application.

### APPROXIMATE WEIGHTS

| | Size | Net, pounds | Export, pounds | Domestic, pounds | Export Volume |
|---|---|---|---|---|---|
| Complete Shaker | NRM-124-S | 1060 | 1375 | 1225 | 70 cu. ft. |
| Complete Shaker | NRM-145-S | 1580 | 2190 | 1925 | 145 cu. ft. |
| Shaker only, no mud tank, flume, motor supports, etc. | NRM-124-S | 580 | 1000 | 800 | |
| | NRM-145-S | 810 | 1425 | 1160 | |
| Auxiliary Mud Tank | NRM-124-S | 655 | 1000 | 800 | |
| | NRM-145-S | 1165 | 1325 | 1465 | |

# HISTORICAL PERSPECTIVE

## Link-Belt Company

### NRM-124-D AND NRM-134-D MODEL 49 DUAL SHALE SHAKERS

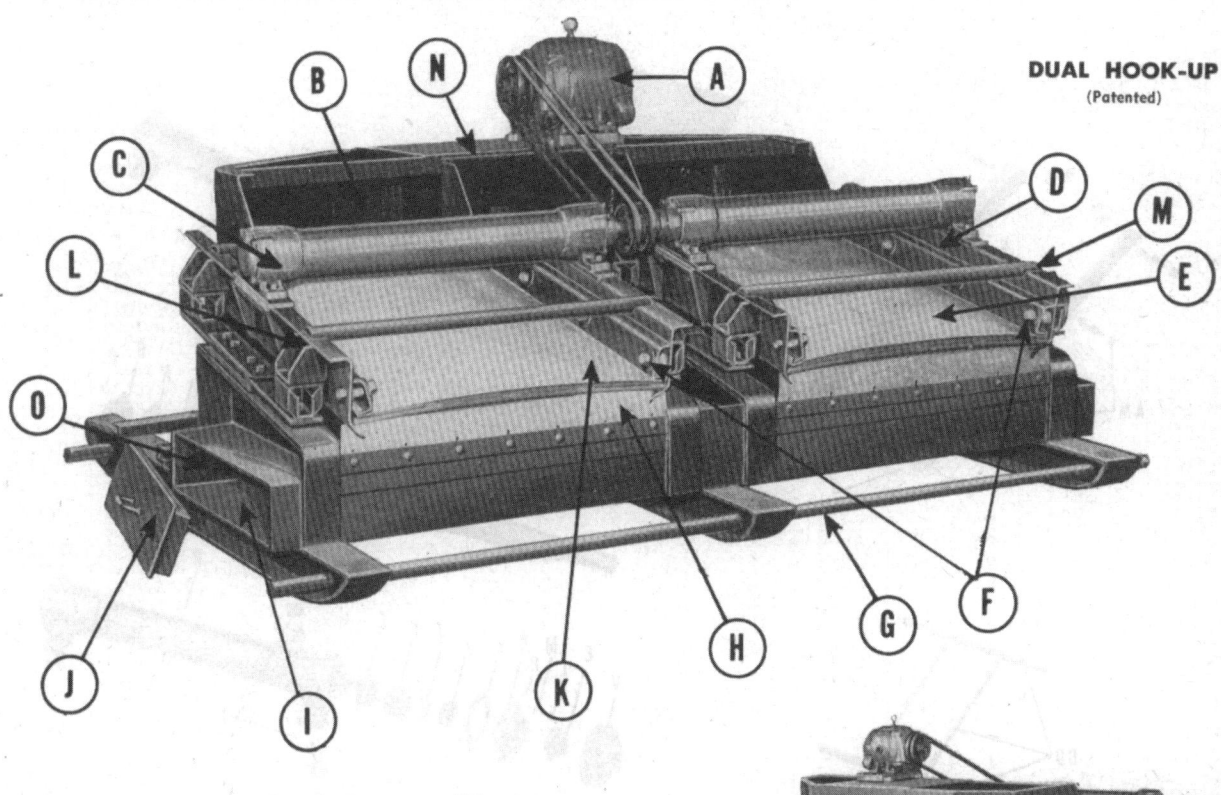

**DUAL HOOK-UP** (Patented)

A. One motor or turbine drives both screens (2 horsepower motor, 5 horsepower turbine or equivalent internal combustion engine).
B. Integral mud receiving box provided with threaded pipe connection for mud inlet and by pass pipes. Weir type feed to screen provides uniform, low velocity distribution of mud to screen.
C. Fully enclosed, roller bearing vibrator imparts full vibration to screen; can be quickly detached by removing only four bolts.
D. Screen box vibrates, producing uniform vibration over entire area of cloth.
E. Stainless steel screen cloth cushioned on flexible strips gives maximum life.
F. Tension members permit quick replacement of cloth.
G. Heavy skids permit easy portability.
H. Rubber Seal to prevent rejects falling back into screened mud.
I. Screened mud outlet or trough connection either or both sides.
J. Mud outlet plug.
K. Two independent vibrating shale shakers, side by side.
L. Combination rubber in shear and steel leaf spring assembly maintain high energy of deck, assuring movement of shale.
M. Sturdy pipe cross bracing stiffens deck.
N. One mud receiving box for both screens.
O. One collecting tank for both screens.

NRM-124-S or NRM-134-D Dual Model 49, shale shakers include as standard equipment: Two screen boxes with 20 x 20 mesh stainless steel cloth and tension members. V-belts with vibrator sheaves; self-contained, skid mounted mud inlet tank with inlet and by pass connections, slide gates and motor supports; mud collecting tank with two way outlet.

Domestic and Canadian prices are F.O.B. designated points. Export prices are F.A.S. designated points, packed for export.

### PRICE AND SHIPPING POINT OF COMPLETE SHAKER—LESS MOTOR

| | | |
|---|---|---|
| F.O.B. | Los Angeles, California<br>Houston, Texas<br>Casper, Wyoming | See Current Price List |
| F.A.S. | New York, N. Y. (Packed for Export)<br>Houston, Texas (Packed for Export) | |

Guard for V-Belt Drive: Price on application.
Extra Screen Cloth: Price on application.
Additional price for galvanizing on application.

### APPROXIMATE WEIGHTS

| | Size | Net, pounds | Export, pounds | Domestic, pounds | Export Volume |
|---|---|---|---|---|---|
| Complete Shaker | NRM-124-D | 2100 | 2860 | 2480 | 175 cu. ft. |
| Complete Shaker | NRM-134-D | 3850 | 4350 | 4000 | 340 cu. ft. |
| Shaker only, no mud tank, flume, motor supports, etc. | NRM-124-D | 1160 | 1660 | 1310 | |
| | NRM-134-D | 1600 | 2200 | 1800 | |

# Link-Belt Company

## REPLACEMENT PARTS, MODEL 49, SHALE SHAKERS

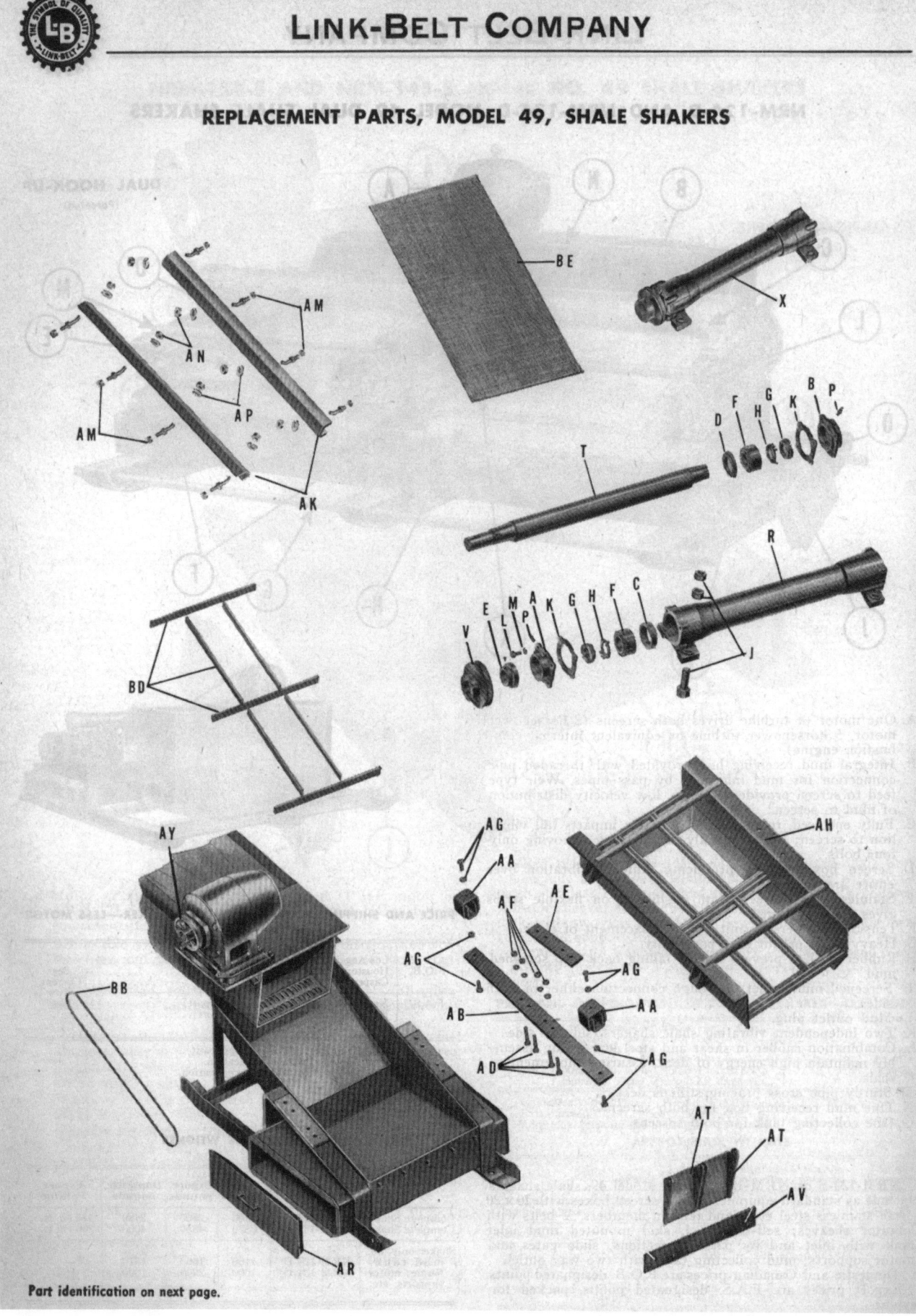

Part identification on next page.

# LINK-BELT COMPANY

## REPLACEMENT PARTS, MODEL 49, SHALE SHAKERS

| Symbol | Part Number | Description of Screen Parts | Quantity per Shale Shaker NRM-124 S | NRM-124 D | NRM-134 D | NRM-145 S | Shipping Weight, Each | List Price, Each |
|---|---|---|---|---|---|---|---|---|
| AA | 364X10-1 | Shear Mounting | 4 | 8 | 8 | 4 | 5 | |
| AB | 436W6-1 | Spring Leaf | 2 | 4 | 4 | 2 | 15 | |
| | 364X9-A | Complete Spring Assembly includes 1 leaf, symbol AB; 2 mountings, symbol AA; 2 nuts, symbol AG; 2 bolts, symbol AG | 2 | 4 | 4 | 2 | 25 | |
| AD | 126W131 ⅝x3¼ | Spring Bolts | 8 | 16 | 16 | 8 | 1 | |
| AE | 167W50-1 | Spring Leaf Hold Down Flat | 2 | 4 | 4 | 2 | 4 | |
| AF | 126W166 ⅝ | Spring Bolt Nut | 8 | 16 | 16 | 8 | ½ | |
| AF | 242W3 ⅝ | Spring Bolt Lock Washer | 8 | 16 | 16 | 8 | | |
| AG | 126W141 ¾x1½ | Shear Mounting Bolt | 4 | 8 | 8 | 4 | 1 | |
| AG | 126W192 ¾ | Shear Mounting Nut | 4 | 8 | 8 | 4 | ½ | |
| AH | 1053Z23-A | Screen Box (Bare Box Only) | 1 | 2 | | | 260 | |
| | 1053Z24-A | Screen Box (Bare Box Only) | | | 1 | | 300 | |
| | 1053Z22-A | Screen Box (Bare Box Only) | | | | 1 | 350 | |
| | 1053Z23-B | Screen Box Assembled includes Box AH; | 1 | 2 | | | 340 | |
| | 1053Z24-B | Tension Plates AK with bolts, washers and nuts; | | | 1 | | 380 | See Current Price List |
| | 1053Z22-B | and Neoprene strips BD | | | | 1 | 450 | |
| AK | 438X4-1 | Tension Plate | 2 | 4 | 4 | | 25 | |
| | 438X5-1 | Tension Plate | | | | 2 | 30 | |
| AM | 126W105-A | Tension Bolt with Nut | 8 | 16 | 16 | 12 | 2 | |
| AN | 126W120 ¾ | Tension Nut | 8 | 16 | 16 | 12 | 1 | |
| AP | 128W52-12 | Tension Washer | 8 | 16 | 16 | 12 | ½ | |
| AR | 440W1-A | Outlet Plug | 1 | | | | 20 | |
| | 440W2-A | Outlet Plug | | 1 | 1 | 1 | 25 | |
| AT | 374W6-B | Feed Gate with Bolts | 2 | | | 5 | 5 | |
| AV | 131W289-1 | Rubber Seal | 1 | 2 | | | 2 | |
| | 131W291 | Rubber Seal | | | 2 | | 2 | |
| | 131W263-1 | Rubber Seal | | | | 1 | 3 | |
| AY | 536X3-1 | V-Belt Pulley 6.2″ P.D. Single Groove bore ¾″ | 1 | | | | 6 | |
| | 536X3-2 | V-Belt Pulley 6.2″ P.D., Single Groove, 1″ bore | | 2 | 2 | | 6 | |
| | 537X3-1 | V-belt Pulley 6.2″ P.D., Two Groove, 1″ bore | | | | 1 | 8 | |
| BB | B 75 | Endless V-Belt | 1 | | | | 1 | |
| | B 90 | Endless V-Belt | | 2 | 2 | 2 | 1 | |
| BD | 364W4 | Neoprene Strips | | | | | ½ lb. per ft. | |
| BE | | Screen Cloth | 1 | 2 | 2 | 1 | | |

| Symbol | Part Number | Description of Vibrator | Quantity per Shale Shaker NRM-124 S | NRM-124 D | NRM-134 D | NRM-145 S | Shipping Weight, Each | List Price, Each |
|---|---|---|---|---|---|---|---|---|
| A | 305X60-1 | Bearing Retainer for Drive End | 1 | 2 | 2 | 1 | 8 | |
| B | 305X62-1 | Bearing Retainer End Opposite Drive | 1 | 2 | 2 | 1 | 5 | |
| C | 280X4-1 | Grease Retainer for Drive End | 1 | 2 | 2 | 1 | 2 | |
| D | 280X5-1 | Grease Retainer End Opposite Drive | 1 | 2 | 2 | 1 | 2 | |
| E | 496X3-A | Collar including Set Screws | 1 | 2 | 2 | 1 | 3 | |
| F | 323W10-3 | Roller Bearing DE-22310T | 2 | 4 | 4 | 2 | 5 | |
| G | 126W217-11 | Bearing Lock Nut N-10 | 2 | 4 | 4 | 2 | 1 | |
| H | 242W13-11 | Bearing Lock Washer W-10 | 2 | 4 | 4 | 2 | 1 | |
| J | 122W4-B | Cap Screw (⅞ x 3½) Nut and Jam Nut | 4 | 8 | 8 | 4 | 2 | |
| K | 131W89-1 | Gasket 8⅞ x 3½ | 2 | 4 | 4 | 2 | | |
| L | 126W115 ½x1¼ | Cap Screw | 8 | 16 | 16 | 8 | | See Current Price List |
| M | 242W1 ½ | Lock Washer | 8 | 16 | 16 | 8 | | |
| P | 147W32-1 | Alemite Fitting (A336) | 2 | 4 | 4 | 2 | | |
| R | 535X1-A | Shaft Housing | 1 | 2 | | | 100 | |
| | 535X2-A | Shaft Housing | | | | 1 | 140 | |
| | 535X29-A | Shaft Housing | | | 2 | | 120 | |
| T | 309X110-A | Shaft Including Key | 1 | 2 | | | 125 | |
| | 309X111-A | Shaft Including Key | | | | 1 | 150 | |
| | 309X205-A | Shaft Including Key | | | 2 | | 140 | |
| V | 536X1-1 | V-Belt Pulley 5.6″ P.D.—1 Groove 1⅞″ bore | 1 | 2 | 2 | | 6 | |
| | 537X1-1 | V-Belt Pulley 5.6″ P.D.—2 Groove 1⅞″ bore | | | | 1 | 5 | |
| X | 427Z14-C | Complete Vibrator Unit Assembly | 1 | 2 | | | 300 | |
| | 427Z14-D | Complete Vibrator Unit Assembly | | | | 1 | 360 | |
| | 427Z14-F | Complete Vibrator Unit Assembly | | | 2 | | 330 | |

# LINK-BELT COMPANY

## SCREEN CLOTH FOR LINK-BELT SHALE SHAKERS

| Mesh | Nominal Opening, Inches | Wire Diameter, Inches | Link-Belt Shale Shaker Size | | | | | |
|---|---|---|---|---|---|---|---|---|
| | | | NRM-124-S NRM-124-D 24½" x 47½" | | NRM-134-D 36½" x 47½" | | NRM-145-S 48½" x 59½" | |
| | | | Symbol | Price | Symbol | Price | Symbol | Price |
| 12 | .060 | .023 | AA | | AAA | | J | |
| 16 | .040 | .023 | AB | | AAB | | Q | |
| 18 | .038 | .018 | AC | | AAC | | R | |
| 18 | .0326 | .023 | AT | | AAT | | AS | |
| 16 x 18 | .0326 | .023 | AX | See Current Price List | AAX | See Current Price List | AW | See Current Price List |
| 20 | .033 | .017 | AD | | AAD | | M | |
| 20 | .033 | .017 | | | | | N RM* | |
| 24 | .027 | .015 | AE | | AAE | | S | |
| 20 x 30 | .018 | .015 | AF | | AAF | | L | |
| 24 x 30 | .020 | .0135 | AG | | AAG | | DD | |
| 24 x 30 | .020 | .0135 | | | | | A RM* | |
| 30 | .020 | .013 | | | | | EE | |
| 40 x 36 | .015 | .0105 | AH | | AAH | | O | |
| 50 | .011 | .010 | AI | | AAI | | BB | |
| 60 x 40 | .008 | .009 | AJ | | AAJ | | CC | |

*Cloths measure 45" between bends instead of 48½"

## OPERATING INSTRUCTIONS

### SPEED
Screen should operate at a steady speed of 1800 to 2000 R.P.M.

The direction of rotation should always be down-hill, that is, the top belt of the drive should always travel towards the discharge end of screen.

### NEOPRENE STRIPS
The screen cloth is supported and cushioned on specially designed sponge Neoprene strips (Part No. 364W4) which are easily inserted in channel.

Neoprene strips and channels are proportioned so that strips will remain in position without bolting or cementing.

### TENSION BOLTS
Tension is applied to the screen cloth by uniformly tightening tension nuts (Part No. 126W120¾) on specially designed tension bolts (Part No. 126W105A). These bolts are fastened rigidly to the screen box as shown on drawings, permitting easy installation of replacement bolts when necessary and also facilitating the removal of tension plates (Parts No. 438X4-1 or 438X5-1) when renewing the screen cloth.

### CARE OF SCREEN CLOTH
Before placing the cloth on the screen be sure all the Neoprene strips are in place. The screen cloth furnished is made of stainless steel to resist corrosion and is bound along both longitudinal edges with special steel binders which engage the tension plates. Lay the screen cloth in place on the Neoprene strips and then add tension to the screen cloth uniformly, tightening the tension nuts until the cloth has the proper tension for satisfactory operation.

Sufficient tension should be applied to remove all sag from cloth.

Because fine mesh cloth as a rule will have an initial stretch shortly after being put in use, it is particularly important that the tension of the cloth be checked at this time. After this, periodic check of the cloth tension is advisable.

The screen is usually furnished with cloth of stainless steel wire to provide a clear opening of .033". This gives best results under average conditions although coarser or finer mesh cloths can be furnished if the individual mud characteristics are of such nature to require them. Care should be used in handling the screen cloth. Do not lay tools or heavy material on it and in cleaning always wash down with a hose; never scrape mud or sand off the cloth. If the meshes show a tendency to blind, a wire brush may be used to clear the openings. Similarly, if oil or pipe compounds clog the openings, they may be cleared by washing with a rag soaked with kerosene.

### REPAIRING SCREEN
Before making repairs read carefully all explanations given under the various parts you are particularly concerned with. Refer to REPAIR PARTS LIST corresponding drawings shown in this folder.

When ordering repair parts we suggest that you use the part number as shown in parts list. In every case refer to our serial number which is stamped upon the name plate fastened on the side of each screen. The serial number is important.

### LUBRICATION
(a) Each bearing is lubricated separately and is furnished with fittings for grease lubrication.

(b) Bearings are lubricated before leaving factory. The usual practice is to lubricate each bearing with one shot from grease gun every 24 hours.

A rise in temperature of as much as 40 degrees F. may be considered normal.

(c) It is not good practice to use kerosene or gasoline for flushing purposes. If care is used to keep lubricant clean and free from grit, there should be no need to flush the bearings. When it is found necessary to clean the bearings, remove the housing covers to expose the rollers and races.

(d) Use only a good lubricant such as recommended for high grade roller bearings. Under normal conditions any of the following greases or their equivalents are suggested. If bearings are to be operated in extreme temperatures, refer your problem to our Engineering Department.

New York & New Jersey Lubricant Co., "No. F-925 Non-fluid oil".

Socony-Vacuum Oil Company, "Gargoyle BRB No. 2".

The Texas Company, "Starfak No. 2".

Gulf Refining Co., "Gulf Anti-Friction No. 1".

Standard Oil Co. of Indiana, "Superla 2X".

Sinclair Refining Co., "No. AF-2".

Standard Oil Co. of New Jersey.

Humble Oil and Refining Co.

HISTORICAL PERSPECTIVE  45

# THOMPSON TOOL CO.          IOWA PARK, TEXAS

## MODEL "DWF" SELF-MOTIVATED TYPE SEPARATORS
### FOR DEEP WELLS — WITH SAMPLE MACHINE

| MODEL "DWF" SPECIFICATIONS | |
|---|---|
| Length Overall | 15' 2" |
| Height Overall | 6' 0" |
| Width | 3' 5" |
| Flow Line Connection Above Base | 2' 7" |
| Size of Flowline Connection | 0' 10" |
| Weight with Sample Machine Attached | 2,400 lbs. |

Many important features are afforded the Thompson "DWF" Model that are not found on any other self-motivated shale separators. It is larger in size to the extent it is capable of handling the flow of mud from the largest mud pumps now in operation. The Power Wheel is larger in size for greater efficiency. Water jets are located on the side of the drum where their function will be unhampered by any wind.

Like other Thompson Shale Separators, the "DWF" Model is simple in construction and operation. It is motivated entirely from the flow of mud . . . no motors or other auxiliary units to maintain, repair, or replace. Driven by V-Belts and gears . . . no chains. So designed that all moving parts are readily accessible . . . easily removed . . . replaceable at low cost.

No matter how much mud is driven into this machine by your mud pumps, the shale and abrasives are removed. By reconditioning your drilling mud, wear and tear on costly drilling equipment is decreased . . . the life of these tools will be prolonged.

The Sample Machine is standard on all self-motivated separators unless otherwise specified. With this amazingly accurate machine, the Thompson separator does a twofold job: (1) it removes the shale and abrasives from the valuable drilling mud (2) it collects sample sections of the cuttings for foot-by-foot analysis. The value and accuracy of the Sample Machine are recognized by drilling and geologic authorities.

### HEAVY DUTY SPIRAL DRUM ON ALL MODELS

• REMOVES SHALE AND ABRASIVES      • THOROUGHLY RECONDITIONS MUD

The drum in all models is constructed from solid tube of metal and cut to form spirals on the inside. Extremely rigid and durable. Adjustable stretcher rods keep the screen tight and smooth. Screen is attached by steel clamp bands.

**CAN BE REPLACED EASILY AND QUICKLY**

## SOLD THROUGH SUPPLY STORES EVERYWHERE

# LINK-BELT COMPANY

## INTAKE MUD BOXES FOR NRM-124-S SINGLE AND NRM-145-S SINGLE MODEL 49-A SHALE SHAKERS

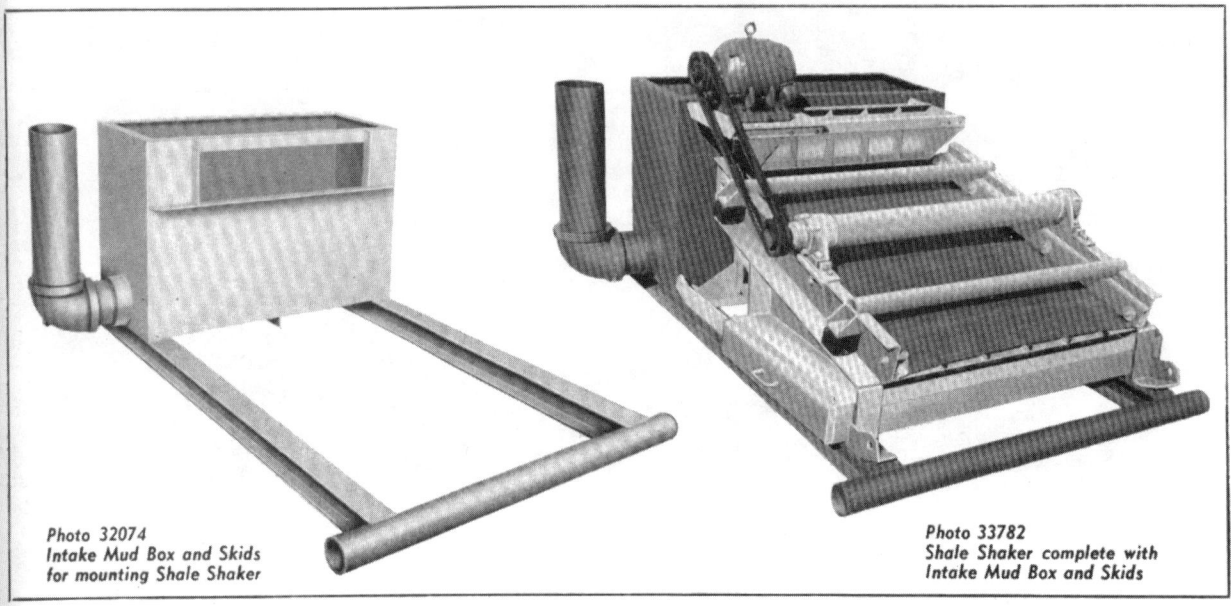

Photo 32074
Intake Mud Box and Skids for mounting Shale Shaker

Photo 33782
Shale Shaker complete with Intake Mud Box and Skids

Shale Shaker is easily mounted on intake mud box skids, and welded to box. This provides a unitized, compact, easy to handle, longer life and more efficient mud cleaning equipment.

Mud enters at bottom of box, which cuts down velocity and results in a smooth flow of mud onto the screen. This permits screen to handle a greater volume.

Eliminates time and expense of building flow line mud trough at each location.

Mounted on "H" Beam Skids for sturdiness and ease of moving between locations.

Mud Box has three 8⅝" OD openings so you may arrange it to suit your rig location. One opening for flow line, one for By-Pass Nipple, and one for Plug. By-Pass Nipple and Plug are furnished at no extra cost.

Stock carried by Link-Belt Company at Houston and Odessa in Texas, and Casper, Wyoming.

Can be purchased through your oil field supply house.

| DIMENSIONS AND WEIGHTS OF INTAKE MUD BOXES AND SKIDS | NRM-124-S | NRM-145-S |
|---|---|---|
| Over All Height | 3' 1" | 3'-7" |
| Over All Width | 4'-0" | 6'-0" |
| Over All Length | 7'-6" | 8'-6" |
| Approximate Weight | 850 lbs. | 1050 lbs. |
| Approximate Weight for Export | 1050 lbs. | 1425 lbs. |
| Approximate Export Volume | 25 cu. ft. | 60 cu. ft. |

## MUD CONVEYORS FOR OFF-SHORE DRILLING OPERATIONS

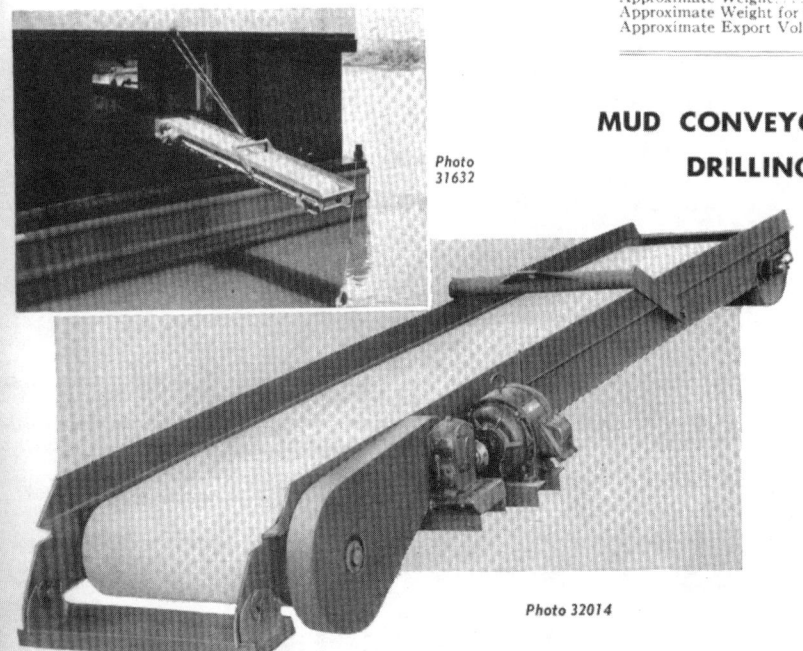

Photo 31632

Photo 32014

Moving bags of mud and cement from the supply boat to drilling barge is done easier, faster and safer with Link-Belt mud conveyors. Built to your requirements in Link-Belt's Houston Plant.

Link-Belt engineers will be glad to help you with your conveyor or power transmission problems. There's no obligation.

# HISTORICAL PERSPECTIVE

Alhambra, Calif. **VERNON TOOL CO., LTD.** Houston, Texas

## VERNON-CORWIN DESANDER

### THE VERNON-CORWIN DESANDER PROVIDES THESE OUTSTANDING FEATURES

- Minimum Sand Content
- Thinner Filter Cake
- Greater Economy in Mud and Chemicals
- Controlled Mud Viscosity
- Effective De-gassing Action
- Easier Finishing Jobs
- Reduced Maintenance Expenses
- Less Pipe, Bit and Pump Wear

The Vernon-Corwin Desander provides a simple and inexpensive means of removing sand and abrasives too fine to be removed from the rotary drilling mud by the conventional shaker type screens. The Desander operates on a velocity-vacuum principle in converting the velocity of the mud entering the machine into centrifugal force while a vacuum is automatically generated by the mud discharging from the machine. This concentrates the sand and solids for discharge into the sump while returning clean mud to the system.

When unitized independently the Desander can be operated continuously for recycling the pit mud whether or not drilling is in process. This further reduces the sand content during idle periods and also keeps the mud agitated and conditioned.

**THE DUAL HEAD DESANDER** — (shown here) is designed to accommodate operators using large capacity mud pumps. It will handle up to 800 g.p.m. The Single Head models handle from 400 to 500 g.p.m. Both the single head and dual head Desanders are available for use with standby pump or unitized with the Vernon Centrifugal Pump and Drive.

### EXPORT BOXING SPECIFICATIONS

Unitized Dual Head Model: Net Weight—3050 lbs.—Box 200 Cubic Feet
Unitized Single Head Model: Net Weight—2400 lbs.—Box 125 Cubic Feet

## VERNON CENTRIFUGAL MUD PUMP

### ... for SUMP CLEAN OUT, MUD TRANSFER, MUD LOADING, SUCTION BOOSTER

The Vernon Centrifugal Pump is designed specifically to handle rotary mud, sand and other abrasives common in the oil fields. Hardened and ground steel sleeve, fully protected bearings and overall rugged construction insure dependable service and long life. Large impeller openings permit passage of rocks and solids up to 2½" in diameter.

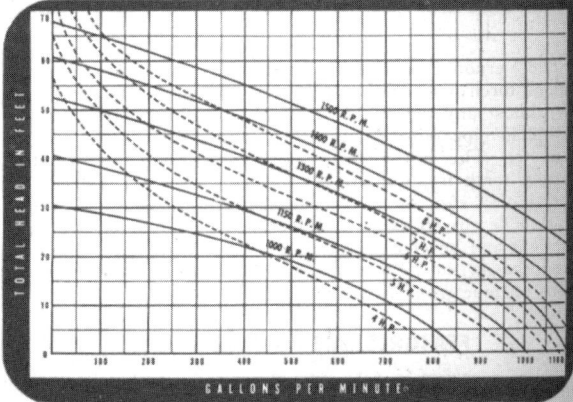

### PERFORMANCE CURVE VERNON CENTRIFUGAL PUMP

Vernon Centrifugal Pumps are available separately or unitized on a sturdy steel skid base and driven by either electric motor or gasoline engine. Pump is easily portable.

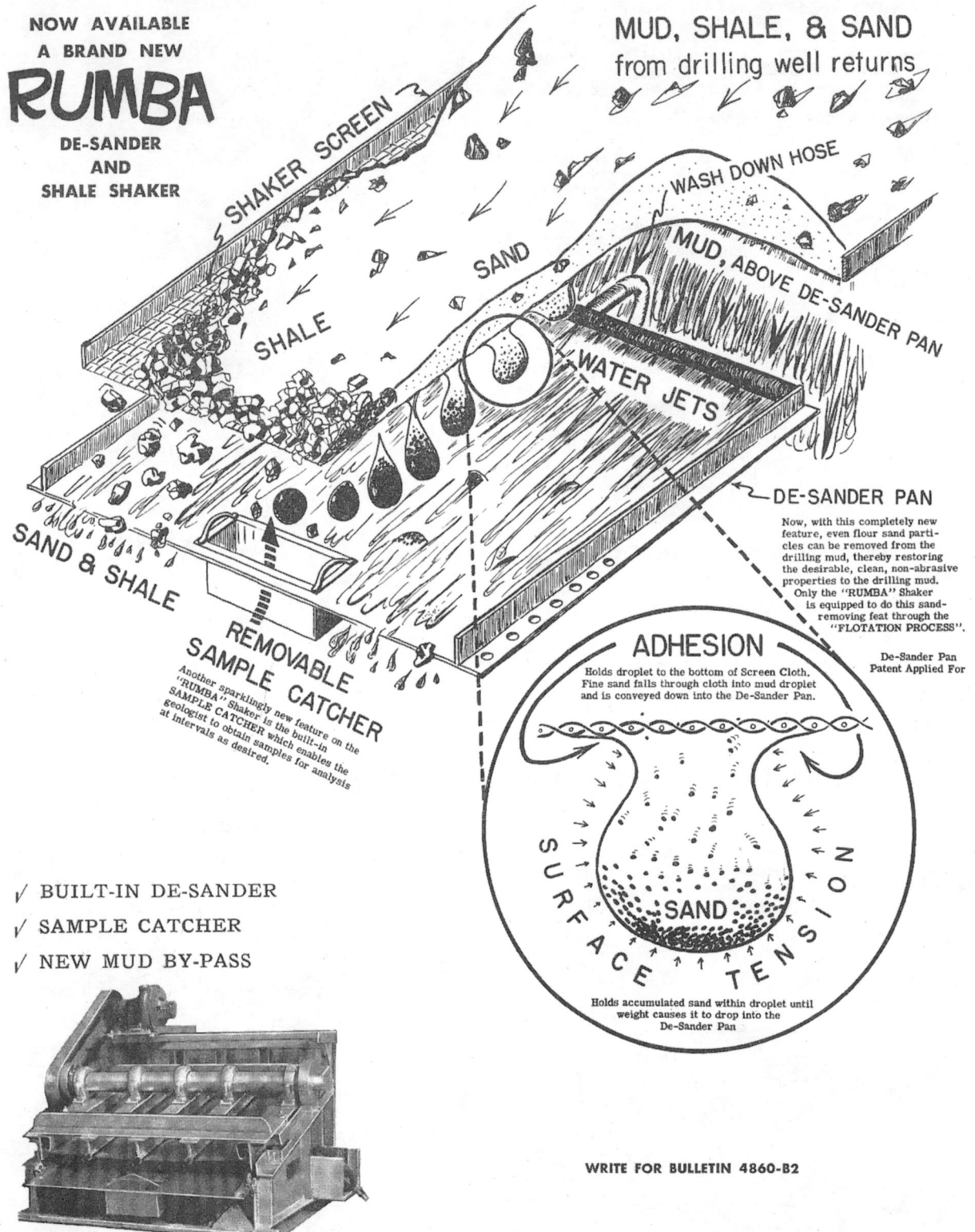

# HUTCHISON MANUFACTURING COMPANY — HOUSTON, TEXAS

## TONS OF SAND REMOVED FROM MUD BY NEW "RUMBA" SHALE SHAKER AND DE-SANDER

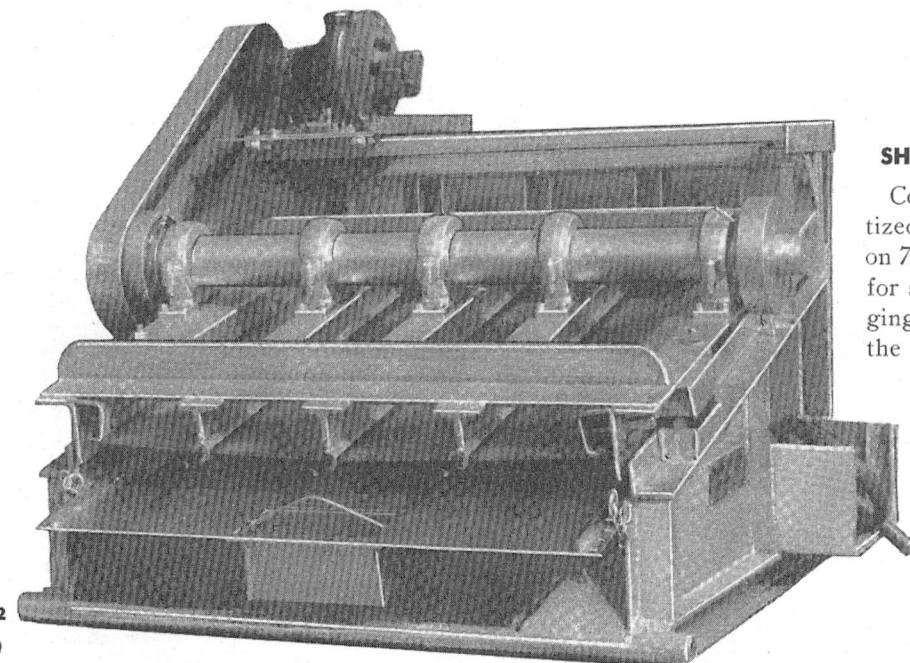

**Large Model Shaker Cat. No. 4860-B2**
(Patent Applied for)

**"RUMBA" SHALE SHAKERS**

Completely unitized and mounted on 7' 10" skid ready for service. No rigging up expense in the field.

**LONG SCREEN CLOTH LIFE:** The screen frame carrying the under-slung screen cloth is made of 3/8" steel, and is driven by a rubber mounted vibrator. Rubber-lined pillow blocks equally distribute the vibration through the five foot lateral frame struts which hold and transmit the vibration to the under-slung screen cloth. This friction hold **prevents screen cloth whipping** as the cloth moves in the same vibrational path as its supporting frame; therefore, when screen cloths need replacement, they are worn out by the abrasive materials going over them, not whipped out.

**SIMPLICITY IN DESIGN:** Fewer exposed parts reduce replacement costs. The heavy steel back tank is 3/16", and the bottom tank is 10 gauge steel to assure long shaker life when corrosive mud chemicals are used.

**SAND REMOVAL BY FLOTATION WITH NEW DE-SANDER:** By using the under-slung screen cloth with the entire frame above the screen, there are no obstructions on the under side of the cloth. A droplet of mud on the under side of the screen cloth is loaded with fine flour-like sand that falls through the meshes of the cloth. This drop is held together on its way to the bottom of the cloth with its load of sand by surface tension, and the molecular force of adhesion holds it to the bottom of the cloth, where it falls clear on the de-sander. Sand entrapped in the droplet falls clear of the mud returns. This patented method of sand removal is very effective and it is obvious that pump repairs will be reduced, and that the hole will be cleaner with less sand in the mud. The SAMPLE CATCHER provides the operator with an easily accessible method of taking samples.

### VIBRATOR WITH RUBBER-LINED PILLOW BLOCKS

Each of the five pillow blocks drives one of the five lateral steel members that make up the vibrating top, assuring uniform movement which prevents metal fatigue of the steel. The screen cloth is driven evenly through the friction hold of these 5 members, resulting in longer screen cloth life, as there is no whipping of this under-slung screen cloth. As the entire vibrating frame is above the screen cloth, it is easily taken care of. Recommended SPEED OF VIBRATOR is 2000 R.P.M. Maximum Speed is 2200 R.P.M. Both Shakers are furnished with 6" two-groove pulleys for B section belts.

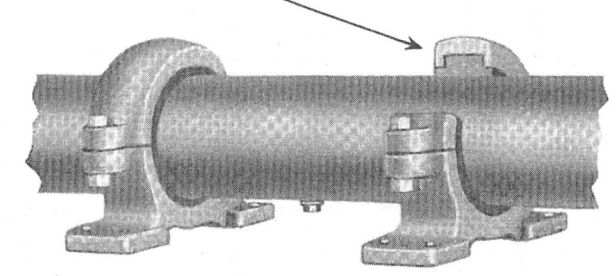

| Size of Shakers | Junior | Large |
|---|---|---|
| Catalog Number of the Shakers | 2760-B2 | 4860-B2 |
| Screening Area of Cloth, Sq. Ft. | 10 ft. | 19 ft. |
| Mud Connection to Well (Pipe) O.D. | 10" | 10" |
| Height to Top of Screen Cloth | 31 1/8" | 31 1/8" |
| Height Over All (Less power unit) | 43 1/2" | 43 1/2" |
| Width Over Mud Troughs | 50 1/2" | 72 1/2" |

| Size of Shakers | Junior | Large |
|---|---|---|
| Length Over Skids | 7'-10" | 7'-10" |
| Shipping Weight (Less Power) | 2200 Lbs. | 2400 Lbs. |
| Export Weight (Less Power) | 3300 Lbs. | 3610 Lbs. |
| Export Width, Boxed | 51" | 79" |
| Export Height, Boxed | 55" | 55" |
| Export Length, Boxed | 106" | 106" |

# THOMPSON TOOL CO.         IOWA PARK, TEXAS

## NEW THOMPSON VIBRATING-TYPE
### SHALE SHAKER AND SAMPLE MACHINE

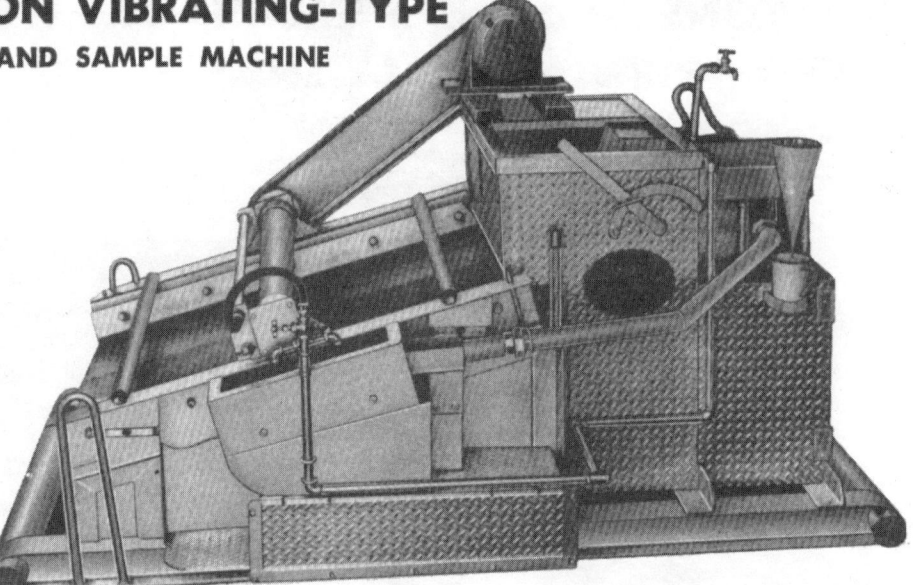

### EXCLUSIVE FEATURES

- **SAMPLE MACHINE** for easy sample taking
- **LIFTING BAILS** for easy pick-up
- **WASH BASIN** for cleaning tools and hands
- **STATIONARY WATER SPRAY** for washing screen
- **SAMPLE, SAMPLE BAG, and TOOL CABINET**

NOW—after years of engineering, development and field testing—it's ready for you!

Thompson, leading manufacturer for many years of the Famous Self-motivated Rotating Shale Separator, now introduce a new Vibrating-type SHALE SHAKER. This Thompson SHALE SHAKER has been proven in the field and many of them are operating throughout Oklahoma and Texas with complete satisfaction. Many NEW and IMPROVED FEATURES—some exclusively Thompson—spearpoint this highly efficient machine.

Here the operator is taking a sample from Sample Machine exclusive on NEW Thompson Shale Shaker.

Above is rear view of NEW Thompson Shale Shaker showing convenient wash basin and tool cabinet.

### OTHER ADVANTAGES

- By-pass built in mud box
- Screen bed sets on Rubber Mounts
- Extra Strong Screen Bed
- Coated Against Corrosion and Rust
- All Heavy Steel—welded construction
- Takes Heaviest Mud Flow
- Uses Electric, Steam, Gas Engine or Oil Drive
- HIGHLY ECONOMICAL OPERATION

### SPECIFICATIONS

**Model A54 Thompson Vibrating Shale Shaker**
Length over all.................................9 ft. 6 in.
Width over all ...................................6 ft. 1 in.
Height..............................................4 ft. 0 in.
Width of Skid....................................4 ft. 11 in.
Flow Line Connection Above Base........2 ft. 10 in.
Size of Flow Line Connection..............8 5/8" O.D. Pipe
Size of Screen..................................4 ft. x 5 ft.
Weight Approx. .................................2300 lbs.

**Model A3 Thompson Vibrating Shale Shaker**
Length over all.................................9 ft. 6 in.
Width over all ...................................5 ft. 3 in.
Height..............................................3 ft. 0 in.
Width of Skid....................................4 ft. 1 in.
Flow Line Connection Above Base........2 ft. 2 in.
Size of Flow Line Connection..............8 5/8" O.D. Pipe
Size of Screen..................................3 ft. x 4 ft.
Weight Approx. .................................1600 lbs.

For best service, vibrator shaft should run between 1950 and 2050 R.P.M. The direction of rotation should be down the screen. The top of the belt should travel towards the discharge of the screen.

**SOLD THROUGH SUPPLY STORES EVERYWHERE**

# HISTORICAL PERSPECTIVE 51

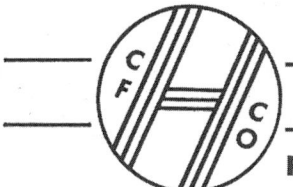

*Originators of LINDA "K" Rotary Shale Extractor*

## C. F. HICKMAN COMPANY

P. O. Box 1224     ALICE, TEXAS     Phone 4-6901

### LINDA "K" ROTARY SHALE EXTRACTOR

WITH LINDA "K" SAMPLE CATCHER

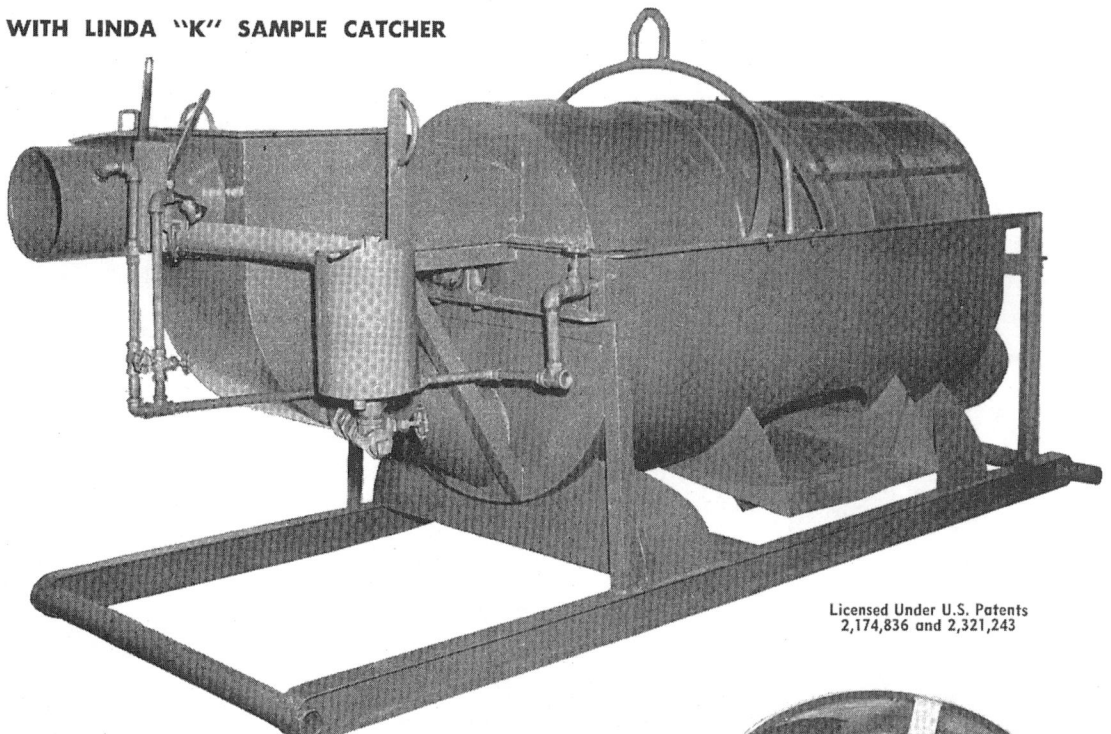

Licensed Under U.S. Patents
2,174,836 and 2,321,243

- **COMPLETELY SELF PROPELLED**, requiring no power cost or power units to maintain
- **SIMPLE IN CONSTRUCTION AND OPERATION**, only two moving parts, reduces maintenance requirements
- **HIGH MUD-CLEANING EFFICIENCY**, reducing wear and tear on pumps, swivel and rotary hose
- **REDUCES MUD WASTE** by keeping mud equalized

The Linda "K" is a completely automatic rotary shale extractor with only two moving parts. Since it is self-propelled by mud flow, it is automatic in operation, with no expensive power units to maintain. Compact, rugged construction reduces maintenance costs.

The long-life rotor drum is feather balanced and rotates at a constant speed on two roller bearings, regardless of the flow of mud. It has no chains, gears, motors, steam turbines or belts to wear and maintain.

The Linda "K" has definitely superior mud capacity, will handle any size pump, and is capable of handling any mud regardless of how heavy the flow. Convenient by-pass and dual mud trough.

10 years of field research, testing and drilling operation in the tough heaving shale regions of the South Texas-Gulf Coast have proved the superiority of the Linda "K" as an efficient, economical shale extractor.

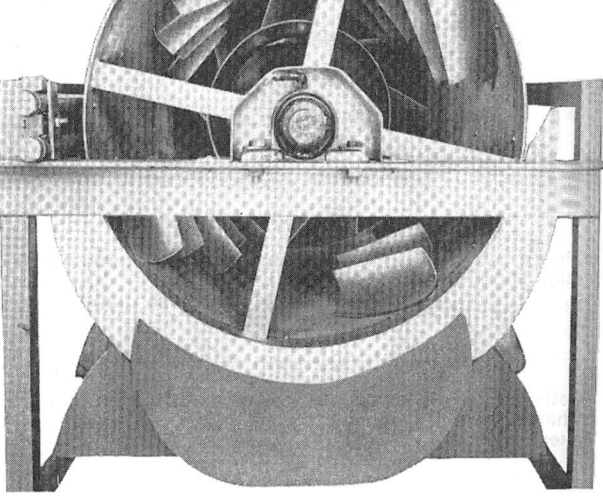

*Illustration showing Tailing Blades*

The above illustration shows the "Patented Tailing Blades" inside the Linda "K" Screen. These blades thoroughly separate and clean any mud, regardless of weight or viscosity, and keep mud equalized at no extra cost. All shale is extracted, reducing wear and tear on pumps, swivel and rotary hose, and water is required on screen only in sticky formations.

**SHALE SHAKERS AND DRILLING FLUID SYSTEMS**

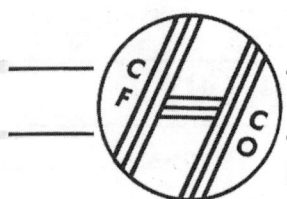

*Originators of LINDA "K" Rotary Shale Extractor*

# C. F. HICKMAN COMPANY

P. O. Box 1224     ALICE, TEXAS     Phone 4-6901

## LINDA "K" ROTARY SHALE EXTRACTOR PARTS DESCRIPTION AND SPECIFICATIONS

Part No. 51—Screen Sample Receptacle

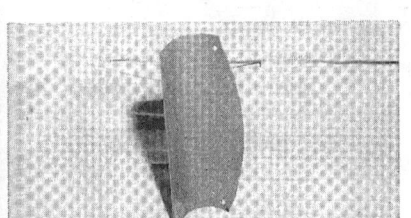

Part No. L or N 56—Tailing Blade, 20 Required

Part No. L or N 59—Rotor and Screen Drum

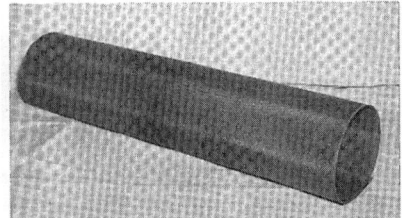

Part No. L or N 50—Monel Screen, 16 Mesh .028 Gage

### SPECIFICATIONS—Model L

Length Overall .............. 11 ft. 8 in.
Width Overall .............. 4 ft. 5 in.
Height Overall .............. 4 ft. 3 in.
Weight Net .............. 1440 lbs.
Screen Area .............. 36 sq. ft.

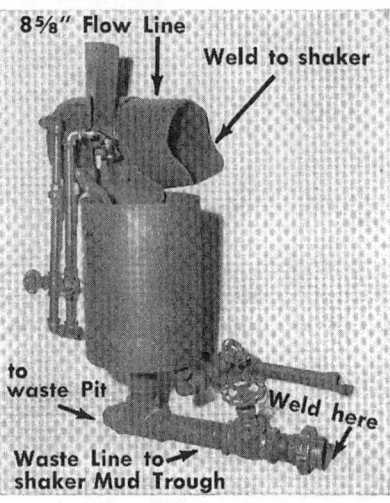

Part No. 57—Sample Catcher

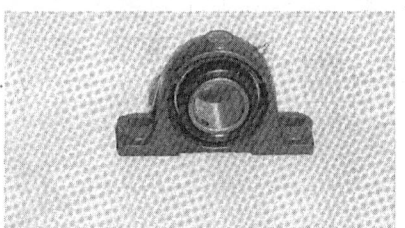

Part No. L or N 52—Bearing, 2 required

### LINDA "K" SAMPLE CATCHER FULLY FIELD TESTED AND PROVEN

Washed cuttings of the finest sand can be taken continuously, or at any desired intervals by the Linda "K" Sample Catcher, optional with the Linda "K" Rotary Shale Extractor, since the flow of mud through the Sample Catcher is continuous as long as the pump is operating.

A removable catcher screen is enclosed in the charged cylinder of the Sample Catcher. This screen is sprayed by jetted water from the inside wall of the charged cylinder. This prevents overflow of the screen, and gives a true washed sample of the cuttings.

Washing fluid can then be returned to the mud system or the waste pit, as desired.

Ruggedly constructed for longer service life, and sold under a 12 month guarantee against all mechanical defects, 275 Linda "K" Shale Extractors are helping cut operating costs for economy conscious drilling contractors.

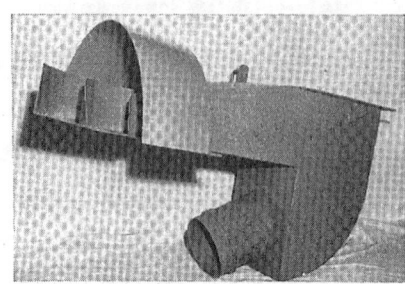

Part No. L or N 55—Intake Chute & By-Pass

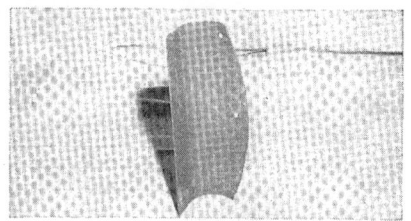

Part No. L or N 58—Tailing Blade, 4 required

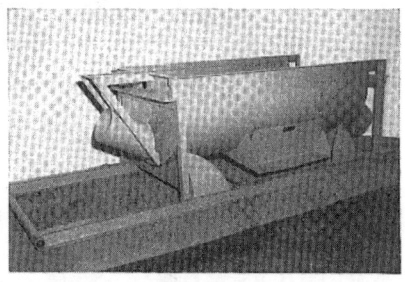

Part No. L or N 60—Dual Mud Trough With Skids

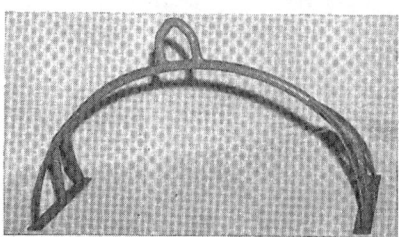

Part No. L or N 54—Hangar

### SPECIFICATIONS—Model N

Length Overall .............. 10 ft. 2 in.
Width Overall .............. 3 ft. 10 in.
Height Overall .............. 3 ft. 9 in.
Weight Net .............. 1260 lbs.
Screen Area .............. 27½ sq. ft.

**WHEN ORDERING LINDA "K" SHALE EXTRACTOR OR PARTS, PLEASE SPECIFY MODEL DESIRED.**

# MERCO CENTRIFUGAL CO.
### San Francisco, California

OFFICES: 150 Green Street
Export Division: 74 Trinity Place, New York City, N. Y.

PLANT: 211 Green Street
Cable Address: LURCO

## MERCO A-24 CONCENTRATOR AND MERCO A-24 DESANDER

The Merco A-24 Separator is a versatile unit with a 20-year history of proven field performance. It has been engineered to deliver continuous service in the oil country, and with changes in speed, and disc-stack the A-24 can be readily adapted for either the conditioning of drilling mud, or in special cases, for the desanding of drilling mud.

In anticipation of the abrasive materials that must be handled on either of these operations, special wear-resistant materials have been used throughout the A-24 to minimize replacement costs.

### MERCO A-24 CONCENTRATOR

The Merco A-24 Concentrator is a high speed centrifugal unit specifically designed for the concentration of weighting material in rotary drilling mud. In actual operation, it may be employed on a drilling well, or may be put to use on mud sumps. For either type of application, operation of the equipment is the same but the objective achieved will differ in this respect: On a drilling well, the purpose of processing the mud through the A-24 is to maintain the mud in proper condition at all times. When salvaging material from the mud sump, the A-24 concentrates the valuable weighting materials into a small volume for re-use on other wells.

Capacity of the A-24 Concentrator will vary according to the conditions encountered on each individual well. However, for purposes of computation the capacity of this unit on mud conditioning will be between 25 and 60 GPM.

### Operational Data

Shown below is a cutaway view of the A-24. Mud from the sump or from the return circulation is introduced into the A-24 either full strength, or diluted with water if this is desirable. It enters the Concentrator at (A); here it is subjected to high centrifugal force which separates the heavier solids from the lighter liquid in separation chamber (B). The concentrated solid underflow occurs at (C); waste material leaves machine through overflow port (D).

### Automatic Control of Separation

An exclusive feature of the Merco A-24 Concentrator, known as the Specific Gravity Control, coupled with the Merco Return Flow Principle permits the operator to regulate the degree of concentration of solids at any desired density. Once set, this regulating device automatically maintains the specific gravity of the mud regardless of the density of the feed.

### MERCO A-24 DESANDER

At lower speeds, with disc-stack removed, the Merco A-24 operates as a high efficiency desander. At capacities up to 350 GPM, less than 1% of plus 100 mesh sand remains in the mud, and less than 3% of plus-200 mesh sand is residual after processing in the A-24.

High efficiency desanding of rotary drilling mud results in multiple savings. It eliminates abrasive wear on drilling equipment; reduces the hazard of sticking-pipe; and minimizes down time when circulating to condition the mud.

For desanding, the Merco A-24 may be placed directly on the mud overflow line on the casing of a drilling well, or may be used to condition mud drawn from a sump or storage pit. If return flow capacity from the well is in excess of the capacity of the unit, the overage may be bypassed to a storage area for later processing.

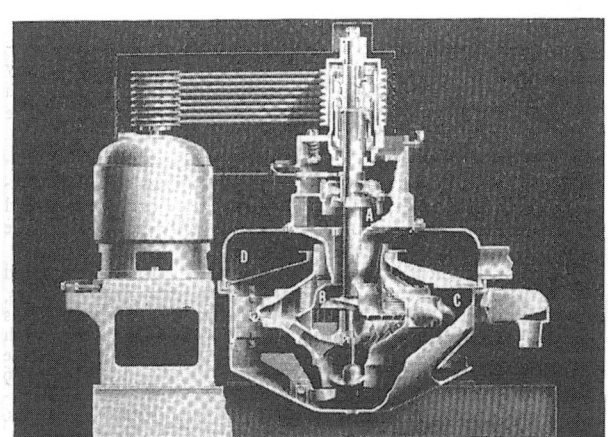

| BRIEF SPECIFICATIONS MERCO A-24 SEPARATOR | | |
|---|---|---|
| | CONCENTRATOR | DESANDER |
| Capacity (Under ordinary conditions) | 25-60 GPM | 250-350 GPM |
| Dimensions | H: 5'; L: 7'; W: 43" | H: 5'; L: 7'; W: 43" |
| *Approx. Shipping Weight | 4,000 to 6,000 lbs. | 4,000 to 6,000 lbs. |
| Horsepower | 40 | 20 |
| Speed of Centrifuge | 700 to 1400 RPM | 900 RPM |
| Underflow Density | Up to 2.4 Sp. Gr. | Up to 2.0 Sp. Gr. |

*Varies with type of drive. Units available with electric motor, turbine or engine drive, as required.

# MEDEARIS *Oil Well* SUPPLY CORP.

## UNITIZATION

The pump manifold is an integral part of the tank. This permits transportation of both tank and manifold in one load, also permits rigging up mud system in a minimum of time. There are no loose valves, manifold sections or pipe fittings to become lost or damaged. Units are built to legal sizes for movement without permits.

*Fig. 3—Type U Tank*

**MEDEARIS COMPLETE MUD SYSTEM INCLUDES SUCTION AND DISCHARGE MANIFOLD, MUD LAB AND DRY MUD STORAGE FOR SMALL WORK OVER RIG.**

*Fig. 4—Type W Tank*

## MEDEARIS UNITIZED MUD SUCTION TANK

This late model Medearis mud suction tank incorporates a number of improvements developed over several years of experience in building and supplying these units to meet the differing needs of many operators. These unitized tanks have been growing steadily more popular, since they incorporate by-passes, shakers and screens, mud ditch and chemical tank into one compact, skid-mounted, easily transported unit.

The tank shown here has Medearis vibrating mud screens installed. These are the screens supplied unless the customer specifies otherwise. Under each shaker is a de-sander, with discharge nozzle and lever-operated dump gate for cleaning, which traps any fine sand which might be passed by the screen cloth. Cleaning gates are also provided at tank bottom for easy cleaning. The chemical tank is mounted on the mud ditch, between the screens.

For positive control, this unit is equipped with Medearis 10-inch suction valves, controlled from the grating at the top of the tank. These special valves are easy to operate, either to open or to close, and provide a positive, leak-proof seal. Dual by-passes allow by-passing the shakers and discharging directly either into the tank or into the sump.

Medearis unitized mud suction tanks are all-steel, all-welded construction, and are completely painted. Catwalks all around are non-slip grating; guard rails are supplied all around to prevent accident, and access stairs are welded on. The unit shown here is a 150-barrel capacity tank, 24 feet long, and weighs 13,000 pounds completely equipped. It is shown only as a typical unit which meets the requirements of many operators, since these unitized mud suction tanks are built by us in any size dictated by the customer's requirements.

# THOMPSON TOOL CO.        IOWA PARK, TEXAS

## ALL NEW DOUBLE VIBRATING SHALE SHAKER

### ENGINEERED FOR OFFSHORE BARGES OR ISLANDS, DEEP WELLS AND WELLS USING HIGH VOLUME AND HIGH PRESSURE MUD PUMPS

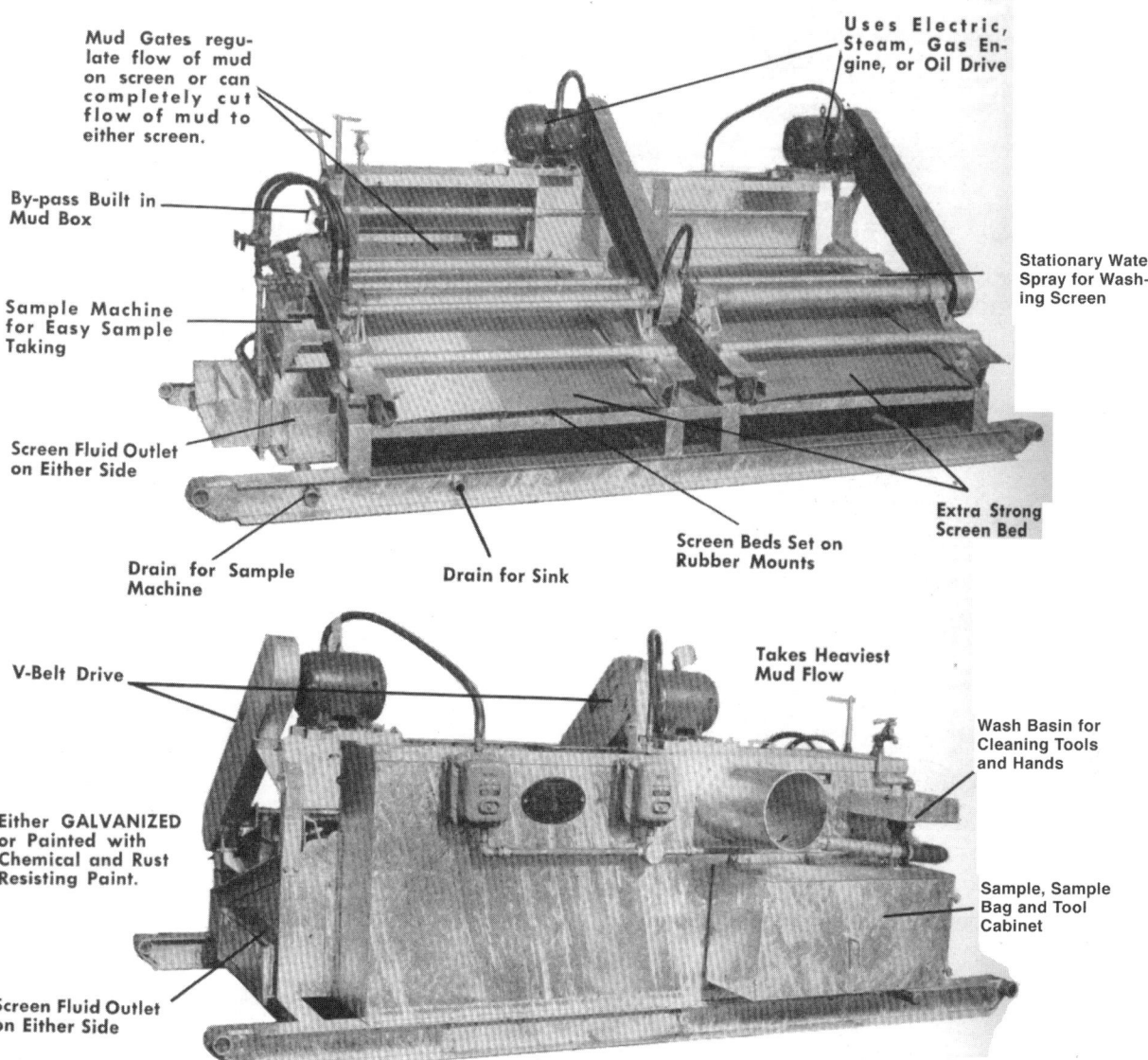

THOMPSON, leading manufacturer of vibrating and rotating shale separators, now provides the drilling industry with its newest and finest shale shakers . . . the THOMPSON DOUBLE VIBRATING SHALE SHAKER. This new twin vibrating model is specially engineered for off-shore barges or islands, deep wells, or wells using big volume and high pressure mud pumps.

THOMPSON Shakers are famous for their highly efficient service at any depth or flow and for their economical first cost. They do one of the most effective jobs . . . of removing shale and abrasives from mud, prolonging drilling tool life, and general cutting down of drilling expense. The THOMPSON Shale Shakers are easily transported and installed, using either electric, steam, gas engine or oil drive. One of the most outstanding features, exclusive to the Model A54RD is the Thompson SAMPLE MACHINE which gives the operator an important foot-by-foot sample of the cuttings.

With THOMPSON on the job, you are assured of getting clean mud.

**SOLD THROUGH SUPPLY STORES EVERYWHERE**

# MEDEARIS *Oil Well* SUPPLY CORP.

## MEDEARIS IMPROVED TYPE X MUD SHAKERS

### SINGLE SHAKER

Newly engineered and redesigned, this new Medearis Single Type X Mud Shaker has fewer moving parts than any other mud shaker. Simplicity of construction and operation reduce maintenance far below what is usually considered normal. (Some operators report savings as great as 50%.) There are no breakable tension boards. The new, simplified rubber strip deck is easy to install and gives longer screen life. All parts are corrosion-protected, and the vibrator housing and screen box are hot-dip galvanized. The vibrator is roller-bearing mounted and housed for complete protection. Tension and vibration are uniform over all parts of the screen.

Weight has been reduced to facilitate mounting and rig-down. The compact design also contributes to easier handling.

Vanes in the feed gate are individually adjustable for even distribution and flow of mud over the whole screen surface. Gates in the integral mud tank allow mud flow in either direction. Depending upon screen mesh opening and pump capacity, the Medearis Single Type X Mud Shaker will handle up to 800 gallons per minute of average drilling mud. Units can be supplied with or without surge box.

### DUAL SHAKER

The Medearis Dual Type X Mud Shaker consists of two Medearis Single Type X Mud Shakers complete with bottom tank and distributing chute with mud wings. These dual shakers are equipped with a surge box in which the flow of mud coming directly from the well is slowed down to ensure an even distribution and flow of mud over each shaker. The dual unit is also equipped with a by-pass for by-passing either or both shakers during mechanical maintenance or in the event the operator is using a lost-circulation material which he does not want to pass over the shakers. A cleanout valve located directly in the bottom of the surge box assures easy cleanout operations.

The dual shakers are mounted on a unitized skid so that the entire assembly can be easily tailboarded and loaded on a truck without the need for disassembling or separating the shakers. Mud handling capacity is equivalent to that of two Medearis Single Type X Mud Shakers.

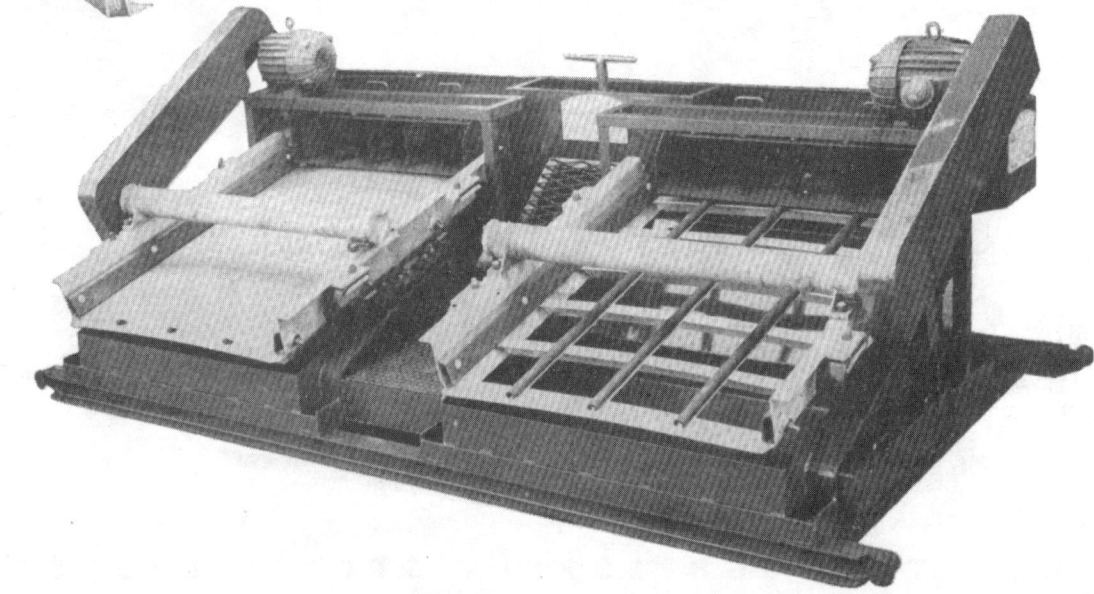

# MEDEARIS *Oil Well* SUPPLY CORP.

*These two views show clearly the design and construction of the Medearis Master Mud Mixer and Degasser. Any type of power may be used—gas, diesel, or electric—at customer's option.*

## MEDEARIS MASTER MUD MIXER AND DEGASSER

### FAST
Proved in the field on rig after rig, this unit is one of the fastest, most efficient mixers ever designed. 250 gallons of mud per minute—thoroughly mixed and put into circulation. Handles bulk material of all types—bentonite, clays, oil-base or oil-emulsion, weight materials, dry chemicals . . . even such hard-to-mix lost circulation materials as fibers, redwood bark, cottonseed hulls and beet pulp.

### EFFICIENT
Material added through the hopper goes directly to the four 13" opposed-pitch mixing propellers at the bottom of the tank. These blades, turning at 1200 rpm, produce thorough, uniform agitation throughout all the 20-barrel tank. An intake ditch along the fluid column leads directly into the center hopper, delivering fluid direct to the mixing blades while, at the same time, mud is being shipped back to the suction tanks.

### ECONOMICAL
Maintenance of the Medearis Master Mixer is simple and easy. The corrosion-resistant mixing blades have long life, and are bolted to the shaft for easy installation or removal. The shaft itself revolves on pressure-lubricated pillow blocks . . . there are no bearings in the tank to wear out!

### A DEGASSER, TOO
For use as a Degasser, the unit is equipped with 2 stacks, each with its own suction fan run by an explosion-proof motor. Thus equipped, this unit does an excellent job of getting the gas out of gas-cut mud, and restoring the mud to normal conditions. Operators report substantial mud savings from this use.

### ACTS AS A THIRD PUMP
The Medearis Combination Mud Mixer and Degasser is equipped with a 4" x 4" impeller shipping pump driven by a 30-hp explosion-proof electric motor, or by a combustion engine which may be diesel, natural gas or butane. Many operators report that the Medearis Master Mud Mixer will replace the third pump on today's larger rigs, or the second pump on small rigs. It is used, for example, to pump out the cellar or sump, and to transfer mud from storage to suction tank.

### EASY TO TRANSPORT
As shown, the entire unit, 20' long, 6' wide, is skid-mounted for easiest moving. This includes the power source, which may be any type specified by the customer—gas, diesel or electric. Removable tank covers are provided, and the non-slip hinged catwalk and demountable stairs may be mounted on either side.

Weight of Unit 7,300 lbs. less prime mover.

## HUTCHISON MANUFACTURING COMPANY

6609 Avenue U  
HOUSTON 11, TEXAS  
Subsidiary of Revere Electric Mfg. Co.  
Phone: WAlnut 6-7471

### "RUMBA" SHALE SHAKERS
### The NEW "RUMBA" THE MOST EFFECTIVE OF ALL SHALE SHAKERS.

Only "RUMBA" provides all the plus features. Check them for yourself. Know why more "RUMBAS" are in use than all other Shale Shakers combined.

### HERE'S WHY - - -

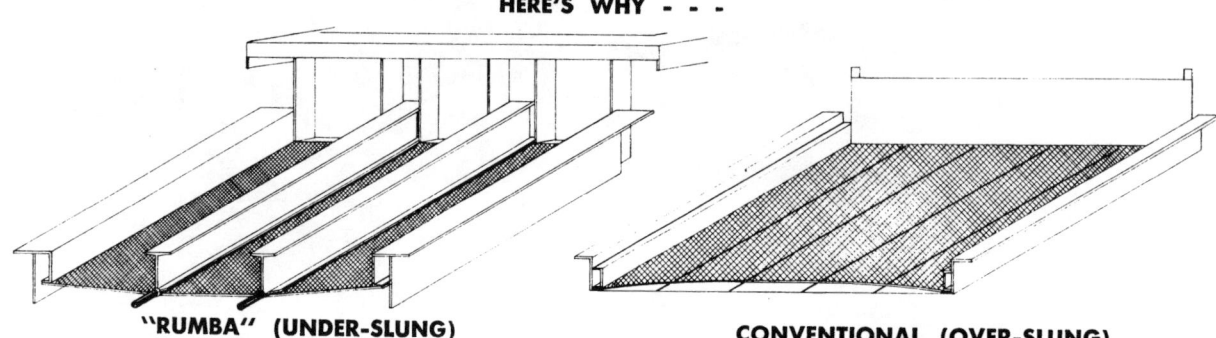

**"RUMBA" (UNDER-SLUNG) SCREEN TOP**

The center struts of the Shaker top are equipped with Rubber Cushion Tubes. This arrangement provides three separate channels which positively control the distribution of Drilling Fluid over the **entire** screening area, providing higher capacity and more effective screening.

**VS.**

**CONVENTIONAL (OVER-SLUNG) SCREEN TOP**

The over-slung Screen has **no** control over distribution. The Drilling Fluid flows to the sides—utilizing only a part of the screening area—providing less effective screening of the Drilling Fluid.

### ONLY "RUMBA" PROVIDES EVEN DISTRIBUTION OF DRILLING MUD OVER THE ENTIRE SCREENING AREA

★ **ONLY "RUMBA,"** through Constant Research, builds more performance into the Shale Shaker. Now available on all "RUMBA" is a NEW and BIGGER Vibrational Orbit to speed cuttings off the screen. The NEW ORBIT permits more effective screening at still higher RATES OF FLOW.

★ **ONLY "RUMBA"** provides the longest screen cloth life. The four heavy struts on the Screen Top drive the screen with TWENTY LINEAL FEET OF RIGID DRIVING SURFACE— positively preventing whipping or flexing of the screen cloth.

★ **ONLY "RUMBA"** has built-in DE-SANDER that removes tons of sand through Flotation.

The under-slung Screen Cloth, with the entire frame above the Screen, leaves no obstructions on the underside of the Cloth. A droplet of mud on the underside of the Screen Cloth is loaded with fine flour-like sand that falls through the meshes of the Cloth. This droplet (with its load of sand) is held together on its way to the bottom of the Cloth by surface tension. The molecular force of adhesion holds it to the bottom of the cloth. Sand entrapped in the droplet falls clear of the mud returns onto the De-Sanding Tray. This patented method of sand removal is very effective and permits substantial savings in pump maintenance, plus producing cleaner Drilling Mud.

# THOMPSON TOOL CO. — IOWA PARK, TEXAS

## THOMPSON MUD AND CHEMICAL MIXER

1. Automatic Feed and Mixer designed to give close control of dry muds and chemicals (with the exception of caustic soda) mixed in mud system.
2. Saves Mud—Delivers mud to system thoroughly mixed. Actual field tests have proven that an increase in Jel yield of 15% to 30% over mud mixed by the usual method of hand feed and jet mixing.
3. Elimination of "light" spots ordinarily found in mud mixed by hand, provide greater safety when drilling in high pressure zones.
4. Mixing and evenly distributing all dry chemicals (except caustic soda) through mud system.
5. When desired a system of mud may be mixed and on hand ready for drilling out from under surface. This may be accomplished by setting up mud system and mixing mud while completing rig-up of remaining equipment.
6. Thompson Mud and Chemical Mixer is a complete unit in itself—Does not require additional pump or power. Unit is ready for use as soon as it is set in place.
7. Elimination of using large expensive pumps and power units for mud mixing.
8. Saves time on one-pump rigs by not having to lower pump pressure due to mixing mud.
9. Labor Saving—When hopper is filled and correct feed set —crewman is free for some other job until hopper is emptied and ready to be refilled.
10. No High Pressure Jets are used in the Thompson Mud and Chemical Mixer therefore there is no cutting action due to high pressure abrasive mud.

### Specifications
Length—11' - 6"
Width—6' - 4"
Height—6' - 7"
Weight—
Horse Power Req.: 36 B.H.P. at 1500 RPM
Volume hopper: 500 LBS. Bentonite
                  1000 LBS Barytes
Hopper will feed from 100 to 3500 LBS per hour
Pump Volume: 250 GPM

## THOMPSON DESANDER

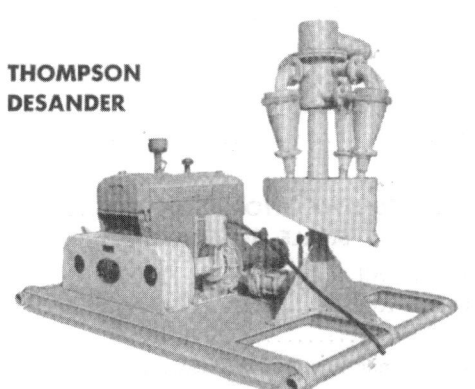

Saves mud and equipment
Keeps sand content of mud as low as ½ of 1 per cent.
The cone is an efficient size.
Equipped with engine or electric motors and centrifugal pump as required or specified. Components with cone cluster are mounted on heavy H-beam skid.

### Capacities
Single Cone—250 GPM
Two Cone—500 GPM
Three Cone—750 GPM
Four Cone—1000 GPM

### Advantages
1. Completely rubber lined 9/16" thickness.
2. Discharge head adjustable to turn in any direction — speeds set up.
3. Sand trap adjustable for discharging in desired direction for faster set up.
4. Quick cleanout.
5. Rate of sand discharge easily controlled with orifice adjustment disc. Easily operated by lever handle.

**SOLD THROUGH SUPPLY STORES EVERYWHERE**

 **BAROID DIVISION NATIONAL LEAD COMPANY** ◄◄◄◄◄◄

## LIQUID MUD

The list of Baroid Liquid Mud Plants is growing almost daily as the demand for this popular service increases. Each plant has storage and mixing facilities for any drilling or completion fluid specified. Such fluids include conventional water muds; water muds modified to operator's specifications; oil muds prepared to specifications for drilling, fishing, setting casing, perforating, gravel parking and coring. They also include brine fluids, gum fluids, completion and packer fluids.

For the fluids needed for the job at hand—made to your specifications as to weight, viscosity, filtration, type and condition—contact any Baroid store, office or representative.

## MISCELLANEOUS MUD MATERIALS

| | | | |
|---|---|---|---|
| Cottonseed Hulls | Cement | Quebracho | Sodium Chromate |
| Bicarbonate of Soda | Gypsum | Salt | Sodium Acid Pyrophosphate |
| Calcium Chloride | Graphite | Soda Ash | Aluminum Stearate |
| Caustic Soda | Lime | Sodium Bichromate | |

Baroid's principal mud products are described in this catalog. Any commonly used mud ingredient can be supplied even though it is not listed in these pages.

## DOUBLE DECK SHAKER

Baroid's new Double Deck Shaker provides the latest advancement in the control of drilled solids. Two screens are employed, usually a 30 mesh on top with an 80 mesh below. The unit employs a gyrating motion rather than the simple shaking motion of regular shale shakers. This motion keeps the solid particles moving toward the discharge end and helps prevent plugging of the screens. Maximum capacity of the unit is 12 barrels of mud per minute. Recommended power supply is three-phase 220 volts, AC, but units can be modified to use 110 volts, single-phase AC or DC.

## PORTABLE LIQUID MUD MIXER

Baroid's Portable Liquid Mud Mixer provides an extra 100-barrel mud mixing capacity at the rig. It has been used in combatting lost circulation, in weighting up rapidly when unexpected formation pressures are encountered, and as a general auxiliary to rig mixing equipment. Power is supplied by a diesel driven pump. The hydraulically driven mixing propeller can be raised or lowered to any desired mixing level. Dry and liquid material can be added directly to the mixing tank, or through a jet mixer on top of the tank. Dry material can also be added through a ground level hopper equipped with a conveyor system. The entire unit is skid-mounted for easy handling.

▶▶▶▶▶▶ BAROID DIVISION NATIONAL LEAD COMPANY

## MUD CENTRIFUGE

The BAROID MUD CENTRIFUGE cuts mud costs by efficiently recovering valuable barite, rejecting unwanted solids and providing better control of viscosity, gel, filtration rate and cake thickness. When used as a desander, on low weight mud, the centrifuge will remove solids without wasting valuable fluid.

A large throughput mud volume can be efficiently processed with accurate control of output weight. Flow-rators allow direct reading of mud and water volume. A tachometer provides continuous speed readings and total operating hours. All equipment has been specially designed to reduce maintenance and increase efficiency.

The BAROID MUD CENTRIFUGE is available in two models: Standard and Compact. Either model can be mounted on any mud pit or tank. The Compact unit can be installed in an area as small as 27″ high, 45″ wide and 90″ long. Both units are powered by a diesel, gasoline, natural gas or butane engine that is separately skid mounted and can be placed anywhere within fifty feet of the centrifuge unit. The mud to be processed is pumped to the centrifuge unit by a small portable pump that can easily be handled by two men. The pump can be placed in the mud tank, pit or reserve pit.

More information is available on request.

## AUTOMATIC GAS DETECTION & ALARM SYSTEM

The Baroid Automatic Gas Detection and Alarm System for compressor station protection was developed by Baroid to meet an expressed need for a reliable, automatic system that will not generate false alarm warnings of gas leaks and gas contamination build-up in gas compressor stations.

The instrument can be built to test air samples from any number of locations up to eighteen. The test cycle does not exceed 15 sec/sampling point regardless of the number of sample points. If gas is detected at any point, the test cycle may be interrupted and a hand sampler used to determine the source of the leak manually. Once the leak is located, the instrument can be switched back to automatic operation. If for any reason this is not done, the instrument returns itself to automatic operation after one hour.

An air-pumping system delivers the samples to the detector element and continuously purges all sample lines. A standby air-pump automatically switches into operation if the primary air-pump fails.

The gas detection unit tests the instrument daily and before registering an alarm. It features automatic zeroing, automatic sensitivity checking, automatic filament burnout detection and automatic switching to a standby detector unit if the primary unit is inoperable.

Four output signals provide: (1) warning when gas content of the air sample reaches a pre-determined level, (2) alarm when gas content reaches the danger level, (3) signals when the system needs service, and (4) when the system is inoperable.

The entire unit is mounted in enameled steel cabinet, 70″ high, 24″ wide and 22″ deep. Panels and chassis are of anodized aluminum. Power requirements: 105 to 130 volts, AC, 60 cycle, 500 watts. Explosion proof units are available.

# MEDEARIS *Oil Well* SUPPLY CORP.

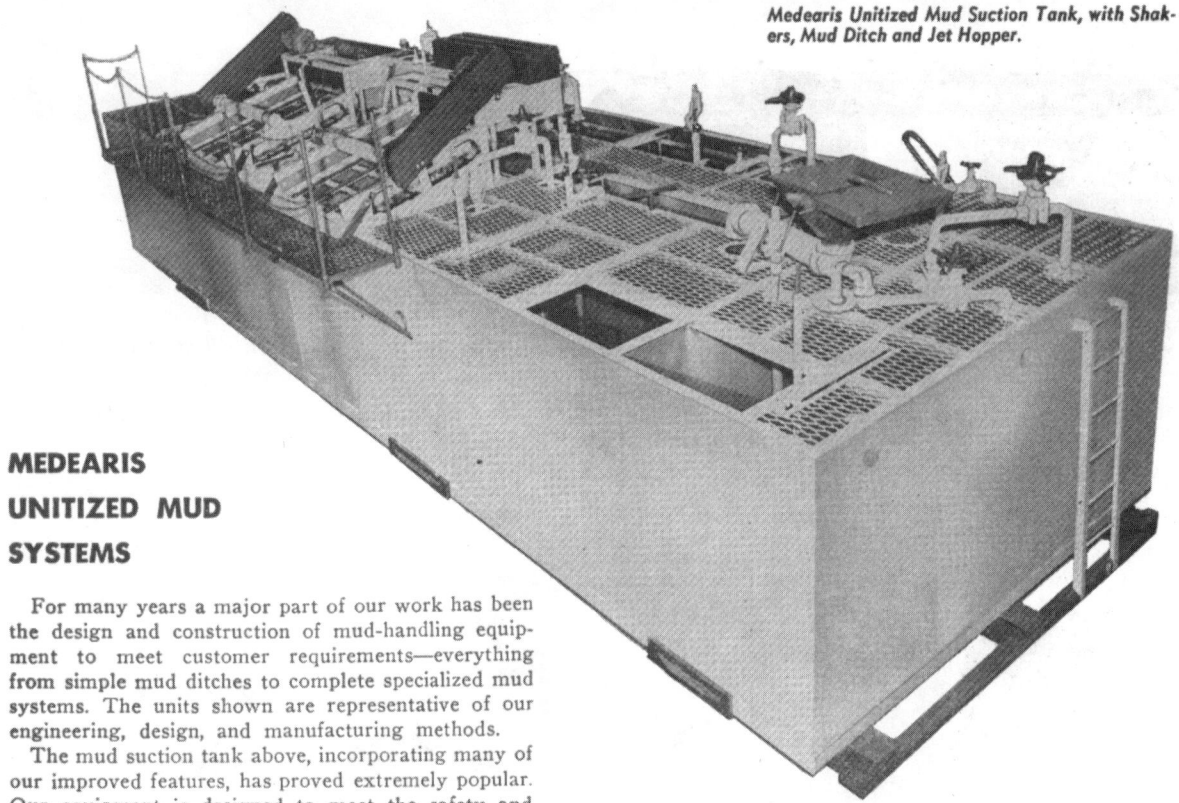

*Medearis Unitized Mud Suction Tank, with Shakers, Mud Ditch and Jet Hopper.*

## MEDEARIS UNITIZED MUD SYSTEMS

For many years a major part of our work has been the design and construction of mud-handling equipment to meet customer requirements—everything from simple mud ditches to complete specialized mud systems. The units shown are representative of our engineering, design, and manufacturing methods.

The mud suction tank above, incorporating many of our improved features, has proved extremely popular. Our equipment is designed to meet the safety and highway regulations of any state or country where it is to be used. Medearis Improved Type "X" Mud Shakers with new style Flow Gates, as shown, are normally sold with our systems.

Sloped bottom desanders, immediately below each shaker, trap fine floury sand and sediment which passes through the screen. The settlings are disposed of through dump gates to outside facilities. Both the desander dump gates and the cleanout gates at the tank bottom are lever-controlled from the tank deck. A Bottom Cam-operated Dump Gate can be supplied if desired.

The mud ditch is equipped with by-pass valves and can include a chemical tank. Medearis Quick-Opening pump suction valves provide positive mud control, and thorough mud agitation is provided by built-in Medearis 360° revolving bottom mud guns.

Top decks and catwalks are made of non-slip grating. Hand rails are instantly demountable, as are the access stairs. The hinged catwalk folds for transportation.

Our mud systems incorporate tanks, ditches, chutes, mud laboratories, storage facilities, pump and tank manifolds, mixing tanks, pill tanks, and other specified equipment, as required. Design and construction can be tailored to fit your needs.

## MEDEARIS CUTTING AND WASHING SYSTEM

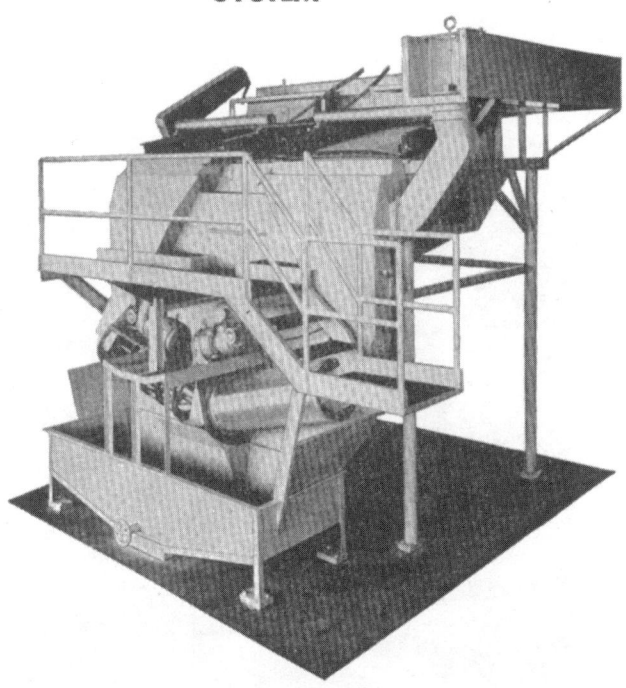

A good example is the cuttings and washing system shown at left. This system, with two new and improved Medearis Type "X" Shakers, is being used by major oil companies on offshore drilling platforms for the control of the Mud Flow, separation of cuttings, return of mud to storage, and washing of oil from the cuttings for disposal in the ocean. This can be used on any water or land operation where oil contamination is a problem.

The speed of the washers can be regulated to obtain the best cleaning results for different areas and conditions.

This system is designed to meet the requirements of the Fish & Game Commission and can be modified to fit individual requirements.

# SLUSH PUMPS & ACCESSORIES

## POWER SLUSH PUMPS

Complete range of reciprocating pumps for all requirements of the oil industry.

General representation of CONTINENTAL-EMSCO Mud Pumps from 225 to 1650 HP.

JANDIN CYCLE PUMPS: from 30 to 225 HP manufactured by S.N. MAREP (see specifications and performance chart below).

UNITIZED PUMPS—The combination of Diesel engines or electric motors and power pumps allows the assembly of unitized pumps—The units can be trailer or skid-mounted.

### SPECIFICATIONS

| TYPE | GE 30 | DD 45 | DB 110 | DB 225 |
|---|---|---|---|---|
| SIZE (Max. Liner Bore x Stroke) | 4½" x 4" | 5" x 6" | 5½" x 8" | 6¾" x 10" |
| Piston Rod API Taper | No. 1 | No. 1 | No. 3 | No. 3 |
| Fluid End Test Pressure (psi) | 1000 | 1700 | 3100 | 3900 |
| Fluid End Suction Flange | 4" | 4" | 6" | 8" |
| Fluid End Discharge Flange | 2" | 2" | 3" | 4" |
| Maximum H.P. Input at Maximum Rated Speed (1) | 30 | 45 | 110 | 225 |
| Maximum Rated Speed of Crank Shaft (rpm) | 180 | 140 | 150 | 125 |
| Corresponding Speed of Input Shaft (rpm) | 1320 | 690 | 750 | 565 |
| Gear Ratio | 7,33/1 | 4,92/1 | 5/1 | 4,5/1 |
| Approximate Weight (lbs.) | 1300 | 2200 | 4000 | 11000 |
| **PERFORMANCE CHART AT MAX. RATED SPEED (2)** | | | | |
| G.P.M. Output With Maximum Size Liners | 145,5 | 210 | 355 | 577 |
| Discharge Pressure With Maximum Size Liners (psi) | 295 | 305 | 435 | 555 |
| G.P.M. Output With Minimum Size Liners | 86,7 | 60 | 98 | 142 |
| Discharge Pressure With Minimum Size Liners (psi) | 495 | 1070 | 1600 | 2260 |

(1) Based on 85% mechanical efficiency  (2) Based on 100% volumetric efficiency

## SLUSH PUMP PARTS

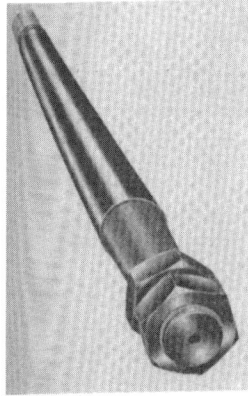

PISTON RODS—PISTONS and RUBBERS—LINERS

MUD PUMP VALVES:
— Z type "Super-Valve" for the highest service pressures
— X and Y types, wing-guided valves, for small pumps

STUFFING BOX with "CONIC" type packing (Patended)

S.N. MAREP supply a wide range of these wear parts for various makes and types of pumps, in order to meet most of the field requirements.

## "ROL" SHALE SHAKERS

S.N. MAREP is World Commercial Agent for the Oil Industry of "ROL" Shale Shakers manufactured by Ets G. CHAUVIN (Grenoble), screen specialist since 1930.

Four models:
— 2 types (with or without mud collecting tank)
— 2 sizes (single or dual screen cloth)

One size screen cloth fits all models (59 1/16" x 23 5/8")

Maximum screening capacity: 450 and 900 US gpm

**THOMPSON TOOL CO.**         **IOWA PARK, TEXAS**

## THOMPSON DESANDER

Keep sand content of mud as low as ¼ of 1% with
**THOMPSON DESANDER**

### ADVANTAGES

1. Operates successfully with over 30 lbs. lost circulation material to the barrel.
2. Removes all large cuttings when used in system without shale shaker.
3. Saves weight material—because weight material is so fine it will be returned with cleaned mud.
4. Control mud to desired weight.
5. Completely rubber lined 9/16 to ¾" thickness.
6. Discharge head adjustable to turn in desired direction... speeds set up.
7. Sand trap adjustable for discharging in desired direction for faster set up.
8. Quick cleanout.
9. Rate of sand discharge controlled with automatic orifice adjustment.

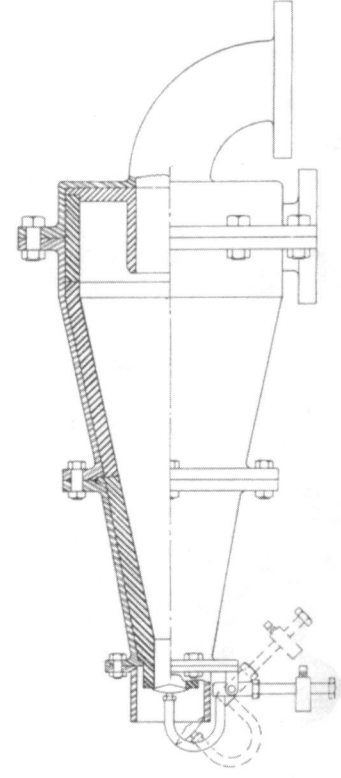

Automatic orifice adjustable from 0 to 1¼" opening.

**SOLD THROUGH SUPPLY STORES EVERYWHERE**

# Centrifuges Inc.

## Dual Screening Machine

**HIGH SPEED DUAL SHALE SHAKER WITH AUTOMATIC LUBRICATOR**

P. O. Box 51853, Phone 318—232-0188
Lafayette, Louisiana 70501 U.S.A.
Write for Catalog on Sales, Lease and Rental

### 3600 REVERSALS PER MINUTE ASSURES ACCURATE SEPARATION

The direction of movement of the screen bed on a Centrifuge machine varies throughout its length. The material deposited on the feed end of the screen is thrown forward and spread over the surface of the wire cloth. High speed action quickly stratifies the load and fines are rapidly eliminated. As particles move toward the center of the screen, their forward motion is slowed and a vigorous circular motion is imparted to the load. This creates finer stratification and encourages the passage of near-size material through the screen. The backward throw at the discharge end of the screen retards movement of the load still further, assuring clean, near-perfect separation of undersize and oversized material.

ACCESSIBILITY: Top-mounted explosion proof motor, equipped with automatic lubricator, permits easy access for servicing and inspection.

**Screen cloth quickly replaced.**

INCREASED SCREEN LIFE: Supported in hoop tension, screen cloth becomes integral part of machine, eliminating flutter and destructive fatigue. Uniform action is maintained over entire screen surface.

### Specifications

CAPACITY:
- 1400 gallons per minute with 50 mesh screen
- 1000 gallons per minute with 60 mesh screen
- 750 gallons per minute with 70 mesh screen
- 550 gallons per minute with 80 mesh screen

SIZE:  Width.................7'11"
       Length...............12'5"
       Height...............5'2½"

ACTUAL SCREEN AREA: 4' x 6' ea.
DRY WEIGHT: 3,500 lbs.
POWER: 1½ H. P. 60 cycle, 3 phase 220 or 440

### Price List

| TWO SCREENS | THREE SCREENS |
|---|---|
| Machine Screens | Machine Screens |
| Lubricator | Lubricator |
| Screens—80 mesh | Screens—80 mesh |
| Screens—50 mesh | Screens—50 mesh |
| Skid | Skid |
| Connections | Connections |
| Sandblasting | Sandblasting |
| Electrical Connections | Electrical Connections |
| **Price:** $7,977.60 | **Price:** $8,261.10 |

### Sale Price

Dual unit with automatic lubricator ............................................................. $14,535

Lease purchase contracts or long term rentals are available upon request.

### Delivery and Installation:

For delivery on all sale units — allow 45 days.

All Transportation is F. O. B. our Lafayette plant.

Centrifuges, Inc. will send a service engineer to instruct installation of the units and operation of the units.

The units can be purchased on a 12 month Lease-Purchase Agreement.

**Centrifuges Inc.**

P. O. Box 51853, Phone 318—232-0188
Lafayette, Louisiana 70501 U.S.A.
Write for Catalog on Sales, Lease and Rental

# Centrifuge...

THE BIRD 18 X 28 DECANTING CENTRIFUGE LEADS THE FIELD IN PROCESSING WEIGHTED MUD FOR VISCOSITY CONTROL.

### THE HIGH VOLUME
## Mud Centrifuge

The high volume mud centrifuge is designed to save the operator and contractor time and money. The unit is equipped with a GM diesel power skid, enabling it to perform for continuous periods without readjustment or maintenance. Benefits are low water use, coupled with 97% barite reclamation. High drilled solids rejection and short operation per day add to the extra economies of the Centrifuge.

High density muds of up to 21 lbs. per gallon are synonymous with deep holes along the Gulf Coast. The Centrifuge's ability to be used as a mud reclamation device eases the contractor and operator's concern with fluid properties of high density muds.

The unit is efficient, dependable, economical and effective. When tested against all other types of centrifuges the unit demonstrates its superiority in barite savings, and is not surpassed in volume handled by any other mechanical centrifuge available.

**271 GM POWER UNIT FOR CENTRIFUGE**

**DIMENSIONS OF POWER SKID:**
8'6" x 4' x 4'6"
Weight approximately 4,000 lbs.

**DIMENSIONS OF UNIT:**
6' x 4' x 3'6"
Weight approximately 4,000 lbs.

**POWER SYSTEM DISTRIBUTION:**
Hydraulic

**Inquire for Rental, Lease, and Purchase**

# DAHLORY, INC.

## CUTTINGS PROCESSING AND WASHING EQUIPMENT

**DAHLORY** standard and custom designed Cuttings Washing Equipment is being used with excellent results in the Gulf of Mexico, off the coast of California — any area where contamination is a problem. Oil and Water Separators can be supplied if required to process the waste water from the Cuttings Washing Unit, the working floor and other hosed down areas.

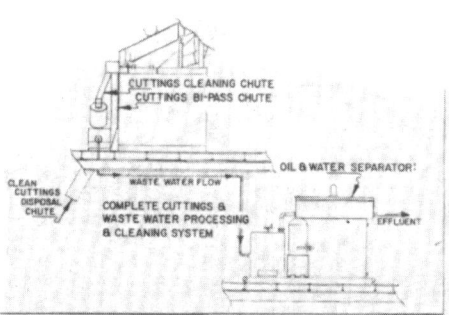

DAHLORY Hi-Angle Shaker, available in single or dual models, incorporates the following advantages and features:

- Vibrator design eliminates V-Belts.
- Hi-Angle and elliptical motion allows faster screening of drilling fluid with less blinding.
- 30 sq. ft. of screening area per deck allows more material to be processed with finer mesh screens.
- Easy access to change the 3 piece screen cloth reduces maintenance cost.
- Ability to use up to 80 Mesh Screen Cloths. 2 x 4 and 4 x 5 Shakers are standard. Other sizes available.
- Purchase or rent agreements.

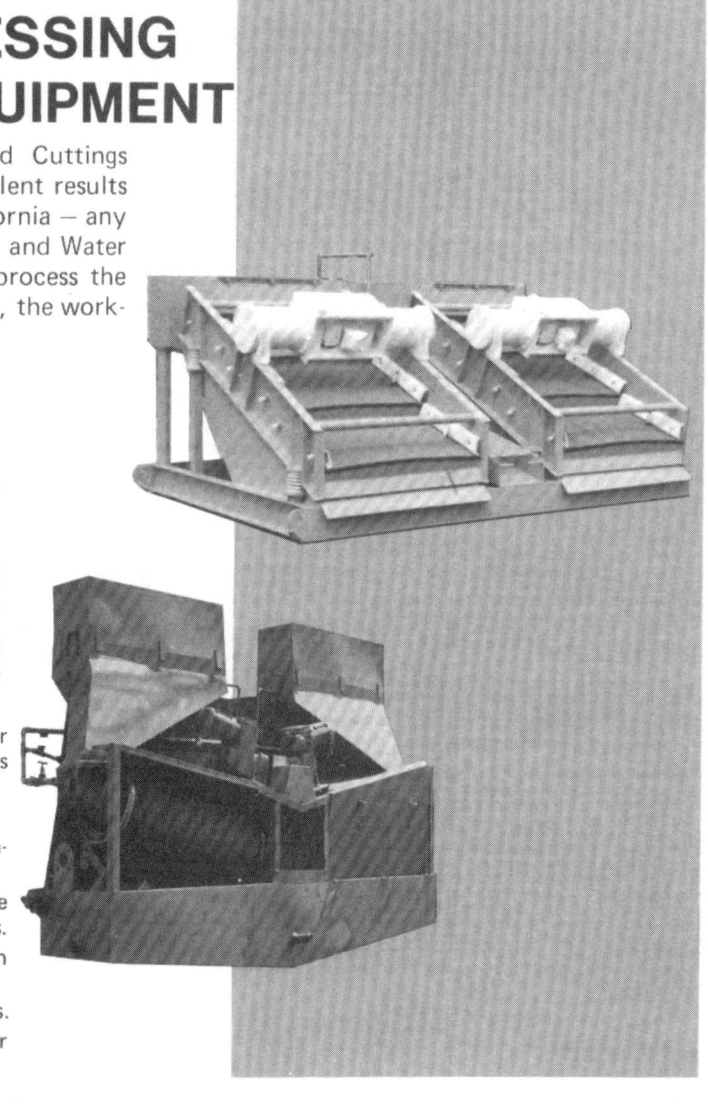

DAHLORY also designs and fabricates Mud Systems incorporating DAHLORY Liquid Mud Agitators • Bug Blowers • Mixing Hoppers • Mud Guns • Tank Leveling Valves • Mud Suctions Valves • Cleanout Gates • Sewage Treatment Plants • Trip Tanks • Mud Boxes • Pipe Wipers • and other integrated equipment. Please write or call for information.

**DAHLORY, INC.**  9928 South Romandel Avenue
Post Office Drawer 2807
Santa Fe Springs, California 90670

# Solids Control

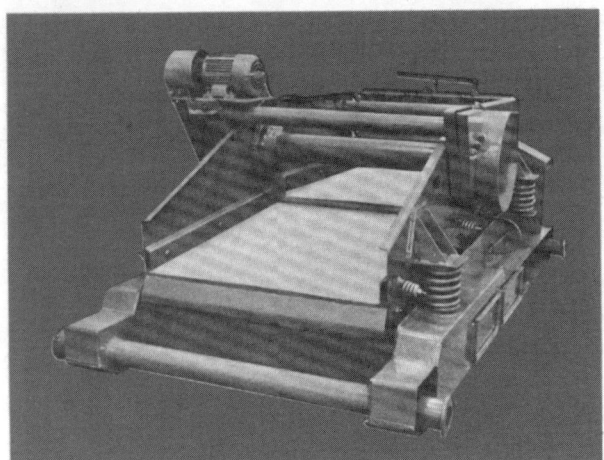

## Super Screen
### Fine Mesh Vibrating Screen

The SWACO Super Screen is the first and only fine mesh vibrating screen designed especially for the drilling industry. It is the result of almost five years of development and field testing. It has been proved superior to any other method of solids screening.

The SWACO Super Screen above has a single deck configuration with two screens at different slopes. The top screen is set flat. Maximum use of direction of rotation is made for conveying solids to the second screen which is set at a 5° down-slope against the axis of rotation. This assures complete dewatering of solids for minimum fluid loss off the end of the screen.

Flow diverters are placed in the entry flume to insure uniform flow across the entire screen surface. Flume height has been adjusted to spend as much of the forward entry velocity of the drilling fluid as possible before it reaches the screen surface.

The drive motor is mounted above the eccentric shaft to eliminate box alignment and belt tension problems. Variations in rotational speed and amplitude (throw) are possible with simple adjustment to fit the Super Screen to the requirements of various muds and conditions. A pneumatic lubricating system has been installed to insure proper lubrication and free the rig crew for other duties.

The fast separation efficiency is proof positive that the Super Screen is returning more cuttings-free fluid to the system than any other model on the market.

Consider these field-proven advantages:
- Improves solids control in oil continuous, inverts, and water-base muds.
- Improves fluid properties.
- Reduces daily fluid maintenance expense.
- Greatly improves pump part life.
- Enhances centrifuge efficiency by removing solids the centrifuge returns with barite.
- Increases D-Silter performance by removing large sand particles.
- Unique design increases screen life, an especially important feature to overseas operators, permitting them to carry smaller screen inventories.

For more information, send for Bulletin # E-S3043-SC.

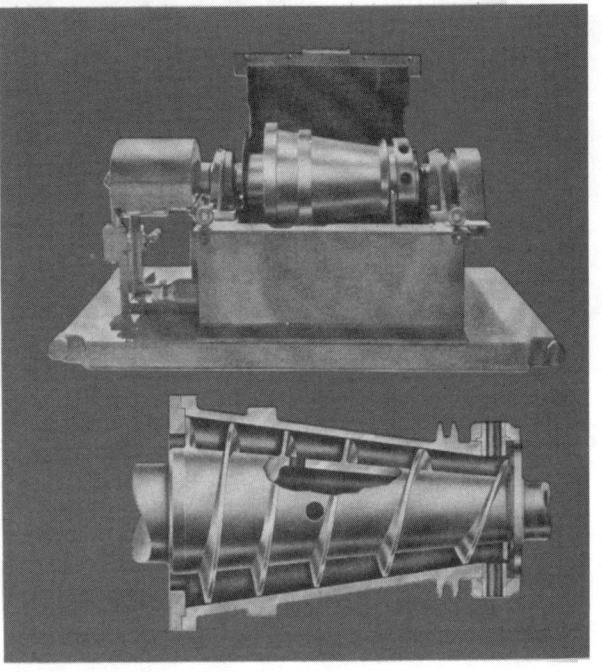

## Solids Control
### Turbo-Flite® Centrifuge

The SWACO Turbo-Flite Centrifuge is the most unique innovation in centrifuging. All previous centrifuge models have maintained a static scroll design over the entire face of the cone. This creates a varied centrifugal torque as the cone becomes smaller toward the apex. In the SWACO Turbo-Flite Centrifuge this variable torque is eliminated by the balanced scroll design. With this design, solids mass can maintain an equal volume between each blade of the conveyor. A constant height of as much as one and one-half inch solids volume can be maintained between flights and held throughout the cycle. This affords almost twice the capacity normally expected at the inlet ports. It is also the only centrifuge to incorporate eight discharge ports instead of the usual four.

No excessive torque caused by compaction will retard separation of the weight material. The balanced conveyor permits free movement of solids mass creating faster material separation. Less force is required to scrape the bowl. A single lead conveyor relieves pressure buildup common in standard conveyor design.

SWACO Turbo-Flite Centrifuges are available on a rental or long term lease basis from any SWACO service center throughout the world. All SWACO Centrifuges are skid-mounted for ease of installation and are available in any desired prime mover power capacity.

For more information, send for Bulletin # E-S3035-SC.

®Registered TM Gilreath Hydraulics.

# Solids Control

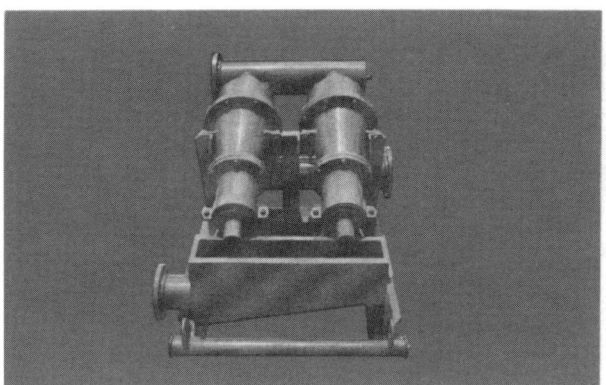

## D-Sander

SWACO was the first in the industry to introduce the 12" cyclone desander. It is designed for continuous removal of sands and abrasive cuttings from rotary drilling fluids. The D-Sander can help limit repair and replacement or parts damaged by abrasive-laden drilling fluids and minimizes sand-caused problems while it increases effective drilling rate and reduces rig downtime.

The SWACO D-Sander can be efficiently designed to handle any combination of sand and rig pump conditions. Any combination of 12" cyclones can be furnished, including single Dorrclone® cyclones, manifolded assemblies or complete units with pumps and prime movers of the driller's choice.

The Swaco 12" D-Sander reduces sand content to a trace in normal drilling fluids. In a water slurry carrying an assortment of sand and silts, the 12" D-Sander will remove about 95% of the 39 micron particles and more than 50% of the 20 micron particles.

The economical multi-sectioned liners of the Swaco D-Sander permit replacement of worn parts without replacing the entire liner. The Swaco apex valve segment includes the area of fastest wear and is provided with a quick-change apex valve holder to minimize downtime. As drilling conditions and sand content of the flowline change, this quick-change feature permits fast valve substitution without the distortion of cyclone design common to other apex valve adjustment systems.

The SWACO D-Sander is installed where it can take suction from the actual mud system as soon as it passes through the shale shaker or leaves the flowline. The front end of the first mud tank should be partitioned in half with an overflow gate. The D-Sander can then take suction from the first half of the tank and return clean mud to the second half of the tank. With this arrangement, virtually no sand settles in the tank, and the D-Sander pressure is in the 30-35 psi range. Back pressure or vacuum should be avoided in the overflow line for maximum efficiency.

The position of the clones in relation to the active mud system and the size and length of all connecting piping should be carefully considered. A water line should be placed at the underflow collecting point to wash sand onto the waste area. This is particularly important if D-Sander underflow is relatively dry.

For more information, send for Bulletin #E-S3034-SC.

®Registered TM of Dorr-Oliver, Inc.

## D-Silter

The SWACO D-Silter is a highly effective and economical unit for removing silt-size particles from oilwell drilling fluids. Combined with the benefits of an effective desanding system, the D-Silter can reduce downtime due to pump wear. It can also increase effective drilling rate by reducing mud weight due to silt particles. Reducing the solids content of the drilling mud due to silt reduces the thickness of the filter cake and leads to less wall-stuck drill pipe. When properly used, the SWACO D-Silter removes practically all silt particles larger than 25 microns and more than half of the particles as small as 15 microns.

The exclusive twin cone design of the SWACO D-Silter consists of a pair of four-inch cyclones mounted in a single case. The clones have a unique 20° taper angle compared to the 15° clone taper on most other units. As a result, the SWACO D-Silter has a GPM capacity of 40-50 per cent greater than other four-inch cyclone desilters and the percentage of mud lost to the underflow is much lower. Therefore, more mud is treated while less fluid is lost to the underflow. Also, the underflow valve, or apex valve, can be larger and is much less likely to plug.

The twin clone assembly is also unique in that it contains four easy-to-use assembled steel sections, and each of the two clones consists of three rubber liner segments and a common feed port. The twin clone is simple and quick to assemble and disassemble, thus limiting downtime. When wear occurs in any of the rubber sections, particularly the apex valve, the worn section alone can be quickly replaced instead of the entire cyclone liner. This provides significant savings in downtime and costs.

For more information, send for Bulletin #E-S3051-SC.

# SHALE SHAKERS AND DRILLING FLUID SYSTEMS

# Blow-Out Detection & Control

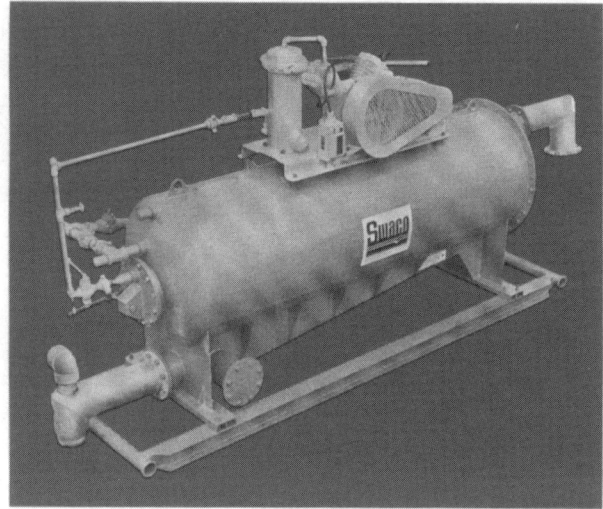

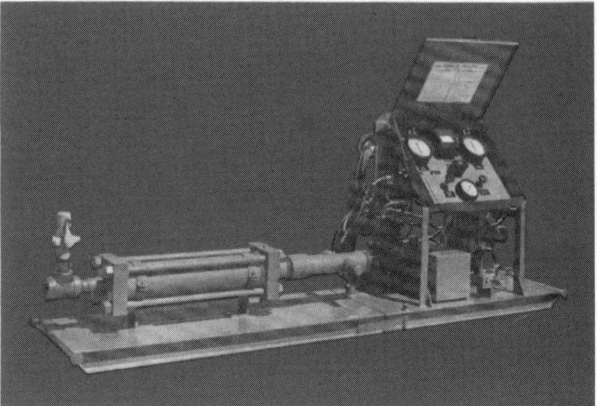

## D-Gasser

Since its introduction in 1951, the efficiency and reliability of the SWACO D-Gasser has been proven on more than 10,000 locations worldwide. It has become the standard against which all degassing equipment performance is measured.

The SWACO D-Gasser removes virtually all gases including air and volatile distillates from the drilling fluid system and vents them at a safe distance from the D-Gasser vessel. It also provides maximum control of gas-cut drilling fluids and effectively reduces the threat of dangerous and costly well blowouts due to recycling of the gas-cut mud or gas locking of the rig pumps.

The SWACO D-Gasser has only two moving elements—the float inside the vacuum vessel and the vacuum pump. The float insures that the system will not overfill or pump dry. The pump is used for positive removal of volatile gases from the vessel and discharges them at a safe distance away from the D-Gasser. With proper return jet sizing, peak flow rates as high as 1000 gpm can be efficiently degassed. The SWACO D-Gasser is ruggedly built and comes skid-mounted for easy installation. Corrosion resistant epoxy inside and outside coatings provide long life with little or no maintenance.

Gas-cut mud is drawn into the degassing vessel by a vacuum created by the discharge jet and vacuum pump. The drilling fluid is distributed over the D-Gasser baffle in a thin, evenly dispersed layer, and the gases are throughly removed by the vacuum pump. The vacuum pump can then discharge the gas at a safe distance from the drilling rig. The degassed mud is returned to the drilling mud system through the exclusive SWACO jet.

While the system appears to be too simple to be so efficient, the design is, however, the result of continued experience and refinement of the basic vessel since 1951. Invariably the SWACO D-Gasser returns the mud to its original weight and essentially removes all of the gas.

For more information, send for Bulletin #E-S3032-BC.

## Blow-out Detection & Control Adjustable Choke—2500 PSI

The SWACO 2500 psi Adjustable Choke, the first remotely controlled adjustable choke designed for rig crew operation, provides complete adjustability of choke size for positive pressure control on flowing systems. Originally built for well killing procedures, particularly with the drill pipe pressure method, it has also found extensive use as a throttling valve while drilling with gas-cut muds.

The adjustable choke is ideally suited for any situation where high pressure flows are to be controlled by restricting the flow and so creating back pressures, especially where the amount of restriction must be determined by progressive steps or must be changed as other conditions change.

The choke was developed specifically to provide accurate surface control of subsurface pressures while circulating highly abrasive drilling mud and/or well fluids. It allows sensitive adjustment of choke size to maintain controlled back-pressure for many hours, even with the most abrasive fluids.

Choke size is controlled by hydraulic pressure against a rubber choke element. Hand regulation of hydraulic pressure allows extremely accurate setting and maintaining of desired differential pressure. If hand manipulation of the relief valve is required to assist in dislodging the particle, the hydraulic regulator will then return the system to the same hydraulic pressure and relationship of choke size to casing pressure.

The remote control panel that regulates choke sizes is mounted on a separate skid and may be placed at any desired distance from the choke valve assembly. It is most often installed on the rig floor near the drilling controls.

The control panel contains all valves, regulators, gauges and pumps required for complete operation. A pump stroke counter and pump rate meter are included for programming certain operations. Relief valves, pulse dampeners and similar aids provide safe positive control of the choke. A standby hydraulic pump is also provided for backup operation.

For more information, send for Bulletin #E-S3033-BC.

# HUTCHISON-HAYES INTERNATIONAL, INC.

## NEW RUMBA® SOLIDS SEPARATORS
### Rumba Solids Control System

The Rumba Solids Control System is the only efficient unit available where high mud volumes and heavy gumbo concentrations are encountered as in the North Sea. Under these conditions conventional shaker systems do not give the required efficiency. This new SCS system developed by Hutchison-Hayes provides the necessary equipment to properly handle mud conditioning when drilling large diameter surface hole in pure gumbo. During the surface drilling phase the primary solids separator section is fitted with a perforated plate screen and is intended to be used for gumbo removal. The gumbo is discharged from the system via shale slides which are equipped with high pressure water sprays to prevent build up of the gumbo. Prescreened mud from the primary separator section is discharged into the back tank of the secondary separator section. The secondary section dual screen deck would be equipped with a 20 x 20 mesh top and 40 x 40 mesh bottom screen. Thus giving a 40 x 40 mesh cut at approximately 38 bbl. P.M. in surface hole conditions. Additionally, incorporated in the unit is a by-pass system that allows the operator to use any combination of screens required. For example, when the surface drilling phase is completed, it is possible to by-pass the primary separator section and screen only on the secondary separator. The reverse is also possible. Should it be necessary to change screens on the primary or secondary sections, drilling may continue as each section may be operated independently.

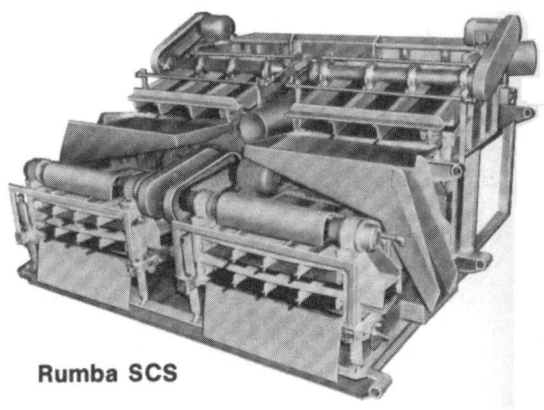

**Rumba SCS**

**Rumba SS101**

### Rumba SS101-102

The Rumba SS101 and 102 Solids Separators provide more solids removal quickly through controlled elliptical motion of the screens. Amplitude and direction of this motion varies down the length of the screen to provide maximum acceleration of solids particles away from the mud stream. In addition, the adjustable basket slope of the 101/102 (from −13° to +3°) provides maximum efficiency from surface to T.D.

The 101/102 is provided with air-mount basket suspension or with hard rubber block mounts where temperature extremes preclude the use of air-mounts.

## RUMBA® SHALE SHAKERS

**SUPER DUAL, MODEL 4860-SU-DU—1600 GPM CAPACITY** An "over-under" four-screen shale shaker for large hole drilling at very high pump volumes. Includes "By-Pass Gate" system to divert flowline discharge of cement, gumbo, etc., direct to waste chute.
SPECIFICATIONS
Max. Capacity, GPM, one side . . . . . . . . . . . . . . . . . . . . . . . . . . 800*
Total Max. Capacity, GPM, both sides . . . . . . . . . . . . . . . . . . . . 1600*
Number & HP of Motors . . . . . . . . . . . . . . . . . . . . . . . . . . . . . (2) 3 HP
Overall Dimensions: . . . . . . . . . . . . . . . . . 13'4" L x 7'3½"W x 58½"H
Coatings:   Galvanized—Dimetcote®—Painted
Net Weight:      6700#       6590#       6400#
*Top Screen 10 x 10, Bottom Screen 20 x 20—10-lb. Mud

4860-SU-DU

**DUAL, MODEL 4860-DU—1000 GPM CAPACITY**
A dual screen shale shaker known the world over for low operating cost. "By-Pass" system as described above is optional accessory.
SPECIFICATIONS
Max. Capacity, GPM, one side . . . . . . . . . . . . . . . . . . . . . . . . . . 500*
Total Max. Capacity, GPM, both sides: . . . . . . . . . . . . . . . . . . . 1000*
Number & HP of Motors: . . . . . . . . . . . . . . . . . . . . . . . . . . . . (2) 3 HP
Overall Dimensions: . . . . . . . . . . . . . . . . . . 13'0"L x 7'0"W x 60½"H
Coatings:   Galvanized—Dimetcote®—Painted
Net Weight:      5110#       5000#       4800#
*Screens 20 x 20—10-lb. Mud

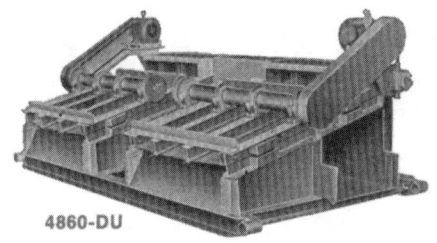

4860-DU

## IMCO Hydrocarbon Logging

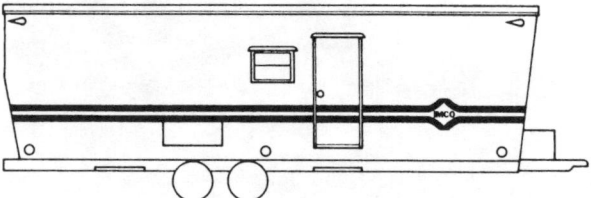

The units employ the most effective techniques and equipment currently being used in the search for hydrocarbons. (Services available only in West Texas and Central U.S.A.).

## IMCO 3-D Shaker

IMCO's 3-D Shaker takes solids control one step further than ordinary equipment. It utilizes three fine mesh screens for more precise solids control and sampling. The 3-D Shaker has hydraulically controlled screen bed tilt which allows a variation from zero to 13 degrees. This affords versatility in handling fluid flow rates with varying viscosities and a faster gravitational pitch of cuttings over screens. Screens may be rapidly changed by one man in 15 minutes or by two men in half this time.

Other features of the 3-D Shaker include a positive bypass gate which can handle any volume of fluid, hinged mud control gates which regulate flow of muds to the screens, Aero-Union type mud box connections which minimize rig-up time for flowline connections, an underslung design feature which extends screen life and heavy duty all-welded construction with a rubber mounted screen bed on an extra heavy base which provides exceptionally quiet and efficient operation.

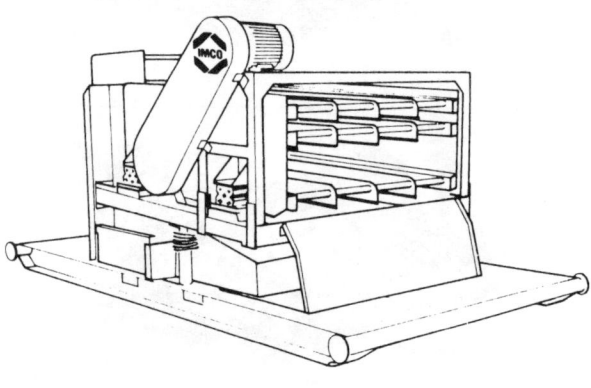

## IMCO Mud Gas Separator

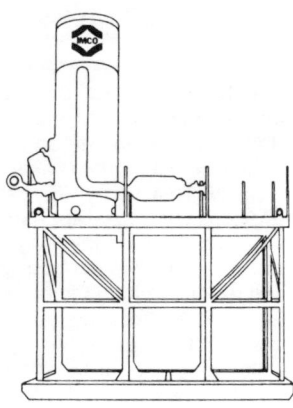

The IMCO Mud Gas Separator is a low pressure vessel similar in design to gas separators used in production. Its function is the removal of large volumes of gas from the drilling fluid during the drilling operations. The gas is then vented away from the drilling rig for burning.

The unit is capable of handling 20 million cubic feet per day and is designed for maximum flow—maximum safety.

The IMCO Mud Gas Separator has a surge pot which allows gas to expand before entering the main vessel and is equipped with adjustable legs for gravity drainage. The unit has a 6 inch flare line.

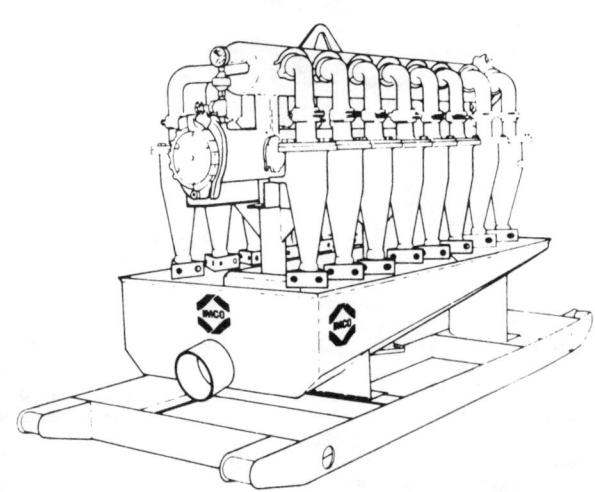

## IMCO Desilter

The IMCO Desilter utilizes 16 four-inch desilter cones for highly efficient removal of silt sized particles from drilling fluids.

The Desilter Unit and the Power Unit are both skid mounted with the Power Unit being available with either diesel or electricity.

The IMCO Desilter, equipped with a centrifugal pump, is designed to operate optimally with rates up to 800 gpm. For additional capacity, a second unit may be added.

# HISTORICAL PERSPECTIVE

*Drilling Fluids*

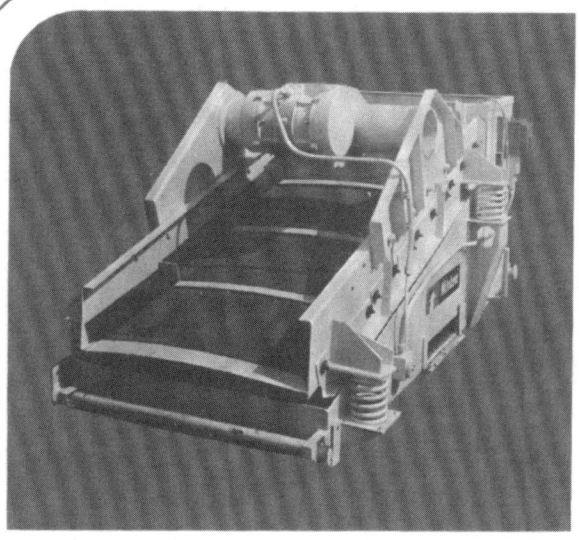

## RVS Shaker

RVS is the new name for the fine screen shale shaker we manufacture and market.

RVS stands for "Rotary Vibrating Shaker." Its unique drive system develops 11,000 lbs. of thrust to handle exceptionally-high mud and cutting volumes. There are no belts, chains, or gears...so there's no down-time to service them. The shaker screens are in a series on a sloped bed where a torn screen can be quickly replaced. The RVS will handle flow rates in excess of 900 gpm with no sacrifice of cleaning efficiency.

Transportation and installation are easy, because the RVS is designed to break down into two sub-assemblies for fly-in rigs. Its rugged welded construction and dimetcoating make the RVS suitable for any environment. From Gulf Coast Offshore to the Arctic Circle—from California to the North Sea and Middle East—from South America to Singapore, customers are using our shaker because from gumbo to hardrock cuttings and sand, the RVS outshakes them all. Available in single, dual, and triple models.

## Rotor Mud Separator

Milchem's Rotor Mud Separator RMS is the only centrifuge on the market designed specifically for use with oilwell drilling muds. The RMS simply, effectively, and economically removes undesirable, low gravity solids from the drilling mud system. It at the same time reclaims high percentages of barite at high flow rates.

The RMS has a greater throughput mud volume capacity than any other centrifuge on the market. Unlike any decanting centrifuge, the RMS does not depend upon gravity to recover the underflow of high density solids. A complete, self-contained pumping system enables the RMS to pick up mud to be centrifuged from a fixed position from as far away as 25 feet. Both the effluent and the underflow are discharged under pressure and can be transported as much as 50 feet from the unit. The RMS is available in a trailer mounted configuration to facilitate ease of transportation.

**The total Milchem concept embodies the RVS (Rotary Vibrating Screen) SHALE SHAKER, the RMS ROTOR MUD SEPARATOR, and the DESILTER...designed to work together to cut costs of rig operation in any environment.**

Drilling Fluids

## Desilter

The Milchem Desilter offers the operator an inexpensive, efficient method of removing undesirable, low gravity solids (as small as 10 microns) from an unweighted drilling fluid system. The average unweighted mud system can be maintained at mud densities of less than 8.8 pounds per gallon if proper application of the desilter is utilized.

A planned program of utilizing the Milchem Desilter to process both the active drilling mud system and the excess mud located in the sump more than justifies the investment in this equipment.

During trips the desilter can be used to process the sump mud. By recycling this mud through the equipment several times, the effluent water can be reduced to a mud weight of as low as 8.6 pounds per gallon. This effluent contains almost an equivalent concentration of chemicals as are contained in the active system. Usually this water, containing drilling mud chemicals, can be added to the active system as needed to build volume. This program can reduce the sump volume by as much as 75%.

The Milchem Desilter is available in sizes ranging from 4 to 16 cone units. Fabricated from Schedule 80 steel, the manifold is designed to last for years. Power skids are available with either diesel or explosion-proof electric motors sized to process maximum volumes.

## Testing Equipment

Milchem Drilling Fluids offers a complete line of testing equipment for use in the investigations of performance characteristics of drilling muds and the evaluation of drilling mud additives. Detailed information on individual equipment, kits and prices is available by contacting Milchem Incorporated, Testing Equipment Dept., P. O. Box 22111, Houston, Texas 77027

## the brandt company

6300 Midvale
Houston, TX 77017

Phone (713) 644-1638

## Brandt Tandem Screen Separators

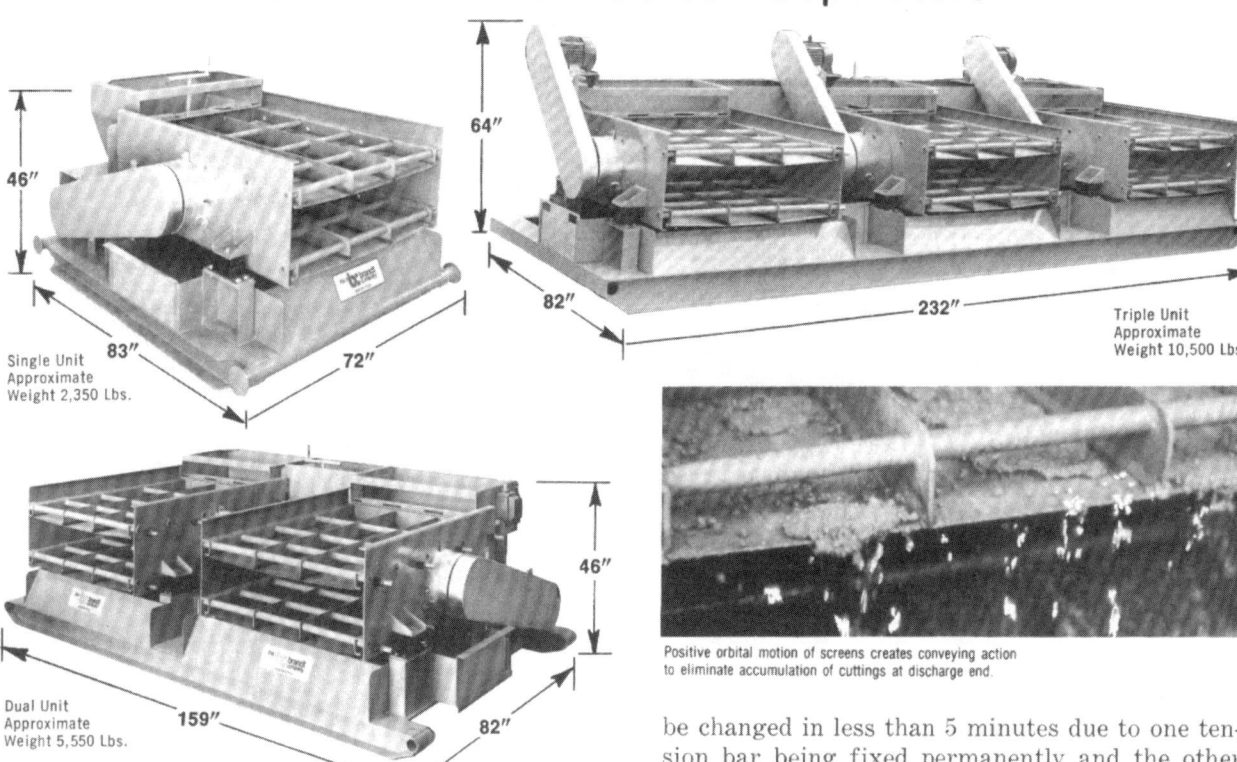

Single Unit Approximate Weight 2,350 Lbs.
Triple Unit Approximate Weight 10,500 Lbs.
Dual Unit Approximate Weight 5,550 Lbs.

Positive orbital motion of screens creates conveying action to eliminate accumulation of cuttings at discharge end.

The high flow rate and positive solids control you get with Brandt Tandem Screen Separators are the result of the preferred horizontal screen attitude and positive orbital motion at all points on the screens. Solids move off at a constant rate. Accumulations are eliminated. A wash system is seldom needed.

Brandt Tandem Screen Separators have demonstrated higher flow rates with all types of mud than other well known separators. The horizontal screens allow longer slurry retention time and provide more volume/sq. ft. of screen than high-angle separators.

The tandem screen feature increases the volume/sq. ft. of the lower screen and maintains a more constant flow rate than other designs over a wide range of drilling rates. The top screen removes large solids which otherwise would reduce the flow rate of the fine mesh lower screen as the drilling rate increases.

The low profile of Brandt Tandem Screen Separators allows units to be used with low substructure rigs. Weir height is only 34".

Standard size screens (4' x 5') can be obtained through oilfield supply stores. Either screen can be changed in less than 5 minutes due to one tension bar being fixed permanently and the other being adjustable.

Standard motors, belts, sheaves are used. No delicate alignments are necessary. Skid-mounted units are sand blasted to white metal, prime-coated with inorganic zinc and finish coated with epoxy.

Junior Tandem units for workover and low volume drilling also are available.

### Flow Rates for Brandt Tandem Screen Separators

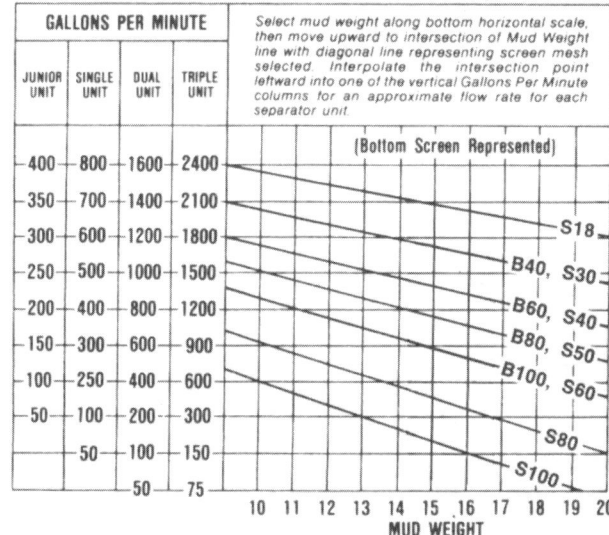

Select mud weight along bottom horizontal scale, then move upward to intersection of Mud Weight line with diagonal line representing screen mesh selected. Interpolate the intersection point leftward into one of the vertical Gallons Per Minute columns for an approximate flow rate for each separator unit.

Phone (713) 644-1638   6300 Midvale
Houston, TX 77017

## Brandt Single Screen Separators

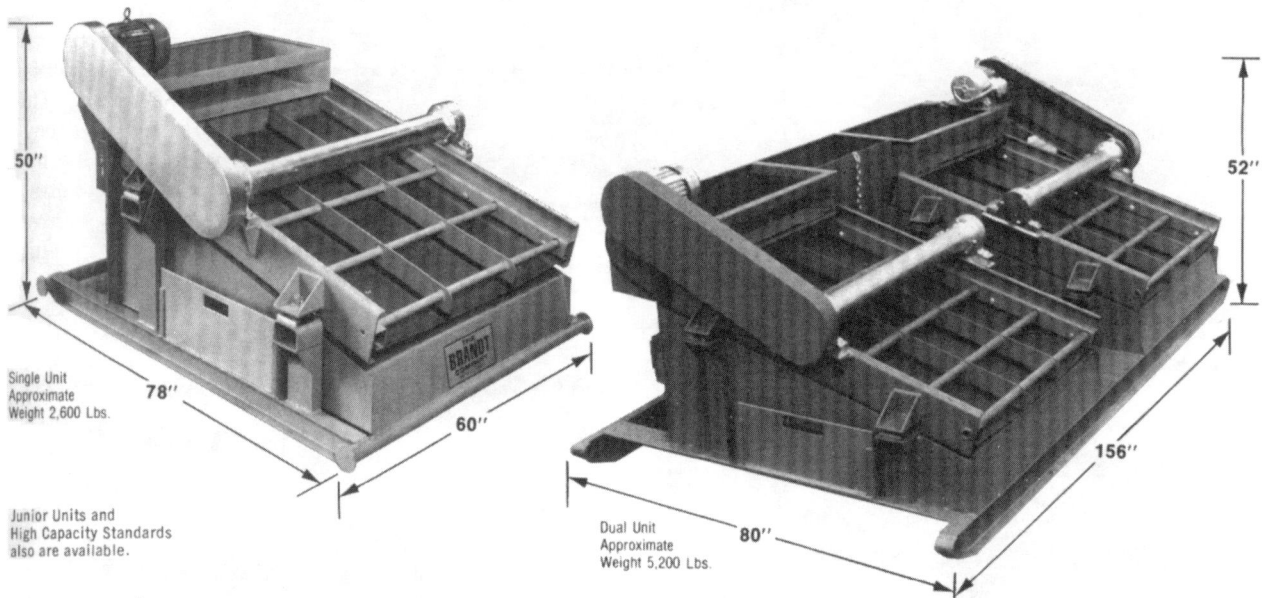

Single Unit Approximate Weight 2,600 Lbs.
Junior Units and High Capacity Standards also are available.
Dual Unit Approximate Weight 5,200 Lbs.

Many operators want high volume performance at minimum cost and prefer separators designed to utilize the traditional high-angle single screen design. They will find Brandt Single Screen Standard Separators to be their most reliable equipment.

The low amplitude screen action of Brandt Standard Separators assures thorough separation of cuttings from the mud without any loss of volume. This speeds drilling operations.

Readily available and easily interchangeable 4' x 5' screens are tension-adjusted from one side only for convenience and simplicity. Standard bearing and shaft components are easy to adjust and maintain, and the belt-driven vibrator mechanism is shielded completely for operator safety. Curved surface structural parts reduce adherence and accumulation of mud on machine surfaces.

### Flow Rates for Brandt Single Screen Separators

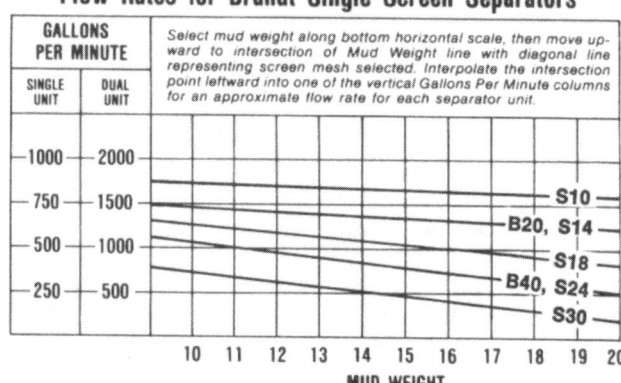

Select mud weight along bottom horizontal scale, then move upward to intersection of Mud Weight line with diagonal line representing screen mesh selected. Interpolate the intersection point leftward into one of the vertical Gallons Per Minute columns for an approximate flow rate for each separator unit.

## Brandt Blowers for Drilling Rigs

Brandt developed these specifically for drilling rigs. They can be positioned to disperse dangerous gases or fumes or bothersome insects. They increase personnel comfort in land and offshore areas. Heavy duty guards meet OSHA specifications for safety. Non-sparking aluminum blades move large volumes of air at minimum noise level. Explosion-proof electric motors, starters. Power input 230v or 460v 3-phase AC.

15,000 CFM BLOWER: Impeller 36", Height 58", Weight 345 Lbs.

25,000 CFM BLOWER: Impeller 48", Height 70", Weight 485 Lbs.

6300 Midvale
Houston, TX 77017

Phone (713) 644-1638

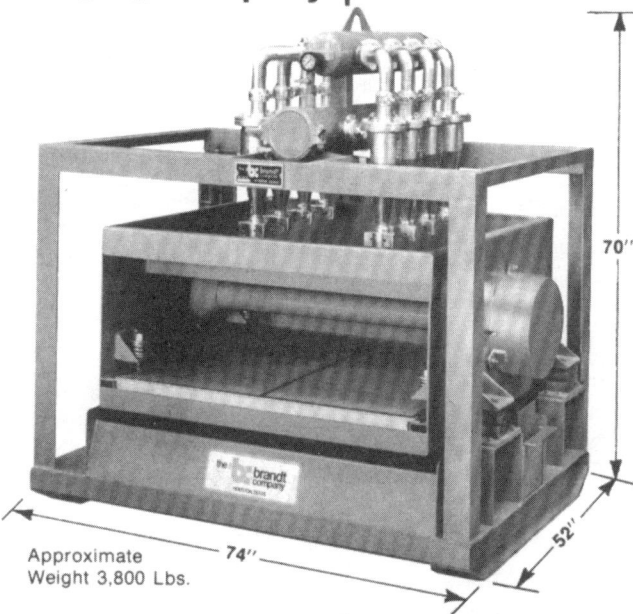

Approximate Weight 3,800 Lbs. — 74" — 52" — 70"

## Brandt Silt Separator For Solids Control

Solids control resulting in a reduction in drilling costs is accomplished by the Brandt Silt Separator. The equipment combines a bank of four-inch hydrocyclones with an ultra-fine-screen separator. The underflow of the desilter cones passes onto a fine mesh screen, and the liquid and majority of the desired barite pass through the screen to return to the mud system. The drilled solids are discarded.

Operation of the Brandt Silt Separator is visible and gives you better knowledge of down-hole conditions and the solids separation efficiency. Screen changes, when necessary, are a fast, one-man operation by the easy removal of four bolts. Maintenance is simplified by Brandt's use of standardized replacement parts which are available at oilfield supply stores and other nearby sources.

## Brandt DS/DT Separator for Gumbo Control

Approximate Weight 10,400 Lbs. — 156" — 88" — 82"

Removal of gumbo in the first separation stage of the Brandt DS/DT Separator permits the use of fine mesh screens in the second and third stages of the separator action. As a result, a high percentage of the drilled solids is removed, and a maximum amount of valuable drilling fluid is retained.

The term DS/DT stands for "Dual Standard over Dual Tandem"— a Brandt standard separator mounted above a Brandt Tandem Screen separator and manufactured into one piece of equipment having high mud-handling capacity but compact dimensions.

An integral water spray system is designed into the DS sections to facilitate the washing down of gumbo as necessary. Controls permit the bypass of any combination of the separate units. With all units in operation, flow rate can be as high as 1,800 gpm, depending on screen mesh and mud weight.

### New Brandt Equipment Cleans Oil From Cuttings in Offshore Drilling

The Brandt Company offers a new packaged *Cuttings Cleaner* to remove oil from drilled cuttings for environmentally safe disposal offshore. The *Cuttings Cleaner* includes a Tandem fine screen separator, solvent and detergent spray equipment, retention washing tank, oil-water separator, and a cuttings disposal system. Skid units are designed for individual installation.

 brute shakers

## BRUTE VIBRATOR

The BRUTE (balanced uniform rotary thrust effort) vibrator is adjustable in thrust from 0 to 11,000 pounds. A standard 5 H.P., 1800 RPM explosion proof motor, frame 184T is used to drive the vibrator shaft. The vee belt drive is constant center, constant tension, eliminating the flopping and severe belt wear that can occur on shakers that mount the vibrator on the bed and the drive motor on the frame.

A single adjustable counterweight accessible through the front of the housing insures that no misadjustment of thrust weights causing the shaker bed to gallop is possible. The vibrator shaft runs on large spherical bearings independent of the motor bearings. Over greasing of the vibrator bearings is permissible, as there is no possibility of grease entering the drive motor cage. In continuous service the vibrator bearings should be greased at least daily.

## THE SHAKER

Brute shakers are available in three and four screen models, in single, double and triple assemblies. These high speed, series screen, fine screen shakers utilize fine screens (2' x 4') with a heavy mesh back up for additional life. The single three screen model gives good fine screen separation characteristics at flow rates of up to 400 GPM, and the single four screen unit at flow rates up to 550 GPM. The low profile three screen unit has a spillway height of only 30 inches making it ideal for low substructure rigs.

All shakers screens are visible from above permitting a torn screen to be detected immediately. A minimum of clearance is required in front of the shaker to facilitate removal of screens. The narrow screens (2 ft.) are relatively easy to tension evenly, an important consideration for maximum screen life. Most parts, with the exception of the vibrator are interchangeable with other popular makes.

## FEATURES

1) Fine screening characteristics at high flow rates.
2) Simple, rugged, adjustable vibrator, standard motor.
3) All screens visible from top.

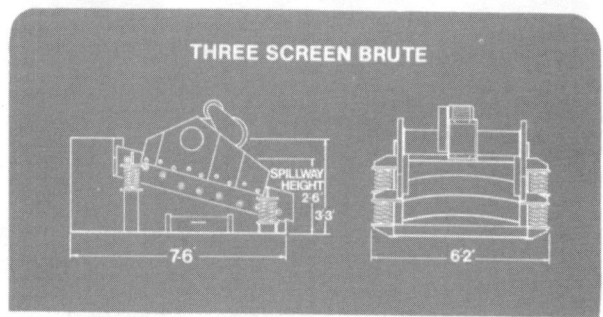

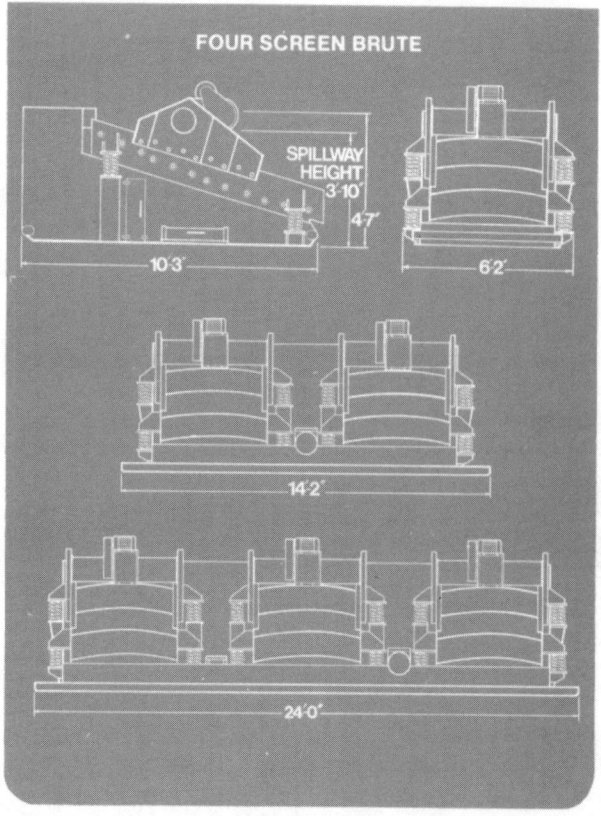

| 3 SCREEN SHAKERS | Painted P | Zinc Rich Coating Z | Epoxy Coating E |
|---|---|---|---|
| Single, Model B3SS—X | $ 7,400 | $ 7,600 | $ 7,650 |
| Double, Model B3SD—X | 13,900 | 14,275 | 14,380 |
| Triple, Model B3ST—X | 20,300 | 20,850 | 20,975 |

| 4 SCREEN SHAKERS | Painted P | Zinc Rich Coating Z | Epoxy Coating E |
|---|---|---|---|
| Single, Model B4SS—X | $ 9,350 | $ 9,570 | $9,620 |
| Double, Model B4SD—X | 17,400 | 17,825 | 17,900 |
| Triple, Model B4ST—X | 25,700 | 26,500 | 26,650 |

Please specify coating denoted X.

### WEIGHTS

| | | |
|---|---|---|
| B3SS — 3,950 # | | B4SS — 4,800 # |
| B3SD — 7,700 # | | B4SD — 9,300 # |
| B3ST — 11,500 # | | B4ST — 13,700 # |

A lightweight version of the Model B4SS (3950 #) is available for helicopter transportability.

# Solids Control

# Superclone Centrifuge Super Solids Separator

## Superclone Centrifuge

The SWACO Superclone Centrifuge is a new generation clay removal-barite recovery tool based on the successful Clayjector that SWACO introduced in 1957. Mechanical fluid treatment through the Superclone Centrifuge's four ceramic-lined, 2-inch cyclones can reduce chemical costs about 20% by permitting reuse of muds that would ordinarily be abandoned.

The new, simplified Superclone Centrifuge requires only about 1/10th as much space on top of mud tanks as decanting centrifuge units, but it is capable of rejecting 60% to 90% of light clays and drilled solids while reclaiming 85% to 90% of the usable barite in a normal mud system. The Superclone Centrifuge rejects particles smaller than 8 microns, including excess barite fines that cause high viscosity and gel in the mud system.

With Superclone Centrifuge's high clay rejection efficiency, the solids in the underflow are 90% to 98% pure, API-standard barite, and are returned in a smooth slurry that mixes readily with the active mud system.

New air-operated controls are manipulated from a conventional control panel. The system includes automatic shutdown of the unit if conditions change enough to adversely affect the mud-treating efficiency. It also includes an automatic washdown cycle to prevent service problems.

The SWACO Superclone Centrifuge consists of two units. The Dilution Unit is placed conveniently near the mud tanks, and includes the 25-hp Diesel or electric prime mover, centrifugal pumps, water tank and piping to mix and feed water-diluted mud to the cyclones. The Separator Unit (approximately 24 inches square x 66 inches high) sits on top of the suction mud tank. Cyclones have quick-connect couplings so one, two, three or four can be easily put into service to suit volume requirements (up to 60 bbls/hr., depending on mud weight).

For more information, send for Bulletin #E-S3062-SC.

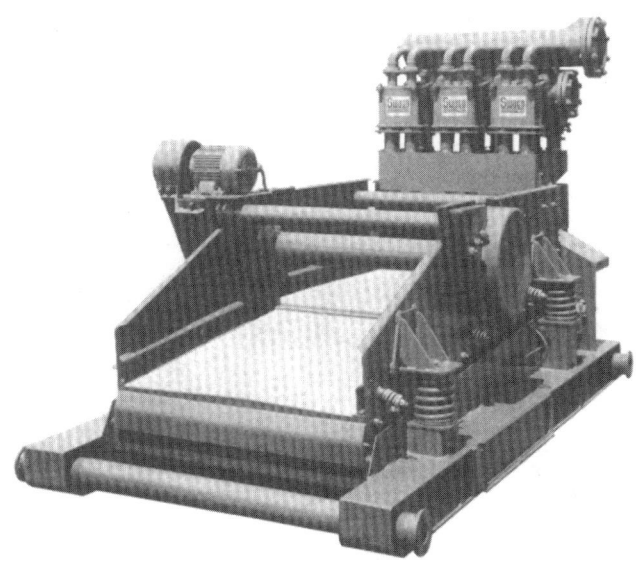

## Super Solids Separator

The SWACO Super Solids Separator combines two proven pieces of solids control equipment into one mud maintenance system. The result is reduced mud maintenance and reduced waste disposal costs.

All of the circulating mud volume is first processed through a bank of 4-inch D-Silter cyclones that are mounted above a modified Super Screen fine mesh vibrating screen. The D-Silter separates the majority of the liquid mud phase and returns it to the active mud system.

The concentration of drilled solids is then fed onto the screen deck, where special fine screens in the range of 120-200 mesh are used to separate out all particles larger than 117 to 74 microns. Since the screens under the D-Silter are handling reduced, dewatered volumes, special high-speed, high-strength screens are used for superior service and reduced costs.

The maximum dewatering accomplished by this combination equipment reduces disposal chores to only the cuttings, not the liquid mud and water.

**G&C ENTERPRISES INC.**

P. O. Box 1867
Monahans, Texas 79756
U. S. A.
Phones as follows:
(915) 943-4140
(915) 943-4292
(915) 943-2746

For total solids control

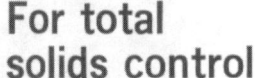

## one SHIMMY-SHAKER replaces shale shaker, de-sander, and de-silter

**IMAGINE** a single, explosion-proof, pneumatic machine capable of total solids control on a flow of 350 gallons per minute of 16-pound mud . . . yet requiring only 28 cubic feet per minute of rig air. No de-sander, centrifuge, or de-silter needed.

**AMAZINGLY QUIET,** the Shimmy-Shaker provides its operator with instant and infinitely-variable control over all solids in the mud-flow . . . from the roughest cuttings down to particle sizes scarcely larger than barium additives.

**UNIQUE DESIGN** produces a super-fast vibrating action (variable from 0 to 5,000 cycles per minute), enabling the use of screens up to 300 mesh to handle large gallonage without danger of overflow. All three screens tilt individually for optimum angle.

**MAINTENANCE** is surprisingly simple. The high-frequency, low-impact vibrating action multiplies screen life many times over that of ordinary shakers. Screens are pre-framed and pre-stretched to the proper tension. Any one screen can be replaced in five minutes without shutting down the other two. Likewise with vibrators, which seldom require changing because they are protected from wear by lubricators and intake air-filters. Otherwise, there are virtually no moving parts, no bearings or mechanical assemblies to wear out or fail.

**ECONOMY,** along with safe, efficient operation, is a primary factor with the Shimmy-Shaker. It overrides any need for secondary or supplemental equipment. It virtually eliminates downtime for replacing components in the solids-control system. And it reduces to an absolute minimum the cost of replenishing additives or fluid bases. In the event of a gas kick, an automatic by-pass prevents mud-loss by shunting unfiltered fluid back to the pits.

**PNEUMATIC VIBRATION** is super-fast, gentle and quiet, multiplying the life of screens up to 300 mesh for maximum solids control. Flow capacity exceeds that of the largest standard shale shaker.

**SIMPLE CONTROLS** govern vibration cycles from 0 to 5,000 per minute, plus the amplitude or impact-strength. Automatic by-pass stops mud-loss during gas-kicks and can, if desired, be controlled from rig floor.

**ADAPTABLE TO ANY RIG,** Shimmy-Shaker employs diversion gates to direct the flow through either side or end. Screens can be changed in minutes without shutting down. Custom-fabrication is available to meet unusual space requirements.

**FULL DETAILS AND PRICES AVAILABLE ON REQUEST**

## MEDEARIS SINGLE DECK AND DOUBLE DECK MODEL 45-E
### SINGLE AND DUAL SHAKER ARRANGEMENTS:

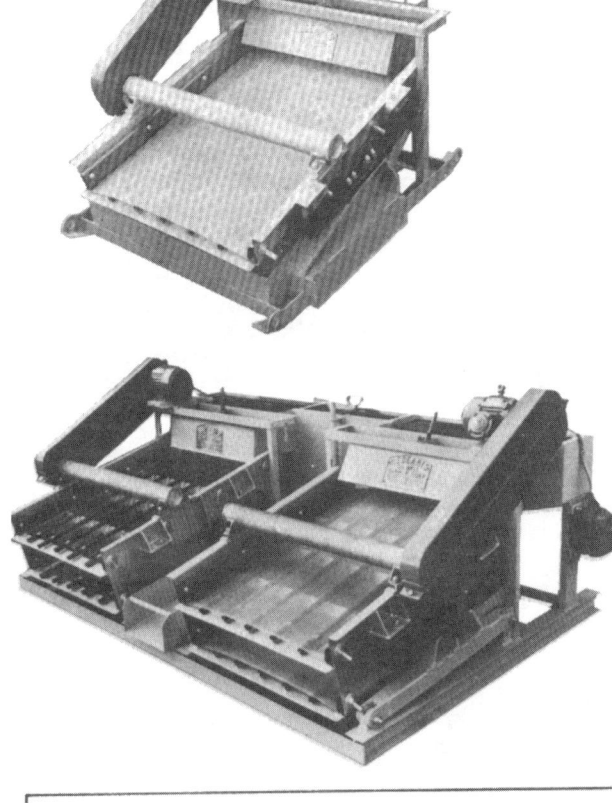

The newly engineered and redesigned Medearis mud shakers have fewer moving parts than any other shale shaker in the industry, and the simplicity of construction and operation reduces maintenance costs far below what is considered normal. New type rubber cushions and screen assembly increases screen life at least 25% over prior models.

All of the shaker units are equipped with a surge box in which the flow of mud coming directly from the well is slowed down, and using internal mud wings and new flow gate control, the mud flow is evenly distributed over the entire screen deck.

All Medearis shaker units are equipped with bypass valves or gates for bypassing the shaker during mechanical maintenance, or when using lost circulation material that should not be screened out of the mud.

All parts are corrosion protected, and the vibrator housing and screen box are hot-dip galvanized. The vibrator is roller-bearing mounted and housed for complete protection. Tension and vibration are uniform over all parts of the screen cloth. The compact design also contributes to easier handling and operation.

Depending upon screen mesh opening and pump capacity, the Medearis single deck shaker will handle up to 800 GPM of average drilling mud; the single deck dual shaker will handle up to 1600 GPM, the double deck single shaker will handle up to 1000 GPM, and the double deck dual shaker will handle up to 2000 GPM of average drilling mud.

Units can be supplied with open bottoms for mounting on an existing tank, or with a bottom collection and distribution tank, as well as with or without surge boxes.

### SHAKER SCREENS AND PARTS
Contact your nearest Medearis office, Geolograph office, or your Medearis-Geolograph agent for shaker screens or repair parts for all brands and models of shale shakers.

## NEW MEDEARIS HI-VOLUME SHAKERS

These newest members of the old family of Medearis Mud Shakers incorporate all of the superior mud distribution characteristics of the Medearis Type "E" and Type "X" Shakers *plus* a new vibrating principle that increases separating ability and efficiency, even with finest of screens. The result is the most high-volume, fine-separation shale shakers in the industry.

With the high volume shaker, the vibrator is coupled directly to shaker bed which eliminates belts, chains and gears. Eight intermediately-cushioned rails under each screen provide optimum support for fine mesh screens, adding significantly to the screen life. The high-volume shaker is available in both dual and single models.

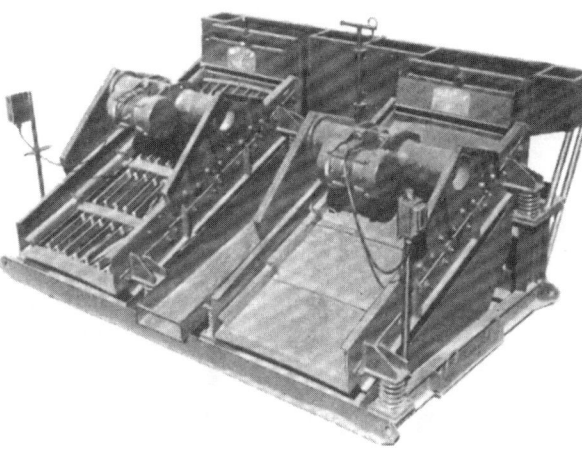

**Medearis Oilwell Supply Corp.**

One of the world's leading designers and manufacturers of solids control equipment

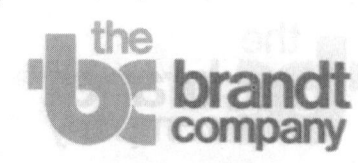

the brandt company

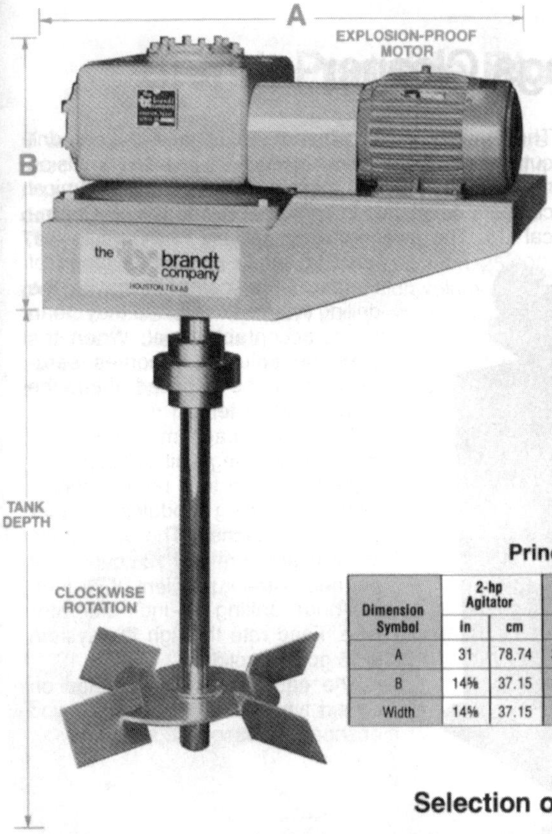


# Brandt Mud Agitators

Brandt mud agitators, available in seven models from 2 through 25 horsepower, are quiet, smooth and vibration-free. The simplicity and strength of their single-reduction worm gear drive provide long, reliable service.

Worm gear drives are 50% more resistant to shock and vibration than gear reductions utilizing twice as many helical and bevel gears. Also, worm gear drives wear *in* rather than wear out. Steady operation continues to form the tooth rather than destroy the tooth form.

Brandt impeller blades are canted at 60° to promote axial as well as radial flow. This induces a bottom-to-top movement and results in a more homogeneous mixture.

### Principal Dimensions of Brandt Mud Agitators

| Dimension Symbol | 2-hp Agitator | | 5-hp Agitator | | 7.5-hp Agitator | | 10-hp Agitator | | 15-hp Agitator | | 20-hp Agitator | | 25-hp Agitator | |
|---|---|---|---|---|---|---|---|---|---|---|---|---|---|---|
| | in | cm | in | cm | in | cm | in | cm | in | cm | in | cm | in | cm |
| A | 31 | 78.74 | 39½ | 100.33 | 51½ | 130.81 | 51½ | 130.81 | 60⅝ | 153.99 | 60⅝ | 153.99 | 68 | 172.72 |
| B | 14⅝ | 37.15 | 16¾ | 42.55 | 24½ | 61.60 | 24¼ | 61.60 | 27⅛ | 68.90 | 27⅛ | 68.90 | 30⅞ | 78.42 |
| Width | 14⅝ | 37.15 | 18¾ | 47.63 | 24½ | 62.23 | 24½ | 62.23 | 30½ | 77.47 | 30½ | 77.47 | 35⅜ | 89.85 |

## Selection of Agitator Size

Select the right size agitator by first locating the tank width on the right side of the graph. A recommended impeller diameter is shown across on the left side. This impeller size is correlated to the mud weight and the required horsepower. Simply follow a horizontal line from the impeller diameter to the curve showing the heaviest anticipated mud weight. Now locate the nearest vertical line to the *right* of this point and note the required horsepower at the top of the graph.

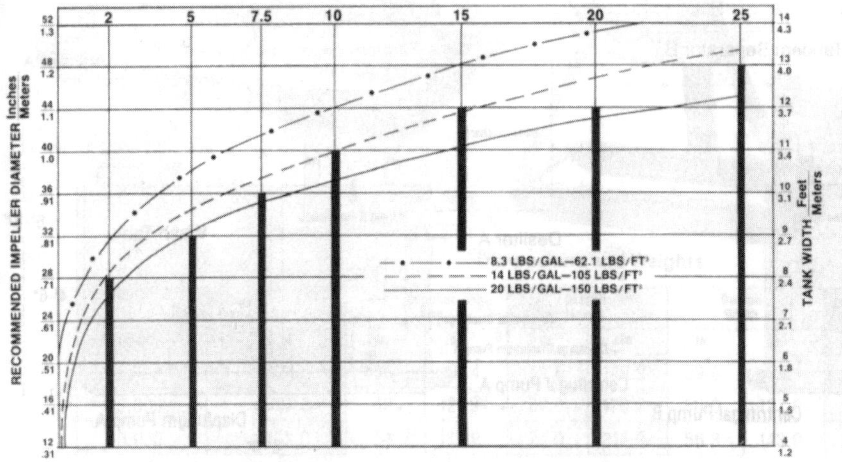


NOTE: Horsepower is based on 57.5 rpm and a four-bladed impeller. Maximum impeller size for given horsepower shown by heavy line.

**EXAMPLE:** Agitators are required for a 10-foot-wide tank, 30 feet long, to maintain weighting materials in suspension for a 12 lbs/gal mud:

Find the tank width (10 ft) and the recommended corresponding impeller diameter (36 in) on the graph. Follow a horizontal line from the impeller diameter to the curve of the given mud weight (12 lbs/gal mud—use the curve on the *next higher* mud weight). From the intersection of the mud weight curve and the impeller diameter, locate the nearest vertical line to the *right* and note the horsepower at the top of the graph.

This particular application will require a 7.5-hp size Brandt Agitator for each 10 feet of tank length—a total of three 7.5-hp agitators.

# HISTORICAL PERSPECTIVE

**One of the world's leading designers and manufacturers of solids control equipment**

6300 Midvale
Houston, Texas 77087
Phone (713) 644-1638
TWX 9108815451

## Brandt Cuttings Cleaner

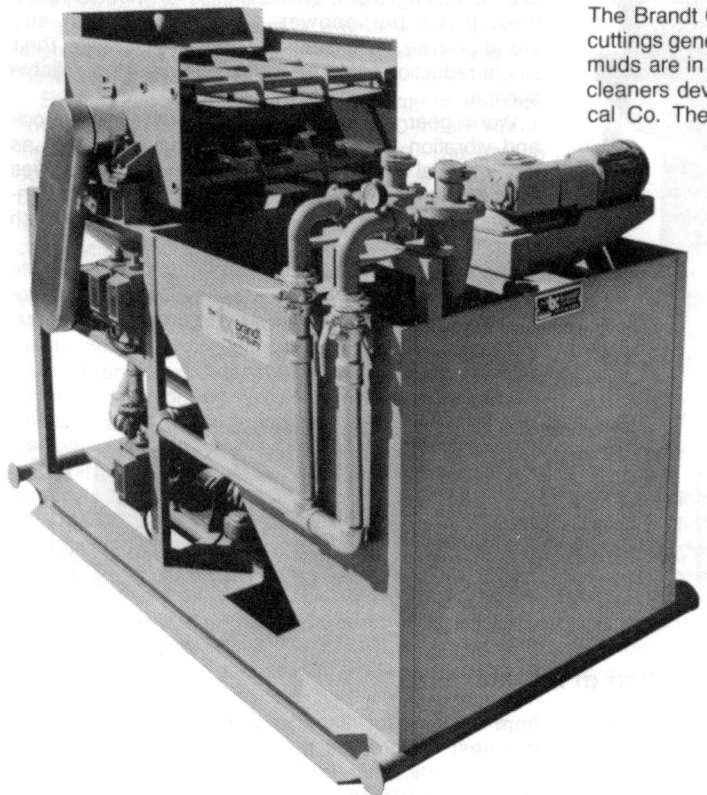

The Brandt Cuttings Cleaner is designed to clean drill cuttings generated from holes where inverted emulsion muds are in use. The cleaning process uses chemical cleaners developed in conjunction with Exxon Chemical Co. The cleaners are relatively non-toxic to life, easy to handle in a wide range of temperatures, compatible with the drilling systems used, and they clean to an acceptable level. When the cleaning solution becomes saturated, it can be disposed of into the active mud system.

The flow diagram shows the modular cleaning unit concept, designed to clean to a predetermined level. By adding modules, cleaner levels are reached. The unit cleans cuttings at the rate of 125 cubic feet per hour—the equivalent of 75 feet-per-hour drilling 15-inch-diameter hole. Feed rate through the system is 15 gpm of solids.

The equipment is assembled on oilfield type skids and can be used onshore or offshore.

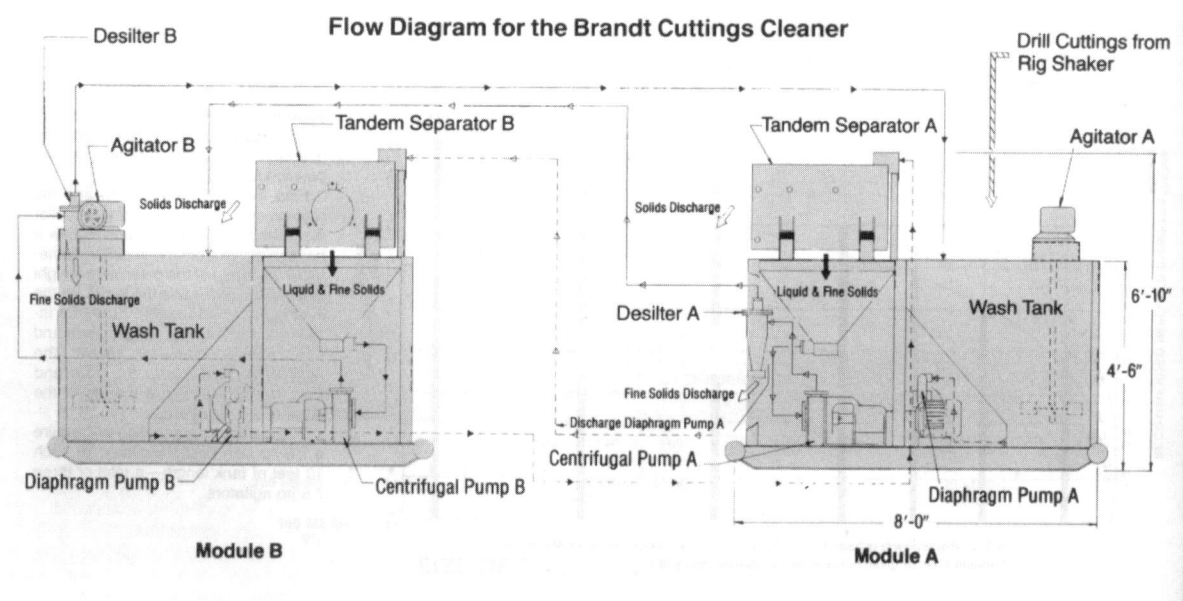

Flow Diagram for the Brandt Cuttings Cleaner

# HYDROCYCLONES

## For Drilling and Production

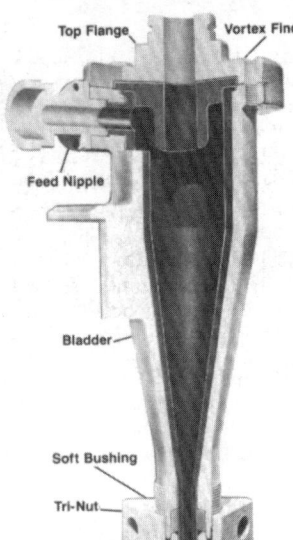

The high centrifugal separation forces in a hydrocyclone are created by the feed stream entering tangentially at a constant head pressure as from a centrifugal feed pump (See Figure 2). A Picenco balanced-design hydrocyclone can be adjusted so clean liquid will spiral toward the apex, reverse axial directions, and spiral to the overflow with no loss of liquid at the underflow.

If solids are present, they will be settled to the outside wall in the down-spiraling stream, and will **discharge at the underflow by inertia** when the liquid stream changes direction. The removed solids take with them free liquid on their surfaces.

If the underflow opening is large enough to handle all the feed solids separating to the underflow, a balanced cyclone will operate in "spray discharge". See Figure 4. If the feed solids separating to the underflow cannot pass through the underflow opening, the solids discharge will "rope", or "sausage discharge", indicating solids overload. See Figure 5. Although a gallon of rope underflow will weigh more, a spray discharge will remove more solids.

Picenco standard hydrocyclones are designed for easy balancing with the largest possible underflow openings per unit of feed rate to obtain the most removal capacity per feed pump horsepower.

### FOR DRILLING AND OTHER APPLICATIONS WITH ATMOSPHERIC DISCHARGE

Picenco mud hydrocyclones are designed to operate with a feed head of 75 feet ± 5 feet. See Figure 3. They are available unitized on skids and equipped with:

- Two quick-opening couplings on each cyclone overflow to permit quick internal inspection without removing the cyclone from the header.
- A shelf-rest for each cyclone for easy one-man removal and replacement in seconds.
- Quick-connections and welding stub ends furnished on all outlets, including solids trough discharge.
- Complete connecting flexibility. (Point the overflow header in the desired direction, enter the feed header from *either* end, and the solids trough is quickly reversible to the proper direction.)

**Picenco Head Gauges** accurately measure feed pressure to centrifuge or cyclone. Readings are in meters/slurry specific gravity or feet/pounds per gallon. Reflects changes immediately. Lower heads reduce throughput and cause abrasion. Varying engine speed or changing the pump impeller diameter will affect the head pressure.

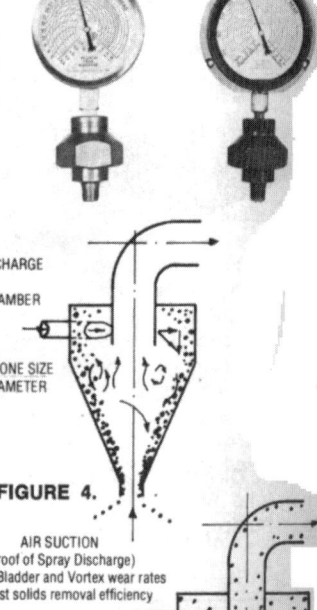

**FIGURE 2.**

**FIGURE 4.** AIR SUCTION (Proof of Spray Discharge) Lowest Bladder and Vortex wear rates Highest solids removal efficiency

**FIGURE 5.** NO AIR SUCTION (Proof of Rope Discharge) High Bladder and Vortex wear rates Low solids removal efficiency Underflow will soon plug completely

**FIGURE 3.** HEAD—DENSITY PRESSURE RELATION Shaded area shows recommended operating range for pioneer hydrocyclones on drilling fluids

# HYDROCYCLONES

**PICENCO INTERNATIONAL INC.**

## The SILTMASTER®
### A 4-INCH HYDROCYCLONE FOR DESILTING MUDS

This is the Picenco hydrocyclone design whose performance in the field led to the term "desilting," and first eliminated wall-sticking, resulted in holes closer to gauge, increased drilling rates, and bit life, etc. when properly rigged and with sufficient total removal equipment to operate in spray discharge.

For many drilling rigs, a set of Picenco SILTMASTERS large enough to process slightly more than the total circulating rate will handle the solids load below good screens. However, they will plug immediately under by-passed screen conditions (unless protected by VOLUMEMASTERS®) and will tend to plug quickly if overloaded (rope discharge) by relatively fast penetration rates in medium and large holes.

In the case of higher drilling rates, SANDMASTERS® or VOLUMEMASTERS® should be used in a stage ahead of the SILTMASTERS. Picenco can advise you on this if you furnish your mud circulation rate and expected maximum bit penetration rate with each size bit in your drilling program.

STANDARD SILTMASTER UNITS-shipped with cast polyurethane bladders at no extra cost, unless Buna N is specified.

Dimensions and weights (uncrated):

| Model No. | Length Ins. | Length Cm. | Width Ins. | Width Cm. | Height Ins. | Height Cm. | Min. Feed Rate @ 75 Ft. Hd.* GPM | Min. Feed Rate @ 75 Ft. Hd.* L/M | Weight Lbs. | Weight Kg. | Headers Outside Dia. Inches | Headers Outside Dia. Cm. |
|---|---|---|---|---|---|---|---|---|---|---|---|---|
| S 4-4 | 62 | 157 | 30 | 76 | 52 | 132 | 200 | 757 | 651 | 295 | 4½ | 11.4 |
| S 6-4 | 69 | 175 | 30 | 76 | 55 | 140 | 300 | 1135 | 975 | 442 | 5½ | 14.0 |
| T 8-4 | 62 | 157 | 40 | 102 | 53 | 135 | 400 | 1514 | 1175 | 533 | 6⅝ | 16.8 |
| T10-4 | 69 | 175 | 40 | 102 | 55 | 140 | 500 | 1893 | 1375 | 625 | 6⅝ | 16.8 |
| T12-4 | 69 | 175 | 40 | 102 | 58 | 140 | 600 | 2271 | 1520 | 690 | 6⅝ | 16.8 |
| T14-4 | 85 | 216 | 40 | 102 | 58 | 147 | 700 | 2650 | 1760 | 800 | 6⅝ | 16.8 |
| T16-4 | 85 | 216 | 40 | 102 | 59 | 150 | 800 | 3028 | 2045 | 928 | 8⅝ | 21.9 |
| T20-4** | 101 | 257 | 40 | 102 | 63 | 160 | 1000 | 3785 | 2400 | 1090 | 8⅝ | 21.9 |
| T24-4** | 117 | 297 | 40 | 102 | 66 | 167 | 1200 | 4542 | 2875 | 1305 | 8⅝ | 21.9 |

\* A new PICENCO head indicator is optional equipment. Specify API or metric.
\*\* Two smaller units offer more flexibility in operation and maintenance, and costs very little more than one of these.

## The SOLIDSMASTER™
### A 3-INCH HYDROCYCLONE FOR SPECIAL APPLICATIONS

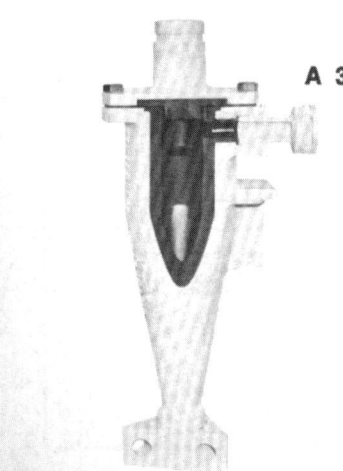

This cyclone is designed to make a very fine separation at a *constantly* high feed solids content.

Although it operates in spray discharge, it is designed on the very wet side of balance. Because the feed opening will plug very easily, it must be used only downstream from screens or scrubbers if coarse material is present.

The standard SOLIDSMASTER hydrocyclone has a basic feed capacity of 22 GPM at 75 feet of head. It is often equipped with special modifications for more or less feed rate for special situations, such as power oil. It is also operated at an unusually high feed pressure equivalent of 125 feet of head when unitized in a special CLAYMASTER® unit for working on weighted water base muds.

# SWECO

SWECO, Inc., Dept. 215-542, 6033 E. Bandini Blvd., P.O. Box 4151,
Los Angeles, CA 90051, (213) 726-1177, Telex: 67-4968

## SAND SEPARATOR

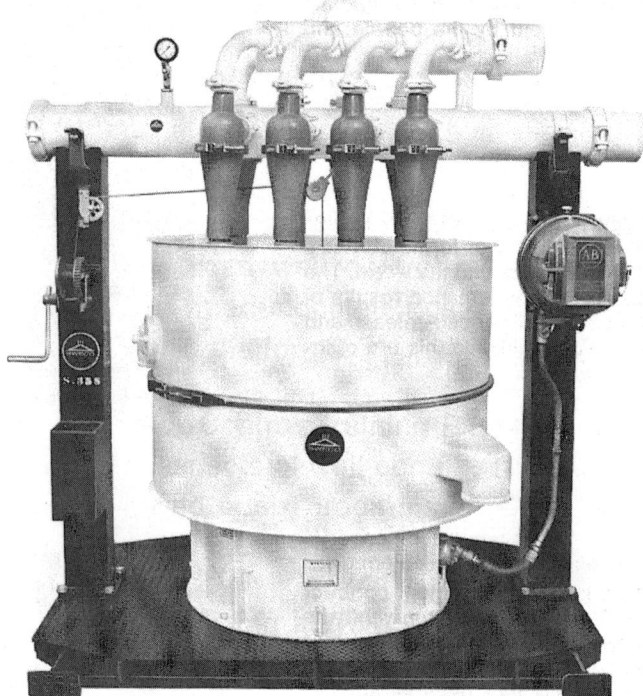

### SAND SEPARATOR SPECIFICATIONS

| | |
|---|---|
| Overall Weight | 2400 lbs. |
| Overall Dimensions | 7'5" high x 6'0" long x 4'0" wide |
| Construction | Epoxy-coated carbon steel |
| **Hydrocyclones** | |
|   Size | 4-inch Polyurethane |
|   Number | 8 |
|   Manifold | 6-inch |
| **Sand Shaker** | |
|   Screen Diameter | 48-inch |
|   Screen Mesh | 150 mesh, 200 mesh or finer for special applications |
|   Motor HP | 2½, Explosion-proof |
|   Motor Voltage | 230/460 dual voltage |
|   Motor Cycles/Phase | 60/3 |

All specifications are subject to change without notice.

The SWECO® Sand Separator provides a practical solution to problems that have plagued the oil industry for years—the economical removal of fine drilled solids from weighted mud, and the ecological disposal of fine drilled solids from unweighted mud.

SWECO's unique solids control equipment uses a two-stage mechanical process to remove fine drilled solids from the mud. By a combination of cycloning and screening, the entire mud circulating volume is stripped of native sand without loss of valuable weighting materials or chemicals. The discarded solids are dry enough for economical disposal.

Field application has proven these benefits: Chemical treatment is reduced and mud costs are lower; bit life and penetration rates are increased; filter cake properties are improved and the danger of differential sticking is minimized; cementing problems are reduced; downtime due to abrasion of pump parts is minimized; waste disposal problems are reduced; the entire drilling operation runs smoother.

## DESANDER

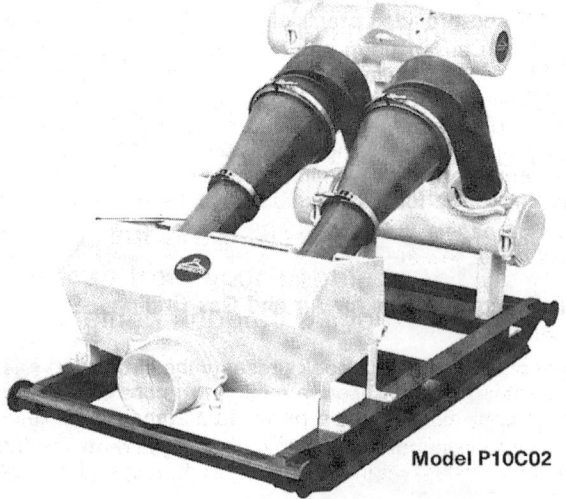

Model P10C02

SWECO Desanders are equipped with 10-inch polyurethane hydrocyclones. They are a low-cost, lightweight, low maintenance alternative to conventional desander units. SWECO Desanders are available in a two-cone model (1000 gpm) and a three-cone model (1500 gpm). Desanders are furnished with oil field "break-over" skid ends, grooved couplings and weld stubs, pressure gauge, and snap-on cone couplings. Units are sandblasted, zinc-epoxy primered and epoxy-painted.

# CHAPTER ONE

# Introduction

This book is primarily about **shale shakers** and their role in a complete **drilling fluid** system. A shale shaker is a device that removes undesirable solids from a drilling fluid. Drill bit cuttings and pieces of formation that have sloughed into the wellbore (collectively called "**drilled solids**") are brought to the surface by the drilling fluid. The fluid flows across a shale shaker before entering the mud pits. Most shale shakers impart a vibratory motion to a wire or plastic mesh **screen**. This motion allows the drilling fluid to pass through the screen and removes particles larger than the openings in the screen. Usually, drilled solids must be maintained at some relatively low concentration. The reason for this control is explained in the next section. The shale shaker is the initial and primary drilled solids removal device and usually works in conjunction with other solids removal equipment. Drilled solids may also be controlled by adding a fluid containing no drilled solids or a lower drilled-solids concentration. This reduces the solids concentration in the **circulating system** and is called "**dilution**" (see Chapter 8).

**Solids control equipment**, also called **solids removal equipment**, is designed to remove drilled solids from the circulating drilling fluid. This equipment includes a **gumbo remover**, **scalper shakers**, **shale shakers**, **dryer shakers**, **desanders**, **desilters**, **mud cleaner**s, and **centrifuges**. Various arrangements of these components are used to remove specific size particles from drilling fluid. The scalper, shale, and dryer shakers are the main subjects of this book. The other solids removal equipment (gumbo remover, desanders, desilters, mud cleaners, and centrifuges) are discussed in Chapter 7.

## WHY CONTROL DRILLED SOLIDS?

Usually, a well has a maximum level of drilled solids that can be tolerated before problems develop. This level depends on the formations drilled, the type of drilling fluid used, and the characteristics of the solids. There are some wells in which solids seem to have very little effect. These cases, which allow high drilled solids content levels to be tolerated, are the exception rather than the rule. Many work-over rigs that occasionally do some drilling, still do not employ solids removal equipment. This seems to work satisfactorily only as long as drilling is confined to short intervals. In any event, rather than experience increased drilling costs, it is best to plan for solids control.

Drilling without removing solids or liquid, the circulating system will contain a constant volume of liquid phase. The pit levels will also remain at a constant level even though the circulating volume increases as an additional hole is drilled. (Pit levels will actually rise by the volume of the drill string added to the hole.) These additional solids from drilling will increase the mud weight. Usually, many problems will arise if too many drilled solids are allowed to remain in the drilling fluid. These effects frequently do not immediately appear after the drilled solids concentrations reach excessive levels. For this reason, many drillers accustomed to drilling shallow wells with unweighted drilling fluids, find it hard to believe that drilled solids are the cause of their problems. Experience has shown them otherwise. Therefore, it is beneficial to take the detrimental effects of drilled solids seriously and remove them from the drilling fluid. Some of the detrimental effects are discussed below.

## FILTER CAKES

As drilled solids are added to a drilling fluid, **fluid loss** decreases but filter cakes become thicker and less compressible. Thick filter cakes can cause many problems:

1. The most common problem caused by poor-quality filter cakes is torque and drag, which

may result in **stuck pipe**. When drill collars are buried in a thick filter cake, the differential pressure between the wellbore and the formation fluid pressure pushes the drill collar against the formation. This is called "**differential sticking**." Drilled solids create a high friction force that can exceed the strength of the drill string.

2. Thick filter cakes can cause high surge and swab pressures. When a drill bit moves through the thick filter cake, it resembles the plunger in a syringe. As the drill bit is raised, formation fluid can be pulled into the wellbore. This can result in a kick. A hole cannot be made while the well is shut-in handling a kick. If the drill bit is lowered rapidly, the pressure below the bit increases dramatically. This can result in lost circulation.

3. Poor filter cakes can create formation damage. Formations that are sensitive to the drilling fluid filtrate should be exposed to as small amount as possible. Fluid invasion should be minimized.

4. Large invasion zones may make interpreting electric logs difficult. This is particularly true in thin beds. Shale barriers create a problem in thin beds because the depth of electric investigation is greatly diminished. Deep invasion of the filtrate means that hydrocarbons may go undetected by conventional logging techniques.

5. Cement will not effectively displace thick filter cakes. If this occurs, cement cannot fill the annulus between the casing and the formation. In this case, several problems may occur: (a) the casing seat will not pressure-test and will require a **squeeze job**, or several squeeze jobs, delaying drilling; (b) formations will not seal, thereby allowing fluid to enter the well behind the casing; and (c) this fluid could charge an upper formation or could eventually broach the surface.

## PLASTIC VISCOSITY

Plastic viscosity is determined by four factors: the viscosity of the liquid phase, the size, the shape, and the number of particles in the fluid. Drilled solids increase the plastic viscosity of a drilling fluid. A pound of colloidal particles (less than 1 to 2 **microns**) will have a greater detrimental effect on drilling fluid performance than a pound of 30 to 40 micron-size particles. The shale shaker removes many large particles that will eventually disintegrate into colloidal size. It is important to remove drilled solids as soon as they reach the surface because they will never again be that large. The following points are important when discussing plastic viscosity:

1. Plastic viscosity of a drilling fluid impacts the effectiveness of cuttings removal from the bottom of a borehole. Fluids with low plastic viscosity can remove cuttings better. This raises the **founder point** of a drill bit and allows for greater bit loading. Higher drill rates are now possible.

2. High plastic viscosities diminish the effectiveness of yield point for hole cleaning in nearly vertical holes. Carrying capacity of a drilling fluid is dependent on the viscosity of the fluid in the annulus. Decreasing the plastic viscosity of a drilling fluid while maintaining a constant yield point increases the low-shear-rate viscosity of a drilling fluid. This is known as the "K" value of the Power Law Rheological Model.

3. High plastic viscosities have a detrimental effect on the performance of all solids removal equipment.

4. High plastic viscosity increases fluid pressure losses in the circulating system, which decreases drilling performance.

## WEAR

Solids, particularly sand-size particles (greater than 74 microns), can cause significant wear of expendables. The effect is similar to grinding pump parts and elastomer seals against coarse sandpaper. Seals in drill bits, pump liners, swabs, valves, swivels, and so forth, will deteriorate much faster if the large drilled solids, such as sand, are not removed from the drilling fluid. Bearing life of roller cone bits with no seals is strongly dependent on the quantity and type of drilled solids in the drilling fluid. The life of conventional diamond bits is also adversely affected by an increase in the quantity of sand-size particles in the drilling fluid.

## CARRYING CAPACITY

One important component of a good solids management philosophy is the drilling fluid carrying capacity. As drilled solids (sloughings and drill bit-generated) enter the wellbore, they should be brought to the surface as soon as possible. If these drilled solids are tumbled and slowly brought to the surface, they have time to disintegrate and increase the low-gravity solids content of a drilling fluid.

Surprisingly, drilled solids also affect the drilling fluid properties that control carrying capacity in vertical and nearly vertical (up to 35°) wells. An empirical relationship has been published relating a carrying capacity index (CCI) to the product of the mud weight (MW), annular velocity (AV), and a characteristic viscosity (K). This can be expressed as:

$$CCI = \frac{(MW)(AV)(K)}{400,000}$$

The "400,000" constant was empirically determined by observing hole cleaning conditions on many rigs over an 8- to 10-year period. A CCI value of "one" seems to indicate good hole cleaning in both water-based and oil-based drilling fluids. The constant is probably not accurate to more than one significant figure.

The "K" viscosity is the viscosity from the Power Law Rheological Model, expressed as equivalent centipoise. It can be related to the Plastic Viscosity (PV) and Yield Point (YP) through the equations:

$$K = (511)^{1-n}(PV + YP)$$

and

$$n = 3.322 \log \frac{2PV + YP}{PV + YP}$$

At a constant yield point value, increasing the plastic viscosity decreases the value of the viscosity. This is clear from the curves in Figure 1-1.

If the cuttings on the shaker screen indicate that they are not arriving at the surface without considerable tumbling, the CCI equation can be used to calculate the yield point needed to clean the wellbore. Because the mud weight and annular velocity will not usually be changed, the K-value needed can be calculated. Generally, the plastic viscosity will not change significantly when the yield point is increased. Consequently, the curve in Figure 1-1 (or a similar curve) can be used to determine the yield point needed to clean the hole.

## DRILLING FLUID COSTS

Drilling fluid density can be increased using drilled solids for the lowest initial weight-up cost. If drilled solids had no other effect, using them for a weighting agent would be cost-effective. However, using drilled solids could potentially increase drilling costs because slow drilling, poor cement jobs, stuck pipe, and lost circulation will frequently follow. Downtime on a rig or reduced penetration rates are far more expensive than the cost of using adequate weighting material.

Removal of drilled solids as they reach the surface usually costs less than any other method of mitigating their effect. Conversely, controlling drilled solids with dilution is usually the most expensive method. Although drilled solids concentration can be reduced by half with dilution by doubling the system volume with clean drilling fluid, this clean fluid is expensive.

**Polymer drilling fluids**, with X-C, PHPA, and so on, require relatively low drilled-solids concentrations. The polymer attaches to all solids in the system whether they are desirable or not. Failure to keep a low drilled-solids concentration will result in using excessive quantities of chemical (polymer). This can make the use of polymer drilling fluids cost-prohibitive. The use of polymer systems requires the utmost in planning solids control equipment for optimum performance.

A word of caution is appropriate here. Neophytes in drilling have a tendency to try to minimize the cost of each category of drilling expenses with the misconception that this will minimize the total cost of the well. It is important to realize that additional expenses can be incurred because of inadvisable decisions to cut costs in easily monitored expenses while drilling wells. When line-items are independent of each other, minimization of each line-item will result in the lowest possible cost. When line-items are interconnected, minimization of each line-item may be very expensive. Drilled solids concentrations and trouble costs (or the cost of unscheduled events) are very closely related.

For example, a common mistake is to allow an initial increase in mud weight to occur with drilled solids. Clearly, less money will be spent on the

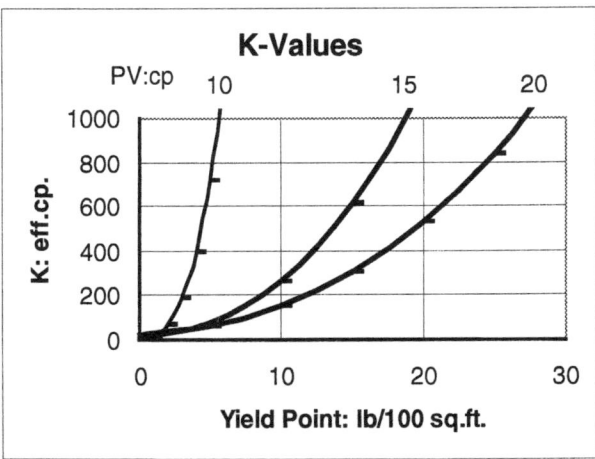

**FIGURE 1-1.** K-viscosity.

drilling fluid if no weighting agents are added. These savings are easily documented. Less apparent, however, will be the additional expenses incurred because of excessive drilled solids in the drilling fluid. These problems will obviously increase the well cost as described previously.

Another common mistake, usually made while drilling with weighted drilling fluid, is to relate the cost of the discarded weighting agent with the drilled solids discard. The cost of discarded weighting agents (barite or hematite) can be relatively small compared to the problems associated with drilled solids. This is particularly true in the expensive offshore environment. Even in less expensive land drilling, a comparison normally justifies discarding weighting agents to eliminate drilled solids.

Solids control equipment properly used, with the correct drilling fluid selection, will usually result in lower drilling costs. Decisions made for specific wells depend on the well depth and drilling fluid density. Shallow, large-diameter, low mud-weight wells can tolerate more drilled solids that more complicated wells. Each well must be evaluated individually with careful consideration of the risk of the problems associated with drilled solids.

## WASTE MANAGEMENT

Failure to remove drilled solids with solids control equipment results in solids control with dilution. This creates excessive quantities of fluid that must be handled as a waste product. It can be very expensive if this excess fluid must be removed from the drilling location. Even if the fluid can be handled on-site, large quantities of fluid frequently increase costs.

Smaller quantities of waste products can significantly decrease the cost of a well. Decreasing the quantity of drilling fluid discarded with the drilled solids will decrease the cost of rig-site cleanup. Dilution techniques for controlling drilled solids concentrations greatly increase the quantity of waste products generated at a rig. This results in an additional expense that ultimately adds to the total cost of drilling.

# CHAPTER TWO

# The Role of Shale Shakers

Shale shaker is a general term for a vibrating device used to screen solids from a circulating drilling fluid. Many configurations have been used, including:

- A square or rectangular screening area with drilling fluid flow down the length
- Revolving, nonvibrating, cylindrical screens with longitudinal flow down the center axis
- Circular screens with flow from the center to the outside

The majority of shale shakers flow the drilling fluid over a rectangular screening surface. Larger solids are removed at the discharge end, with the smaller solids and drilling fluid passing through the screen(s) into the active system. All drilled solids above 74 microns are considered especially undesirable in any drilling fluid. For this reason, 200-mesh (74 microns) screens are extremely desirable on shale shakers. Weighting materials that meet API specifications may still have 3% by weight larger than 74 microns. Screens this size may remove large quantities of barite, which may significantly impact the drilling fluid cost, but may decrease well cost.

Shale shakers are the most important and easiest to use solids removal equipment. In most cases, they are highly cost-effective. However, if shale shakers are used incorrectly, the remaining solids removal equipment will not perform properly.

A shale shaker can be used in all drilling applications where liquid is used as the drilling fluid. Screen selection is controlled by the circulation rate, shaker design, and drilling fluid properties. Variation in drilling fluid properties control screen throughput to such a large degree that shaker capacities are not included in this handbook.

Most operations involved in drilling a well can be planned in advance because of experience and engineering designs for well construction. Therefore, well planners expect to be able to determine from statistical data the size and number of shale shakers required to drill a particular well, as well as the mesh size of the shaker screens used for any portion of any well. However, due to the large number of variables involved, these data simply do not exist. Many shale shaker manufacturers, because of customer demand, publish approximate flow charts that indicate their shakers can process a certain flow rate of drilling fluid through certain size screens. These charts are usually based on general field experience with a lightly treated, water-base drilling fluid and should be treated as approximations at best.

Rheological factors, fluid type, solids type and quantity, temperature, drilling rates, solids/liquid interaction, hole diameters, hole erosion, and other variables dictate actual flow rates that can be processed by a particular screen. Drilling fluid without any drilled solids can pose screening problems. Polymers that are not completely sheared tend to blind screens and/or appear in the screen discard. Polymers that increase the low-shear-rate viscosity or gel strength of the drilling fluid also pose screening problems. Polymers, like starch, that are used for fluid-loss control are also difficult to screen through a fine mesh screen (such as 200 mesh). Oil-base drilling fluids without adequate shear and proper mixing are difficult to screen, as are oil-base drilling fluid through screens finer than 100 mesh without sufficient oil-wetting additives.

Screen selection for shale shakers is dependent on geographical and geological location. Screen combinations that will handle specific flow rates in the Middle East or Far East, for example, will not necessarily handle the same flow rates in Norway or the Rocky Mountains. The best method to select shale shaker screens and/or to determine the number of shale shakers for a particular drilling site is to first use the recommendations of a qualified solids control advisor from the area. Screen use records should be established for further guidance.

## HOW A SHALE SHAKER SCREENS FLUID

Shale shakers should remove as many drilled solids and as little drilling fluid as possible. These dual objectives require that cuttings (or drilled solids) convey off the screen while simultaneously separating and removing most of the drilling fluid from the cuttings. Frequently, the only stated objective of a shale shaker is to remove the maximum quantity of drilled solids. Disregarding the need to conserve as much drilling fluid as possible defeats the ultimate objective of reducing drilling costs.

Cutting sizes greatly influence the quantity of drilling fluid that tends to adhere to the solids. As an extreme example, consider a golfball-size drilled solid coated with drilling fluid. Even with a viscous fluid, the volume of fluid would be very small compared with the volume of the solid. If the solids are sand-sized, the fluid-film volume increases as the solids surface area increases. For silt-size or ultra-fine solids, the volume of liquid coating the solids may even be larger than the solids volume. More drilling fluid returns to the system when very coarse screens are used than when screens as fine as 200 mesh are used.

Drilling fluid is a rheologically complex system. At the bottom of the hole, faster drilling is possible if the fluid has a low viscosity. In the annulus, drilled solids are transported better if the fluid has a high viscosity. When the flow stops, a gel structure slowly builds to prevent cuttings or weighting agents from settling. Drilling fluid is usually constructed to perform these functions. This means that the fluid viscosity depends on the history and shear within the fluid. Typically, low-shear-rate viscosities of drilling fluids range from 300 to 400 centipoise up to 1000 to 1500 centipoise. As the shear rate (or usually the velocity) increases, drilling fluid viscosity decreases. Even with a low-shear-rate viscosity of 1500 centipoise, the plastic viscosity (or high-shear-rate viscosity) could be as low as 10 centipoise.

Drilling fluid flows downward, on and through shaker screens. If the shaker screen is stationary, a significant head would need to be applied to the drilling fluid to force it through the screen. For example, imagine pouring honey onto a 200-mesh screen (Figure 2-1). Honey at room temperature has a viscosity around 100 to 200 centipoise. The flow through the screen would be very slow. If the screen is moved rapidly upward through the honey, more fluid would flow in a given period of time (Figure 2-2). The introduction of vibration to this process applies upward and downward forces to the honey. The upward stroke moves the screen rapidly through the honey. These same forces of

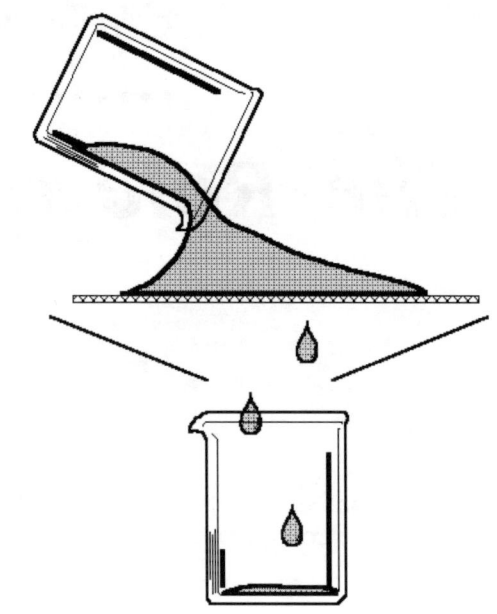

**FIGURE 2-1**

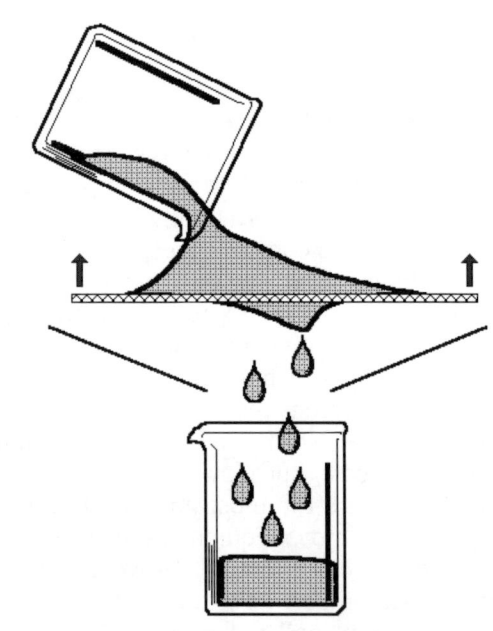

**FIGURE 2-2**

vibration affect drilling fluid in a similar manner. The upward stroke moves drilling fluid through the screen. Solids do not follow the screens on the downward stroke and, therefore, are propelled from the screen surface.

The upward motion of the shaker screen forces fluid downward through the shaker openings and moves solids upward. When the screen moves on the downward stroke, solids do not follow the screen. They are, instead, propelled forward along the screen. This is the theory behind elliptical, circular, and linear motion screens.

Screens are moved upward through the fluid with the elliptical, circular, and linear motion shale shakers. The linear motion shaker has an advantage because solids can be transported out of a pool of liquid and discharged from the system. The pool of liquid creates two advantages: it provides an additional head to the fluid and also provides inertia, or resistance, to the fluid as the screen moves upward. This significantly increases the flow capacity of the shaker.

The movement of the shaker screen through the drilling fluid causes the screen to shear the fluid. This decreases the effective viscosity and is an effective component to allow shakers to process drilling fluid.

The upward movement of the shaker screen through the fluid is similar to pumping the drilling fluid through the screen opening. If the fluid gels on the screen wires, the effective opening is decreased. This is the same as pumping drilling fluid through a smaller diameter pipe. With the same head applied, less fluid flows through a smaller pipe in a given period of time than a larger pipe. If a shaker screen becomes water-wet while processing an oil-base drilling fluid, the water ring around the screen opening effectively decreases the opening size available to pass the fluid. This, too, would reduce the flow capacity of the shaker.

## SHAKER DESCRIPTION

The majority of shale shakers use a **back tank** (commonly known as a **possum belly** or a **mud box**) to receive drilling fluid from the flow line (Figure 2-3). Drilling fluid flows over a weir and is evenly distributed to the **screening surface (deck)**. The screen(s) are mounted in a **basket** that vibrates to assist the throughput of drilling fluid and the movement of separated solids. The basket rests on **vibration isolation** members, such as helical springs or rubber float mounts. The vibration isolation members are supported by the **skid**. Below the basket, a **collection pan** (or **bed**) is used to channel the screen underflow to the active system.

Shale shaker performance is affected by the type of motion, stroke length of the deck, and the rotary speed of the motor. The shape and axial direction of the vibration motion along the deck is controlled by the position of the **vibrator(s)** in relation to the deck and rotation direction of the vibrator(s). There are many commercially available basket and deck configurations. The deck may be horizontal (Figures 2-4C and D) or mounted at a slope (Figures 2-4A, B, E, and F), and its surface may be tilted up or down in the basket. The basket may be horizontal or have a fixed or adjustable angle. An adjustable basket angle allows the deck to be tilted up or down.

On sloped deck units (**cascade** or parallel flow), the screens may be continuous with one screen covering the entire deck length (Figures 2-4A and

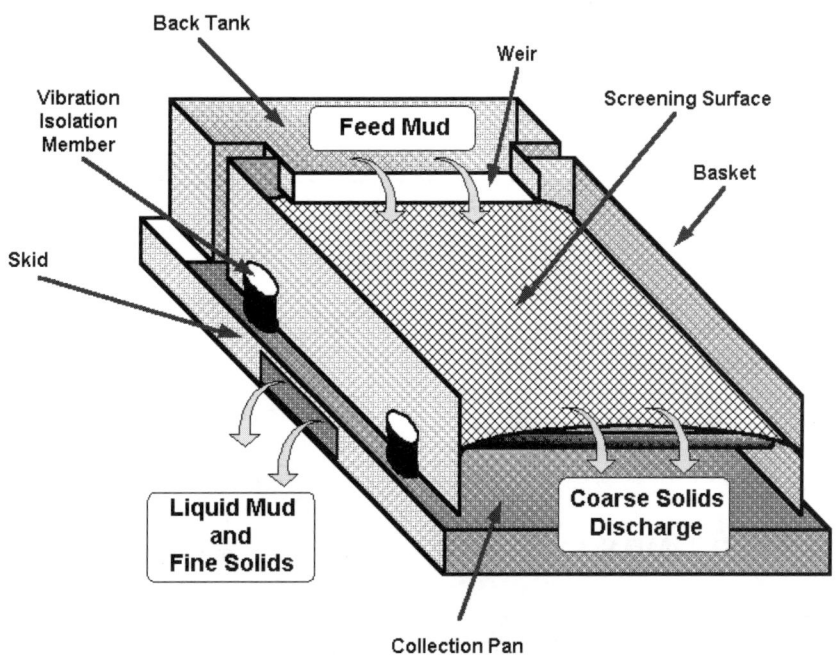

FIGURE 2-3

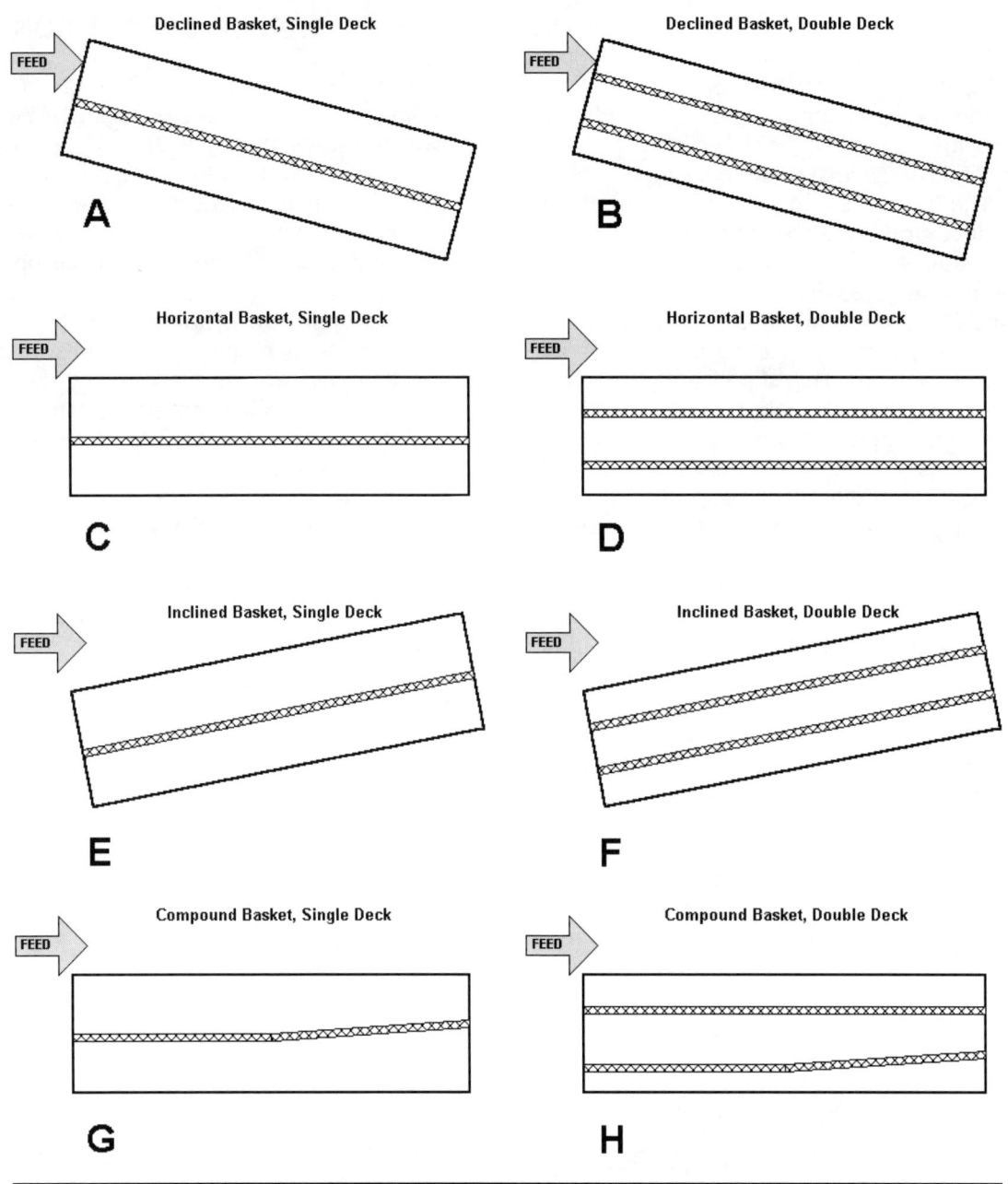

**FIGURE 2-4**

E); or have a divided deck which uses more than one screen to cover the screening surface (Figures 2-4B and F); or with individual screens mounted at different slopes (Figures 2-4G and H). On multiple deck units, fluid passes through the upper deck before flowing to the next deck (Figures 2-4B, F, and H).

## SHALE SHAKER LIMITS

A shale shaker's **capacity** is reached when excessive drilling fluid (or the drilling fluid liquid phase) first begins discharging over the end of the shaker. The capacity is determined by the combination of two factors:

1. The fluid limit is the maximum fluid flow rate that can be processed through the shaker screen.
2. The solids limit is the maximum amount of solids that can be conveyed off the end of the shaker.

These two limits are interrelated in that the amount of fluid that can be processed will decrease as the amount of solids increases.

Any shale shaker/screen combination has a fluids-only capacity (i.e., no solids are present that can be separated by the screen) that is dependent on the characteristics of the shaker ("G"-factor, vibrational frequency, type of motion, and angle of the screen deck), the screen (area and conductance), and the fluid properties (density, rheology, additives, and fluid type). The fluid-only capacity is the fluid limit with zero removable solids. Although not true in many instances, the following assumes the drilling fluid to be a fluid with solids no larger than the openings in the shaker screen.

The screen cloth can be considered a permeable media with a permeability and thickness (conductance) and an effective filtration area. The fluid capacity will decrease as the fluid viscosity increases (plastic viscosity is most important but yield and gel strengths can have a significant impact as well). Capacity will also increase as the fluid density increases due to increased pressure on the screen surface acting as a force to drive fluid through the screen.

The fluid-only capacity will generally be reduced when certain polymers are present in the fluid. Partially-hydrolyzed polyacrylamide (PHPA) is most notable in this respect as it can exhibit an effective solution viscosity in a permeable media higher than that measured in a standard viscometer. At one time, this effective viscosity of PHPA solutions was determined by flowing the solution through a set of 100-mesh screens mounted in a standard capillary viscometer. PHPA drilling fluids typically have a lower fluid-only capacity for a given shaker/screen combination than similar drilling fluids without PHPA because of this higher effective viscosity. This decrease in fluids-only capacity can be as much as 50% compared to a gel/water slurry. Adsorption of PHPA polymer may decrease effective opening sizes (as it does in porous media), thereby increasing the pressure drop required to maintain constant flow. This makes the PHPA appear to be much more viscous than it actually is. This effect also occurs with high concentrations of XC in water-base fluids, drilling fluids with high concentrations of starch, newly prepared oil-base drilling fluids, and polymer-treated viscosifiers in mineral oil-base fluids.

The solids limit can be encountered at any time, but occurs most often when drilling large diameter holes, soft, sticky formations, or during periods of high penetration rates. A relationship exists between the fluid limit and solids limit. As the fluid flow rate increases, the solids limit decreases. As the solids loading increases, the fluid limit decreases. Internal factors that affect the fluid and solids limits are discussed in Chapter 3.

Major external factors that affect the solids and fluid limits are:

- *Fluid rheological properties*—Literature indicates that the liquid capacity of a shale shaker screen decreases as the plastic viscosity of a drilling fluid increases. Plastic viscosity (PV) is the viscosity that the fluid possesses at an infinite shear rate.[1] Drilling fluid viscosity is usually dependent on the shear rate applied to the fluid. The shear rate through a shale shaker screen depends on the opening size and how fast the fluid is moving relative to the shaker screen wires. Generally, shear rates through the shaker screen vary significantly. The exact capacity limit, therefore, will depend on the actual viscosity of the fluid. This will certainly change with plastic viscosity and yield point.

- *Fluid surface tension*—Although drilling fluid surface tensions are seldom measured, high surface tensions decrease the ability of the drilling fluid to pass through a shale shaker screen, particularly fine mesh.

- *Wire wettability*—Shale shaker wire screens must be oil-wet when drilling with oil-base drilling fluids. Water adhering to a screen wire decreases the effective opening size for oil to pass through. Frequently, this results in the shaker screens being incapable of handling the flow of an oil-base drilling fluid. This is called "sheeting" across the shaker screen and often results in discharging large quantities of drilling fluid.

- *Fluid density*—Drilling fluid density is usually increased by adding a weighting agent to the drilling fluid. This increases the number of solids in the fluid and makes it more difficult to screen the drilling fluid.

- *Solids: type, size, and shape*—The shape of solids frequently make screening difficult. In single-layer screens, particles that are only slightly larger than the opening size and can become wedged in the openings. This effectively plugs the screen openings and decreases the area available to pass fluid. Solids that tend to cling together, such as **gumbo**, are also difficult to screen. Particle size has a

---

[1] The Bingham Plastic Rheological Model may be represented by the equation: Shear Stress = (PV) Shear Rate + YP, where YP is the Yield Point. By definition, viscosity is the ratio of shear stress to shear rate. Using the Bingham Plastic expression for shear stress: Viscosity = [(PV) Shear Rate + YP]/Shear Rate. Performing the division indicated, the term for viscosity becomes (PV) + (YP/Shear Rate). As shear rate approaches infinity, viscosity becomes PV.

significant effect on both solids and liquid capacity. A very small increase in **near-size particles** usually results in a large decrease in fluid capability for any screen, whether they be single, double, or triple layer.

- *Quantity of solids*—Solids compete with the liquid for openings in the shaker screen. Fast drilling can produce large quantities of solids. Usually, this requires coarser mesh screens to allow most of the drilling fluid to be recovered by the shale shaker. Fast drilling is usually associated with shallow drilling. The usual procedure is to start with a coarser mesh screen in the fast drilling of larger holes near the top of the well and to then use finer mesh screens as the well becomes deeper. Finer mesh screens can be used when the drilling rate decreases.

  Boreholes that are not stable can also produce large quantities of solids. Most of the very large solids that arrive at the surface come from the side, and not the bottom, of the borehole. Drill bits usually create very small cuttings.

- *Hole cleaning*—One factor frequently overlooked in the performance of shale shakers is the carrying capacity of the drilling fluid. If the cuttings are not brought to the surface in a timely manner, they tend to disintegrate into small solids in the borehole. If they stay in the borehole for a long enough period before arriving at the surface, the plastic viscosity and the solids content of the drilling fluid increases. This makes it appear that the shale shaker is not performing adequately, when actually the solids are disintegrating into those that cannot be removed with the shale shaker.

## SHAKER DEVELOPMENT SUMMARY

Shale shakers have undergone many improvements since the last shale shaker handbook was published in the early 1970s. The current linear motion shakers, introduced in the 1980s, have become widely used because of their improved solids conveyance and fluid throughput. Balanced elliptical and linear motions make it possible to convey solids toward the discharge end of the deck while the screen is tilted uphill. The uphill tilt of the deck creates a pool of fluid at the feed end of the deck, which, in combination with the balanced elliptical or linear motion, exerts greater pressure on the fluid flowing through the screen openings. This allows the use of a finer mesh screen than with previous shaker designs. The normal component of the acceleration controls the liquid throughput. The normal component of a circular and a linear motion shaker may have the same acceleration (or "G"-factor) but the linear motion shaker can process a greater flow rate. The linear motion conveys solids uphill, whereas circular motion does not. The uphill solids conveyance allows the linear motion to process a greater flow rate.

The use of linear motion shakers has become feasible with the development of improved screen designs. The life of shaker screens has been extended with the introduction of repairable bonded and pretensioned screen panels. Other design improvements are available in wire cloth, non-metallic screens, and three-dimensional screen surfaces, which have improved the solids separation capabilities of all shakers.

Although linear motion shale shakers have made a significant impact in solids removal concepts, the other shale shakers have many advantageous features. Circular motion is easier on the shale shaker structure, easier on the shaker screens, and conveys gumbo. Linear motion and elliptical motion machines can convey gumbo downhill but not upward. Linear motion shakers may use bonded screens where 30% to 50% of the area is forfeited. The liquid pool at the back of the linear motion screens can cause solids to disintegrate into many smaller particles and forced through the shaker screens. This liquid pool also gives solids slightly finer than the screen openings more of a chance to pass through the screen. The various types of motions will be discussed more thoroughly in Chapter 3.

# CHAPTER THREE

# Shale Shaker Design

The purpose of a shale shaker is to induce drilling fluid to flow through a screen, transport solids across a screen surface, and discharge solids off the end of the screen. Screening is the result of using the energy developed by a rotating eccentric mass and applying that force to a porous surface. The energy causes the screen to vibrate in a fixed orbit or path.

The elements of shale shaker design focuses on several aspects of the machine:

- Shape of motion (orbit or path)
- Deck design
- "G"-factor
- Power systems

All of these parameters contribute to the results achieved by a machine and each will be discussed in this chapter.

## SHAPES OF MOTION

Historically, the progression of shale shaker design has been to allow the use of finer mesh screens. They have developed through the years from relatively simple, uncomplicated designs to today's more complex models. This evolutionary process encompasses several distinct eras of shale shaker technology and performance. These developmental timeframes can be divided into four main categories:

1. **Unbalanced elliptical** motion
2. **Circular** motion
3. **Linear** motion
4. **Balanced elliptical** motion

The eras of oil field shaker (and screening) development may be defined by the types of motion(s) produced by the **vibrators** and their associated machines.

If a single rotating vibrator is located away from the center of gravity of the basket, the motion is elliptical at the ends of the deck and circular below the vibrator (Figure 3-1). This is an unbalanced elliptical motion. If a single rotating vibrator is located at the center of gravity of the basket, the motion is circular (Figure 3-2). Two counter-rotating vibrators attached to the basket are used to produce linear motion (Figure 3-3). When placed at an angle to the basket, two counter-rotating vibrators will produce a balanced elliptical motion (Figure 3-4).

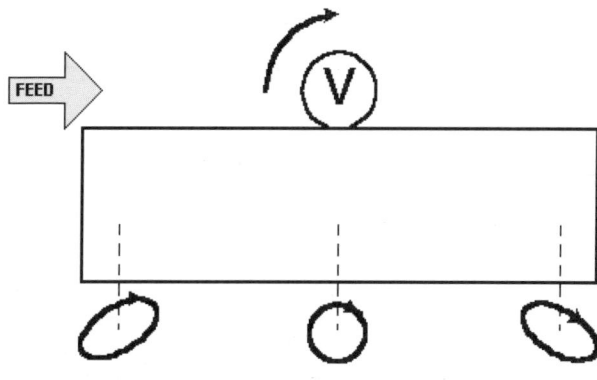

**FIGURE 3-1.** Unbalanced elliptical motion.

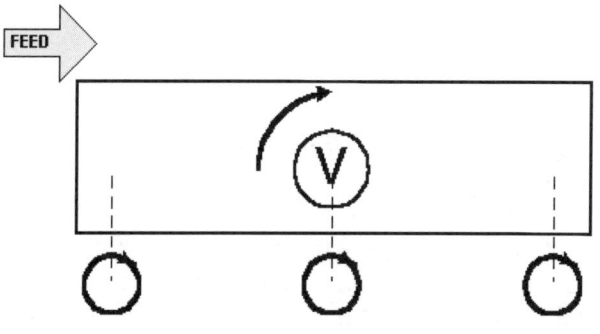

**FIGURE 3-2.** Circular motion.

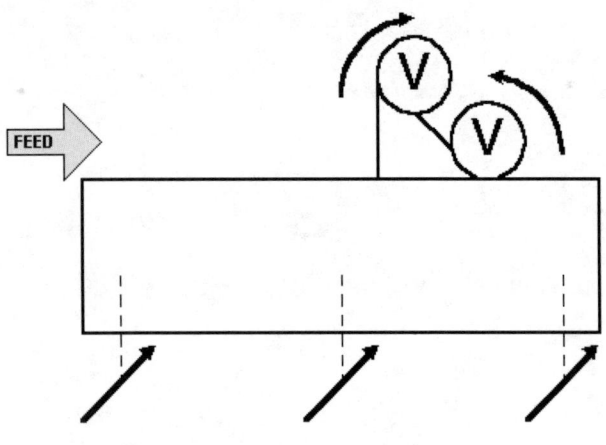

**FIGURE 3-3.** Linear motion.

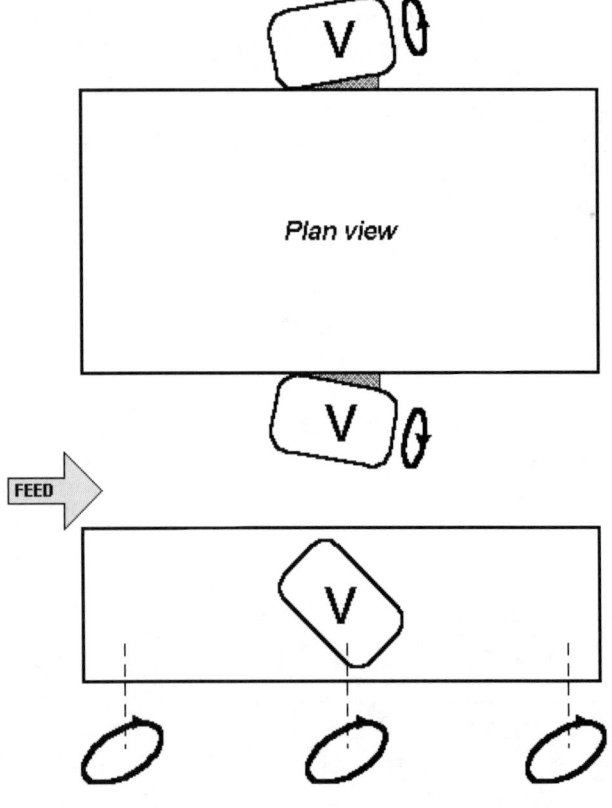

**FIGURE 3-4.** Balanced elliptical motion.

## Unbalanced Elliptical Motion Shale Shakers

In the 1930s, **elliptical (unbalanced)** shale shakers were adapted to the oil field. These first shakers originated from the mineral ore dressing industries (e.g., coal, copper, etc.) with little or no modifications. These machines were basic, rugged, and mechanically reliable, but were generally limited to 20-mesh and coarser screens.

In an unbalanced elliptical motion shaker (Figure 3-1), the movement of the shaker **deck/basket** is accomplished by placing a single vibrator system above the shaker deck. In other words, the mechanical system of spinning counterweights (or an elliptical-shaped drive shaft) is installed above the center of gravity of the deck. The resulting motion imparted to the bed is a combination of elliptical and circular. Directly below the vibrator the motion of the basket is circular, while at either end of the deck the motion is elliptical.

The orientation of the major axes of the ellipses formed at the feed end and solids discharge end of the basket has a major impact on solids **conveyance**. Specifically, it is desirable for the major axis of the ellipsoidal trace to be directed toward the solids discharge end. However, the orientation of the major axis of the ellipse formed at the solids discharge end is exactly the opposite; it is directed backward toward the feed end. This discharge end **thrust** orientation is undesirable since it makes discharging solids from the shaker more difficult (Figure 3-5). To assist in solids conveyance, the deck or last screen is tilted downward (Figure 3-6), or the vibrator is moved to the discharge end. Moving the vibrator toward the discharge end significantly reduces the fluid capacity and the screen life of the end screen. This also reduces the **residence time** of the feed slurry on the screening surface.

Early elliptical motion shale shakers used **hook strip screens** that were manually tensioned. A series of **tension rails** and tension bolt spring assemblies were used to pull the screens tightly over the support bars to ensure proper tightening. **Pretensioned screens** and pretensioned screen panels were not introduced until the 1970s and even then were not commonly used on elliptical motion units.

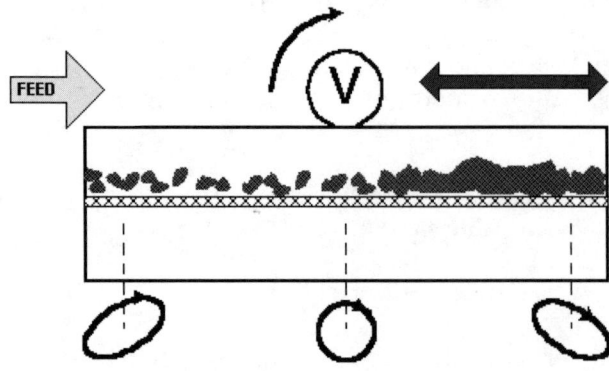

**FIGURE 3-5.** Undesirable discharge end thrust orientation.

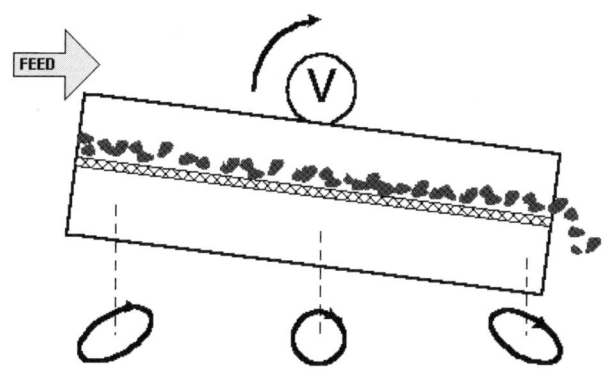

**FIGURE 3-6.** Tilted orientation.

Like most engineered products, compromises are made to achieve an acceptable balance between the amount of feed slurry the shale shaker can process and its ability to effectively convey solids along the screen deck. The early elliptical motion shakers typically had one screen surface driven by a motor sheaved to the vibrator with a belt drive. More recent models employ additional screen area and/or integral vibrators to increase flow capacity. These shakers are capable of processing drilling fluid through 80- to 100-mesh screens.

Unbalanced elliptical motion shale shakers are compact, easy to maintain, and inexpensive to build and operate. They use relatively coarse screens (80 to 100 mesh) and, for this reason, are frequently used as **scalping shakers**. Scalping shakers remove large solids, or gumbo, and reduce solids loading on downstream shakers. The use of scalping shakers in **cascading systems** is discussed in Chapter 4.

## Circular Motion Shale Shakers

**Circular (balanced) motion shakers** were introduced in 1963. These shakers have a single vibrator shaft located at the center mass of the basket. A motor drives a concentric shaft fitted with counter-weights, which provides pure circular motion along the entire length of the vibrating deck. This feature improves solids conveyance off the end of the deck compared to unbalanced elliptical designs. The circular motion transports solids along a horizontal screen, thus reducing the loss of liquid without sacrificing solids conveyance.

Circular motion units often incorporate multiple, vertically stacked decks. Coarse mesh screens mounted on the top deck separate and discharge the larger cuttings, thereby reducing solids loading on the bottom screens. These multiple-deck units allowed the first practical use of 80- to 100-mesh screens.

**Flow-back trays** (Figure 3-7), introduced in the late 1970s, direct the slurry onto the feed end of the finer mesh screen on the lower deck. The tray allows full use of the bottom screen surface to achieve greater cuttings removal with less liquid loss. Even with these units, screens are limited to approximately 100 mesh by the available screening area, the vibratory motion, and screen panel design. If bonded screens are used, screens as fine as 150 mesh have been used.

Screens on the circular motion units are installed either **overslung** or **underslung**. The **open-hook strip screen** is tensioned across the longitudinal support members. Both designs have advantages and disadvantages. Overslung screens have reasonable screen life but the drilling fluid tends to channel to the sides. On underslung screens, drilling fluid tends to congregate around and beneath the longitudinal support members. Grinding this accumulation of drilled solids between the rubber support and the screen tends to reduce screen life. To overcome this problem, rubber supports with flatter cross-sections are used, and strips are installed between the rubber support and the screen.

In the 1980s, some circular motion machines began being fitted with repairable bonded, underslung screens that increased screen life and fluid throughput. Even though the use of repairable bonded screens reduced the net nonblanked area, the detrimental effect on fluid capacity was more than offset by the use of higher conductance screen cloths and larger bonded openings. **Repairable bonded screens** will be further discussed in Chapter 6.

## Linear Motion Shale Shakers

The introduction of **linear motion shale shakers** in 1983, combined with improved screen technology, resulted in the practical use of 200 and

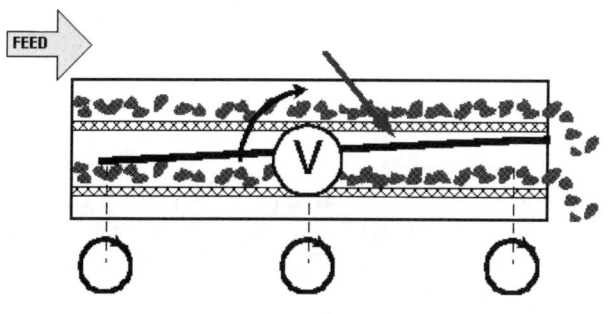

**FIGURE 3-7.** Flow-back tray.

finer mesh screens. Linear motion is produced by a pair of eccentrically weighted, counter-rotating, parallel vibrators. This motion provides cuttings conveyance when the screen deck is tilted upward.

Linear motion shakers have overcome most of the limitations of elliptical and circular motion designs. Straight-line motion provides superior cuttings conveyance (except for **gumbo**) and superior liquid throughput capabilities with finer mesh screens. These shakers can effectively remove gumbo if they are sloped downward toward the discharge end; however, they generally do not convey gumbo uphill. The increased physical size of these units (and an accompanying increase in deck screen surface area) allows the use of even finer screens than those used on circular or elliptical motion shakers.

Screening ability is the result of applying the energy developed by a rotating eccentric mass to a porous surface or screen. The energy causes the screen to vibrate in a fixed orbit. This transports solids across the screen surface, off the discharge end, and induces liquid to flow through the screen.

In conventional unbalanced elliptical and circular motion designs, only a portion of the energy transports the cuttings in the proper direction, toward the discharge end. The remainder is wasted, due to the peculiar shape of the screen bed orbit, and is manifested as solids that become nondirectional or traveling the wrong direction on the screen surface. Linear motion designs provide positive conveyance of solids throughout the vibratory cycle because the motion is straight-line rather than elliptical or circular. The heart of a linear motion machine is its ability to generate this straight-line or linear motion and transmit this energy in an efficient and effective manner to the vibrating bed.

As shown in Figure 3-8, a linear motion system consists of two eccentrically weighted, counter-rotating shafts. The net effect of each equal eccentric mass being rotated in opposite directions is that resultant forces cancel at all positions along the vibratory trace, except at the very top and bottom of each stroke, resulting in a thrust (vibration) along a "straight-line." Hence, the term "linear" or "straight-line" motion.

To achieve the proper relationship between the rate of solids conveyance and liquid throughput, the drive system must be mounted at an angle to the horizontal bed. A thrust angle of 90° relative to the screen surface would simply bounce solids straight up and down. Taken to the other extreme, a thrust angle of 0° would rapidly move solids but yield inadequate liquid throughput and solids that were discharged very wet. On most units the appropriate angle is approximately 45° to the horizontal (Figure 3-8).

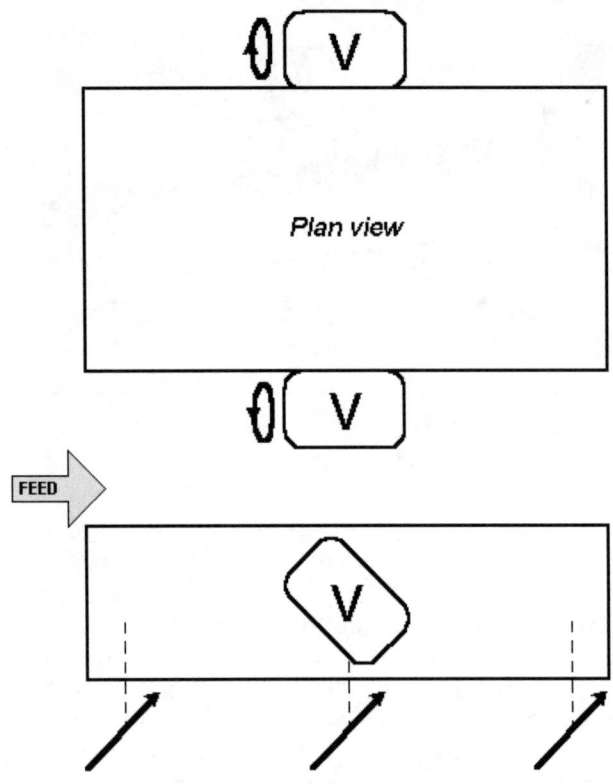

**FIGURE 3-8.** Linear motion system with two eccentrically weighted, counter-rotating shafts.

Some machines have adjustable angle drive systems that can be changed to account for various process conditions (Figure 3-9). If the thrust angle is decreased (for example, to 30° to the horizontal), the "X" component of the resultant vibratory thrust (force) would increase and the "Y" component would decrease. Conversely, building a greater angle would cause the "X" component to decrease and the "Y" component to increase.

A larger "X" vector component of thrust will move solids along the deck faster. A larger "Y" component vector increases liquid throughput and increases residence time of material on the screen. Most manufacturers choose a fixed angle near 45° that yields near equal values for each vector. This is a logical approach since the shaker must simultaneously transmit liquid through the screen as it conveys solids off the screen.

The ability to create linear motion vibration allows the slope of the bed to vary up to a +6° incline (which impacts residence time and, therefore, shaker performance) and to create a liquid pool at the flow line end of the machine. This allows a positive liquid pressure head to develop, which helps "drive" liquid and solids through the finer wire cloths. The deck on most linear motion shale shaker designs can be adjusted up to a maximum

# SHALE SHAKER DESIGN

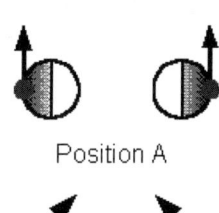

Position A

Position B

Position C

Position D

### How linear motion is created:

When two equal weights are rotating, each about its own center as shown on the left, the vertical velocity is a maximum value in position A. At this point, the weights start moving vertically more slowly. For example, in position B the vertical velocity is smaller as the weights start moving in a more lateral direction. This means that the vertical acceleration is a negative value. The circular rotation indicates that the velocity is changing as a sinusoidal function of time.

At position C there is no vertical velocity left. The vertical acceleration reaches zero. At this position, clearly the horizontal velocities are equal and opposite. The horizontal accelerations cancel each other. The vertical accelerations are additive.

The final position D shows that the vertical velocity is increasing in a negative direction. Since the acceleration is time rate of change velocity, the acceleration is a positive value. Again the horizontal components cancel each other.

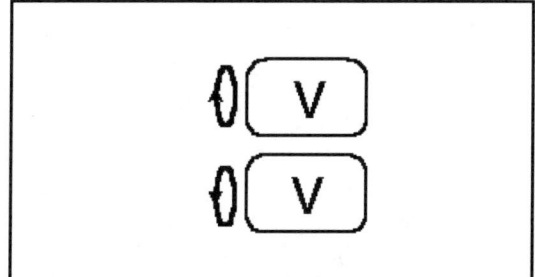

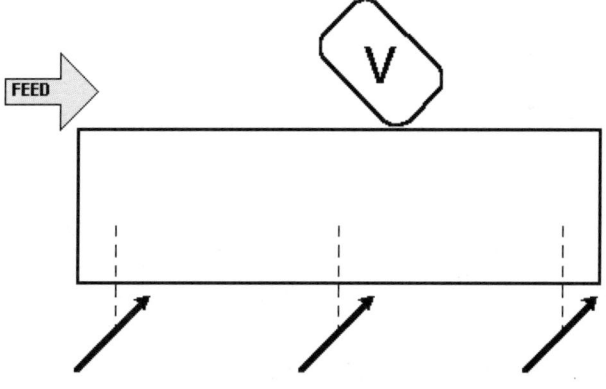

**FIGURE 3-9.** Adjustable angle drive system.

of +6°. In some cases, the beds can be tilted downward to help in situations where gumbo is encountered. These movements of bed on skid can be accomplished with mechanical, hydraulic, or combination mechanical/hydraulic systems. On some units these adjustments can be made while the unit is running. In general, fine mesh screens require an upward tilt to prevent flooding.

The ability of linear motion to convey uphill allows the use of finer shaker screens. The finer the screen, the smaller the particles that can be removed from the drilling fluid. Hence, a solids control system that uses fine screen linear motion shakers will better maintain the drilling fluid and improve efficiency of downstream equipment such as hydrocyclones and centrifuges. When screens are tilted too much uphill, many solids are ground to finer sizes as they are pounded by the screen. This tends to increase—not decrease—the solids content of the drilling fluid.

When linear motion shale shakers were introduced, other solids removal equipment was sometimes erroneously eliminated. While for a short time this appeared to be a solution, particle size analyses proved the need for downstream equipment. Linear motion shale shakers should not be expected to replace the entire solids removal system).

## Balanced Elliptical Motion

**Balanced elliptical motion** was introduced in 1992 and provides the fourth type of shale shaker motion. With this type of motion, all of the ellipse axes are sloped toward the discharge end of the shaker screen. Balanced elliptical motion can be produced by a pair of eccentrically weighted, counter-rotating parallel vibrators of different masses. This motion can also be produced by a pair of eccentrically weighted, counter-rotating vibrators that are angled away from each other (Figure 3-10). The ellipse aspect ratio (major axis to minor axis) is controlled by the angle between vibrators or by different masses of the parallel vibrators.

The greater the minor axis angle, or angle of vibrators relative to each other, the broader the ellipse and the slower the solids conveyance. A thin ellipse with a ratio of 3.5 will convey solids faster than a fat ellipse with a ratio of 1.7. Typical operating range is 1.5 to 3.0, with the lower numbers generating slower conveyance and longer screen life.

Balanced elliptical motion shale shakers can effectively remove gumbo if they are sloped downward toward the discharge end, just like linear motion. The increased physical size of these units (and an accompanying increase in deck screen surface area) allows the use of even finer screens than those used on circular or elliptical motion shakers.

In conventional unbalanced elliptical and circular motion designs, only a portion of the energy transports the cuttings in the proper direction, toward the discharge end. Balanced elliptical motion continues transporting the cuttings toward the discharge end of the screen in the same manner as linear motion. Balanced elliptical motion provides positive conveyance of solids throughout the vibratory cycle.

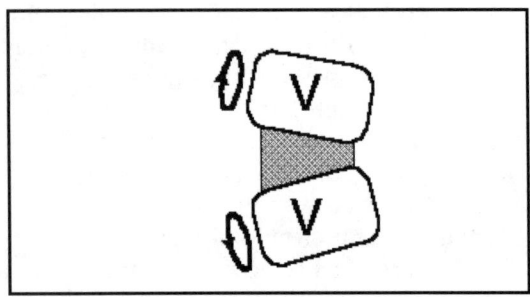

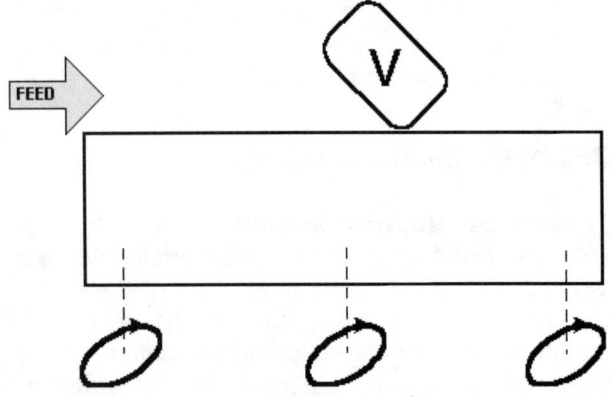

**FIGURE 3-10.** Balanced elliptical motion system.

## VIBRATING SYSTEMS

The type of motion imparted to the shaker depends on the location, orientation, and number of **vibrators** used. In all cases, the correct direction of rotation must be verified.

- Circular motion shakers use a single vibrator mounted at the shaker's center of gravity. Belt-driven vibrators and hydraulic-drive vibrators are used for this shaker design.

- Unbalanced elliptical motion shakers use a single vibrator mounted above the shaker's center of gravity. Integral vibrators, enclosed vibrators, and belt-driven vibrators are used for this shaker design.

- Linear motion shakers use two vibrators rotating in opposite directions and mounted in parallel, but in such a manner that the direction and angle of motion desired is achieved. Integral vibrators, enclosed vibrators, belt-driven vibrators, and gear-driven vibrators are used for this shaker design.

- Balanced elliptical motion shakers use two vibrators rotating in opposite directions but at a slight angle to each other so that they are not parallel. These vibrators must be oriented in such a manner that the desired direction and angle of motion is achieved. The motion elliptical traces must all lean toward the discharge end and not backwards toward the possum belly. If two vibrators of different masses are mounted in the same manner as the linear motion vibrators (i.e., parallel), a balanced elliptical motion may also be achieved.

There are various vibrating systems used on shakers. These systems include:

1. *Integral vibrators*—the eccentrically-weighted shaft is an integral part of the rotor assembly in that it is entirely enclosed within the electric motor housing.

2. *Enclosed vibrator*—a double-shafted electric motor that has eccentric weights attached to the shaft ends. These weights are enclosed by a housing cover attached to the electric motor case.

3. *Belt-driven vibrator*—the eccentrically-weighted shaft is enclosed in a housing and a shieve is attached to one end. A shieved electric motor is used to rotate the shaft with a drive belt. The electric motor may be mounted along side,

above, or behind the shaker depending on the model. It may also be mounted on the shaker bed along with the vibrator assembly.

4. *Dual-shafted, belt-driven vibrator*—this system is similar to the above description except that it has two vibrator shafts rotating in opposite directions and driven by one electric motor with a drive belt.

5. *Gear drive*—a double-shafted, electric motor drives a sealed gear box, which in turn rotates two vibrator shafts in opposite directions.

6. *Hydraulic drive*—a hydraulic drive motor is attached directly to a vibrator shaft, which is enclosed in a housing. The hydraulic motor must have a hydraulic power unit that includes an electric motor and a hydraulic pump. The hydraulic-drive motor powers the vibrator shaft.

## DECK DESIGN

Hook strip screens have been mounted with both underslung and overslung supports. Some previous generations owf oil field shaker designs used screens that were "underslung," or "pulled" up, from the bottom of a group of support or "bucker" bars (Figure 3-11). These support bars would divide the flow of material down the screen. Problems were occasionally experienced when solids were trapped in areas beneath the rubber bar supports.

Some linear motion shale shakers used "overslung" screens (Figure 3-12). With this approach, screens were attached to the bed of the shaker by pulling the screen down onto the bed from the top. This resulted in a screening area completely free of obstacles. Modern shale shaker bed design has also increased the number of support ribs located beneath the screen to aid in fine mesh support and to reduce the amount of "**crown**," or "**bow**," necessary to properly tension screen panels. Occasionally, fluid traveled from the high, center of the screen and flowed down the sides.

Most circular motion shale shakers were built with a double deck, which means fluid flowed over and through the top screen onto a finer mesh screen immediately below. This design led to some operational problems because the bottom screen was not easily visible. Generally, a flashlight was needed to inspect the lower screen, and a torn screen could remain in operation for a long time before it was noticed and changed. This created problems with solids removal because the bottom screen did not provide the intended finer screening. Some manufacturers installed backflow pans beneath the top screen to direct the flow through the entire surface area of the lower screen, which made it even more difficult to see the bottom screen.

Most manufacturers of linear motion shakers have adopted a single-deck design. These units have clear visibility for ease of care and maintenance. This unobstructed approach also makes screen changing much easier. The fluid pool tends to obscure any torn screens until drill pipe connections are made. Therefore, a torn screen on a single-deck shaker reduces solids removal efficiency until a new screen is installed.

Crews need to be alert to torn screens regardless of what type shaker is used. This is especially true during slow drilling when drill pipe connections are infrequent. When riser-assist pumps are used, flow should be periodically directed to different shakers during connections. This allows screens to be properly inspected and replaced, if needed.

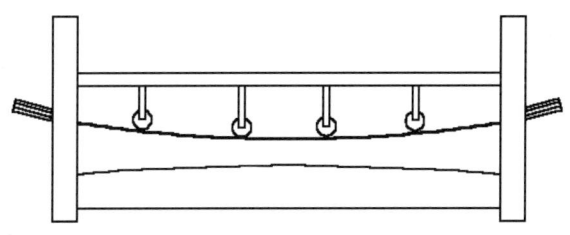

**FIGURE 3-11.** "Underslung" or "pulled up" deck support system.

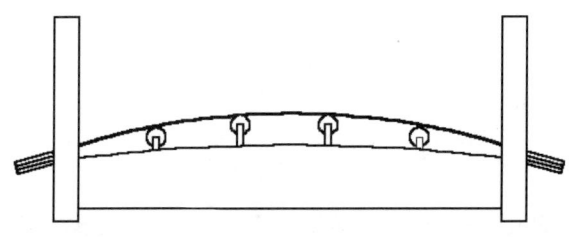

**FIGURE 3-12.** "Overslung" deck support system.

## "G"- FACTOR

The "G"-factor refers to the ratio of an acceleration to Earth's gravitational acceleration. For example, a person on Earth that weighs 200 pounds would weigh 600 pounds on Jupiter because Jupiter is much larger than Earth. A person's mass remains the same on Earth or Jupiter, but weight is a force and depends on the acceleration of gravity. The gravitational acceleration on Jupiter is three times the gravitational acceleration on Earth. Therefore, the "G"-factor is 3.

The term "G"-force is sometimes used incorrectly to describe a "G"-factor. In the above example, the "G"-force on Earth would be 200 pounds and the "G"-force on Jupiter would be 600 pounds.

## Calculating the "G"-Factor

Accelerations are experienced by an object or mass rotating horizontally at the end of a string. A mass rotating around a point with a constant speed has a centripetal acceleration ($C_a$), which can be calculated from the equation:

$$C_a = r\omega^2$$

where r is the radius of rotation
w is the angular velocity in radians per second

This equation can be applied to the motion of a rotating weight on a shale shaker to calculate an acceleration. The centripetal acceleration of a rotating weight in a circular motion, with a diameter (or stroke) of 2r in inches, rotating at a certain rpm (or $\omega$), can be calculated from the above equation.

$$(C_a) = \left(\frac{1}{2}\right)(\text{diameter})(\omega)^2$$

$$C_a = \frac{1}{2}(\text{stroke, in.})\left(\frac{1\,\text{ft}}{12\,\text{in.}}\right)$$

$$\left[\text{RPM}\left(\frac{2\pi\,\text{radians}}{\text{revolution}}\right)\left(\frac{1\,\text{min}}{60\,\text{sec}}\right)\right]^2$$

Combining all of the conversion factors to convert the units to feet per second squared:

$$C_a(\text{in ft/sec}^2) = \frac{(\text{stroke, in.})}{(\text{rpm})^2}$$

Normally this centripetal acceleration is expressed as a ratio of the value to the acceleration of gravity:

$$\text{Number of G's} = \frac{C_a}{32.2\,\text{ft/sec}^2}$$

$$= \frac{(\text{stroke, in.}) \times (\text{rpm})^2}{70490}$$

Shale shakers are vibrated by rotating eccentric masses. A tennis ball rotating at the end of a 3-foot string, and a 20-pound weight rotated at the same RPM at the end of a 3-foot string, will have the same centripetal acceleration and the same "G"-factor. Obviously, the centripetal force, or the tension in the string, will be significantly higher for the 20-pound weight.

The rotating eccentric weight on a shale shaker is used to vibrate the screen surface. The vibrating screen surface must transport solids across its surface for discard and allow fluid and solids smaller than the screen openings to pass through to the mud tanks. If the rotating weights rotate at a speed or vibration frequency that matches the natural frequency of the basket containing the screen surface, the amplitude of the basket's vibration will continue to increase and the shaker will be destroyed. This will occur even with a very small rotating eccentric weight. For example, consider a child in a swing on a playground. The application of a small force each time the swing returns to full height (amplitude), soon results in a very large amplitude. In this case, the "forcing function" (the push each time the swing returns) is applied at the natural frequency of the swing.

When the forcing function is applied at a frequency much larger than the natural frequency, the vibration amplitude depends on the ratio of the product of the unbalanced weight (w) and the eccentricity (e) to the weight of the shaker's vibrating member (W). This can be expressed as:

$$\text{Vibration Amplitude} = \frac{we}{W}$$

The vibration amplitude is one half of the total stroke length.

The peak force, or maximum force, on a shaker screen can be calculated from Newton's second law of motion:

$$\text{Force} = \frac{W}{g}a$$

where a is the acceleration of the screen.

For circular motion, the displacement is described by the equation:

$$x = X \sin \frac{Nt\pi}{30}$$

The velocity is the first derivative of the displacement $dx/dt$ and the acceleration is the second derivative of the displacement $d^2x/dt^2$. This means that the acceleration would be:

$$a = \left(\frac{N}{30}\right)^2 (\pi)^2 X \sin\left[(\pi)^2 \frac{Nt}{30}\right]$$

The maximum value of this acceleration occurs when the sine function is one. Since the displacement (X) is proportional to the ratio of we/W for high vibration speeds, the peak force, in pounds (from the peak acceleration), can be calculated from the equation:

$$F = \frac{weN^2}{35,200}$$

Therefore, the force available on the screen surface is a function of the unbalanced weight (w), the eccentricity (e), and the rotation speed (N).

Stroke length for a given design depends on the amount of eccentric weight and its distance from the center of rotation. Increasing the weight eccentricity and/or the rpm increases the **"G"-factor**. The "G"-factor is an indication of only the acceleration of the vibrating basket and not necessarily its performnce. Every shaker design has a practical "G"-factor limit. Most shaker baskets are vibrated with a five horsepower or smaller motor and produce 2 to 7 G's of thrust to the vibrating basket.

**Conventional shale shakers** usually provide a "G"-factor of less than 3; **fine screen shale shakers** usually provide a "G"-factor between 4 and 6. Some shale shakers can provide as much as 8 G's. The higher the "G"-factor, the greater the solids separation possible and, generally, the shorter the screen life. The higher the solids capacity, the less tendency there is for the screens to blind.

In unbalanced elliptical and circular vibration motion designs, only a portion of the energy transports the cuttings in the proper direction. The remainder is lost, due to the peculiar shape of the screen bed orbit, and is manifested as solids becoming nondirectional or traveling the wrong direction on the screen surface. Linear motion designs provide positive conveyance of solids throughout the vibratory cycle because the motion is straight-line rather than elliptical or circular.

Generally, the acceleration forces perpendicular to the screen surface are responsible for the liquid and solids passing through the screen, or the liquid capacity. The acceleration forces parallel to the screen surface are responsible for the solids transport, or the solids capacity.

On a linear motion shaker, the motion is generally at an angle to the screen. Usually, the two rotary weights are aligned so that the acceleration is 45° to the screen surface. The higher liquid capacity of linear motion shale shakers for the same mesh screens on unbalanced elliptical or circular motion shakers is primarily related to the fact that a pool of drilling fluid is created at the entry end of the shale shaker. The linear motion moves the solids out of the pool, across the screen, and off the end of the screen.

On a linear motion shaker with a 0.13-in. stroke at 1500 rpm, the maximum acceleration is at an angle of 45° to the shale shaker deck. The "G"-factor would be 4.15. The acceleration is measured in the direction of the stroke. If the shale shaker deck is tilted at an upward angle of 5° from the horizontal deck, the stroke remains the same. The component of the stroke parallel to the screen transports the solids up the 5° incline.

## Relationship of "G"-factor to Stroke and Speed of Rotation

An unbalanced rotating weight vibrates the screen deck. The amount of unbalanced weight, combined with the speed of rotation, will yield the "G"-factor imparted to the screen deck. The stroke is determined by the amount of unbalanced weight and its distance from the center of rotation and the weight of the shale shaker deck. (This assumes that the vibrator frequency is much larger than the natural frequency of the shaker deck.) The stroke is independent of the rotary speed.

The "G"-factor can be increased by increasing the stroke, rpm, or both. It can also be decreased in the same manner. The stroke must be increased by the inverse square of the rpm reduction to hold the "G"-factor constant. Examples are given in Table 3-1 to hold 5 G's constant while varying the stroke length at different values of rpm. The shaker manufacturers select the combination that appears to work best for their machine.

## POWER SYSTEMS

The most common power source for shale shakers is the rig electrical power generator system.

**TABLE 3-1**

| | | | |
|---|---|---|---|
| 5 G's | @ 0.44" stroke at 900 rpm | 4 G's | @ 0.35" stroke at 900 rpm |
| 5 G's | @ 0.24" stroke at 1200 rpm | 4 G's | @ 0.20" stroke at 1200 rpm |
| 5 G's | @ 0.16" stroke at 1500 rpm | 4 G's | @ 0.13" stroke at 1500 rpm |
| 5 G's | @ 0.11" stroke at 1800 rpm | 4 G's | @ 0.09" stroke at 1800 rpm |

The rig power supply should provide constant voltage and frequency to all electrical components on the rig. Most drilling rigs generate 460 VAC, 60 Hz, 3-phase power or 380 VAC, 50 Hz, 3-phase power. Other common voltages include 230 VAC, 190 VAC, and 575 VAC. Through transformers and other controls, a single power source can supply a variety of electrical power to accommodate the requirements of different rig components.

Shale shakers should be provided with motors and starters that match the rig generator output. Most motors are **dual-wound**. These may be wired to accommodate two voltages and starter configurations. For example, some dual-wind motors operate at 230/460 VAC, while others operate at 190/380 VAC. Dual-wind motors allow the shaker to operate properly with either power supply after relatively simple rewiring. Care must be taken, however, to make certain that the proper voltage is used. Additionally, electrical motors are designed to rotate with a specific speed. Typically the rotational speed is 1800 rpm for 60 Hz applications and 1500 rpm for 50 Hz applications.

Shale shakers use a vibrating screen surface to conserve the drilling fluid and reject drilled solids. The effects of this vibration are described in terms of the "G"-factor, or the function of the angular displacement of a screen surface and the square of the rotational speed. Angular displacement is achieved by rotating an eccentric mass. Most shale shakers are designed to be operated at a specific, fixed "G"-factor by matching the stroke to a given machine rotational speed. It follows that any deviation in speed will affect the "G"-factor and influence performance.

Deviations in speed may be caused by one or more factors. Typically, they are caused by fluctuations in voltage or the frequency of the alternating current. If the voltage drops, the motor cannot produce the rated horsepower and may not be able to sustain the velocity needed to keep the eccentric mass moving properly. Low voltage also reduces the life of electrical components. Deviations in frequency result in the motor turning faster (frequencies higher than normal) or slower (frequencies lower than normal). This directly influences rpm and shaker performance.

Slower rpm reduces the "G"-factor, causes poor separation, and poor conveyance. Faster rpm increases the "G"-factor and, although it may improve conveyance and separation, often increases screen fatigue failures. In extreme cases, higher rpm may cause structural damage to the shaker. Thus, it is important to provide proper power to the shale shaker.

For example, a shale shaker is designed to operate at 4 G's and has an angular displacement, or stroke, of 0.09 inch. This shaker must vibrate at 1750 rpm to produce 4.1 G's. At 60 Hz, the motor turns at 1750 rpm, which yields a "G"-factor of 4.1, just as designed. If the frequency drops to 55 Hz, the motor speed reduces to 1650 rpm, resulting in a "G"-factor of 3.5. Further reduction of frequency to 50 Hz results in 1500 rpm and a "G"-factor of 2.9.

Most rigs provide 460 VAC, 60 Hz power and most shale shakers are designed to operate with this power supply. However, many drilling rigs are designed for 380 VAC, 50 Hz electrical systems. To provide proper "G"-factors for 50 Hz operations, shale shaker manufacturers rely on two methods: increasing stroke length or through the use of voltage/frequency inverters (transformers).

A motor designed for 50 Hz applications rotates at 1500 rpm. At 0.09 inch stroke a shale shaker will produce 2.9 G's. Increasing the stroke length to 0.13 inch provides 4.1 G's, similar to the original 60 Hz design. However, the longer stroke length and slower speed will produce different solids separation and conveyance performance. At the longer stroke lengths, shakers will probably convey more solids and have a higher fluid capacity. If the stroke length is not increased, some manufacturers use voltage inverters to provide 460 VAC, 60 Hz output power from a 380 VAC, 50 Hz supply.

Constant electrical power is necessary for proper, consistent shale shaker performance. Tables 3-2 and 3-3 below assist in designing a satisfactory electrical distribution system. Alternating current (AC) motors are common on most shale shakers. The motor rating indicates the amount of electrical current required to operate the motor. The values in Table 3-2 provide some guidelines for various motors. The manufacturer's recommendation should always take precedence over the generalized values in these tables. The amount of electric current that a conductor (or wire) can carry increases as the diameter of the wire increases. Common approximate values for currents are presented with the corresponding wire size designation in Table 3-3. Conductors, or even relatively large-diameter wire, still have some resistance to the flow of electric current. This resistance results in a line voltage drop. When an electric motor is located in a remote area relative to the generator, the line voltage drop may decrease the motor voltage to unacceptably low values. Wire diameter guidelines necessary to maintain the voltage drop to 3% are presented in Table 3-4.

> **WARNING**
>
> **ELECTRICAL HAZARD**— Follow ALL National Electrical Codes, Local Electrical Codes, and Manufacturer's Safety and Installation Instructions. Always conform to regulatory codes, as applicable, regionally and internationally.

**TABLE 3-2.** Electric Current Required by Motors Running at Full Load

| Motor Rating (HP) | Single Phase | | Three Phase | | Three Phase | |
|---|---|---|---|---|---|---|
| | 115 V | 230 V | 190 V | 230 V | 460 V | 575 V |
| 1 | 16 | 8 | 8 | 3.6 | 1.8 | 1.4 |
| 1.5 | 20 | 10 | 10 | 5.2 | 2.6 | 2.1 |
| 2 | 24 | 12 | 12 | 5.8 | 3.4 | 2.7 |
| 3 | 34 | 17 | 17 | 9.6 | 4.8 | 3.9 |
| 5 | 56 | 28 | 28 | 15.2 | 7.6 | 6.1 |
| 7.5 | 80 | 40 | 40 | 22 | 11 | 9 |
| 10 | 100 | 50 | 50 | 28 | 14 | 11 |
| 15 | | | | 42 | 21 | 17 |
| 20 | | | | 54 | 27 | 22 |

**TABLE 3-3.** Maximum Allowable Electric Current for Various Wire Sizes

*These values are to be used as general guidelines only. Many factors, including insulating material and temperature, control the values.*

| Current amps | 35 | 50 | 70 | 80 | 90 | 100 | 125 | 150 | 200 | 225 | 275 |
|---|---|---|---|---|---|---|---|---|---|---|---|
| Wire size AWG | 10 | 8 | 6 | 5 | 4 | 3 | 2 | 1 | 0 | 00 | 000 |

**TABLE 3-4.** Copper Wire Size Required for Limiting Line Voltage Drop to 3%

**120V, SINGLE PHASE**

| Current Amps | Wire Length (ft) | | | |
|---|---|---|---|---|
| | 50 | 100 | 150 | 200 |
| 10 | 12 | 12 | 10 | 8 |
| 20 | 12 | 8 | 6 | 6 |
| 40 | 8 | 6 | 4 | 3 |
| 60 | 6 | 4 | 2 | 1 |
| 80 | 6 | 3 | 1 | 0 |
| 100 | 4 | 2 | 0 | 00 |

**190V, THREE PHASE**

| Current Amps | Wire Length (ft) | | | |
|---|---|---|---|---|
| | 50 | 100 | 150 | 200 |
| 10 | 12 | 12 | 12 | 12 |
| 20 | 10 | 10 | 8 | 8 |
| 40 | 8 | 6 | 6 | 4 |
| 60 | 6 | 4 | 4 | 3 |
| 80 | 4 | 4 | 3 | 2 |
| 100 | 4 | 3 | 2 | 1 |

**230V, THREE PHASE**

| Current Amps | Wire Length (ft) | | | |
|---|---|---|---|---|
| | 150 | 200 | 250 | 300 |
| 10 | 12 | 12 | 12 | 12 |
| 20 | 10 | 10 | 8 | 8 |
| 40 | 8 | 6 | 6 | 4 |
| 60 | 6 | 4 | 4 | 3 |
| 80 | 4 | 4 | 3 | 2 |
| 100 | 4 | 3 | 2 | 1 |

# CHAPTER FOUR

# Shaker Applications

Most drilling rigs are equipped with at least one shale shaker. The purpose of a shale shaker, as with all drilled solids removal equipment, is to reduce the drilling cost. Most drilling conditions require limiting the quantity and size of drilled solids in the drilling fluid. Shale shakers remove the largest drilled solids that reach the surface, which are the ones that create many wellbore problems if they remain in the drilling fluid.

## SELECTION OF SHALE SHAKERS

The first consideration in selecting a shale shaker is to decide whether or not a gumbo slide, or gumbo-removal device, will be needed. This is often necessary when drilling recent sediments.

A scalping shaker must be considered next. Scalping shakers are usually needed when large quantities of drilled solids or gumbo reach the surface. Usually, long intervals of $17\frac{1}{2}$-in.-diameter holes, with flow rates above 1000 gpm, require scalping shakers in front of fine mesh screens (see Chapter 7).

The final consideration is to decide on the type and quantity of **main shakers** necessary for processing all the drilling fluid. The goal should be to sieve an unweighted drilling fluid through the finest mesh screen possible. For weighted drilling fluids, the goal should be to screen all drilling fluid through 200-mesh screens (finer screens may remove too much weighting material).

Many factors affect the liquid capacity of specific shale shaker and screen combinations. While no publication accounts for all of these variables, some manufacturers publish curves relating the fluid flow capacity to screen sizes as a function of one or two parameters. These curves are usually generated without a comprehensive testing program. Many manufacturers use generalizations to gauge the number of shakers needed based on the maximum flow rate anticipated. For example, flow rates between 300 gpm and 500 gpm can probably be processed through a 200-mesh screen on most linear motion shale shakers.

## SELECTION OF SHAKER SCREENS

Proprietary computer programs are available that reportedly allow estimating screen sizes used on specific shale shakers. Most of these computer programs are verified with data obtained from laboratory-prepared drilling fluid with limited property variation. Different drilling fluid ingredients can reduce the capacity of a shaker system and, therefore, predicting screen-mesh sizes that will handle certain flow rates is very difficult. For example, a drilling fluid containing starch is difficult to screen because starch, acting as a good filtration control additive, tends to plug fine mesh screens. Drilling fluids with high gel strengths are also difficult to screen through fine meshes. For these reasons, screen selection for various shale shakers is primarily a trial and error evaluation. The best advice is to contact the manufacturer for recommendations for various geographic areas.

### Cost of Removing Drilled Solids

Few wells can be drilled without removing drilled solids (see Chapter 1). Even for 3,000- to 4,000-foot wells, problems created by drilled solids, such as lost circulation, stuck pipe, or well control, are more than enough reason to properly process the drilling fluid. In expensive drilling operations, the proper use of solids removal equipment will significantly reduce costs.

Although drilled solids can be maintained by simply diluting the drilling fluid to control their acceptable levels or concentrations of drilled solids, the expense and impracticality of this approach are evident using the following example. A $12\frac{1}{4}$-in.-diameter hole, 1,000 feet deep, will contain approximately 146 bbls of solids. If these solids are to be reduced to a 6% volume target concentration, they must be blended into 2400 bbl ($\frac{146}{0.06}$) of slurry. To create the 2400 bbl of slurry, 2256 bbl of clean drilling fluid must be added to the 146 bbl of solids [146 bbl/2256 + 146 bbl)] = 6% volume. Not only would the cost of the clean drilling fluid be prohibitive, but most drilling rigs do not have the necessary surface volume to build 2256 bbl of clean drilling fluid for every 1,000 feet of hole drilled. (See Chapter 8 for a more complete discussion of dilution calculations.)

As demonstrated, it is important to remove as many drilled solids as possible with the shale shaker. Shakers are an important component of this process but they are only one portion of a complete drilled solids removal system. Careful attention to details is the key to developing the most efficient drilled solids removal operation. Complete processing will decrease the cost of accumulating excess drilling fluid, thereby contributing to the ultimate goal of reducing the costs associated with oil well drilling (Chapter 8).

## Specific Factors

Specific factors that should be considered when designing the shale shaker system include: flow rate, fluid type, rig space, configuration/power, available elevation, and discharge dryness (restrictions).

Most programs extrapolate laboratory-generated performance curves to predict field performance. Unfortunately, laboratory-manufactured drilling fluid does not duplicate properties of drilling fluid that has been used in a well. High shear rates through drill bit nozzles at elevated temperatures produce colloidal-size particles that are not duplicated in surface-processed drilling fluid.

***Flow rate.*** The flow rate that a particular shaker/screen combination can handle greatly depends on the flow properties of the drilling fluid. The lower the values of plastic viscosity, yield point, gel strength, and mud weight, the finer the mesh size that can be used on a shale shaker. The conductance of the shaker screen provides a guide for the fluid capability but does not reveal how the screen will actually perform. Screens with the same conductance may not be able to handle the same flow rate if used on different shale shakers.

Shaker screen selection programs have been developed to predict the quantity of solids that can be removed from a drilling fluid by various shaker screens on specific commercial shakers. Many programs start by assuming that the flow rate of drilled solids reaching the surface is identical with the generation rate of the drilled solids. Unfortunately, many drilled solids are stored in the well-bore and do not reach the surface in the order in which they are drilled. Frequently, in long stretches of open hole, as many drilled solids enter the drilling fluid from the sides of the wellbore as are generated by a drill bit.

One proposed relationship shows that the maximum flow rate that can be handled by a shaker (Q), is inversely proportional to the product of the plastic viscosity (PV), mud weight (MW), and proportional to the screen conductance (K). This relationship answers the question: If a linear motion shale shaker is handling 1250 gpm of a 10.3 ppg drilling fluid, with a PV of 10 cp on a 120-square MG mesh screen, what flow rate could be handled on a 200-square MG mesh screen if the mud weight is increased to 14.0 ppg and the PV becomes 26 cp?

$$Q_2 = \frac{(K_2)(PV_1)(MW_1)}{(K_1)(PV)(MW_2)} Q_1$$

$$Q_2 = \frac{(1.24 \text{ kd/mm})(10 \text{ cp})(10.3 \text{ ppg})}{(0.68 \text{ kd/mm})(26 \text{ cp})(14.0 \text{ ppg})} \times 1250 \text{ gpm}$$

$$Q_2 = 645 \text{ gpm}$$

The problem with this equation is that it fails to account for other rheological variables. For example, if the gel strength of the 10.3 ppg drilling fluid significantly increased, the shaker could no longer handle the fluid. To further demonstrate, take a shaker that handles 750 gpm of an 11.0 ppg drilling fluid with a certain plastic viscosity (PV). If the yield point is significantly increased, or additives such as PHPA or a high concentration of starch are added to this fluid, the shaker capacity might be only 350 gpm. In both these cases, the PV would change very little but there would be a significant effect on the screening capability. Therefore, the above equation should only be used to predict the flow rate if no other properties in the drilling fluid change other than the mud weight and plastic viscosity. The equation should be used with caution.

***Rig configuration.*** On some drilling rigs, the derrick rig floor is not high enough to allow some shale shakers to be used because the flow line is not high enough. Whichever shaker is used, consideration must be given to providing sufficient safe power to the shaker motors. It is best to check with the manufacturer concerning the electrical requirements for individual shakers.

***Discharge dryness.*** In some areas, drilled solids and drilling fluid cannot be discarded at the rig location. This applies to both land and offshore rigs. The cost of handling discarded material may require drying the discard. The fine mesh screens discharge much wetter solids than those discarded from very coarse screens (see Chapter 13).

## CASCADE SYSTEMS

The first **cascade system** was introduced to the drilling industry in the mid-1970s. A **scalper shaker** received fluid from the flow line and removed gumbo, or large drilled solids, before the fluid passed through the main shaker using a fine mesh screen (at this time, 80- to 120-mesh screens were the practical limits). The first unit combined a single-deck, elliptical motion shaker mounted directly over a double-deck, circular motion shaker (Figure 4-1). This combination was especially successful offshore where space is limited.

One advantage of multiple-deck shale shakers is their ability to reduce solids loading on the lower, fine mesh screen deck. This increases both shaker capacity and screen life. However, capacity may still be exceeded under many drilling conditions. The screen mesh and, thus, the size solids returned to the active system, is often increased to prevent loss of whole mud over the end of the shaker screens.

Processing drilling fluid through shale shaker screens, centrifugal pumps, hydrocyclones, and drill bit nozzles can cause degradation of solids and aggravated problems associated with fine solids in the drilling fluid. To remove drilled solids as soon as possible, additional shakers are installed at the flow line—sometimes, as many as six to ten parallel shakers—so that the finest mesh screen may be used. With the finer mesh screens and additional shakers in place, downstream equipment is often erroneously eliminated. It is important to remember that the improved shale shaker still remains only one component of a complete drilled solids removal system (see Chapter 7).

A system of **cascading shale shakers** is designed to use one set of screens (or shakers) to scalp large solids and gumbo from the drilling fluid, and another set of lower screens (or shakers) to receive the upper shaker underflow fluid for removal of fine solids. This combination increases the solids removal efficiency of high performance shakers, especially during fast, top-hole drilling or in gumbo-producing formations. The cascade system is used where the solids' loading exceeds the capacity of the fine mesh screen. The advantages of this cascade arrangement include:

1. Higher overall solids loading on the system
2. Reduced solids loading on fine mesh screens
3. Finer screen separations
4. Longer screen life
5. Lower fluid well costs

There are three basic designs of cascade shaker systems:

1. The separate unit concept
2. The integral unit with multiple vibratory motions
3. The integral unit with a single vibratory motion

The choice of which design to use depends on many factors, including space limitations, performance objectives, and overall cost.

### Separate Unit

The separate unit system mounts usable rig shakers (elliptical or circular motion) on stands above newly installed linear motion shakers (Figure 4-2). Fluid from the rig shakers (or **scalping shakers**) is routed to the possum belly (back tank)

FIGURE 4-1

FIGURE 4-2

of a linear motion shaker. Line size and potential head losses must be considered with this arrangement to avoid overflow and loss of drilling fluid. This design may reduce overall cost by using existing equipment. This concept, where space is available, has the advantages of highly visible screening surfaces and ease of access for repairs. The disadvantage of separate individual shakers is the space needed for mounting the shakers and the extra piping needed to tie the scalping shakers to the flow-line shakers.

### The Integral Unit with Multiple Vibratory Motions

This design type combines the two units of the separate unit system into a single, integral unit mounted on a single skid. Usually, a circular motion shaker is mounted above a linear motion shaker on a common skid (Figure 4-3). The underflow (liquid mud) from the top scalping shaker is directed through a flowback, sheetmetal pan onto the back of the bottom flow-line shaker. The primary advantages of this design are reduced installation costs and space requirements. The internal backflow pan eliminates the manifold and piping needed for the two separate units. By mounting the units above one another, normally the flow line entry height and the weir height of the mud flow on the scalping shaker, is lower than that of separate units.

The disadvantage of this design is the limited visibility of the upper and lower screens, due to the height of the scalper shaker and the minimum clearance between the upper and lower shakers. On some models, it is also difficult to perform routine maintenance and replace vibratory drive components.

### The Integral Unit with a Single Vibratory Motion

This design consists of an integral unit with a single vibratory motion, as shown in Figure 4-4. Typically, the units use linear motion shakers and incorporate a scalping screen in the upper part of the basket. The lower bed consists of a fine screen mesh, flow-line shaker unit and the upper scalper section is designed with a smaller-width bed, using coarse mesh screens. Most of the available single vibratory motion shakers use hook strip-type screens for the upper scalping deck and the lower section deck designs vary from hook strip to pretension framed screens. Compared to the cascade shaker units, this design significantly lowers the weir height of the mud inlet to the upper screening area.

Conversely, as with the cascade shaker units, the visibility and access to the fine screen deck of the single vibratory unit is limited by the slope of the upper scalping deck.

### Summary

Cascading systems use one set of shakers to scalp large solids and/or gumbo from the drilling fluid and another set of shakers to remove fine

**FIGURE 4-3**

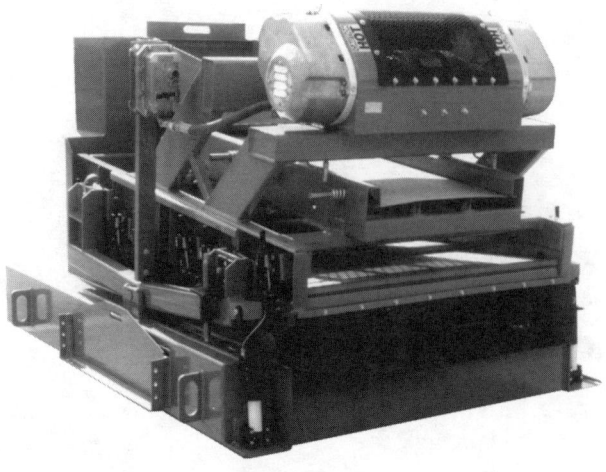

**FIGURE 4-4**

solids. Their application is primarily during fast, top-hole drilling, or in gumbo formations. This system was designed to handle high **solids loading**, which occurs when rapidly drilling large-diameter holes or when gumbo arrives at the surface.

The introduction of linear motion shale shakers has allowed development of fine screen cascade systems capable of 200-mesh separations at the flow line. Scalping screens, as coarse as 10 mesh, may be used to avoid dispersing solids before arriving at the linear motion shaker. This is particularly important in areas where high circulation rates and large amounts of drilled solids are encountered.

After either the flow rate or solids-loading is reduced in deeper parts of the bore hole, the scalping shaker should be used only as an insurance device. When the linear motion shaker, with the finest mesh screen available, can handle all of the flow and solids arriving at the surface, the need for the cascade system disappears. When this occurs, the inclination may be to discontinue the use of the scalping screen unit. However, even when the fine mesh screen can process all of the fluid, screens should be maintained on the scalper shaker for use as needed. These scalping screens can be changed to a relatively medium-coarse mesh (12 or 40 mesh) and still be used to protect the finer mesh on the main shaker.

## DRYER SHAKERS

The dryer shaker, or **dryer**, is a linear motion shaker used to minimize the volume of liquid associated with drilled cuttings discharged from the main rig shakers and hydrocyclones. The liquid removed by the dryers is returned to the active system. Because of environmental efforts to reduce liquid waste disposal, dryers were introduced with **closed loop mud systems**. Two methods are available to minimize liquid discharge: chemical and mechanical. (The chemical system, called a dewatering unit, is described in Chapter 13.) Linear motion shakers are used to mechanically reduce the liquid discard. These systems may be used independently or in combination.

The dryer is used to deliquify drilled cuttings initially separated by another piece of solids separation equipment. These drilled solids can be the discharge from a main shaker or a bank of hydrocyclones. Dryers recover liquid discharged with solids in normal liquid-solids separation that would have been previously discarded from the mud system. This liquid contains colloidal solids and the effect on drilling fluid properties must be considered since dewatering systems are frequently needed to flocculate, coagulate, and remove these solids.

The dryer family incorporates equipment long used as independent units: the main linear motion shaker, the desander, and the desilter. They are combined in several configurations to discharge their discard across fine mesh screens of a linear motion shaker to capture the associated liquid. These units, formerly identified as mud cleaners, are mounted on the mud tanks, usually in line with the primary linear motion shaker. They can also be connected to the flow line to assist with fine screening when not being used as dryers. Their pumps obtain suction from the same compartments as desanders and desilters and discharge their overflow into the proper downstream compartments.

A linear motion dryer may be used to remove the excess liquid from the main shaker discharge. The flow rate across a linear motion dryer is substantially smaller than the flow rate across the main shaker. This lower flow rate permits the use of finer mesh screens and removal of the excess fluid by the linear motion dryer. This dryer is usually mounted at a lower level than other solids separation equipment to allow gravity to transport the solids. Whether using a slide or conveyor, the cuttings are deposited into a large hopper located above the screen that replaces the **back tank** or **possum belly**. As the cuttings are transported along the screen, they are again deliquified. This excess fluid, together with the fine solids that passed through the screens, is collected in a shallow tank that takes the places of a normal sump. This liquid is pumped to a catch tank, that acts as the feed for a centrifuge, or back to the active system.

A dryer unit can be used to remove the excess fluid from the underflow of a bank of hydrocyclones (desanders or desilters). This arrangement resembles a mud cleaner system. When used in this configuration, the dryer unit may be used on either a weighted or unweighted mud system. The liquid recovered by the linear motion shaker under the hydrocyclones can be processed by a centrifuge as previously described.

The perfection of the linear motion shaker for drilling fluid use, coupled with advanced fine screen manufacturing technology, has made these dryers very efficient. In most configurations, the dryers use the same style of screens, motors, and/or motor/vibration combinations as other linear motion shakers.

Depending on the fluid, saving previously lost liquid from the discard may be financially advantageous. The dryer discard is relatively dry and, therefore, may be removed with a backhoe and dump truck rather than a vacuum truck.

It is important to remember that drilling fluid properties must be properly monitored when the recovered liquid is returned to the active system. Large quantities of colloidal solids may be recovered with the liquid and this could affect plastic viscosity, yield point, and gel strengths of a drilling fluid.

## NON-OILFIELD DRILLING USES OF SHALE SHAKERS

**Trenchless drilling** is one of the fastest growing areas for shale shaker use other than drilling oil and gas wells. Trenchless drilling encompasses several areas of use:

- *Micro-tunneling*—Micro-tunneling has become very popular in Europe and is being used more frequently in the United States. Micro tunneling is the horizontal boring of a large-diameter hole (from 27 inches up to 10 feet) while simultaneously laying pipe. Typically, this technique is used in cities when laying or replacing water and sewer pipe beneath buildings and heavily traveled roads.

  To prepare for these operations, large-diameter, vertical holes, or "caissons," are excavated allowing the drilling equipment and hydraulic rams to be set up at the desired depth. The caisson is excavated slightly below the equipment level creating a sump for the returned drilling fluid and associated drilled solids. The returns are pumped to the surface by a submersible pump to a compact solids removal system, which typically consists of a shale shaker and mud cleaner mounted over a small tank.

- *River crossing*—This technique facilitates running a pipeline under a river. First, a small-diameter hole is directionally drilled under the riverbed. The pipe for the pipeline is then attached to the end of the drill string and pulled back under the river while a larger hole is back-reamed to accomodate the pipe.

  When laying large-diameter pipelines, a substantial solids control system must be established with multiple shakers, desanders, desilters, and centrifuges. Additionally, the use of mud cleaners will reduce drilling fluid disposal volumes.

- *Road crossing*—Pipelines or cables often need to be placed beneath roads. Drilling beneath a road does not require disrupting traffic or destroying the road surface. Frequently, the hole volume is small enough that solids removal equipment is not necessary. Should drilling fluid accumulation cause a problem, or in the case of large-diameter holes and wide roadbeds, a shaker or mud cleaner is used.

- *Fiber-optic cables*—Fiber-optic cables are often required in residential or business areas where drilling fluid and drilled solids must be contained. Since these cables do not require large-diameter holes, solids control systems usually consist of only a small tank, pump package, and a small shaker.

# CHAPTER FIVE

# Shaker User's Guide

All solids removal systems should have enough shale shakers to process 100% of the drilling fluid circulating rate. For expensive wells, an evaluation should be conducted to determine the drilling fluid processing system needed to minimize drilling and disposal costs. On the basis of this evaluation, the number, type, and configuration of shaker(s) can be chosen (see Selecting Shaker Screens discussed in Chapter 4). The following guidelines address the installation, operation, and maintenance of shale shakers.

## INSTALLATION

In all cases, the owner's manual should be consulted for proper installation procedures. If unavailable, the general guidelines below may be helpful:

1. Low places in the flow line will trap cuttings. The flow-line angle should be such that solids settling does not occur. In general, establish a 1-inch drop for every 10 feet of flow line.
2. When using a back tank, also known as a possum belly, the flow line should enter the bottom to prevent solids from settling and accumulating. If the flow line enters the top of the back tank, it should be extended to within one pipe diameter of the flow line from the bottom.
3. Rig up with sufficient space and approved walkways around the shaker(s) to permit easy maintenance.
4. Branch tees, Figure 5-1, should be avoided. Solids preferentially travel in a straight path resulting in uneven solids distribution to the shale shakers.
5. Ensure equal fluid and solids distribution when more than one shaker is used (Figure 5-2).
6. Options shown in Figures 5-2 and 5-3 are better than the distribution system shown in Figure 5-1.

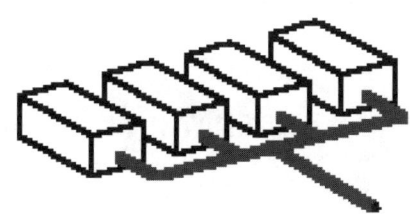

**FIGURE 5-1**

**FIGURE 5-2**

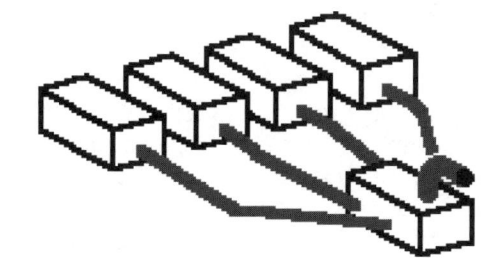

**FIGURE 5-3**

7. An optional top delivery (Figure 5-3) prevents cuttings from settling in the back tank.
8. A cement bypass that discharges outside the active system is desirable.
9. Mount and operate the shale shaker where it is level. Otherwise, both the solids and fluid limits will be reduced.
10. Motors and starters should be explosion-proof. Local electrical codes must be met. Be sure the proper sized starter heaters are used.

11. Provide the proper electrical voltage and frequency. Low line voltages reduces the life of the electrical system. Low frequency reduces the motion and lowers the capacity of the shale shaker.
12. Check for proper motor rotation.
13. Check for proper motion of the shale shaker deck.
14. Check drive belts for proper tension according to manufacturer's instructions.
15. Screens should be installed according to manufacturer's instructions.
16. Provide a wash-down system for cleaning.
17. Water-spray bars, if installed, should provide only a mist of water—not a stream of water.

## OPERATION

1. For double-deck shale shakers, run a coarser mesh screen on the top deck and a finer mesh screen on the bottom The coarser screen should be at least two standard mesh sizes coarser. Watch for a torn bottom screen and replace or patch torn screens at once. During normal drilling operations, cover at least 75% to 80% of the bottom screen with drilling fluid to maximize utilization of available screen area. Properly designed flow-back pans may improve shaker performance. (Gumbo shakers mounted above as an integral part of linear shale shakers are not called double-deck shale shakers—although the operation guidelines above still apply.)
2. For single-deck shale shakers with multiple screens on the deck, run the same mesh screens throughout. If coarser screens are necessary to prevent drilling fluid loss, run the finer mesh screens closest to the back tank. All screens should have approximately the same size openings. For example, use a combination of **market grade** (MG) 100 mesh (140 microns) and MG 80 mesh (177 microns), instead of MG 100 mesh (140 microns) and MG 50 mesh (279 microns). Under normal drilling conditions, cover at least 75% to 80% of the screen area with drilling fluid to properly utilize the screen surface area.
3. Spray bars (mist only) may be used for sticky clay to aid conveyance, which reduces whole drilling fluid loss. High-pressure washers should not be used on the screen(s) while circulating because they disperse and force solids through the screen openings. Spray bars are not recommended for weighted or oil-based drilling fluids.

4. Do not bypass the shale shaker screens or operate with torn screens; these are the main causes of plugged hydrocyclones. This results in an accumulation of drilled solids in the drilling fluid. Also, do not empty the back tank into the pits (for cleaning purposes, etc.) since this is considered a form of bypassing the shale shaker.
5. All drilling fluids that are not processed by solids removal equipment and are intended to be added to the active system, should be screened by the shale shakers to remove undesirable solids. Specifically, this includes drilling fluid delivered to a location from remote sources.

## MAINTENANCE

1. For improved screen life with nontensioned screens, ensure the components of the screen tensioning system, including any rubber supports, nuts, bolts, springs, and so forth, are installed and properly operating. Install screens according to the manufacturer's recommended installation procedure.
2. Lubricate and maintain the unit according to the manufacturer's instructions. Some units are **self-lubricating** and, therefore, this step should be skipped.
3. Check the screen tension one, three, and eight hours after installation and hourly thereafter.
4. Check the tension of drive belts and make any adjustments according to the manufacturer's instructions.
5. If only one deck of a multiple-deck shaker is used, make sure other tension rails are secured.
6. Wash screen(s) at the beginning of a trip so as not to allow fluid to dry on the screen(s). Do not empty the back tank into the active system or the sand trap below the shaker. This results in plugged hydrocyclones downstream and/or an increase in drilled solids concentration in the drilling fluid.
7. Inspect screens frequently for holes and repair or replace any that are torn.
8. Check the condition of **vibration isolators** and **screen support rubbers** and replace if they show signs of deterioration or wear.
9. Check the fluid bypass valve and the area around the shaker screens for leaks.
10. Remove drilling fluid accumulation from vibrating bed, vibrators, and motors. *Caution:*

Do not spray electrical equipment or motors with oil or water.

## GENERAL GUIDELINES

1. Shale shakers should run continuously while circulating. Cuttings cannot be separated from drilling fluid if the shaker is not in motion.
2. Drilling fluid should cover most of the screen. If the drilling fluid only covers one-fourth to one-third of the screen, the screen mesh is too coarse.
3. Torn screens should be repaired or replaced at once. Holes in panel screens can become plugged preventing cutting removal. Install screens according to manufacturer's recommended installation procedures.
4. Shaker screen replacements should be made as quickly as possible. This will decrease the amount of cuttings remaining in the drilling fluid because the shale shaker is not operational. Locate and arrange tools and screens prior to actual replacement. If possible, replace the screens during a connection. In critical situations, pumps may be shut down (or slowed) and drilling stopped while screens are being replaced.
5. Water should be added downstream and not in the possum belly (or back tank) or onto the shale shaker screen.
6. Except in cases of lost circulation, the shale shaker should not be by-passed, even for a short time.
7. The possum belly (or back tank) should not be emptied into the sand trap or mud tank system before making a trip. This will eliminate cuttings moving down the tank system and plugging desilters as the next drill bit starts drilling.

## COMPARISON AND ANALYSIS OF SHALE SHAKER PERFORMANCE

Evaluation and comparison of two or more shale shakers, although seemingly a straightforward process, is difficult to accomplish and may yield false results. The process requires a methodical procedure, accurate sampling, and control of all variables involved in the analysis.

Many reasons exist for analyzing shale shaker performance, ranging from legal analysis to objective comparison of different shakers. Regardless of the reason for analysis, the goal is to determine how the shaker and screen combination perform in a particular application. The data can be used to optimize performance of a given machine and assist in determining the number of shale shakers needed. By far the most frequent reason for analyzing performance is to compare one brand of shale shaker with another. This requires that test conditions remain constant and representative of field performance.

### Comparison Methods

Four different comparison techniques are available as shown below. Each technique has advantages and disadvantages.

|  | Laboratory | Field |
|---|---|---|
| *Offset* | ✔ | ✔ |
| *Side-by-Side* | ✔ | ✔ |

Offset comparison combines information from more than one test to determine relative performance. While this data may provide some usable information, erroneous conclusions can result because conditions between tests are difficult, if not impossible, to replicate between tests either in the laboratory or in the field.

Field tests increase the likelihood of varifiable data. Accurate sampling and analysis are essential for these evaluations. Field tests, however, are subject to the uncontrollable nature of drilling conditions. Solids arriving at the surface, even when drilling rates and formations are relatively constant, exhibit a significant variation in quantity during any particular time interval. Data replication must be demonstrated by sequentially testing the same machine three or four times, at the same flow rate, under the same conditions. If two machines are to be compared, data should alternately be collected from each machine three or four times. Experience shows significant variations exist in the rate of discarded solids between tests even though all other conditions appear constant.

***Laboratory testing.*** Laboratory conditions may increase the reproducibility of shale shaker performance since the variables are more easily controlled than in the field. Shale cuttings are virtually impossible to use in the laboratory because they deteriorate with repeated use. This deterioration of particles (shale) is an important variable in shale shaker performance. As an alternative, sand can be used, however, sand does not have similar particle sizes as shale. Additionally, the aspect ratio of sand is different than shale, therefore, their movement across the shaker will differ. Another concern with using laboratory testing is the problem with reproducing the rheological properties of drilling fluids. These properties can substantially vary from field drilling fluids even if the plastic viscosity and yield point are the same. Field drilling fluids generally have a higher percent volume of drilled solids. For this reason, field drilling fluid particle-size distributions will be skewed toward the smaller sizes when compared to laboratory-prepared drilling fluids, which will affect fluid movement across the shaker screen. Consequently, laboratory measurements of shale shaker performance can be reproduced but may have no significant relationship to field performance.

*Field testing.* Field testing requires careful planning and awareness of some variations in solids behavior. When comparing the performance of two shakers, an attempt to split the flow from the well in two equal streams for a side-by-side comparison is difficult to obtain even under the best conditions. Even a slight degree of unequal flow will cause uneven solids loading on the shakers. As an alternative, the entire circulation can be directed from one shaker to the other shaker but this procedure has its share of problems.

## Example Problems

*Problem #1.* Drilled cuttings are transported to the surface by the drilling fluid, however, the rate at which the cuttings are generated and the rate they arrive at the surface are not the same. This is caused by variations in annular flow patterns, mud temperature, cutting size and density, hole deviations, drilling methods, pipe and hole geometry, riser boost systems, and so forth, resulting in a different rate of cuttings being presented to the machines.

*Problem #2.* In most cases, the hole diameter does not equate the bit diameter. Hole enlargement is dependent on many factors, ranging from fluid type and chemistry to hydraulic and mechanical factors. Although the drilling rate may be relatively constant, this does not ensure that the cuttings volume reaching the shale shakers will be the same as the rate when the cuttings were generated.

*Problem #3.* If the flow is split equally between two shale shakers, care must be taken to ensure that both shakers receive fluid with the same amount and type of solids. Turns in the flow line or other sources of turbulence can create flow disturbances (such as pipe constrictions due to barite or drilled cuttings settling) that alter the solids distribution in a moving stream. Solids distribution not only controls the solids concentration but also the rheological properties of the drilling fluid. In the field, accurate measurements of drilled solids arriving at the shakers indicate a significant variation in quantity. Changing all of the flow from one shaker to the other eliminates many of the problems associated with the division of solids. If this method is attempted in the field, measurements must confirm that the flow stream and the fluid rheological properties are relatively the same during the testing interval. Each machine should be sampled alternately three or four times to determine the effect of the uncontrolled variation in cuttings arriving at the surface.

*Problem #4.* It is often assumed that the fluid lost with the cuttings is the same as the fluid in the circulating system. With high-speed shakers, variable deck angles, and fine screens (flat and corrugated), the fluid component discharged is generally different than the drilling fluid in the circulating system. The fluid type, pool depth, and dry beach area on the last screen panel are important factors that determine the fluid composition associated with the cuttings. All discharge from the screens must be divided into low-gravity solids, weighting material, water, and oil. The specific gravity of each must be known to use the gravimetric cuttings analysis in Chapter 10.

Because of the problems associated with cuttings collection and analysis, regular sampling and analysis of the shaker screen discharge can only be considered a worthwhile measurement when used to determine trends of solids removal. The equipment used for the collection of shale shaker discharge ranges from the very sophisticated and complex to the relatively simple and straightforward. Because of the variables discussed previously, the general consensus among participants in solids evaluation is that the results achieved with complex, expensive equipment yield as much accuracy as those achieved with simple procedures (i.e., a split PVC pipe is as accurate as an automatic sampler).

## Evaluation

Usually, tests are performed to determine the quantity of drilled solids discarded by a shale shaker. However, there are many important facets of the actual "solids" discard. If the shaker is turned off, all of the flow across the shale shaker will be discarded, as well as all of the drilled solids arriving at the shale shaker. Of course, it also discards all the drilling fluid. This defeats the purpose of the separation. The quantity of drilled solids discarded, as well as the quantity of drilling fluid, should be the criteria for evaluating shale shakers. The quantity of low-gravity solids (drilled and commercial), weighting agents (high-gravity, commercial solids), and drilling fluid should be determined when testing shale shakers.

## CONCLUSIONS

Laboratory evaluation and comparison has the advantage of maintaining duplicate conditions between tests, but usually does not replicate field conditions. Caution must be exercised when extrapolating these results to field applications.

Field evaluation and comparison of two shale shakers is extremely difficult. Any comparison requires a methodical, invariant procedure that validates the consistency of the conditions during the test.

# CHAPTER SIX

## Shale Shaker Screens

Shale shakers remove solids by processing solid-laden drilling fluid over the surface of a vibrating screen. Particles smaller than the screen openings pass through the screen along with the liquid phase of the drilling fluid. Larger particles and trapped finer particles are separated into the shaker overflow for discard.

The original criterion for early shale shaker screens was a long screen life. This demand for screen life was consistent with shaker designs and solids removal philosophies of that period. Early shale shakers could only remove large solids from the drilling fluid. The sand trap, reserve and settling pits, and downstream hydrocyclones (if used) removed the bulk of the drilled solids. Today's shale shakers are capable of using finer mesh screens that remove more solids.

For any particular shale shaker, the size and shape of the screen openings have a significant effect on solids removal. For this reason, the performance of any shaker is largely controlled by the screen cloth used. Desirable characteristics for shaker screens are:

1. Economical drilled solids removal
2. Large liquid flow rate capacity
3. Plugging and blinding resistance
4. Acceptable service life
5. Easy identification

The first four items in the above list are largely controlled by the actual screen cloth used and the screen panel technology. Improvement in shale shaker performance are a direct result of improved screen cloth and panel fabrication.

### COMMON SCREEN CLOTH WEAVES

Some of the common cloth **weaves** available to the petroleum industry are shown in Figure 6-1. The plain square weave, plain rectangular weave, and the modified rectangular weave are the only screen cloths commonly used in the drilling industry. The twilled square weave is used for coarser screens in the mining industry (Figure 6-1B). The plain dutch weave, mostly used as a filter cloth, has small, triangularly shaped screen openings that allow very little fluid to flow through (Figure 6-1C).

The plain square (Figure 6-1A) and rectangular weaves (Figure 6-1D) are simple over/under weaves in both directions. These weaves can be made from the same diameter wire in one or both directions. The **square weave** is made by making the spacing between the wires the same in both directions. The **rectangular** or **oblong weave** is made by spacing the wire in one direction longer than the wire in the opposite direction. The advantage of plain square and rectangular weaves is that they provide a flow path that has low resistance to flow.

A specialty weave screen is available (Figure 6-1E) that consists of coarse wires in the long direction and multiple groups of fine wires in the narrow direction. The long, narrow openings provide low flow resistance and remove spherical and chunky solids.

Layered screens were introduced to the industry in the late 1970s. They are often chosen because they provide a high liquid throughput and a resistance to blinding from drilled solids lodging in the openings. A layered screen is the result of two or more wire cloths overlaying each other (Figure 6-2). Both square and rectangular cloths can be layered, and reducing the diameter of the wires increases liquid throughput. A large assortment of opening sizes and shapes are produced by the multiple screen layers and the diameter of the screen wire. Because of this, a wide variety of particle sizes pass through the screen.

In 1993, a three-dimensional surface screen was introduced. The screen surface is corrugated, supported by a rigid frame for use primarily on linear motion shale shakers. As drilling fluid flows down these screens, solids are transported in the

SHALE SHAKER SCREENS 121

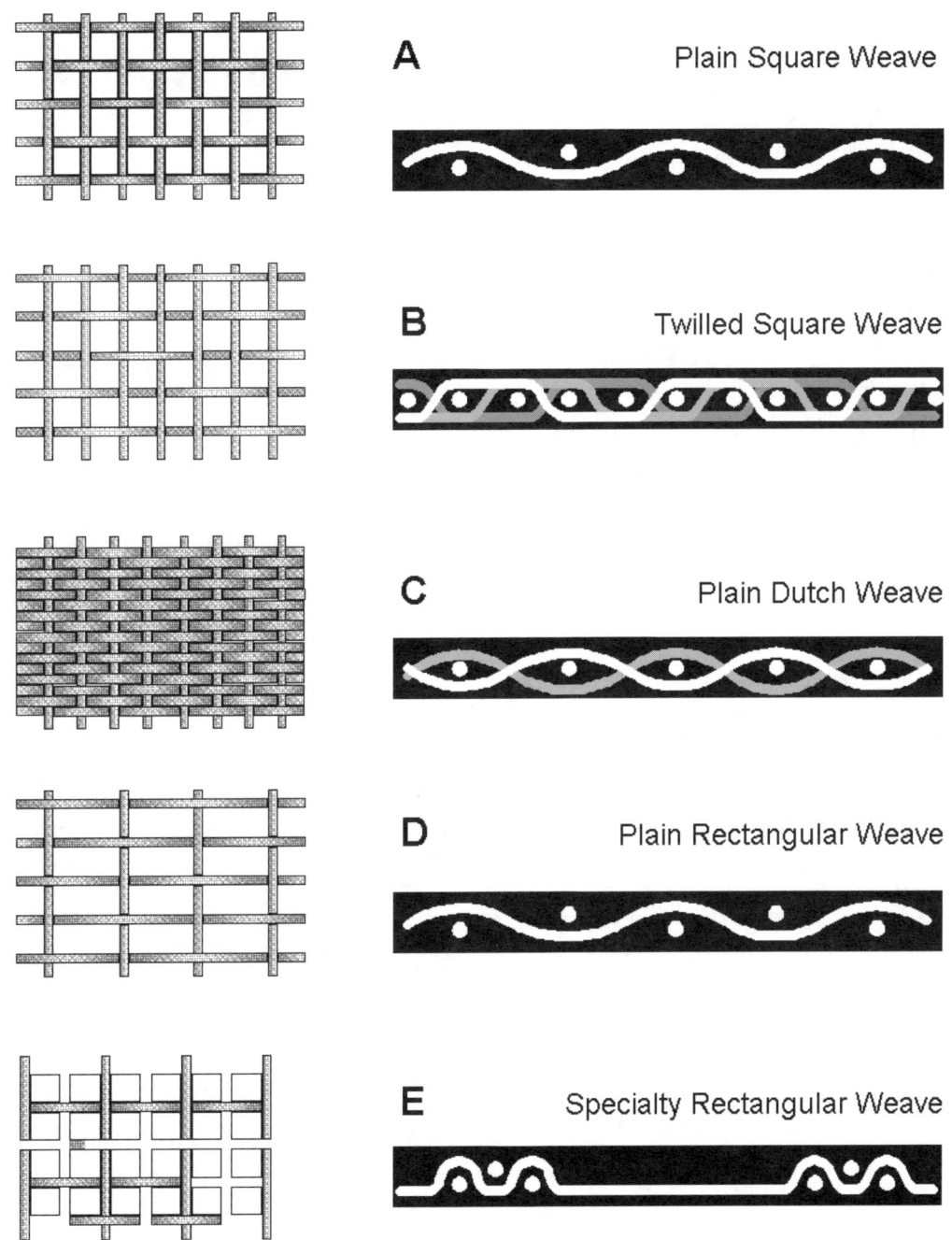

FIGURE 6-1

valleys and the vertical surfaces provide additional area for drilling fluid to pass. This increases the fluid capacity of a particular mesh size when compared with a flat surface screen.

## Screen Identification

The nomenclature used to discuss screens is important in obtaining an accurate representation of the screen performance. The following describes terms presently used within the industry, which resulted in a better screen designation system.

Plain square and rectangular weaves are often referred to by the number of wires (or openings) in each direction per linear inch. This is known as the **mesh** count. The mesh count is determined by starting at the center of one wire and counting the number of openings along the screen grid to the next wire center, one linear inch away. For

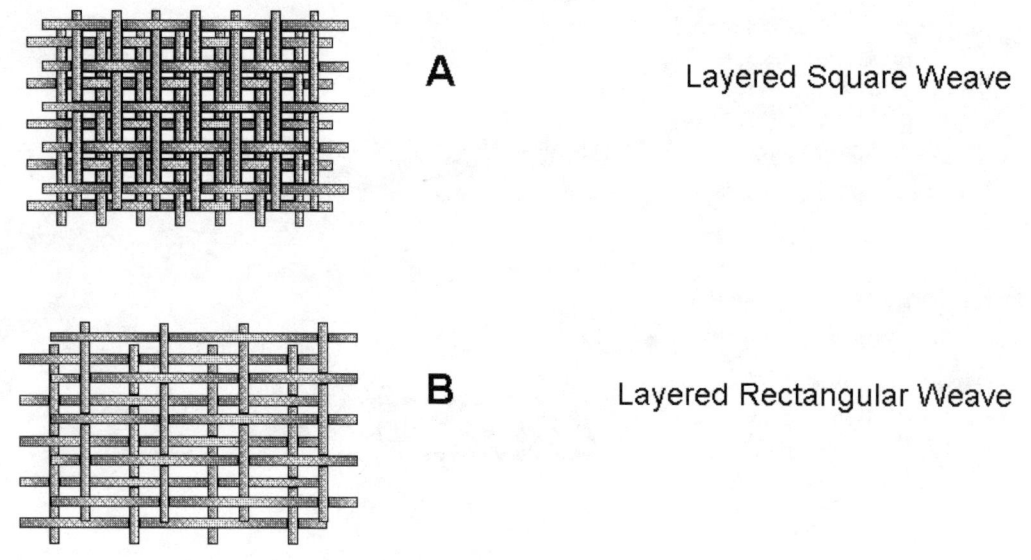

**FIGURE 6-2**

example, an 8-mesh screen has 8 openings in two directions at right angles to each other (Figure 6-3). When counting mesh, a magnifying glass scale designed specifically for this purpose is helpful.

Use of a single number for describing screens implies a square mesh. For example, "20 mesh" usually describes a screen having 20 openings per inch in either direction along the screen grid. Oblong mesh screens are generally labeled with two numbers. A "60 × 20 mesh," for example, is usually understood to have 60 openings in one direction and 20 openings per inch in the perpendicular direction. Referring to a "60 × 20 mesh" screen as an "oblong 80 mesh" is confusing and inaccurate.

The actual separation that a screen is capable of is largely determined by the size of the openings in the screen. The **opening size** is the distance between wires measured along the screen grid and is expressed in either fractions of an inch or in microns, although it is most often stated in microns. One inch equals 25,400 microns. Keep in mind, specifying the mesh count does not specify the opening size. This is because both the number of wires per inch and the size of the wires determines the opening size. If the mesh count and wire diameter are known, the opening size can be calculated as follows:

$$D = 25,400 \, [(1/n) - d]$$

where $D$ = Opening size (microns)
$n$ = Mesh count, number of wires per inch (1/inch)
$d$ = Wire diameter (inch)

The above equation indicates that screens with the same mesh count may have different size openings depending on the diameter of the wire used to weave the screen cloth. Smaller-diameter wire results in larger screen openings, thereby allowing larger particles to pass through the screen. Such a screen will pass more drilling fluid than an equivalent mesh screen made of larger-diameter wires.

In summary, specifying the mesh count of a screen does not indicate screen separation performance since screen opening size, not mesh count, determines the particle sizes separated by the screen. Because there are almost an infinite number of mesh counts and wire diameters, screen

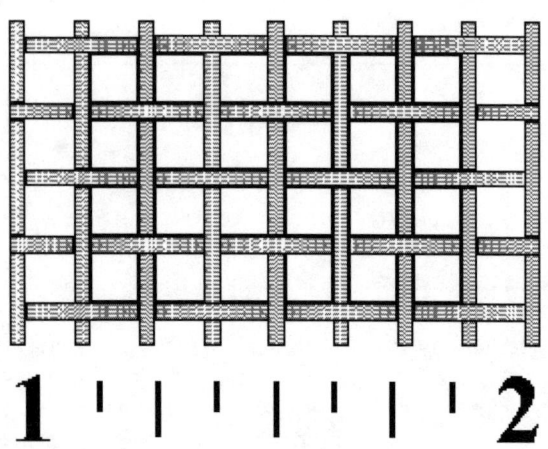

**FIGURE 6-3**

manufacturers have simplified the selection by offering several cloth series such as **market grade** (MG), **tensile bolting cloth** (TBC) and **extra-fine** (or ultra fine) **wirecloth** (XF) as shown in Tables 6-1 through 6-3. Notice in Table 6-1 that a market grade 80 cloth (MG80) has an opening size of 178 microns, whereas in Table 6-2 a tensile bolting 80 cloth (TBC80) has an opening size of 224 microns. The MG80 cloth has a smaller opening size than the TBC80 because the market grade cloths' heavier wires take up some of the opening space. As a result, a MG80 cloth can remove smaller solids than a TBC80. Furthermore, as a result of the larger wires, the MG80 cloth will be more resistant to abrasion and will last longer. The major drawback is that the MG80 cloth will allow about half the flow rate of the TBC80. As shown in Table 6-4, the screen conductance (ability to transmit fluid) for the TBC80 cloth is 3.88 kilodarcy/mm, whereas for the MG80 it is 1.91 kilodarcy/mm. Similar comparisons can be made between the separation/fluid conductance of the TBC cloths relative to the XF cloths. For instance, a single layer of XF180 screen cloth has almost the same opening size as a single layer of TBC165.

This means the XF180 screen could pass more flow. The screen life of the XF180 will most likely be shorter than the TBC165 since the wire diameter of the XF180 cloth is 30.5 microns and the TBC165 is 48.3 microns.

The National Bureau of Standards has a sieve series that is often used to describe screen opening sizes. The opening size of this test series plots uniform increments on semi-log paper, making it ideal for use in plotting particle-size distributions. Shaker screens used in the industry may be assigned an Equivalent National Bureau of Standards Sieve mesh count according to their opening sizes as shown in Table 6-1.

From the discussion above, it should be clear that mesh count alone does not specify the screen opening size. As a result, if mesh count is used, it must be accompanied by a designation of wire diameter, such as Market Grade (mesh count) + mesh count, Tensile Bolting Cloth + mesh count, or National Bureau of Standards Test Sieves equivalent mesh count.

Opening size is used to measure separation performance, but **percent open area** of a single-layered screen does not indicate liquid throughput.

**TABLE 6-1.** Market Grade Wire Cloth

| Mesh | Wire Diameter | | Opening | | Open Area (%) |
|---|---|---|---|---|---|
| | Inches | Mm. | Inches | Microns | |
| 8 × 8 | .028 | .710 | .0970 | 2464 | 60.2 |
| 10 × 10 | .025 | .640 | .0750 | 1905 | 56.3 |
| 12 × 12 | .023 | .584 | .0600 | 1524 | 51.8 |
| 14 × 14 | .020 | .508 | .0510 | 1295 | 51.0 |
| 16 × 16 | .018 | .457 | .0445 | 1130 | 50.7 |
| 18 × 18 | .017 | .432 | .0386 | 980 | 48.3 |
| 20 × 20 | .016 | .406 | .0340 | 864 | 46.2 |
| 30 × 30 | .012 | .305 | .0213 | 541 | 40.8 |
| 40 × 40 | .010 | .254 | .0150 | 381 | 36.0 |
| 50 × 50 | .009 | .229 | .0110 | 279 | 30.3 |
| 60 × 60 | .0075 | .191 | .0092 | 234 | 30.5 |
| 80 × 80 | .0055 | .140 | .0070 | 178 | 31.4 |
| 100 × 100 | .0045 | .114 | .0055 | 140 | 30.3 |
| 120 × 120 | .0037 | .0940 | .0046 | 117 | 30.9 |
| 150 × 150 | .0026 | .0660 | .0041 | 105 | 37.4 |
| 180 × 180 | .0023 | .0584 | .0033 | 84 | 34.7 |
| 200 × 200 | .0021 | .0553 | .0029 | 74 | 33.6 |
| 230 × 230 | .0018 | .0457 | .0026 | 66 | 34.3 |
| 250 × 250 | .0016 | .0406 | .0024 | 63 | 36.0 |
| 325 × 325 | .0014 | .0356 | .0017 | 44 | 30.0 |

**TABLE 6-2.** Tensile Bolting Cloth

| Mesh | Wire Diameter | | Opening | | Open Area (%) |
|---|---|---|---|---|---|
| | Inches | Mm. | Inches | Microns | |
| 16 × 16 | .009 | .229 | .0535 | 1359 | 73.3 |
| 18 × 18 | .009 | .229 | .0466 | 1184 | 70.2 |
| 20 × 20 | .009 | .229 | .0410 | 1041 | 67.2 |
| 24 × 24 | .0075 | .191 | .0342 | 869 | 67.2 |
| 28 × 28 | .0075 | .191 | .0282 | 716 | 62.4 |
| 30 × 30 | .0065 | .165 | .0268 | 681 | 64.8 |
| 32 × 32 | .0065 | .165 | .0248 | 630 | 62.7 |
| 36 × 36 | .0065 | .165 | .0213 | 541 | 58.7 |
| 38 × 38 | .0065 | .165 | .0198 | 503 | 56.7 |
| 40 × 40 | .0065 | .165 | .0185 | 470 | 54.8 |
| 42 × 42 | .0055 | .140 | .0183 | 465 | 59.1 |
| 44 × 44 | .0055 | .140 | .0172 | 437 | 57.4 |
| 48 × 48 | .0055 | .140 | .0153 | 389 | 54.2 |
| 50 × 50 | .0055 | .140 | .0145 | 368 | 52.6 |
| 54 × 54 | .0055 | .140 | .0130 | 330 | 49.4 |
| 58 × 58 | .0045 | .114 | .0127 | 323 | 54.6 |
| 60 × 60 | .0045 | .114 | .0122 | 310 | 53.3 |
| 64 × 64 | .0037 | .094 | .0111 | 282 | 50.7 |
| 70 × 70 | .0037 | .094 | .0106 | 269 | 54.9 |
| 72 X 72 | .0037 | .094 | .0102 | 259 | 53.8 |
| 74 × 74 | .0037 | .094 | .0098 | 249 | 52.7 |
| 76 × 76 | .0037 | .094 | .0095 | 241 | 51.7 |
| 80 × 80 | .0037 | .094 | .0088 | 224 | 49.6 |
| 88 × 88 | .0035 | .089 | .0079 | 201 | 47.9 |
| 90 × 90 | .0035 | .089 | .0076 | 193 | 47.8 |
| 94 × 94 | .0035 | .089 | .0071 | 180 | 45.0 |
| 105 × 105 | .0035 | .076 | .0065 | 165 | 45.0 |
| 120 × 120 | .0025 | .064 | .0058 | 147 | 47.3 |
| 145 × 145 | .0022 | .056 | .0047 | 119 | 46.4 |
| 165 × 165 | .0019 | .048 | .0042 | 107 | 47.1 |
| 180 × 180 | .0018 | .046 | .0038 | 96 | 45.7 |
| 200 × 200 | .0016 | .041 | .0034 | 86 | 46.2 |
| 230 × 230 | .0014 | .036 | .0029 | 74 | 46.0 |

The percent open area is the portion of screen surface not blocked by wire and can be calculated as follows:

$$P = \frac{(O)(o)(100)}{(O + D)(o + d)}$$

where P = Percentage of open area
O = Length of opening in one direction along the screen grid (inches)
o = Length of opening along screen grid perpendicular to the O direction (inches)
D = Diameter of wire perpendicular to O direction (inches)
d = Diameter of wire perpendicular to the o direction (inches)

Comparing the open area with the ability of a screen to transmit fluid, a better measure is the screen's conductance (or equivalent permeability of the screen cloth). Conductance takes into account both the openings and the drag of the fluid on the wires and will be further discussed later in this chapter (Table 6-5).

**TABLE 6-3.** Ultra-Fine Bolting Cloth

| Mesh | Wire Diameter | | Opening | | Open Area (%) |
|---|---|---|---|---|---|
| | Inches | Mm. | Inches | Microns | |
| 32 | .0046 | .117 | .0267 | 678 | 72.7 |
| 38 | .0046 | .117 | .0217 | 551 | 68.1 |
| 50 | .0039 | .099 | .0161 | 409 | 64.8 |
| 58 | .0036 | .091 | .0136 | 345 | 62.6 |
| 70 | .0030 | .076 | .0113 | 287 | 62.4 |
| 84 | .0025 | .064 | .0094 | 239 | 62.4 |
| 100 | .0022 | .056 | .0078 | 198 | 60.8 |
| 130 | .0017 | .043 | .0060 | 152 | 60.7 |
| 160 | .0014 | .036 | .0049 | 124 | 60.2 |
| 180 | .0012 | .030 | .0044 | 112 | 61.5 |
| 220 | .0011 | .028 | .0034 | 86 | 57.5 |
| 250 | .0011 | .028 | .0029 | 74 | 52.6 |
| 270 | .0011 | .028 | .0028 | 71 | 49.4 |
| 325 | .0011 | .028 | .0020 | 51 | 41.3 |

For many years, there has been a great deal of confusion over screen designations. Mesh count and percent open area did not adequately quantify screen cloth performance, and deceptive marketing practices were common. Furthermore, with the advent of the layered cloths that have a range of hole sizes, there were no standards for comparison.

## Cut Points

In general, screens on shale shakers reject solids larger than their opening sizes and retain the drilling fluid and smaller solids. Drilling fluid properties, as well as screen conditions, may affect screen performance. For example, high gel strengths and high surface tensions tend to bridge small screen openings and prevent screens from passing small solids and liquid; filtration-control additives, such as starch, tend to plug screen openings and prevent small solids and liquid from passing; and in an oil-based drilling fluid, water-wet, fine-mesh screens may reject a large portion of the drilling fluid flowing onto the screen.

When 50% of the mass of a particular solid size is found in the underflow of a screen and 50% of the mass of that size is found in the overflow, that size is said to be the D50 or 50% **cut point**. Cut point curves, or a percent separated curve, is a graphical representation of the actual measured separation of solids made by the screen. For example, a D20 cut point would be the size where 20% of the mass of solids of that size are returned to the drilling fluid (pass through the shaker screen) and 80% of the mass of that size solid is rejected from the system (discarded). The measurements required to determine cut point curves are flow rate, percent solids, and particle-size distribution on the feed and discard flow streams. Alternatively, the underflow and discard flow streams can be used. The mass, or weight of solids, in these two streams can be combined to mathematically represent the input stream. In order to ensure that the measurements are accurate, it is best to measure all three flow streams and check continuity. Chapter 9 provides the calculations that must be made to obtain percent separated, or cut point, curves.

Drilled solids enter the wellbore from the sides of the hole (represented by the largest solids) where the drill bit is drilling the rock (represented by the smaller solids), and all drilled solids are subjected to deterioration as they move up the annulus (represented by the various ranges of sizes).

For illustration purposes, assume that six different size spheres reach the surface and present to a very coarse mesh screen. The dimensions discussed are in inches instead of microns to assist visualization. The number of spheres in each size category is shown in Table 6-6.

The total volume of spheres in Table 6-6 represents approximately 252 gal of solids. This is equivalent to less than 100 feet of a 10-inch diameter hole. In a wellbore, the mass of the smaller sizes can and usually do exceed the mass of the larger sizes. Table 6-6 shows the proposed distribution of spheres to be presented to a very coarse

**TABLE 6-4.** Typical Market Grade and Tensile Bolting Cloth Shaker Screen Characteristics

|  | Screen Designation | | Separation Potential in Microns | | | Conductance |
|---|---|---|---|---|---|---|
|  | Mesh | $d_{16}$ | $d_{50}$ | $d_{84}$ | | Kd/mm |
| Market Grade | 10×10 | 1678 | 1727 | 1777 | | 49.68 |
|  | 20×20 | 839 | 864 | 880 | | 15.93 |
|  | 30×30 | 501 | 516 | 531 | | 8.32 |
|  | 40×40 | 370 | 381 | 392 | | 4.80 |
|  | 50×50 | 271 | 279 | 287 | | 2.88 |
|  | 60×60 | 227 | 234 | 241 | | 2.40 |
|  | 80×80 | 172 | 177 | 182 | | 1.91 |
|  | 100×100 | 136 | 140 | 144 | | 1.44 |
|  | 120×120 | 114 | 117 | 120 | | 1.24 |
|  | 150×150 | 102 | 105 | 108 | |  |
|  | 200×200 | 72 | 74 | 76 | | 0.68 |
|  | 250×250 | 59 | 62 | 63 | | 0.58 |
|  | 325×325 | 43 | 44 | 45 | | 0.44 |
| Tensile Bolting Cloth | 20×20 | 1011 | 1041 | 1071 | | 40.93 |
|  | 30×30 | 662 | 681 | 700 | | 24.33 |
|  | 40×40 | 457 | 470 | 483 | | 11.63 |
|  | 50×50 | 357 | 368 | 379 | | 7.94 |
|  | 60×60 | 301 | 310 | 319 | | 5.60 |
|  | 70×70 | 261 | 269 | 277 | | 5.25 |
|  | 80×80 | 218 | 224 | 230 | | 3.88 |
|  | 94×94 | 175 | 180 | 185 | | 2.84 |
|  | 105×105 | 160 | 165 | 170 | | 2.77 |
|  | 120×120 | 143 | 147 | 151 | | 2.51 |
|  | 145×145 | 116 | 119 | 122 | | 2.03 |
|  | 165×165 | 104 | 107 | 110 | | 1.86 |
|  | 200×200 | 84 | 86 | 88 | | 1.49 |
|  | 230×230 | 72 | 74 | 76 | | 1.30 |

**TABLE 6-5.** Comparison of Open Area with Conductance

| Market Grade Screen Mesh | Conductance (Kd/mm) | Open Area (%) |
|---|---|---|
| 50 × 50 | 2.88 | 30.3 |
| 100 × 100 | 1.44 | 30.3 |
| 150 × 150 | 0.68 | 33.6 |
| 200 × 200 | 0.58 | 36 |

**TABLE 6-6**

| Sphere diameter (inches) | 2.0 | 1.5 | 1.0 | 0.5 | 0.25 | 0.125 |
|---|---|---|---|---|---|---|
| Number of spheres | 28 | 130 | 880 | 14,100 | 226,000 | 3,610,000 |
| Volume of spheres | 4 | 8 | 16 | 32 | 64 | 128 |
| Mass of solids (lb) | 88 | 177 | 354 | 707 | 1414 | 2828 |

screen. The mass distribution of these solids will be used to create a cut point curve for this screen. Four gallons of 2.0-inch diameter spheres would have a mass of 88 lb. The 128 gal of $\frac{1}{8}$-inch diameter spheres would represent 2828 lb of solids. The number of spheres increases from 28 to over 3 million. This also is similar to a distribution of solids in a wellbore. During transport to the surface, solids grind smaller increasing plastic viscosity and sometimes yield point. Plastic viscosity indicates liquid phase viscosity and size, shape, and number of solids. Plastic viscosity can increase without an increase in solids content because of drilled solids degradation. (Usually, this indicates the need for a centrifuge.)

The solids concentration traveling up an annulus is assumed to be 10% volume, these 252 gal of spheres would be contained in 2520 gal of drilling fluid. For illustration purposes, the flow rate onto the screen will be assumed to be 400 gal/min and the flow rate off the screen will be assumed to be 20 gal/min. The screen would be presented with all of the solids in 6.3 min. For this illustration, the screen will be assumed to have $\frac{3}{8}$-inch square openings. The screen discard would consist of spheres larger than the opening size as well as liquid that wet these spheres. Drilling fluid clings to the screen discard and small solids also adhere to the large solids discarded from a rig shaker. Actually, some of the large solids find their way through or around shaker screens and appear in the retained drilling fluid. (The hydrocyclone underflow opening is frequently plugged even downstream from a fine-mesh shaker screen. This indicates that large solids are entering the drilling fluid system.)

To obtain a cut point curve for the spheres arriving at the surface, the mass of each size of spheres in the discard would be compared to the mass of that size presented to the screen. Table 6-7 presents the mass captured in the discard.

The cut point curve, plotted in Figure 6-4, indicates that the D 50 cut point would be around $\frac{3}{8}$-inch. This means that 50% of that size particle would be found in the discard and 50% would be found in the retained fluid. Some spheres smaller than the screen mesh are discarded even though they could pass through the screen. In reality, screen discards contain many solids that could pass through the shaker screen. Liquid drilling fluid (whole mud) is carried off the shaker screen with the discard. Finer mesh screens (150 or 200 mesh screens) discard more drilling fluid (or whole mud) than coarser mesh screens (such as 40 to 80 mesh).

A detailed procedure for determining cut points is presented in Chapter 9.

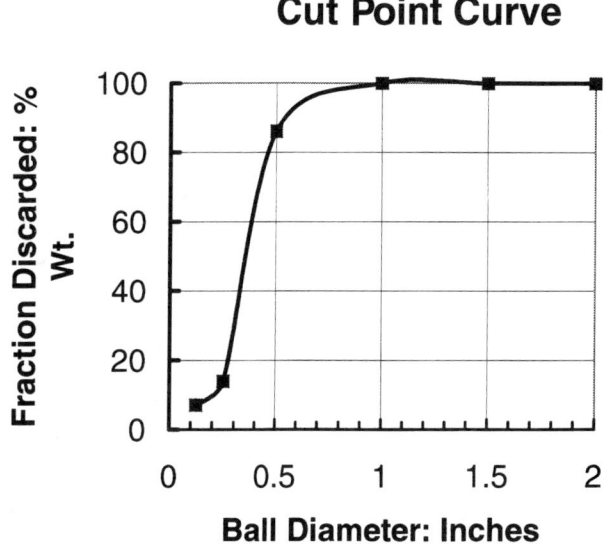

**FIGURE 6-4**

**TABLE 6-7.** Data for Cut Point Curve

| Size (in.) | Mass Presented (lb) | Mass Discarded (lb) | Fraction Discarded (% wt) |
|---|---|---|---|
| 1/8 | 2828 | 226 | 8 |
| 1/4 | 1414 | 212 | 15 |
| 1/2 | 707 | 600 | 85 |
| 1 | 354 | 354 | 100 |
| 1.5 | 177 | 177 | 100 |
| 2.0 | 88 | 88 | 100 |

## Causes of Premature Screen Failure

Several factors may contribute to premature screen failure. Most failures result from improper screen installation or damage to the shaker itself. Cracked or warped shaker beds, which may result from many years of continuous use or improper maintenance, will cause poor vibration patterns. This may cause improper solids conveyance, which, in turn, may cause solids to gather on certain areas of the screen, wearing holes in that section. Damaged beds may also affect the tensioning ability of the tension system, inducing flexure in the screen. This increase in flex causes the screen itself to vibrate separately from the basket against the screen support stringers, damaging the spot on the screen where this is occurring.

An increase in screen flexure ultimately results in most cases of early screen fatigue. All screen tensioning components must be in proper working order to eliminate screen flexure and maximize screen life. Some of the screen tensioning system materials that must be maintained include the cross and side supports, channel rubbers, and tension bolts. As prolonged use of the shakers continues, the support rubbers—rubber liners that cover the support stringers—will begin to wear. In order for the support rubbers to tension the screens properly, they must all be the same thickness; however, this is rarely the case once these rubbers begin to wear. Flexure develops in the areas where the greatest amount of wear has occurred on the rubbers, reducing the screen life. The side and cross supports—fiberglass strips on which the screens rest along the inside of the shaker bed—will wear in a similar manner. This interferes with the ability of the bolts to apply the proper amount of tension on the screens, which will again cause loose screens and rapid failure. Bent steel supports that interlock with hook strips on the screens to fasten the screens directly to the shaker bed—will not allow tension to be applied evenly throughout the full length of the screen, also resulting in early screen failure.

To achieve maximum screen life, all tension bolts must be operating properly. If early screen failure occurs, check to make sure that one or more of the tension bolts are not missing and that they are tightened correctly. The tension bolts should be tightened to manufacturer's recommendations.

Before installing any screens, the shaker bed must be washed clean of any debris. Proper tensioning of the screen cannot be achieved if any substance comes between the screen and the bed. Improper installation or maintenance of the tensioning devices results in premature screen failure.

Excessive solids accumulation in conjunction with poor solids conveyance, causes increased wear on the screens where the accumulation is occurring. This problem may arise, particularly with the three-dimensional screen, if the screen is not in alignment. Where misalignment occurs, solids tend to accumulate and wear screens in that area.

Another possible cause of improper solids conveyance, and therefore, screen wear, is the linear motion vibrators running in the same direction, which causes an improper vibration pattern. This results in massive amounts of solids to accumulate on the first screens, causing them to wear quickly. This can easily be remedied by reversing the electric wiring to the motors. The vibrators of the shakers should be tested before spudding the well. Also, be certain that both vibrators are operating. If not, replace the inoperable vibrator.

## API Recommended Practice for Designations of Shale Shaker Screens

Shale shaker screens made of two or three layers of screen cloth of different mesh size presents openings that historically have not been easy to characterize. A technique to describe these openings

---

### Screen Failure Causes

**A. Within one hour**
1. Heavy solids loading
2. Vibrators operating improperly
3. Coarse formations on fine screen cloths

**B. Within several hours**
1. Missing tension bolts or Ramp-loks®
2. Vibrators operating improperly
3. Damaged shaker bed
4. Missing channel rubbers

**C. Within one day**
1. Damaged shaker bed
2. Inoperable vibrator
3. Extremely heavy mud weights
4. Misalignment of screens
   (with three-dimensional screens)

**D. Within several days**
1. Hematite used as a weighting agent
2. Worn channel rubbers
3. Worn side or cross supports
4. Bent drawbars

has been adopted by the API as a Recommended Practice for Designations of Shale Shaker Screens, API RP 13E, 3rd edition, May 1, 1995. This recommended practice supersedes the second edition (1985) which was valid only for single-layer screens.

The new designation system conveys information on the screen-opening size distribution and the ability of a nonvibrating screen to pass fluid. The following information is stamped on a tag attached to the screen panel, which is visible and legible after the screen is installed on the shale shaker:

>  Manufacturer's Designation
>  Separation Potential
>  $d_{50}$, $d_{16}$, $d_{84}$
>  Flow Capacity
>  Conductance
>  Total Non-Blanked Area

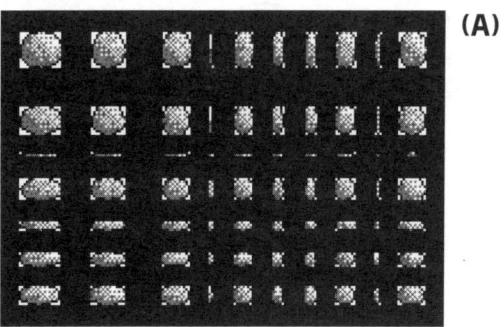

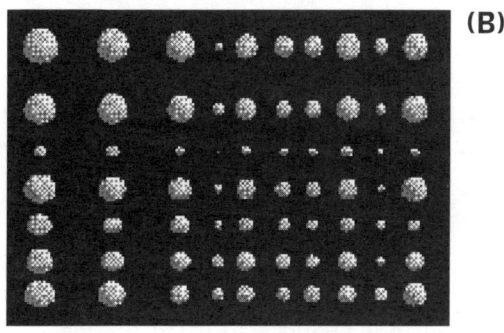

**FIGURE 6-5**

## Manufacturer's Designation

The screen manufacturer may name a particular screen in any manner desirable. The manufacturer's designation is used when ordering a shaker screen with particular characteristics.

## Separation Potential

The **separation potential** analysis determines an ellipsoid volume that fits in each of the screen openings. The length and width of 1,500 openings in multiple-layered screens and 750 openings in single-layered screens are measured, and an ellipsoid volume is calculated for each opening with the following equation (Figure 6-5A):

$$Ve = (3.14 \times la \times lb^2)/6$$

where Ve = Volume of the ellipsoid (microns³)
   la = Length of major axis (microns)
   lb = Length of minor axis (microns)

The diameter of a sphere having the same volume (Figure 6-5B) is then calculated for each of these ellipsoidal volumes as follows:

$$x = (6 \times Ve / 3.14)^{1/3}$$

where x = Diameter of sphere (microns)

Each of these observations is ranked from smallest to largest size and a volume distribution curve is plotted as shown in Figure 6-6. This curve represents the volume size distribution of the open-

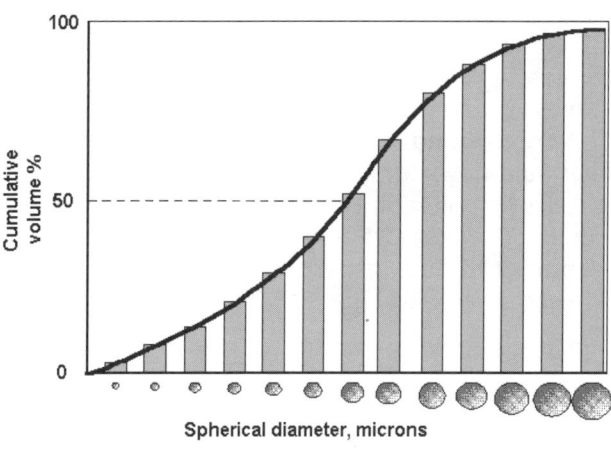

**FIGURE 6-6**

ings in the screen and is designated as the **potential separation curve**. Note that the volume, and not the area, of the various openings is used, and that this curve is obtained strictly from a knowledge of the screen openings. As a result, this curve does not include the effects of solids shape, agglomeration, machine vibration, the inability of a solid that is finer than the screen hole openings to pass through because of interference from other solids, fluid properties effects, and so forth.

Three points—$d_{16}$, $d_{50}$, $d_{84}$—may be obtained from the potential separation curve (Figure 6-7). The lower-case "d" designates the potential separation curves as opposed to the upper-case "D" values of actual cut points, determined by measurements as previously discussed. The $d_{50}$ represents the measured median opening equivalent diameter. This is the size where half (50%) of the apertures in the screen are smaller than this size (i.e., $d_{50}$ = 143 microns). The $d_{16}$ represents the potential size where 16% of the apertures in the screen are smaller than this size (i.e., $d_{16}$ = 113 microns). The $d_{84}$ represents the potential size where 84% of the apertures in the screen are smaller than this size (i.e., $d_{84}$ = 154 microns).

The $d_{16}$, $d_{50}$, and $d_{84}$ values for the separation potential curve for most oil field screens are shown in Table 6-8.

## Screen Conductance and Flow Capacity

A screen that enables an extra-fine separation is useless in the drilling industry if it will not pass a high volume flow rate. The amount of fluid that a screen will process is dependent on the screen construction, as well as solids conveyance, solids loading, pool depth, deck motion and acceleration, drilling fluid properties, and screen blinding. Although it is difficult to calculate the expected fluid processing capacity of a shaker, screens can be ranked according to their ability to transmit fluid.

**Screen conductance** is a measure of the ease that fluid flows through a screen cloth. It is analogous to permeability per unit thickness of the screen, C = k(darcy)/l(mm). To calculate the **flow capacity** through a porous medium, Darcy's law is used as follows:

$$V = K \times \Delta p / (\mu \times l)$$

Conductance, C, can be calculated where $Q = V \times A$ as follows:

$$C = K/l = V \times \mu/\Delta p = Q \times \mu/(A \times \Delta p)$$

where C = Conductance (darcy/cm)
K = Permeability (darcy)
l = Screen thickness (cm)
V = Velocity (cm/sec)
μ = Fluid viscosity (centipoise)
Δp = Pressure drop across screen (atm)
Q = Volume flow rate (cm³/sec)
A = Screen area (cm²)

Higher conductances indicates that for a given pressure drop across the screen, more fluid is able to pass through the screen. Conductance for a screen can be calculated by the equations reported in API RP 13E, 3rd edition, May 1, 1995, Section 1.1. These equations are based on experimental data using a water-glycerol mixture. Conductance is measured experimentally using the following equation, where 0.014375 is required for unit compatibility:

$$C = 0.014375 \times Q \times \mu/(A \times \Delta p)$$

where Q = Volume flow rate (gpm)
μ = Fluid viscosity (centipoise)
A = Screen area (in.²)
Δp = Pressure drop across screen (lb/in.²)

Most screens used by the drilling industry consist of one or two fine screening cloths in combination with a coarse backing cloth. The conductance of these screens can be calculated as follows:

$$1/C_t = 1/C_1 + 1/C_2 + 1/C_3 \ldots$$

where $C_1$ = Conductance of first screen (kilodarcy/mm)
$C_2$ = Conductance of second screen (kilodarcy/mm)
$C_3$ = Conductance of third screen (kilodarcy/mm)

The conductance of commonly used oil field screens and backing cloths is shown in Table 6-8. The conductance of common oil field screens in the later section on Screen Panel Technology may be calculated using the above equation in conjunction with Table 6-8.

## Non-Blanked Area

Continuous cloth screens present all available screen area to the drilling fluid to remove solids.

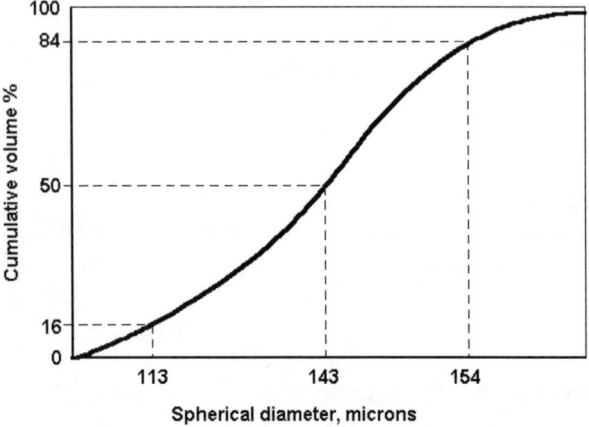

**FIGURE 6-7**

**TABLE 6-8.** Typical Market Grade and Tensile Bolting Cloth Shaker Screen Characteristics

|  | Screen Designation | | Separation Potential in Microns | | | Conductance |
|---|---|---|---|---|---|---|
|  | Mesh | $d_{16}$ | $d_{50}$ | $d_{84}$ | | Kd/mm |
| Market Grade | 10×10 | 1727 | 1777 | 1678 | | 49.68 |
|  | 20×20 | 864 | 889 | 839 | | 15.93 |
|  | 30×30 | 516 | 531 | 501 | | 8.32 |
|  | 40×40 | 381 | 392 | 370 | | 4.80 |
|  | 50×50 | 279 | 287 | 271 | | 2.88 |
|  | 60×60 | 234 | 241 | 227 | | 2.40 |
|  | 80×80 | 177 | 182 | 172 | | 1.91 |
|  | 100×100 | 140 | 144 | 136 | | 1.44 |
|  | 120×120 | 117 | 120 | 114 | | 1.24 |
|  | 150×150 | 105 | 108 | 102 | | 1.39 |
|  | 200×200 | 74 | 76 | 72 | | 0.68 |
|  | 250×250 | 62 | 63 | 59 | | 0.58 |
|  | 325×325 | 44 | 45 | 43 | | 0.44 |
| Tensile Bolting Cloth | 20×20 | 1041 | 1071 | 1011 | | 40.93 |
|  | 30×30 | 681 | 700 | 662 | | 24.33 |
|  | 40×40 | 470 | 483 | 457 | | 11.63 |
|  | 50×50 | 368 | 379 | 357 | | 7.94 |
|  | 60×60 | 310 | 319 | 301 | | 5.60 |
|  | 70×70 | 269 | 277 | 261 | | 5.25 |
|  | 80×80 | 224 | 230 | 218 | | 3.88 |
|  | 94×94 | 180 | 185 | 175 | | 2.84 |
|  | 105×105 | 165 | 170 | 160 | | 2.77 |
|  | 120×120 | 147 | 151 | 143 | | 2.51 |
|  | 145×145 | 119 | 122 | 116 | | 2.03 |
|  | 165×165 | 107 | 110 | 104 | | 1.86 |
|  | 200×200 | 86 | 88 | 84 | | 1.49 |
|  | 230×230 | 74 | 76 | 72 | | 1.30 |

Screen panels are popular because screen tears are minimized and limited to only one small area of the screening surface. The screen panels, however, reduces the amount of available screening area that, in comparison, is totally available with continuous cloth screens. The non-blanked area allows an evaluation of the surface area available for liquid transmission through the screen.

## Factors Affecting Percent Separated Curves

Some of the factors that affect the actual separation made by a screening operation are as follows:

- The relationship between the size and shape of the particles being separated and the size and shape of the screen openings will influence how fine a separation is made. This is reflected in the percent separated curve. If all drilled solids are spherical, then the distribution of the narrowest dimension of the screen openings will establish the percent separated curve. Conversely, if the particles being separated are shaped like the screen openings, then the separation curve will be closer to an ellipsoid, which will barely pass through the screen openings. If a range of solids, from blocky to slivers, is drilled, then the separation curve, again, will be closer to an ellipsoid that will barely pass through the screen openings.

- Solids have mobility in a pool of fluid to seek a screen opening large enough to pass through. As a result, the conveyance velocity, contact time with the screen, and presence of other solids all affect the ability of the solids to pass through the holes in the screen. These variables, therefore, affect the percent separated curve.

- Surface tension of the fluid causes solids to agglomerate together as they exit a pool of fluid. If solids finer than the screen openings leave the pool of fluid, then they are held by the surface tension and have very little chance of traveling through the screen. Adding a spray wash to the last screen panel, disperses these solids and allows finer solids to be washed through the screen.

- Blinding or plugging of the screen cloth dramatically affects both the amount of fluid passing through the screen and the separation the screen makes. Many of the screen openings effectively become smaller and fewer solids will pass through. The screen then makes a much finer separation than originally intended, and the screen capacity descreases significantly.

- Fine solids are also trapped in the moving solids bed. Many of the solids rejected by the screens are finer, or smaller, than the screen openings.

- Reported values for percent separated curves are also affected by how the measurement was made in the laboratory. The greatest error is often the measurement of particle-size distribution. Particle sizing by sieve analysis is the best way to characterize solids being screened since the sieving process is similar to screening. Unfortunately, sieving is a tedious and slow process. Forward laser light scattering particle size analyzers, such as the Malvern and Cilas Granulometer, tend to report size distributions somewhat larger than sieve analysis. These instruments report particle sizes in terms of equivalent spherical diameters. Some drilled solids may be more rectangular in shape, therefore, the equivalent spherical diameter may not exactly agree with the sieve analysis.

In summary, the percent separated curve represents the fraction of solids rejected by the screen as a function of size. From the above discussion, it may be noted that the percent separated curve is dependent on the conditions that exist when the data are taken. As a result, in actual drilling conditions, the percent separated curve will most likely vary as drilling fluid properties change, as the shapes of the solids change, and as the screen blinds.

## Comparison of Separation Potential and Percent Separated Curves

The **separation potential curve** is a representation of the screen opening ellipsoidal volume size distribution expressed as equivalent spherical diameters. The **percent separated curve** (cut point curve) is a representation of the actual separation of solids made by a screen, based on measurements of a particular shale shaker.

Separation potential curves have the same shape as percent separated curves for many types of screens. The separation potential curve provides a way to compare a range of complex screen types in a uniform manner. Since the separation potential curve can be obtained for all screens that have a straight-through flow path, it can be calculated for all oil field screens and provide some information on the confusion of openings in layered screens.

Screens with large length-to-width ratio openings (3:1 or larger) are not well represented with the Image Analysis method. Particles in the drilling fluid normally have a maximum length-to-width ratio of less than 3:1. Ellipsoids that barely fit through the screen openings are not a fair representation of the particle size that the screen's openings will reject. Thus, the opening sizes may not necessarily match the actual performance of a shaker as measured by cut point curves.

A comparison of calculated Image Analysis (IA) $d_{50}$ values and actual measured cut points ($D_{50}$) for a particular shaker is shown in Table 6-9.

## Comparison of Separation Potential Curves for Various Screens

The separation potential curves of several woven, wire screen cloths are shown in Figure 6-8 for market grade 80 × 80 mesh, rectangular 80 × 40 mesh, and layered 84 mesh screens. The following observations may be made:

- The area or region to the left of these curves represents solids returned to the active system. The area or region to the right of these curves represents solids rejected by the shale shaker.

- The market grade 80 × 80 cloth has a narrow distribution of openings, with the smallest openings equaling 165 microns and the largest openings, 193 microns. This type of

**TABLE 6-9.** Comparison of Image Analysis of Separation Potential with Measured Cut Points

| API Screen Designation | Aspect Ratio | Measured $D_{50}$ Cut Point (microns) | Calculated I.A. $d_50$ Separation Potential (microns) | Percent Variation (%) |
|---|---|---|---|---|
| 60 × 40 (200 × 406, 31.1) | 2.03/1 | 249 | 253 | 1.6 |
| 40 × 20 (310 × 90, 36.8) | 2.94/1 | 447 | 444 | 0.7 |
| 80 × 40 (140 × 460, 35.6) | 3.29/1 | 204 | 208 | 2.0 |
| 70 × 30 (178 × 660, 40.3) | 3.71/1 | 236 | 276 | 16.9 |
| 60 × 24 (200 × 830, 41.5) | 4.15/1 | 264 | 321 | 21.6 |
| 35 × 12 (320 × 1700, 42.0) | 5.31/1 | 493 | 558 | 13.2 |

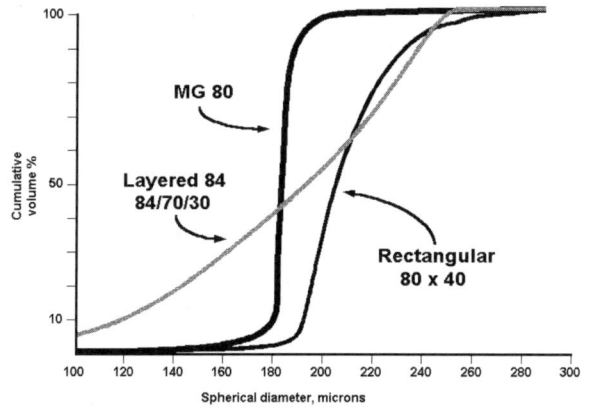

**FIGURE 6-8**

potential separation curve is typical of square mesh screens. Such a curve is often referred to as an "ideal classification" or "narrow classification" because there is little difference between the $d_{50}$, $d_{16}$, and $d_{84}$. If the percent separated curve is the same as the distribution of screen openings, then all solids larger than 193 microns will be discarded and all solids finer than 168 microns will be retained in the mud. In actual practice, some of the finer solids may be discarded along with the larger solids. Notice that the range of opening size is only 28 microns (193 – 165). The median opening size, or $d_{50}$, for the MG80 cloth is 183 microns. This means that the potential exists for half the 183 micron solids to be discarded and the other half to be retained. The analysis indicates that 27% of the openings will pass 180 micron particles and 73% of this size will be retained in the mud.

- The rectangular 80 × 40 cloth has a broader distribution of openings, ranging from 180 microns to 280 microns. Again, if the separation potential curve shown is analogous to the percent separated curve, all solids larger than 280 microns will be rejected and solids smaller than 180 microns will be retained. This is a separation range of 100 microns (280 – 180). The separation potential $d_{50}$ is 205 microns.

- The layered 84/70/30 cloth has a broad distribution of openings, with the smallest openings less than 100 microns and the largest openings, 240 microns. Again, if the separation potential curve shown is analogous to the percent separated curve, all particles larger than 240 microns will be rejected and all particles finer than about 100 microns will be retained. This is a separation range of 140 microns (240 – 100). The potential separation $d_{50}$ is 181 microns (Figure 6-9).

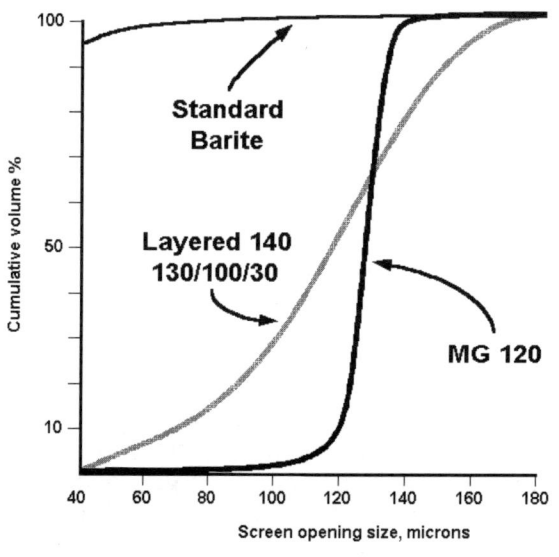

**FIGURE 6-9**

- When the layered 84 curve is compared with the market grade 80, both screens have a $d_{50}$ of approximately 183-micron particles. The layered screen will discard fewer coarser particles between 183 and 245 microns. Both screens will discard all solids larger than 245 microns. Conversely, the layered screen discards more smaller solids between 100 and 183 microns. Both screens will allow most of the solids smaller than 100 microns to pass through.

- The area to the left of the MG80 and to the right of the layered 84 screens indicate a large quantity of smaller solids would be rejected by the layered 84 that would be returned to drilling fluid by the MG80.

- The areas to the right of the MG80 and to the left of the layered 84 indicate that a large quantity of large "sand" size particles would be returned to the drilling fluid by the layered 84 that would be rejected by the MG84.

- The square opening cloth makes a size separation somewhat independent of particle shape. Since the opening size is the same in both directions, particles that pass through the openings must be smaller than the opening in both directions. Furthermore, long slivers that might pass through the screen if turned on end, probably will not because they will tend to remain flat on the screen surface.

- The rectangular opening cloth will have a percent separated curve, or an actual separation, which is a function of particle shape. For instance, if all of the particles being separated are round, the percent separation curve will fall on the narrow dimension of the hole opening. Since the separation potential curve represents the largest ellipsoid that will barely fit in the screen openings, the separation potential curve approximates the separation this rectangular screen makes of oblate spheroids. Particles with a long length can be separated even though the equivalent spherical diameter is much less than the equivalent ellipsoid that will pass through the screen.

- Layered screens have a broad range of opening sizes, which is typical of layered screens. The separation potential curve indicates that many openings are both larger than the $d_{50}$ (or median) cut point and smaller than the $d_{50}$. The layered screen has the greatest potential to have an actual percent separated curve that differs from the separation potential curve.

- The pool of fluid on most linear motion shale shakers may significantly alter the separation made by the layered screen. Solids in the pool have a greater opportunity to find a hole in the screen to pass through. If every particle in the slurry had the same chance to seek out all of the screen openings, then the actual percent separated curve would be the dimension of the largest hole. In reality, even in a pool of fluid, all particles do not have a chance to find the largest hole opening. Layered screens without a pool (circular and unbalanced elliptical motion shale shakers) may provide solids a better chance to agglomerate in patties. Agglomeration severely limits the mobility of the individual solid particles so that they are unable to find openings to pass through.

Separation potential curves for a layered 140 (130/100/30) screen compared with a MG120 screen are shown in Figure 6-9. The cumulative distribution of equivalent spherical sizes in a barite sample, which has been marked with API approval, has been added to this curve. Eighty-three volume percent of this barite is finer than 44 microns, which means that 17% of this barite is coarser than 44 microns. The layered 140 screen has openings in the 44 to 185 micron range. For this reason, some barite may be discarded by this screen. The MG120 should eliminate very little of this barite since all openings are larger than the largest barite particle. (API specifications allow three weight percent barite to be larger than 75 microns, or 200 mesh.) The barite shown in Figure 6-9, contains more than allowed by API standards. Generally, 3 lb from every 100-lb sack would be discarded if presented to a 200-mesh screen. If the slurry presented to the screen is dehydrated, such as hydrocyclone underflows, both screens may remove substantial amounts of barite.

## Review

The separation potential curve is similar in shape to the actual separation made by a screen. Shale shakers classify and separate solid particles based on size and shape. Particles coarser than the screen opening are removed and particles finer than the screen opening are retained. Particles have many chances to find openings to pass through as they are conveyed from the feed end to the discharge. This is particularly true while they are in a pool of fluid. When they exit the pool, they tend to agglomerate with other solids. This agglomeration is caused by surface tension between the air and the liquid wetting phase on the solids. Frequently, a mist or a spray of drilling fluid will cause the agglomeration (or patties) to disperse with better separation of solids, but also returns fine solids to the drilling fluid.

As previously discussed, the openings in a square mesh screen are fairly uniform and vary with a layered mesh screen. Some particles of a given size pass through the screen and some are discarded. If the amount (% by weight) of discarded particles is plotted against the size of the particles, a curve is obtained that is similar to the one shown in Figure 6-8. This curve is called the percent separated curve, cut point curve, tromp curve, or partition curve. As stated earlier, the percent separated curve for common oil field screens has the same shape as the potential separation curves. The **cut point** (also known as the median cut point) is the point on the curve where 50% of the particles of this size are discarded. The area beneath and to the right of the curve represents the mass of solids removed by the shaker. The area to the left of the curve represents the mass of solids retained in the mud. Chapter 9 discusses a method to determine cut points.

The separation potential curves are reported in terms of the $d_{50}$, $d_{16}$, and $d_{84}$. These sizes represent 16%, 50%, and 84% of the opening sizes in the screen and are used to compare various shaker screens. One half of the openings are larger than the $d_{50}$, or median point, and one half of the openings are smaller. The $d_{16}$ size indicates that 16% of the openings are smaller than this size; whereas the $d_{84}$ size indicates that 84% of the openings are smaller than that size. These points are statistically significant for Gaussian distributions because they each represent one standard deviation from the $d_{50}$, or median, point. In a Gaussian distribution, the openings sizes would be uniformly distributed around the $d_{50}$ median value.

The percent separated curve, which is determined experimentally, incorporates all of the test conditions. It is the result of the following variables:

- Size and shape of the openings in the screen
- Solids shape
- Characteristics of solids
- Drilling fluid properties
- Screen wettability
- Solids and liquid loading
- Agglomeration of solids
- Shaker vibration pattern, screen angle, screen tightness, etc.

The separation potential curve, which is determined by examining the screen, is designed to identify only screen parameters. Experimentally, solids discarded from a shale shaker may or may not have the same distribution as the potential separation curve. The variables listed above determine the results of the actual shaker performance. The potential separation curve is designed to eliminate all variables except those relating to the shaker screen openings or **screen cloth**.

### Screen Blinding

Screen blinding occurs when grains of solids being screened lodge in a screen hole. This often occurs when drilling fine sands, such as in the Gulf of Mexico. The following sequence is often observed during screen blinding:

1. When a new screen is installed, the circulation drilling fluid falls through the screen in a short distance.
2. After a time, the fluid endpoint travels to the end of the shaker.
3. Once this occurs, the screens are changed to eliminate the rapid discharge of drilling mud off the end of the shaker.
4. After the screens have been washed, fine grains of sand that are lodged in the screen surface can be observed. The surface of the screen will resemble fine sandpaper because of the sand particles lodged in the openings.

Most every screen used in the oil field is blinded to some extent by the time it needs replacing. For this reason, when the same screen size is reinstalled, the fluid falls through the screen closer to the feed.

One common solution to screen blinding is to change to a finer or coarser screen than the one being blinded. This tactic is successful if the sand that is being drilled has a narrow size distribution. Another solution is to change to a rectangular screen, although rectangular screens can also blind with multiple grains of sand. Unfortunately, the process of finding a screen size that will not blind is expensive.

In the late 1970s, the layered screen was introduced to avoid screen blinding. This hook strip screen was mounted on a downhill sloping, unbalanced elliptical motion shale shaker vibrating at 3600 rpm. The two fine layers of screening cloth, supported at 4-inch intervals, tended to dislodge fine grains of sand and would only blind about 25% in severe laboratory tests (Figure 6-10), leaving 75% of the screen unblinded. The non-blinding feature is assumed to be the result of the deceleration of the two screens. The wire diameter is in the range of 0.002 inch and the opening sizes

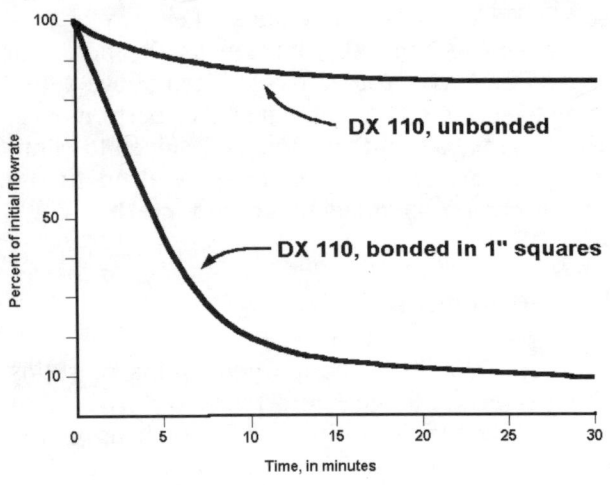

**FIGURE 6-10**

are in the range of 0.004 inch. During the upward thrust of a layered screen, the screens must come to a stop at the upward end of the motion. The screens tended to have inertia that prevented them from stopping at exactly the same time. This created an opening size that was slightly larger than the original opening size of the layered screen during the upward thrust. Solids were then expelled from the screen. On the downward thrust of the motion, the two layers remained together until the screen began decelerating. At the bottom of the stroke, again, the inertial forces caused the screens to slightly separate allowing larger solids to pass through. This may explain why the separation cut point curve shows poorer separation characteristics for a layered screen than for a single, square mesh screen. Many particles larger and smaller than the median opening size are found in the discard from a layered screen.

Unfortunately the downhill sloping basket and high frequency limited the amount of liquid that could pass through the screen. Furthermore, lost circulation material had a high propensity to become lodged in the screen due to the high-frequency, short-stroke vibration. These problems were reduced by limiting the vibration to 1800 rpm and flattening the basket slope. In the early 1980s, linear motion was introduced so that solids could travel up an incline out of a pool of liquid. This fluid pool provided additional pressure to force fluid through the screen. Unfortunately, linear motion, combined with marginal support, tore layered screens apart. The only way to obtain satisfactory screen life on a linear motion machine was to support the layered screen in one-inch squares.

## Materials of Construction

The material used to weave the cloth screens is quite varied. Screens are made from metal wires, plastic wires, and molded plastic cloths.

***Metals.*** The alloys that are most weavable and resistant to corrosion are nickel/chrome steels: 304, 316, and 316L. These alloy wires are available in sizes as low as 20 microns. The finest wire available is 304L, which is available as low as 16 microns. Other materials, including phosphor bronze, brass, copper, monel, nickel, aluminum alloys, plain steel, and plated steel are also available. Within the drilling industry, 304 stainless cloth is most widely used.

***Plastics.*** Two types of synthetic screens are available: woven synthetic polymer and a molded one-piece cloth called a platform.

Conventional looms can be used to weave synthetic polymer screens. Polymers, such as polyesters, polypropylene, or nylon, are drawn into strings having diameters comparable to wire gauges and woven into screen cloth. Synthetic screens exhibit substantial stretch when mounted and used on shale shakers. For this reason, plastic screen openings are not as precise, although this variability is not nearly as great as layered-metal steel screens.

One-piece, injection-molded synthetic cloths are typically made from urethane compounds. These synthetic cloths have limited chemical and heat resistance but display excellent abrasion resistance. The designs range from simply supported molded parts having very little open areas, to complex structures with up to 55% open area. Molded cloths are very popular in the mining industry where abrasion resistance is important. These screens make a coarser separation than screens used in the oil field. Development of molded cloth screens capable of making a fine separation, which have heat and chemical resistance necessary for oil field application, is underway.

Cloth selection for shale shaker screens involves compromises between separation, throughput, and screen life.

## Screen Panels

Shale shaker screens changed as demands on the shale shaker increased. Shaker screens have three primary requirements:

1. High liquid and solids handling capacity
2. Acceptable life
3. Ability to be easily identified and compared

Early shale shaker screens required durability. This demand was consistent with the shaker designs and solids removal philosophies of their period. These shakers could only remove the large coarse solids from the drilling fluid while the sand trap, reserve pit, and downstream hydrocyclones removed the bulk of the drilled solids.

Changes in drilling fluids, environmental constraints, and a better understanding of solids/liquid separation have modified the role of the shale shaker. Generally, the more solids removed at the flow line, the higher the effectiveness of downstream equipment. The results include reserve pits that can be smaller (or eliminated altogether), lower clean-up costs, and increased drilling efficiency.

As important as the mechanical aspects of newly designed shale shakers may be, improvements in screen panels and screen cloth have also significantly increased shaker performance. Older shakers have benefited from these improvements, as well. Two design changes have been made to extend the economic limit of fine-screen operation:

1. A coarse backing screen to support the fine mesh cloth(s), and
2. Tensioned cloth bonded to a screen panel (**pretensioned screen panel**)

## Coarse Backing Screen

Today, fine screen cloth(s) is reinforced with a coarse backing screen. This backing screen protects the fine mesh screen from being damaged and extends life. The backing screen also provides additional support for heavy solids loading.

## Panels

The most important advance in screen panel technology has been the development of pretensioned screen panels. Although similar panels have been used on mud cleaners since their introduction, earlier shakers did not possess the engineering design to allow their use. With the advent of linear motion machines, the pretensioned screen panels extended screen life and permitted more routine use of 200+ mesh screens.

These panels consist of a fine screen layer(s) and a coarse backing cloth layer bonded to a support grid. The screen cloths are pulled tight, or tensioned, in both directions during the fabrication process. This ensures that every screen begins with the proper tension. Correct installation procedures and post-run retightening of screen panels can significantly increase shaker performance and screen life.

Manufacturers employ different geometric-shaped apertures in their screen panel design. Some of the more common panels are square, rectangular, hexagonal, or oval shapes. The apertures in the panels can vary from 1-in. to 3-in. squares, 2-in. × 6-in. ovals, 7-in. × 33-in. rectangles, to 1.94-in. hexagons.

The panels can be flexible (built of thin gauge metal or plastic) so they can be stretched over **crowned** shakers. They can also be flat and built from heavy-gauge mechanical tubing for installation as panels on **flat-decked** (non-crowned) shakers.

Regardless of configuration, the function of the panel is to provide mechanical support for the fine mesh screen cloth bonded to it, and at the same time occlude as little potential flow area as possible with the supporting grid structure.

Some screened panels are made with no support grid at all, but simply by bonding the finer mesh(es) cloths directly to a coarser backing wire using a heat-sensitive adhesive. Essentially this becomes a **hook strip** design, with certain support refinements.

Pretensioned screen panels help address two of the three original design goals: capacity and screen life. The remaining goal of easy identification/comparison is a function of the screen cloths used in today's panel designs and better labeling techniques to display important performance variables.

## Hook Strip Screens

Hook strip screens are also available. Because of the superior life characteristics of the **panel mount units**, they have been relegated to a minor role on linear motion machines, although they are used extensively on circular and elliptical motion machines. Proper tensioning (and frequent re-tensioning) of all screen types is good screen management and can significantly increase screen life. Individual manufacturer's operation manuals should be consulted to obtain the proper installation methods and torque requirements, where applicable, for specific screens/panels.

## Bonded Screens

Several types of **bonded screens** are available. The repairable **perforated plate screen** has one or more layers of fine mesh cloth bonded to a sheet of metal or plastic with punched, patterned holes. Perforated plate designs are available in various opening sizes and patterns. Additional designs include a special application where backing and fine screen(s) materials are bonded together, eliminating the need for perforated plates. Flat-surfaced, pretensioned screen panels are becoming popular because of their even tensioning, easy installation, and the even distribution of liquids and solids across the screen deck.

## Three-Dimensional Screen Panels

Three-dimensional screen panels were introduced in the mid-1990s. These typically offer more screening area than flat-panel, repairable plate screens while retaining the ability to be repaired.

This type of screen panel adds a third dimension to the previous two-dimensional screens. The screen surface is rippled and supported by a rigid frame. Most three-dimensional screen panels resemble the metal used in a corrugated tin roof. Construction consists of a corrugated, pretensioned screen cloth and bonded to a rigid frame.

Like bonded flat screens, the three-dimensional screen panel needs only to be held firmly in place with a hook strip or other means to prevent separation between the shaker bed and the screen panel during vibration. Three-dimensional screen panels can be used to support any type or style of wire cloth and can be used with any type of motion.

Three-dimensional screen panels allow solids to be conveyed down into the trough sections of the screen panel. When submerged in a liquid pool, this preferential solids distribution allows for higher fluid throughput than is possible with flat screen panels by keeping the peaked areas clear of solids. A three-dimensional screen panel improves distribution of fluid and solids across the screen panel. This reduces the characteristic "**horseshoe effect**" caused by shakers using crowned screen beds.

# CHAPTER SEVEN

# Solids Control Equipment

Adequate solids control should be instituted at the beginning of any drilling operation due to the detrimental effects of drilled solids discussed in earlier chapters. Drilled solids will erode, disperse, and subdivide into smaller and smaller sizes as they recirculate in the system. Smaller drilled solids create many problems and are increasingly more difficult to remove. Once this process begins, it is impossible to eliminate the solids unless a large quantity of drilling fluid is jettisoned from the system and clean drilling fluid is introduced.

Proper drilled solids management has many beneficial effects on drilling operations:

- Increased penetration rates
- Decreased drilling fluid costs
- Reduced water (or oil) requirements
- Lower torque and drag
- Reduced differential sticking problems
- Reduced pressure losses in the circulating system
- Lower equivalent circulating densities
- Reduced lost circulation tendencies
- Improved primary cementing operations
- Less remedial cementing
- Less waste or disposal volumes

Drilled solids are separated from the drilling fluid with a variety of different types of equipment. Equipment selection is governed by circulation rates in the wellbore, as well as the concentration and size of drilled solids generated in the wellbore. Large particles are more easily removed than small particles and should be removed as soon as possible to prevent degradation. As particle size decreases, separation becomes increasingly more difficult and expensive. Efficient drilled solids management begins with adequate hole-cleaning of the wellbore. Drilled solids should be brought to the surface with as little degradation as possible.

Most drilling rigs are equipped with many elements of good solids control such as shale shakers, degassers, desanders, desilters, agitators, and sometimes mud guns, mud cleaners, and centrifuges. Surface tank arrangements, including piping and centrifugal pumps, are integral parts of a drilling rig. Regardless of the drilling fluid system used, certain fundamental tank arrangements are common to all types of drilling fluid. Drilling engineers must make certain the arrangements are correct and decide what additional equipment is needed for specific locations.

Some of the standard principles for proper solids control are presented in this chapter and include pit arrangements, piping, pumps, equipment arrangements, and sizing. The drilling fluid type (fresh water-base, saline water-base, oil-base, or synthetic), drilling fluid density (weighted or unweighted), drill bit types, formation reactivity, and disposal costs will dictate additional specific solids control equipment and procedures needed to assure the lowest well cost.

Attention to details such as agitation, proper centrifugal pump and impeller selection, proper plumbing for drilling fluid in the tank system, minimization of piping manifolds, and elimination of leaks through valves will significantly improve overall solids removal and accrue substantial economic benefits.

The purpose of a drilling rig surface fluid processing system is to provide a sufficient volume of properly treated drilling fluid for drilling operations. The surface volume should be large enough to allow enough properly conditioned drilling fluid above suction and equalization lines. This will allow maintenance of drilling fluid properties and keep the wellbore full during a wet trip. The surface system should consist of three clearly identifiable sections (Figure 7-1):

- Suction and testing section
- Additions section
- Removal section

## SUCTION AND TESTING SECTION

The suction and testing section is the last part of the surface system. Most of the usable surface

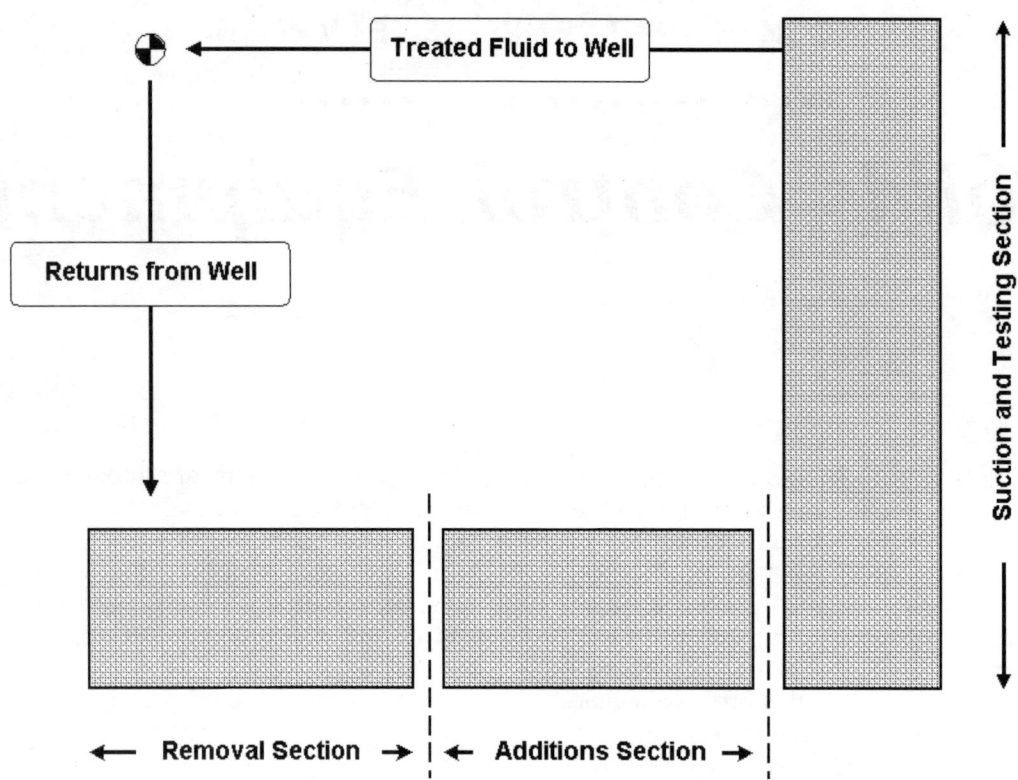

**FIGURE 7-1**

volume should be available in this section. Processed and treated fluid is available for various evaluation and analysis procedures just prior to the fluid recirculating downhole. This section should be mixed, blended, and well stirred. Sufficient residence time should be allowed so that changes in drilling fluid properties may be made before the fluid disappears downhole. Vortex patterns from agitators should be inhibited to prevent entraining air in the drilling fluid.

A slug pit, or small 20- to 50-barrel compartment, should be part of the suction section. This compartment is isolated from the active system and is available for small volumes of specialized fluid. Some drilling fluid systems may have more than one of these small compartments. They are manifolded to a mixing hopper so that solids and chemicals may be added and are used to create a heavier slurry to be displaced partway down the drillpipe before trips. This prevents drilling fluid inside the pipe from splashing on the rig floor during trips. These compartments are also used to mix and spot various pills, or slurries, in a wellbore. The main pump suction must be manifolded to the slug pit(s).

A trip tank should also be a component of the tank system. This tank should have a well-calibrated, liquid-level gauge to measure the volume of drilling fluid entering or leaving the tank. The volume of fluid that replaces the volume of drill string is normally monitored on trips to make certain that formation fluids are not entering the wellbore. When one barrel of steel (drill string) is removed from the borehole, one barrel of drilling fluid should replace it to maintain a constant liquid level in the wellbore. If the drill string is not replaced, the liquid level may drop low enough to permit formation fluid to enter the wellbore. This is known as a "kick." Fluid may be returned to the trip tank during the trip into the well. The excess fluid from the trip tank should be returned to the active system through the shale shaker.

## ADDITIONS SECTION

All commercial solids and chemicals are added to a well-agitated tank upstream from the suction and testing section. New drilling fluid mixed on location should be added to the system through this tank. Drilling fluid arriving on location from other sources should be added to the system through the shale shaker. To assist homogenous blending, mud guns may be used in the addition section and the suction and testing section.

# SOLIDS CONTROL EQUIPMENT

## REMOVAL SECTION

Undesirable drilled solids and gas are removed in this section before new additions are made to the fluid system. Drilled solids create poor fluid properties and cause many of the costly problems associated with drilling wells. Excessive drilled solids can cause stuck drill pipe, bad primary cement jobs, or high surge and swab pressures, which can result in lost circulation and/or well control problems. Each well and each type of drilling fluid has a different tolerance for drilled solids.

Each piece of solids control equipment is designed to remove solids within a certain size range. Solids control equipment should be arranged to remove sequentially smaller and smaller solids. A general range of sizes are presented in Table 7-1.

## PIPING AND EQUIPMENT ARRANGEMENT

The most common problem on drilling rigs is inadequate fluid routing, which allows drilling fluid to be processed through the equipment in a sequential manner. When a substantial amount of drilling fluid bypasses a piece of solids removal equipment, drilled solids cannot be removed. Many factors contribute to inadequate fluid routing including ill-advised manifolding of centrifugal pumps for hydrocyclone or mud cleaner operations, leaking valves, improper mud gun use in the removal section, and routing drilling fluid incorrectly through mud ditches.

Each piece of solids control equipment should be fed with a dedicated, single purpose pump—with *no* routing options. Hydrocyclones and mud cleaners have only one correct location in tank arrangements and, therefore, should have only one suction location. Routing errors should be corrected and equipment color-coded to eliminate alignment errors. If worry about an inoperable pump suggests manifolding, money can be saved by making easy access to the pumps and having a standby pump in storage.

Suction and discharge lines on drilling rigs should be as short and straight as possible. Sizes should be such that the flow velocity within the pipe is kept between 5 and 10 ft/sec. Higher velocities are usually turbulent and cause erosion where the pipe changes direction; lower velocities may result in settling problems. The flow velocity may be calculated with the equation:

$$\text{Velocity, ft/sec} = \frac{\text{Flow rate, gpm}}{2.48 \times (\text{Inside diameter, in.})^2}$$

Pump cavitation may result from improper suction line design, such as inadequate suction line diameter or lines that are too long. The suction line should have no elbows, tees, or pipe reducers within three pipe diameters of the pipe suction flange, and their total number should be kept to a minimum.

## EQUALIZATION

Most compartments should have an equalizing line, or opening, at the bottom. Only the first compartment, if it is used as a settling pit, and the second compartment, the degasser suction tank, should have a high (or weir) overflow to the compartment downstream. Equalizing pipes should be 9 inches in diameter or larger. Equalizing lines that are too large will generally fill with settled solids until the flow velocity in the pipe is adequate to prevent settling.

An adjustable equalizer is preferred between the solids removal and additions sections. The lower end of an "L"-shaped, adjustable equalizer, usually field fabricated from $13\frac{3}{8}$-inch casing, is connected to the bottom of the last compartment in the removal section. The upper end discharges fluid into the additions section and can be moved up or

**TABLE 7-1**

| Equipment | Size | Median-Size Removed Microns |
|---|---|---|
| Shale Shakers | 80-mesh screen | 177 |
| | 120-mesh screen | 105 |
| | 200-mesh screen | 74 |
| Hydrocyclones | 8-inch diameter | 70 |
| | 4-inch diameter | 25 |
| | 3-inch diameter | 20 |
| Centrifuge | | 5 |

down. This controls the liquid level in the removal section and still permits most of the fluid in the suction section to be used.

## SURFACE TANKS

Most steel pits for drilling fluid are square or rectangular with flat bottoms. Each tank should have adequate agitation except for settling tanks. Additionally, each tank should have enough surface area to allow entrained air to leave the drilling fluid. A rule of thumb for calculating the minimum active surface pit area is:

$$\text{Area, ft}^2 = \frac{\text{Maximum circulation rate, gpm}}{40}$$

For example, if the active circulating rate is 650 gpm, the surface area of each active compartment should be 16 square feet.

## Sequential Treatment

Solids control equipment should be arranged so that each piece removes successively finer solids as demonstrated in Figure 7-2. Although all the equipment in the following list may not be needed, the most common arrangements are:

**Unweighted Drilling Fluid**
- Gumbo removal
- Scalper shakers
- Main Shale shakers
- Desanders
- Desilters
- Centrifuge
- Dewatering

**Weighted Drilling Fluid**
- Gumbo removal
- Scalper shakers (seldom needed)
- Main Shale Shaker
- Mud cleaner
- Centrifuge
- Dewatering (seldom needed)

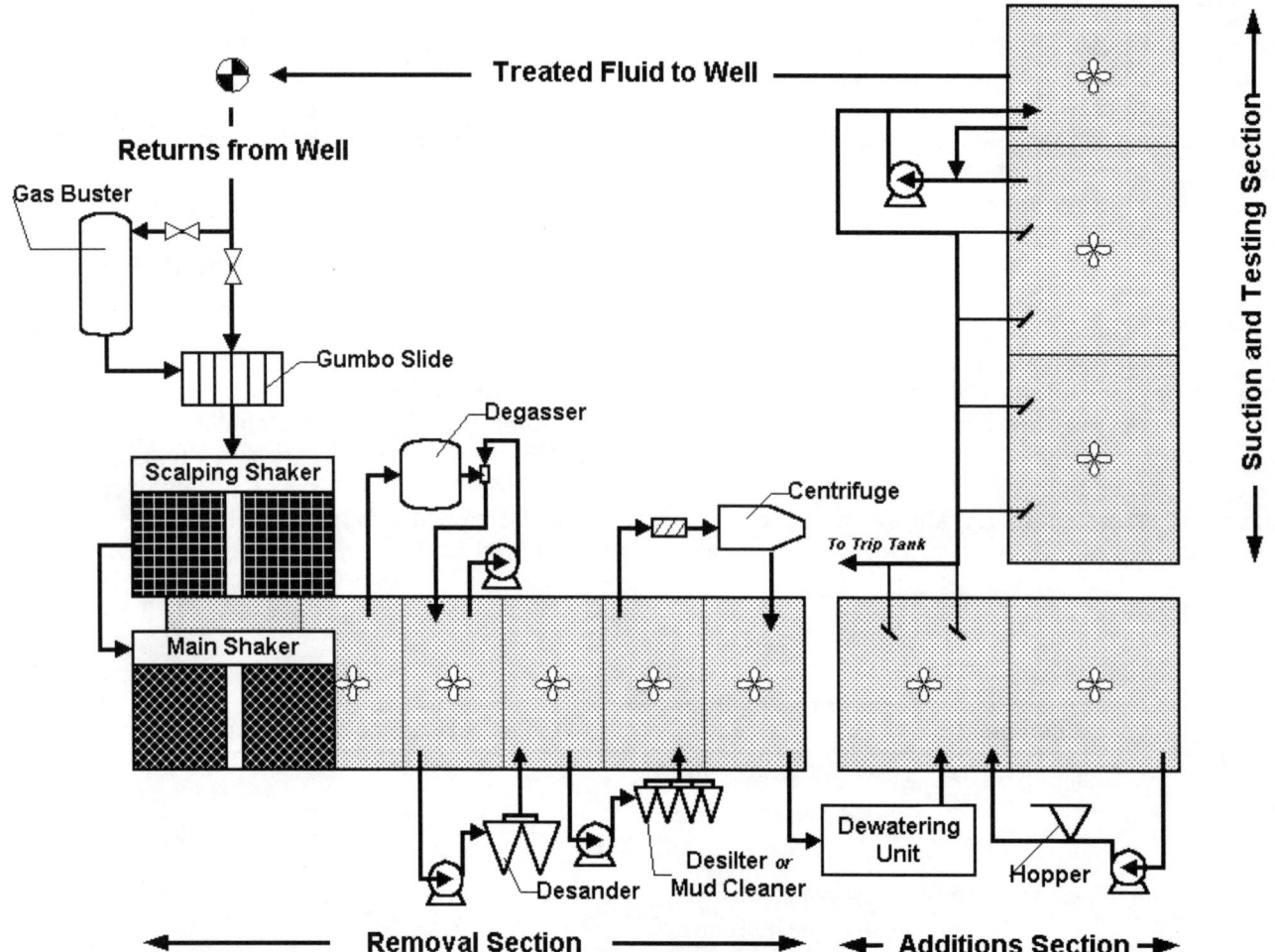

FIGURE 7-2

## GUMBO REMOVAL

**Gumbo** is formed in the annulus from the adherence of sticky particles to each other. It is usually a wet, sticky mass of clay, but finely ground limestone can also act as gumbo. Enough gumbo can arrive at the surface to lift a rotary bushing from the rotary table. This sticky mass is difficult to screen and, in areas where gumbo is prevalent, it is sometimes removed before it reaches shale shakers.

Most gumbo removal devices are fabricated at the rig site, have many different shapes, and are usually slides. Because gumbo does not stick to stainless steel, one effective device is a series of $\frac{3}{4}$- to 1-inch diameter stainless steel rods arranged to slope downward from the end of the flow line. The rods are separated by one to two inches and are about four to six feet long. Gumbo leaving the flow line slides down the rods and is sent to disposal waste. Drilling fluid easily passes through the bars and is sent to the scalping or the main shakers.

Devices and machines designed specifically to remove gumbo are available from several manufacturers. One of these machines uses a series of steel bars formed into an endless belt. The bars are separated by a space of 1–2 inches and are disposed perpendicular to the fluid flow. The unit moves gumbo to the discharge end of the machine. Another machine uses a 5 or 10 mesh synthetic belt run at an uphill pitch to convey gumbo from a pool of drilling fluid. A counter rotating brush is used to clean gumbo from the underside of the belt.

Some shale shakers are also used as gumbo removal devices. Shakers are now available that combine gumbo removal (for example, a scalping shaker) and a main shaker all on one skid.

## SAND TRAPS

After the drilling fluid passes through the main shaker, it enters the mud pit system. When 80-mesh screens and coarser were routinely used, the sand trap performed a very useful function. Large, sand-size particles would settle and could be dumped overboard.

The bottom of a sand trap should be sloped at about 45° to facilitate quick dumping. The trap should not be agitated and should overflow into the next compartment. Linear and balanced elliptical motion shale shakers have all but eliminated this technique. Small drilled solids generally do not have sufficient residence time to settle. When inexpensive drilling fluid was used, sand traps were dumped once or twice per hour. Today, with the use of fine-mesh screens and expensive waste disposal, such dumping is cost prohibitive.

## DEGASSERS

Drilling fluids usually encounter hydrocarbon gasses. At the bottom of a borehole, these gasses may partially dissolve in the drilling fluid or compress to occupy very small volumes. As the fluid rises to the surface, the hydrostatic pressure is reduced and these gasses expand and evolve from the drilling fluid. They must be removed from the surface mud system or pump operations will become erratic.

Shale shakers are not effective mechanisms to separate gas bubbles from a viscous drilling fluid. Degassers should be installed immediately downstream of the shale shakers, and gas should be removed before drilling fluid enters centrifugal pumps. Hydrocyclone performance requires a continuous fluid volume and head generated by the centrifugal pumps. A gaseous drilling fluid reduces centrifugal pump performance and may even "vapor-lock" the pump so that it prevents the movement of fluid. Even positive displacement rig pumps are affected by gaseous drilling fluid.

Two types of degassers are available: atmospheric and vacuum-type. Atmospheric degassers have a submerged centrifugal pump integral with the unit. It is placed into a spray chamber through a disc valve where it strikes the inside wall of the chamber. The thin spray, combined with the impact of the fluid on the wall of the chamber, separates the gas from the fluid.

Vacuum-type degassers separate gas from drilling fluids by spreading the gas-cut fluid into thin layers in a reduced atmosphere. The fluid usually flows over a series of baffles, or plates. Degassed drilling fluid is pumped through an eductor to remove drilling fluid from the vacuum chamber.

Equalization between degasser suction and discharge compartments is through a high weir at the top of the tanks. Degasser suctions should be located at the bottom of the compartments.

## HYDROCYCLONES

**Hydrocyclone** is a general term used to describe a device where liquid swirls inside of a cone. The centrifugal force of the swirling liquid moves the solids to the outside wall. In drilling operations, hydrocyclones use these centrifugal forces to separate solids in the 15- to 80-micron range from the drilling fluid. This solids-laden fluid is discharged from the lower **apex** of the cone, and the cleaned drilling fluid is discharged from the overflow discharge.

Hydrocyclones consist of an upper cylindrical section fitted with a tangential feed section, and a lower conical section that is open at its lower apex allowing for solids discharge (Figure 7-3). The closed, upper cylindrical section has a downward protruding **vortex finder** pipe extending below the tangential feed location.

Fluid from a centrifugal pump enters the hydrocyclone tangentially, at high velocity, through a feed nozzle on the side of the top cylinder. As drilling fluid enters the hydrocyclone, centrifugal force on the swirling slurry accelerates the solids to the cone wall.

The drilling fluid, a mixture of liquid and solids, rotates rapidly while spiraling downward toward the apex. The higher-mass solids move toward the cone wall. Movement progresses to the apex opening at the cone bottom. At the apex opening, the solids along the cone wall, together with a small amount of fluid, exit the cone. The discharge is restricted by the size of the apex. Fluid and smaller-mass particles, which have been concentrated away from the cone wall, are forced to reverse flow direction into an upward spiraling path at the center of the cone to exit through the vortex finder.

The **vortex finder** is a hollow tube that extends into the center of the cone. It diverts drilling fluid from flowing directly to the overlow outlet, causing the drilling fluid to move downward and into the cone. The swirling liquid is forced inward and, still rotating in the same direction, reverses the downward flow and moves upward toward the center of the vortex finder. In a **balanced** cone, the inner cylinder of swirling fluid surrounds a cylinder of air that is pulled in through the cone apex. Solids and a small amount of liquid spray out the lower apex of the cylinder. The apex opening relative to the diameter of the vortex finder will determine the dryness of the discharged solids.

Most balanced cones are designed to provide maximum separation efficiency when the inlet head is 75 feet. Fluid will always have the same velocity within the cone if the same head is delivered to the hydrocyclone inlet. Pressure can be converted to feet of head with the equation frequently used in well-control calculations:

$$\text{Head, in feet} = \frac{\text{Pressure, in psi}}{0.052 \times \text{Mud Weight, in ppg}}$$

The relationship between manifold guage pressure and drilling fluid weight at a constant 75 feet feed head is summarized in Table 7-2.

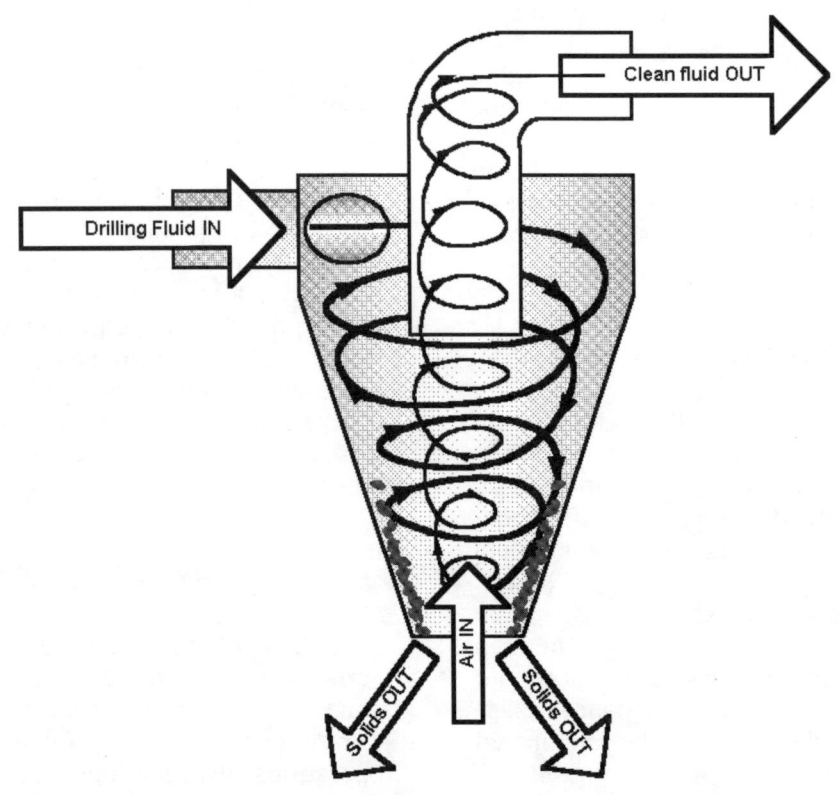

**FIGURE 7-3.** Hydrocyclone

**TABLE 7-2.** Pressure for 75 feet of Head for Various Mud Weights

| Pressure (psig) | Feed Head (ft) | Mud Weight (ppg) |
|---|---|---|
| 32.5 | 75 | 8.34 |
| 35 | 75 | 9.0 |
| 37 | 75 | 9.5 |
| 39 | 75 | 10.0 |
| 41 | 75 | 10.5 |
| 43 | 75 | 11.0 |
| 45 | 75 | 11.5 |
| 47 | 75 | 12.0 |
| 49 | 75 | 12.5 |
| 51 | 75 | 13.0 |

Hydrocylones separate solids according to mass, a function of both density and particle size. However, in unweighted drilling fluids, the solids density has a comparatively narrow range and size has the greatest influence on their settling. Centrifugal forces act on the suspended solids particles, so those with the largest mass (or largest size) are the first to move outward toward the wall of the hydrocyclone. Consequently, large solids with a small amount of liquid concentrate at the cone wall, and smaller particles and the majority of liquid concentrate in the inner portion.

Larger-size (higher-mass) particles, upon reaching the conical section, are exposed to the greatest centrifugal force and remain in their downward spiral path. The solids sliding down the wall of the cone, along with the bound liquid, exit through the apex orifice. This creates the **underflow** of the hydrocyclone.

Smaller particles are concentrated in the middle of the cone with most of the drilling fluid. As the cone narrows, the downward spiraling path of the innermost layers is restricted by the reduced cross-sectional area. A second, upward vortex forms within the hydrocylone and the center fluid layers with smaller solids particles turn toward the overflow. At the point of maximum shear, the shear stress within a 4-inch desilter is on the order and magnitude of 1,000 reciprocal seconds.

The upward moving vortex creates a low-pressure zone in the center of the hydrocyclone. In a balanced cone, air will enter the lower apex in counterflow to the solids and liquid discharged from the hydrocyclone. In an unbalanced cone, a rope discharge will emerge from the cone, resulting in excessive quantities of liquid and a wide range of solids in the discard.

There are two countercurrent spiraling streams in a hydrocyclone; one spiraling downward along the cone surface, and the second spiraling upward along the cone center axis. The countercurrent directions, together with turbulent eddy currents, concomitant with extremely high velocities, result in an inefficient separation of particles. The two streams tend to co-mingle within the contact regions and particles are incorporated into the wrong streams. Hydrocyclones, therefore, do not make a sharp separation of solid sizes.

Hydrocyclone sizes are designed arbitrarily by the inside cone diameter at the inlet. By convention, desanders have a cone diameter of 6 inches and larger; desilters have internal diameters smaller than 6 inches. Normally, discharges from the apex of these cones are discarded when used on unweighted drilling fluids. Prolonged use of these cones on a weighted drilling fluid results in a significant mud weight reduction caused by the discard of weighting material. When these cones are used as part of a mud cleaner configuration, the cone underflow is presented to a shaker screen. The shaker screen returns most of the barite and liquid to the drilling fluid system, rejecting solids larger than the screen mesh. This is a common application of unbalanced cones since the cut point is determined by the shaker screen and not the cone.

Since most hydrocyclones are designed to operate with 75 feet of head at the input manifold, the flow rate through the cones is constant and predictable from the diameter of the cone (Table 7-3). Obviously, manufacturers may select different orifice sizes at the inlet of the cone. The orifice size determines the flow rate through the cone at 75 feet of head.

**TABLE 7-3.** Flow Rates through Hydrocyclones

| Designation | Cone Diameter (in.) | Flow Rate through Each Cone (gpm) |
|---|---|---|
| Desilter | 2 | **10**–30 |
| Desilter | 4 | **50**–65 |
| Desilter | 5 | **75**–85 |
| Desander | 6 | **100**–120 |
| Desander | 8 | **200**–240 |
| Desander | 10 | **400**–500 |
| Desander | 12 | **500**–600 |

*Note:* Numbers in bold indicate the most common flow rate.

The $D_{50}$ **cut point** of a solids separation device is usually defined as the particle size at which one-half of the weight of those particles go to the underflow and one-half of the weight of those particles go to the overflow. The cut point is related to the inside diameter of the hydrocyclone. For example, a 12-inch cone is capable of a $D_{50}$ cut point around 60 to 80 microns, a 6-inch cone is capable of around 40 to 60 microns, and a 4-inch cone is capable of around 20 to 40 microns. These cut points are representative for a fluid that contains a low solids content. The cut point will vary according to the size and quantity of solids in the feed and the flow properties of the fluid. Cut point determination procedures are explained in Chapter 9.

When hydrocyclones are mounted above the liquid level in the mud tanks, a siphon breaker should be installed in the overflow manifold from the cones.

## Discharge

Most hydrocyclones are designed to be **balanced**. A properly adjusted, balanced hydrocyclone has a spray discharge at the underflow outlet and exhibits a central air suction core. A balanced cyclone can be adjusted so that when water is fed under pressure, nothing discharges at the apex. Conversely, when coarse solids are added to the feed slurry, wet solids are discharged at the apex. Even with this adjustment, there still should be a large opening in the bottom of the cyclone. This will confirm that the cyclone is hydraulically balanced and discharges at the bottom (apex) only when solids, which the cyclone can separate, are in the feed slurry (drilling fluid).

A balanced cyclone should be operated with "**spray discharge**." Here, coarser solids separate to the outside in the downward spiral and pass over the lip of the apex as an annular ring. The apex is actually a weir or dam, not a choke or valve.

The high-velocity return stream spinning upward near the center of the cone into the vortex finder generates a column of lower pressure, which sucks air inward through the center of the apex opening.

To set a cone to balance, slowly open the apex discharge while circulating water through the cone. When a small amount of water is discharged and the center air core is almost the same diameter as the opening, the cone is "balanced."

With spray discharge, the device removes the maximum amount of solids and discarding of whole mud is minimized. The umbrella-shaped spray discharge indicates that a uniform solids loading is presenting to the cone, with proper separation occurring. The pattern of the apex discharge provides a good indication of cone operation. The discharge should have a hollow center and appear as a cone spray. A wide cone spray may indicate the apex orifice is too large. When the apex orifice is larger than required, an excess amount of liquid will exit, carrying with it finer feed solids, thereby reducing sharpness of separation and underflow density.

Several conditions restrict separations and exiting of solids that have spiraled along the cone wall. These include:

- Excessive solids concentration
- Excessive volumetric feed rate per cone
- Excessive fluid viscosity
- Restricted (too small) apex
- Inadequate feed pressure

A greater number of larger solids are entrained within the central vortex stream to exit with the overflow. The discharge pattern changes from spray to "**rope discharge**"—characterized by a cylindrical or "ropy" appearance. With the rope discharge, no air core occurs through the center of the cone.

In this case, the apex acts as a choke that restricts flow rather than a weir.

Rope discharge is a process where material pours from the cone apex as a slow moving cylinder (or rope). The hydrocyclone effects only inefficient solids-liquid separations. The apex velocity in rope discharge is far less than that in spray discharge. Separations are less efficient and, because of the lower velocity, fewer solids are discarded.

A rope discharge can create a false sense of success as the heavier rope stream appears to contain more solids than a spray discharge. In reality, a rope discharge indicates that not all solids that have been separated inside the cone can exit through the apex opening. Solids become crowded at the apex and and cannot exit the cone freely. The exit rate is slowed significantly and some solids, which would otherwise be separated, become caught in the inner spiral and are carried to the overflow. Dry discharge can also produce cone plugging.

With rope discharge, the exiting solids stream is heavier than under spray discharge conditions. All discharged solids will have a surface film of "bound" liquid. Since finer solids have a greater surface area to volume (size or mass) ratio, finer solids' streams involve greater volumes of bound water. More bound water causes a less dense underflow stream (the finer the particle separation, the wetter the apex stream). This explains why spray discharge stream densities are less than rope dicharge stream densities.

The amount of fluid lost in cone underflow is important. A hydrocyclone operating with spray discharge gives solids a free path to flow (exit the cone). Rope discharge is a dry discharge. Therefore, spray discharge removes significantly more solids than rope discharge. More fluid may be lost in spray discharge, but the greater solids separation efficiency makes the additional fluid loss insignificant. If fluid loss is a concern, the underflow can be screened (see Mud Cleaners) or centrifuged for liquid recovery.

Rope discharge should be immediately corrected to re-establish the higher volumetric flow and greater solids separation of spray discharge. A rope discharge indicates equipment is overloaded and additional hydrocyclones may be necessary.

## Hydrocyclone Capacity

Since most hydrocyclones are designed to operate at a constant 75 feet of head at the input manifold, and flow rate through any cone is constant at constant inlet pressure, flow rate through any cone is predicatable.

Hydrocyclones are rated from 40 to 80 gpm of liquid removal. The normal 4-inch cones will remove 4 gpm of solids, or 5.7 barrels per hour of solids, per cone. Therefore, the standard 16 cone arrangement will accommodate removal of 510 cubic feet of solids per hour. For a $17\frac{1}{2}$-inch hole, this equates to penetration rates averaging 297 feet per hour. Clearly, if design and operational characteristics are adequately maintained, more than ample solids separation can be effected.

The accelerated gravitational forces generated in hydrocyclones are inversely proportional to the radius of the cyclone cylinder. Thus, the larger the diameter, the coarser the separation. In general, the larger the hydrocyclone, the coarser its cut point and greater its throughput. The smaller the cone, the smaller size particles the cone will separate. In other words, the median particle size removed decreases with cone diameter. Median particle size also increases with increasing fluid viscosity and density, but decreases as particle-specific gravity increases. Oil-field hydrocyclones range between 4 and 12 inches, based on the inner diameter of the intake cylinder A small hydrocyclone diameter is used for ultra-fine separations.

## Hydrocyclone Tanks and Arrangements

Hydrocyclones are arranged with the larger cone-size unit upstream of the smaller unit. A separate tank is needed for each size unit. Generally, a desander size and a desilter size are available as part of the rig equipment.

Suction is taken from the tank immediately upstream of the discharge tank. The number of cones in use should process 100% of the flow rate of all fluids entering the suction tank for the hydrocyclone. A backflow between the hydrocyclone discharge and suction compartment of at least 100 gpm usually ensures adequate processing. Estimations based on rig circulation rates are usually inadequate if the plumbing is not arranged properly. For example, if a 500 gpm hydrocyclone overflow is returned to the suction compartment and a 400 gpm rig flow rate enters the suction compartment, adequate processing is not achieved even though more fluid is processed than is pumped downhole. In this case, the flow entering the hydrocyclone suction compartment is 900 gpm. The fraction of drilling fluid processed would be 500 gpm/900 gpm, or 56%.

## Desanders

Desanding units are designed to separate drilled solids in the 50 to 80 micron range and barite in the 30 to 50 micron range. They are primarily used to remove high solids volume associated with fast drilling of large-diameter top holes. In water-based drilling fluids, desanders make a median separation cut of 2.6 specific gravity solids in the 50 to 80 micron size range. The desander removes sand-size and larger particles that pass through the shale shaker screens.

Desanders are installed immediately downstream of the shaker and degasser. Suction is taken from the immediate upstream tank, usually the degasser discharge tank. Discharge from the desander is made into the tank immediately downstream. Suction and discharge tanks are equalized through valves located on the bottom of each tank.

Desanders are used continuously while drilling. Plumbing can be arranged to process all total surface pit volume after beginning a trip.

Use of desanders is generally discontinued after barite and/or expensive polymers are added to the drilling mud because a desander discards such a high proportion of these materials. Use of desanders is generally not cost effective with an "oil-based" drilling fluid because the larger cones discharge a significant amount of the liquid phase (see Mud Cleaners discussed later in this chapter).

## Desilters

Desilter cones are manufactured in a variety of dimensions, ranging from 2 to 6 inches, and make drilled solids separations in the 12 to 40 micron range. They will also separate barite particles in the 8 to 25 micron range. Desilters are installed downstream from the shale shaker, sand trap, degasser, and desander.

Desilter cones differ from desander cones only in dimensions and operate on exactly the same principles. Common desilter cone sizes are between 2 and 5 inches. Desilter cones should be fed by a centrifugal pump dedicated only to providing fluid to the desilter.

These units make the finest particle-size separations of any full-flow solids control equipment—down to 12 microns of drilled solids. The desilter, therefore, is an important device for reducing average particle size and eliminating drilled solids.

Desilter suction is also taken from the immediate upstream tank, usually the desander discharge tank. Desilter suction and discharge tanks are, again, equalized through a valve(s) located on the bottom of each tank. Suction should not be taken from the tank into which chemicals and other materials (barite and bentonite) are added because valuable treating materials may be lost.

## Summary

Desanders should be used in unweighted mud when shakers are unable to screen down to 140 mesh (100 micron). The role of desanders is to reduce loading downstream on desilters. Installing a desander ahead of the desilter relieves a significant amount of solids loading on the desilter and improves its efficiency. High rates of penetration, especially unconsolidated "surface hole" where the largest diameter bits are used, results in generating larger concentrations of drilled solids. This may place desilters in rope discharge. For this reason, desanders, which have greater volumetric capacity and can make separations of coarser drilled solids, are placed upstream of desilters. Desanders remove a higher mass (coarser drilled solids) during periods of high solids loading. Desilters can then efficiently process the reduced solids content overflow of the desanders.

If the drill rate is slow, generating only a few hundred pounds per hour of drilled solids, the desander may be turned off and the desilter used to process the entire circulating system.

Desilters should be used on all unweighted, water-based mud. These units are not used on weighted muds because they discard an appreciable amount of barite. Most barite particles fall within the silt-size range.

Desilter operation is important for all unweighted fluids, however, in oil-based muds with high viscosities (as found in deep water drilling) the apex discharge may be centrifuged for "oil-phase" salvage.

Hydrocyclones should process all drilling fluid entering their suction compartments independent of the drilling fluid circulation rate.

## Median ($D_{50}$) Cut Points

In spray discharge, for any set of cone diameter, feed slurry compositions, flow properties, volumetric flow rates, and pressure conditions, some particles size (mass) is 100% discarded from the apex.

For every size and design of cone operating at a given pressure with feed slurry of a given viscosity, density, and solids distribution, there is a certain size (mass) of particle that shows no preference for either top or bottom discharge. As a result, 50% of this particular size exits through the vortex and 50% exits through the apex. This particle size is termed the "median cut," "median size particle," or more frequently in drilling operations, the $D_{50}$ cut point.

The median cut, or $D_{50}$ cut point, does not mean that all larger particles exit at the apex and smaller particles exit at the vortex. The $D_{50}$ **cut point** of a solids separation device is defined as that particle size at which one-half of the weight of specific size particles go to the underflow and one-half of the weight go to the overflow. For example, a $D_{30}$ cut point references a particle size which is 30% concentrated in the underflow and 70% in the overflow.

As stated earlier, the cut point is related to the inside diameter of the hydrocyclone. For example, a 12-inch cone has a $D_{50}$ cut point for low-gravity solids in water of approximately 60 to 80 microns, a 6-inch cone around 30 to 60 microns, and a 4-inch cone around 15 to 20 microns (Table 7-4). However, the cut point will vary with the size and amount of solids in the feed, as well as fluid viscosity.

For comparative purposes, consider a 50-micron equivalent drilled solid diameter. Relatively speaking, the percent discharge is as follows:

- 6-inch cone discharges 80% at underflow
- 4-inch cone discharges 95% at underflow
- 3-inch cone discharges 97% at underflow

Now consider a 10-micron equivalent drilled solid diameter:

- 6-inch cone discharges 7% at underflow
- 4-inch cone discharges 11% at underflow
- 3-inch cone discharges 17% at underflow

If a graph of particle size versus percent of particles recovered to underflow is plotted, the portion of the curve near the $D_{50}$, or 50%, recovery point (median cut point) is very steep when separations are efficient.

Particle separations in hydrocyclones vary considerably. In addition to proper feed head and the cone apex setting, drilling fluid properties including density, percent solids (and solids distribution) and viscosity, all affect separations. Any increase in these mud properties will increase the cut point of a separation device.

## Stokes' Law

Stokes Law defines the relationship between parameters that control the settling velocity of particles in viscous liquids, not only in settling pits but also in equipment such as hydrocyclones and centrifuges.

Separations in a settling pit are controlled by the force of gravity and the viscosity of the suspending fluid (drilling mud). A large, heavy particle settles faster than a small, lighter particle. This settling process can be increased by reducing the viscosity of the suspending fluid, increasing the gravitational forces on the particles, or by increasing the effective particle(s) size with flocculation or coagulation.

Hydrocyclones and centrifuges increase settling rates by applying increased centrifugal force, which is equivalent to higher gravity force.

Stokes' Law for settling spherical particles in a viscous liquid is expressed as:

$$V_s = \frac{CgD_E^2(\rho_s - \rho_l)}{\mu}$$

where $V_s$ = Settling or terminal velocity, feet/sec
$C$ = Units constant, $2.15 \times 10^{-7}$
$g$ = Acceleration (gravity or apparatus) ft/sec$^2$
$D_E$ = Particle equivalent diameter, microns
$\rho_s$ = Specific gravity of solids (cutting, barite, etc.)
$\rho_l$ = Specific gravity of liquid phase
$\mu$ = Viscosity of media, centipoise

Various size particles with different densities can have the same settling rates. That is, there exists an **equivalent diameter** for every 2.65 specific gravity drilled solid, be it limestone, sand, or shale, which cannot be separated by gravimetric methods from barite particles of a corresponding equivalent diameter. Presently, it is not possible to separate desirable barite particles from undesirable drilled solid particles that settle at the same rate.

**TABLE 7-4.** Hydrocyclone Size versus $D_{50}$ Cut Point

| Cone Diameter (inches) | $D_{50}$ Cut Point in Water | $D_{50}$ Cut Point in Drilling Fluid |
|---|---|---|
| 2 | 8–10 | 15+ |
| 4 | 15–20 | 35–70 |
| 6 | 30–35 | 70–100 |
| 12 | 60–70 | 200+ |

Generally, a barite particle (specific gravity = 4.25) will settle at the same rate as a drilled solids particle (specific gravity = 2.65) that is $1\frac{1}{2}$ times the barite particle's diameter. This may be verified by applying Stokes' Law.

***Example #1.*** A viscosified seawater fluid with a specific gravity of 1.1, PV = 2.0 centipoise, and YP = 12.0 lbs/100 ft², is circulated to clean out a cased wellbore. What size low-gravity solids will settle out with 5-micron barite particles? With 10-micron barite particles, what is the settling velocity in rig tanks?

Using Stokes' Law, the settling velocity is:

$$V_s = \frac{Cg(\rho_s - \rho_l)D_E^2}{\mu}$$

For equivalent settling rates, $V_s = V_s$. And for $\mu_1 = \mu_2$ (the same fluid, therefore, the same viscosity).

$$D_1^2(\rho_1 - \rho_l) = (\rho_2 - \rho_l)$$

$\rho_1 = 2.65$
$\rho_2 = 4.25$
$\rho_l = 1.1$

$(D_1^2 \times 1.55) = (D_2^2 \times 3.15)$

$D_1^2/D_2^2 = 3.15/1.55 = 2.03$

$D_1/D_2 = 1.42$

Thus, a 5-micron barite particle will settle at the same rate as a 7-micron low-gravity particle, and a 10-micron barite particle will settle at the same rate as a 14-micron low-gravity particle.

Settling velocity for a 5-micron barite (or 7-micron drilled solid) particle is:

$$V_s = \frac{(2.15 \times 10^{-7}) \times 32.2 \text{ ft/sec}^2 \times (4.25 - 1.1) \times 25}{16 \text{ cp}}$$

$= 3.4 \times 10^{-5}$ ft/sec (or 0.24 in./min)

and for a 10-micron barite (or 14-micron drilled solid) particle:

$$V_s = \frac{(2.15 \times 10^{-7}) \times 32.2 \text{ ft/sec}^2 \times (4.25 - 1.1) \times 100}{16 \text{ cp}}$$

$= 13.6 \times 10^{-5}$ ft/sec (or 0.98 in./min)

***Example #2.*** What are the equivalent diameters of barite (specific gravity = 4.25) and drilled solids (specific gravity = 2.65) in an 11.5 ppg mud with PV = 20 cp and YP = 12 lbs/100 ft²?

Using Stokes' Law, the settling velocity is:

$$V_s = \frac{Cg(\rho_s - \rho_l)D_E^2}{\mu}$$

For equivalent settling rates, $V_s = V_s$. And for $\mu_1 = \mu_2$ (the same fluid, therefore, the same viscosity).

$$D_1^2(\rho_1 - \rho_l) = D_2^2(\rho_2 - \rho_l)$$

$\rho_1 = 2.65$
$\rho_2 = 4.25$
$\rho_l = 1.38$

$(D_1^2 \times 1.27) = (D_2^2 \times 2.87)$

$D_1^2/D_2^2 = 2.87/1.27 = 2.26$

$D_1/D_2 = 1.50$

Thus, a 10-micron barite particle will settle at the same rate as a 15-micron low-gravity particle, and a 50-micron barite particle will settle at the same rate as a 75-micron low-gravity particle.

Stokes' Law shows that as fluid viscosity and density increase, separation efficiency decreases.

If the drilling fluid weight is 14.0 pounds per gallon (specific gravity = 1.68):

$$\frac{d^2_{ds}}{d^2_b} = \frac{4.25 - 1.68}{2.65 - 1.68} = \frac{2.57}{0.97} = 2.65$$

or

$$\frac{d_{ds}}{d_b} = 1.63$$

Therefore, in drilling fluid weighing 14 pounds per gallon, a 10-micron barite particle will settle at the same rate as a 16-micron drilled solid particle, and a 50-micron barite particle will settle at the same rate as an 80-micron (or 81.4) drilled solid particle.

It is important to remember that the efficiency of a separator is viscosity dependent. The median cut, or $D_{50}$ cut point, increases with viscosity as shown by Stokes' Law:

$$V_s = \frac{CgD_E^2(\rho_s - \rho_{fl})}{\mu}$$

***Example #3.*** A 4-inch cone will separate half of the 12-micron low-gravity (specific gravity = 2.6) particles in water (that is, the $D_{50}$ cut point is 12 microns). What is the $D_{50}$ cut point in a 50-cp viscosity fluid of the same density?

For constant settling velocity, if fluid density is unchanged and other parameters remain constant:

$$V_s = C\rho D_E^2 / \mu$$

For $\rho_1 = \rho_2$

$$D_1^2 / \mu_1 = D_2^2 / \mu_2$$

$D_1$ = 12 micron
$\mu_1$ = 1 cp
$\mu_2$ = 50 cp

$(12)^2/1 = D_2^2/50$

$D_2^2 = 7,200$

$D_2$ = 84.8 microns

Thus, if fluid specific gravity remains 1.1 and viscosity is 50 centipoise, the $D_{50}$ is raised to 85 microns.

Similarly, for a 4-inch hydrocyclone:

| Fluid viscosity (cp) | $D_{50}$, microns (sp. gr. = 2.6) |
|---|---|
| 1.0 | 12.0 |
| 10.0 | 37.9 |
| 20.0 | 53.7 |
| 30.0 | 65.8 |
| 40.0 | 75.8 |
| 50.0 | 84.8 |

Cut point performance can be further projected by dividing by the projected specific gravity at various viscosities. Thus, in the above example, for 20 centipoise viscosity and 1.4 (11.7 ppg) density fluid, the $D_{50}$ would be:

$[(53.7 \text{ micron})^2 \times 1.1/1.4]^{0.5}$ = 47.5 microns

The $D_{50}$ cut point for 6-inch and 8-inch hydrocyclones (common desilter and desander sizes) are usually given as 25 and 60 microns, respectively.

The variation of $D_{50}$ cut point with viscosity for a 6-inch desilter hydrocyclone (specific gravity = 1.0) is as follows:

| Fluid viscosity (cp) | $D_{50}$, microns (sp. gr. = 2.6) |
|---|---|
| 1.0 | 25.0 |
| 10.0 | 79.0 |
| 20.0 | 112 |
| 30.0 | 137 |
| 40.0 | 158 |
| 50.0 | 176 |

Again, cut point performance can be further projected by dividing by the projected specific gravity at various viscosities. Thus, in the above example, for a 6-inch hydrocyclone, 20 centipoise viscosity, and 1.4 (11.7 ppg) density fluid, the $D_{50}$ would be:

$[(112 \text{ microns})^2 \times 1.1/1.4]^{0.5}$ = 99.3 microns

## Hydrocyclone Operating Tips

- Other than cone and manifold plugging, improperly sized or operated centrifugal pumps are by far the greatest source of problems encountered with hydrocyclones. Centrifugal pump and piping sizes are critical to efficient hydrocyclone operation.

- Hydrocyclones should always have a pressure gauge installed on the inlet manifold to quickly determine if proper feed head is supplied by the centrifugal pump.

- Hydrocyclones are usually mounted in the vertical position but may be mounted horizontally. Cone orientation is irrelevant because the separating force is supplied by the centrifugal pump.

- Feed slurry must be distributed to a number of hydrocyclones operating in parallel. A "radial manifold" provides each cyclone with the same slurry (in terms of feed solids concentration and particle size distribution) and at the same pressure. An "in-line" manifold guarantees the higher mass (larger diameter) particles pass the first cyclones and, instead, enter the last cones. Because these particles have a higher energy, they resist entering the first cones. Thus, the last cones in an in-line manifold receive a higher concentration of coarse feed particles. Cone performance in an in-line manifold will not be identical since feed concentrations and particle size distributions differ for various cones. Further, if the last cyclone(s) in an in-line manifold is taken off-line, the end of the manifold has a tendency to plug.

- To minimize loss of head along the feed line and backpressure on the overflow (top) discharge line, keep all lines as short and straight as possible with a minimum of pipe fittings, turns, and elevation changes. Pipe diameters should be 6 or 8 inches.

- Operate cones in spray discharge with a central air suction, and check cones regularly to ensure the apex discharge is not plugged.

- When balanced cones no longer operate with spray discharge, either too many solids are being presented for design processing, large solids have plugged the manifold or apex, or the feed pressure is incorrect. If the feed pressure is correct at 75 feet of hydraulic head, often the inability to maintain spray discharge can be attributed to the shale shakers. Check for torn screens, open bypass, or improperly mounted screens. Additionally, ensure that there are sufficient hydrocyclones to process the total fluid being circulated by the mud pumps.

- Install a low ("bottom") equalizer to permit backflow from the discharge tank into the suction compartment. Removable centrifugal pump suction screens reduce plugging problems.

- Operate hydrocyclones at the recommended feed head of approximately 75 feet. Efficiency is decreased by operating at too low a feed head. Operating at too high a head wears hydrocyclone parts and yields insignificant separation improvements.

- Do NOT use the same pump to feed the desilter and desander. Each unit should have its own centrifugal pump(s).

- Run the desilter continuously while drilling with unweighted mud and process 100% of the total surface pit volume after beginning a trip.

- Operate the desander when shale shakers will not screen down to 140 mesh (100 microns).

- Install a guard screen with approximately half-inch openings to prevent large debris from entering and plugging the inlets and/or cone apexes.

- Regularly check cones for bottom plugging or flooding. Desilter cones will plug more often than desander cones. Plugged cones may be cleaned with a welding rod. A flooded cone indicates a partially plugged feed or a worn cone bottom section.

- Between wells, or when drilling is interrupted, flush manifolds with water and examine the inside surfaces of the cones.

- Keep the shale shaker well maintained and never bypass it.

- Hydrocyclones discard absorbed liquid with the drilled solids. Discharge solids dryness is a function of the apex opening relative to the diameter of the vortex finder.

- Mud cleaners and/or centrifuges can be used to process the cones underflow.

## Hydrocyclone Installation Tips

- Hydrocyclones should be located so that underflow can be removed with a minimum of trouble and wash-down water. They should be located so they are accessible for maintenance and evaluation.

- Discharge overflow should return to the circulating system into a compartment immediately downstream of the suction compartment. At the same time, the discharge overflow compartment must be bottom equalized with the suction compartment and drilling fluid should backflow from the discharge tank into the suction tank at all times. The hydrocyclones should process all drilling fluid entering its suction compartment. Ensure that all drilling fluid entering the discharge compartment has been processed by the hydrocyclones.

- Centrifugal feed pumps should be located so that they have flooded suction, minimum suction line length, with few elbows and turns to minimize friction losses.

- A centrifugal pump may stir its suction compartment with mud guns. Mud jet mixers should not be supplied with fluid from other parts of the drilling fluid system. Preferably, mechanical agitators should be used in the removal section.

- Keep the end of the discharge line above the surface of the mud in the receiving tank to avoid creating a vacuum. Overflow should be introduced into the next compartment downstream at approximately a 45° angle so lines will be kept full and a siphoning vacuum, which would pull more solids into the overflow discharge, avoided. When hydrocyclones are mounted more than five feet above the liquid level of the mud tanks, a siphon breaker should be installed in the overflow manifold from the cones.

## Conclusions

Hydrocyclones are simple, easily maintained mechanical devices without moving parts. Separation is accomplished by transfer of kinetic input, or feed energy, into the centrifugal force inside the cone. The centrifugal force acts on the drilling fluid slurry to rapidly separate drilled solids and other solid particles in accordance with Stokes' Law.

The solids generated while drilling in some formations are too fine for shale shakers to remove.

## Hydrocyclone troubleshooting:

| Symptom: | Probable Cause(s) and/or Action: |
|---|---|
| Some cones continually plug at apex | • Partially plugged feed inlet or outlet; remove cone and clean out lines<br>• Check shaker for torn screens or bypassing<br>• Possibly increase apex size |
| Some cones losing whole mud in a stream | • Plugged cone feed inlet allowing backflow from overflow manifold |
| Low feed head | • Check centrifugal pump operation—RPM, voltage, etc.<br>• Check shaker for torn screens or bypassing<br>• Check for line obstructions, solids settling, or partially closed valve |
| Cones discharge dry solids | • Increase apex size and/or install more cones |
| Vacuum in manifold discharge | • Install anti-siphon tube |
| Increasing solids concentration in drilling fluid | • Insufficient cone capacity; solids may be too small; use finer shaker screens |
| Heavy discharge stream | • Overloaded cones; increase apex size and/or install more cones |
| High drilling fluid losses | • Cone apex too large; reduce size<br>• Check for centrifuge cyclone discharge<br>• Reduce cone sizes |
| Unsteady cone discharge, varying feed head | • Air or gas in feed line |
| Aerated mud downstream of hydrocyclone | • Route overflow into trough to allow air break-out |

Hydrocyclones must be relied on to remove the majority of these solids. In this case, the shale shaker protects the hydrocyclones from oversize particles that may cause plugging.

Hydrocyclones should be designed to provide maximum removal of solids with a minimum loss of liquid. There should be sufficient hydrocyclones arranged in parallel to process all drilling fluid arriving into the additions region of the mud tank system.

Hydrocyclones produce a wet discharge compared to shale shakers and centrifuges. Underflow density alone is not a good indicator of cone performance since finer solids have more associated liquid and the resultant slurry density is lower than for coarser solids. As the solids content increases, separation efficiency decreases and the size of particles that can be separated increases.

The advantages of hydrocyclones are simple design, no moving parts, easy maintenance, and good separation ability.

The disadvantages are the inability to handle flocculated materials and a limited separation of ultra-fines.

It is impractical to desand or desilt a mud containing appreciable amounts of barite. Silts and barite have about the same size range. The majority of barite particles are between 2 and 44 microns; however, some range between 44 and 74 microns and, unfortunately, 8 to 15% range between 0 and 2 microns.

A desander median cut falls between 25 and 30 microns. A desilter median cut falls between 10 and 15 microns. Since much of the barite falls above these cuts, it is discarded along with the silt and sand.

> **Benefits of hydrocyclones:**
>
> 1. Replacement of pump fluid end parts is reduced and pumps operate more efficiently.
> 2. Less drill string torque and drag equates to less wear on string and less key-seating (a major potential for stuck pipe). Casing is run easier.
> 3. Bit life is extended due to less abrasion.
> 4. Penetration rates increase.
> 5. Water dilution to maintain low mud weights is reduced. This is reflected in smaller waste pits and smaller drilling fluid volumes.
> 6. Material additions are decreased.
> 7. Additions of weight material are made with little or no difficulty.
> 8. Downhole tools set and release with little or no interference from drilled cuttings.

Generally hydrocyclones are most efficient when solids have a diameter greater than 10 microns and are spherical in shape. If the solids are flat (such as mica), movement tends to be random dependent on whether the flat surface or edge is toward the gravitational force created in the vortex.

Since separation efficiency somewhat depends on the freedom and velocity of the solid moving through the liquid phase, it is logical to use the lowest viscosity fluid possible.

Because fine solids have more surface per unit volume ("specific area"), the amount of liquid discharged per pound of solids is higher with fine solids than with coarse solids. Therefore, the difference between the feed and underflow densities is not a reliable indicator of hydrocyclone performance.

Pressure drop is a measure of the energy being expended in the cone. Thus, a higher pressure drop results in a finer separation. If the $D_{50}$ is increased to 75 microns, 25% of the 100-micron particles are retained and only 25% of the 55-micron particles are discharged.

The purpose of a hydrocyclone is to discharge maximum abrasive solids with minimal fluid loss. Larger particles have a greater probability of discharging through the bottom underflow (apex), while smaller and lighter particles have a greater probability to discharge through the top, or overflow, opening. Cone diameter, cone angle, underflow diameter, feed head, and plastic viscosity have the greatest impact on hydrocyclone performance.

Barite particles 3 microns and smaller have a deleterious effect on drilling fluid viscosity due to a surface charge imbalance resulting from unsatisfied broken bonds on the ultra-fine's surface. Therefore, if a centrifuge is set so that its median ($D_{50}$) cut in 14-pound per gallon mud is a 3-micron barite particle, it's median cut for drilled solids will be 5 (or 4.9) microns.

## MUD CLEANERS

Until the development of linear motion shale shakers, an 80-mesh screen was about the finest screen that could process drilling fluid at a rig. With this limitation, drilled solids smaller than 177 microns remained in the drilling fluid. The **mud cleaner** was originally designed to remove drilled solids from a weighted drilling fluid in sizes ranging from 177 to 74 microns (200 to 80 mesh). Because barite is generally ground so that the majority is smaller than 74 microns, most of it should pass through the screen on a mud cleaner.

The mud cleaner is a combination of hydrocyclones and a fine-mesh screen shaker. Weighted drilling fluid is first processed through hydrocyclones. Enough cones should be used to ensure that all fluid entering the desilter suction compartment is treated. The underflow of the hydrocyclones contains drilled solids, weighting material, and some whole mud. The shaker sieves the fluid through a 150- to 200-mesh screen. This allows most of the weighting material to pass through the screen. The solids retained on the fine-mesh screen are discarded. If the fluid on the screen leaves the material too soon, the discarded material contains appreciable quantities of weighting material. A light mud-spray on the screen helps the separation process.

One advantage of the mud cleaner is that the shaker screen only processes about 1 to 2 gallons for every 50 gallons passing through 4-inch desilters. The $D_{50}$ cut point of a 4-inch hydrocyclone processing a weighted drilling fluid may be as high as 60 to 70 microns. Therefore, most of the weighting material is not presented to the mud cleaner screen.

Usually, the screen discharge from a 200-mesh mud cleaner screen has less whole mud associated with it than a primary shale shaker discharge. This relatively dry discharge can give the appearance of being primarily weighting material. Measurements show, however, that the discharge from a 200-mesh shale shaker screen and a properly treated 200-mesh mud cleaner screen are similar.

The underflow of the desilters does not contain a significant quantity of **viscosifiers**. The material that passes through the screen cannot transport weighting material. Care must be exercised to return the material to a well-agitated location in the pit.

With the advent of 200-mesh screens on linear motion shale shakers, mud cleaner usage significantly decreased. Slowly, they have returned to the removal system because trial usage indicates that they still remove a large quantity of solids. Plugging of the lower apex of mud cleaner hydrocyclones indicates that very large solids frequently bypass the shale shaker screens. Solids large enough to plug a 4-inch cone indicate the presence of a wide range of other solids. Even when linear motion shale shakers are operating properly, mud cleaners remove a significant quantity of drilled solids.

In unweighted drilling fluids, the mud cleaner can be used to salvage an expensive liquid phase, such as oil or a high concentration of KCl. Usually, better results are obtained if the hydrocyclone underflow is fed to a decanting centrifuge instead of being screened. A centrifuge processing the underflow from hydrocyclones is a better choice to return the liquid phase. A mud cleaner can be used but not as effectively.

Currently, the hydrocyclones and screens are used to diminish the liquid discharge from a system. This combination is called a "**dryer**" and is discussed in Chapter 4.

Most cut point curves of 4-inch desilters show that only about 5% of the underflow from the cones is less than 5 microns. Viscosifiers are in this size range. The underflow that passes through the mud cleaner screen contains such a small amount of viscosifier and collodial material that it will not support the other solids. For this reason, the mud cleaner should be viewed as another method of removing drilled solids larger than the fine mesh size on the shaker screen. If this material remains in the drilling fluid, it will eventually become undesirable ultra-fine solids. Colloidal material will increase at an even faster rate because of the deterioration of larger solids.

Mud cleaners do not compete with centrifuges. Centrifuges remove colloidal-size particles, usually smaller than the weighing agent, from the drilling fluid. Mud cleaners remove drilled solids larger than the weighing agent. If improperly used, mud cleaners will also discard large quantities of barite. The slurry on the screen must not "dewater" too soon or the barite will "clump" together and be discharged. Frequently, a small stream of drilling fluid from the desilter overflow line can prevent the loss of too much barite.

As with desanders and desilters, mud cleaner hydrocyclones typically receive drilling fluid through a centrifugal pump. Although there have been reports of solids degradation due to the centrifugal pump, particle size measurements at a variety of rig sites fail to confirm this. Solids will eventually degrade, however, this occurs slowly and in a variety of other places as well. These include the liquid pool of linear motion shale shakers (particularlily deep pools), the turbulent zone of decanting centrifuges, mud guns, crushing by the drillpipe, and the high-shear through drill bit nozzles.

## CENTRIFUGES

A centrifuge works on the principle of accelerated settling. By imparting additional "G"-forces to the content of a centrifuge, solids (in a solids-laden fluid) will settle much quicker.

To understand how a decanting centrifuge works, first look at a simple sedimentation vessel (Figure 7-4). Solids in this fluid will settle to the bottom of the vessel over a period of time. One way to speed the process of settling is to reduce the height of the vessel so the solids do not have as far to drop. If a specific volume is required, the vessels dimensions can be lengthened or widened (Figure 7-5). Depending on the rheological properties of the fluid and the size and density of the solids, settling time can still prove quite slow.

### Stokes' Law

Particles will settle in a given fluid according to Stokes' Law, which is expressed as follows:

$$V_t = \frac{aD^2(\rho_s - \rho_l) \times 10^{-6}}{116\mu}$$

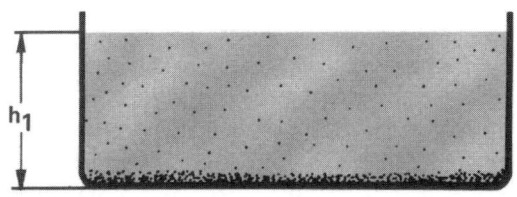

**FIGURE 7-4.** Simple sedimentation vessel.

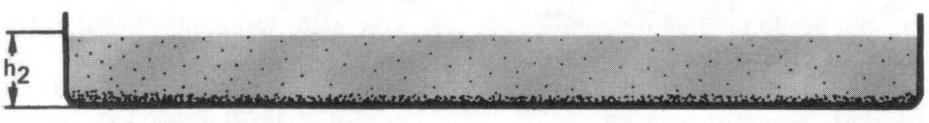

**FIGURE 7-5.** Vessel with small sedimentation height.

where $V_t$ = Terminal or settling velocity
  $a$ = Bowl acceleration, in./sec² = $0.0054812 \times$ bowl diameter $\times$ rpm²
  $D$ = Particle diameter, micron
  $\rho_s$ = Solid (particle) density, grams/cm³
  $\rho_l$ = Liquid density, grams/cm³
  $\mu$ = Liquid viscosity, centipoise (dyne-sec/100 cm²)

Stokes' Law shows that as fluid viscosity and density increases, the settling rate decreases.

It is impossible to separate a drilled solid particle of equivalent mass by settling.

$$\frac{d_{ds}^2}{d_{ds}^2} = \frac{(\rho_s - \rho_l)}{(\rho_{ds} - \rho_l)}$$

where $d_{ds}$ = Diameter of drilled solids particle
  $d_b$ = Diameter of barite particle
  $\rho_{ds}$ = Density of drilled solids particle
  $\rho_b$ = Density of barite
  $\rho$ = Density of liquid

Assuming barite has a specific gravity of 4.25 and drilled solids 2.65, the equivalent diameter ratio for settling in a 14-pound per gallon mud (specific gravity = 1.68) is:

$$\frac{d_{ds}^2}{d_{ds}^2} = \frac{4.25 - 1.68}{2.65 - 1.68} = \frac{2.57}{0.97} = 2.65$$

or

$$\frac{d_{ds}}{d_b} = 1.63$$

In a drilling fluid weighing 14 ppg, a 50-micron barite particle will settle at the same rate as an 81-micron drilled solid particle. All solids, including low-gravity and barite particles 2 microns and smaller (colloidal), can have a detrimental effect on drilling fluid viscosity. That is, a low, specific-gravity particle that has an equivalent spherical diameter that is 1.6 times that of a barite particle, will settle at the same rate as the barite particle. The low-gravity solid will have the same mass as the barite particle. This is the reason that a centrifuge does not separate barite from low-gravity solids.

The particles in a vessel experience an acceleration of 1 "G." The force on a particle is the product of the particle mass and the acceleration. This is called a "G"-force. One way to increase the "G"-force is to rotate the vessel about an axis. In doing so, a simple centrifuge is created as shown in Figure 7-6.

One problem with this design is that as the solids accumulate on the walls of the centrifuge, there is no way to remove them while rotating the centrifuge. Therefore, only small batches of fluid can be treated at one time.

One method that would enable the continuous removal of settled solids is to design a tank as shown in Figure 7-7. This unit uses a drag chain system to remove the settled solids. As the solids are conveyed out of the pool and up the ramp, or beach, they are partially dried prior to discharge. As new fluid is poured into the tank, cleaner fluid may spill out of the weirs. Unfortunately, only 1G is applied to the particles so settling will be very slow.

Another way to remove solids from the tank would be to use an auger, or conveyor, in the tank as shown in Figure 7-8. However, this would not remove the solids that settle away from the conveyor. This could be solved by wrapping the tank around the conveyor (Figure 7-9). This process results in the creation of a complete centrifuge (Figure 7-10).

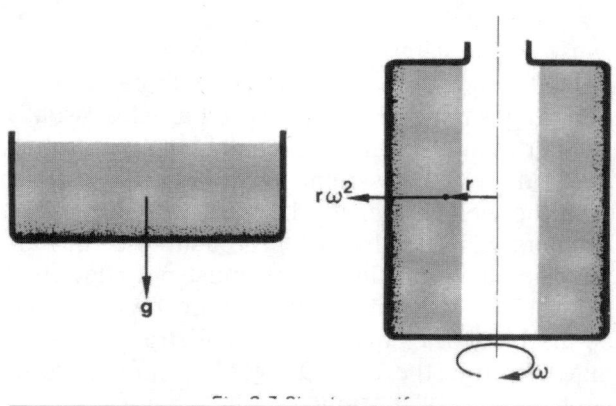

**FIGURE 7-6.** A simple centrifuge.

SOLIDS CONTROL EQUIPMENT **157**

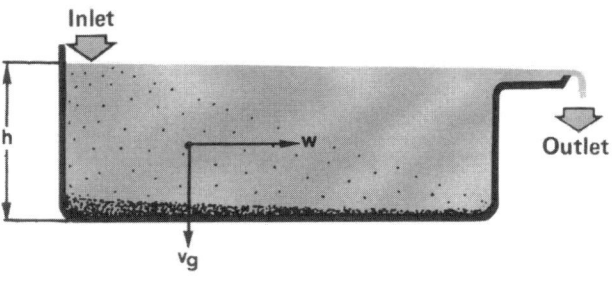

**FIGURE 7-7.** Tank for continuous removal of solid particles from a process liquid.

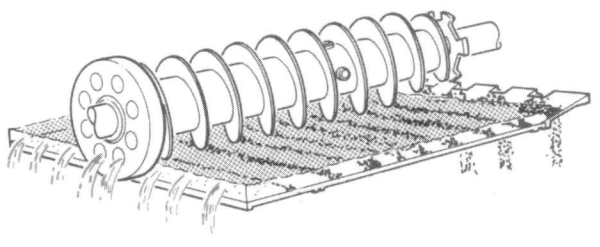

**FIGURE 7-8**

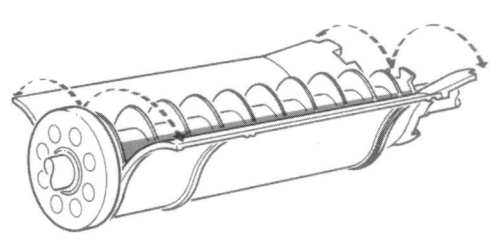

**FIGURE 7-9**

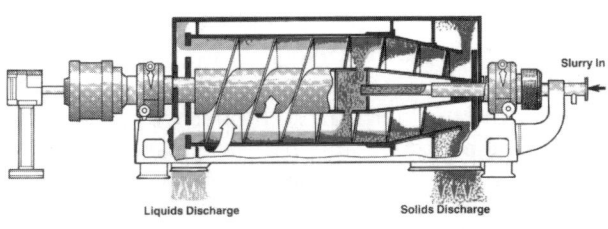

**FIGURE 7-10.** Cross section.

The entire assembly is rotated while increasing the "G"-force on the solids, which accelerates settling. The fluid moves with the outer cylinder of the centrifuge so there is no shear within the fluid. This is the reason that dilution fluid is normally added to the input stream of a decanting centrifuge. The low-shear-rate viscosity of most drilling fluids is increased to aid hole-cleaning and to provide weighting agent support. This low-shear-rate viscosity elevation will also inhibit settling within a centrifuge.

To convey the solids out of the centrifuge, the conveyor and bowl must rotate at slightly different speeds. This is accomplished using a planetary gearbox for belt drive centrifuges. Typically, the entire assembly rotates in the same direction, but the conveyor rotates at a slightly slower speed. The conveyor moves the solids to the solids discharge end and the liquid, or effluent, emptys out of the weirs at the liquid discharge end.

To calculate the "G"-factor a centrifuge imparts to solids, the formula is as follows:

$$G-\text{factor} = \frac{(\text{bowl diameter, in.}) \times (\text{rpm}^2)}{70422}$$

where G is the ratio of the centripetal acceleration of the bowl compared to the acceleration of gravity. (Note, this is the same equation used to calculate the "G"-factor of shale shaker vibrators.)

A centrifuge provides a method of increasing the settling force on particles suspended in liquid. The force depends on the mass of the particle and not the chemical composition. Particles with the same mass, whether they are barite, low-gravity solids, gold, iron, or silver, will settle at the same rate. Centrifuges are able to separate solids above and below the 2- to 10-micron size range. In weighted drilling fluids, centrifuges are capable of eliminating very small particles that can cause dramatic increases in both the low- and high-shear-rate viscosities. In unweighted drilling fluid they are used as "super desilters."

## Types of Centrifuges

Two types of **centrifuges** are generally used to process drilling fluids: decanters and rotary mud separators.

A decanting centrifuge (Figure 7-10 or 7-11) has a bowl that rotates at a high speed (from about 1200 to 3600 rpm). The solids are thrown to the wall of the bowl with a force that depends on their mass. A conveyer transports the solids along the inner wall of the bowl so that they are discharged

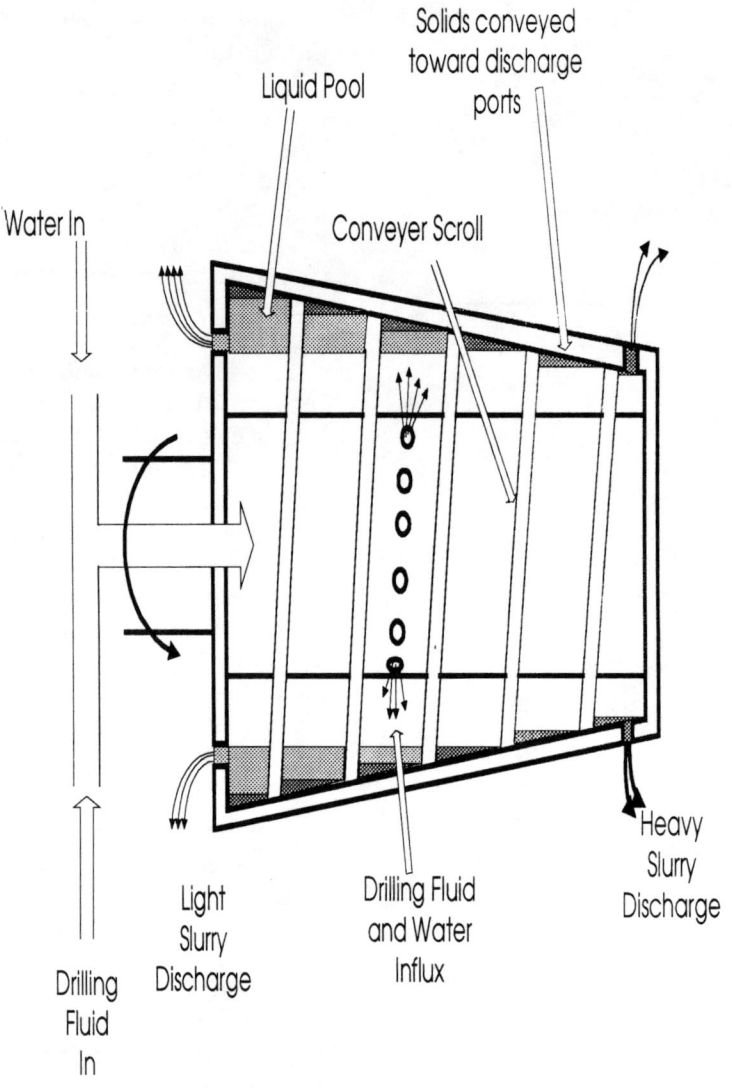

**FIGURE 7-11.** Decanting centrifuge.

as a heavy slurry. The lighter solids and most of the liquid are discharged as a light slurry.

Dilution fluid is added to the input drilling fluid stream in the centrifuge. This reduces the viscosity and lowers the gel structure to permit more rapid solids settling. The fluid inside of a decanting centrifuge moves with the same speed as the bowl so there is only a very small shear-rate applied to the drilling fluid within the machine. In non-Newtonian fluids, the viscosity at low-shear-rates is much higher than the viscosity at high-shear-rates. For this reason, decanting centrifuges are used to remove flocculated and coagulated solids in "closed-mud systems." A desilter, on the other hand, introduces significant shear in the drilling fluid and will usually destroy flocs. Therefore, it cannot be used to remove flocculated and coagulated solids.

A rotary mud separator has a perforated cylinder that rotates inside of an outer cylindrical housing (Figure 7-12). Drilling fluid flows along the outside of the rotating cylinder and smaller solids pass through the perforations. The rotor is attached to a hollow, perforated shaft, which serves as a conduit for the overflow or light slurry. Larger solids are thrown outward to continue passage along the annular space and finally exit from a port on the nonrotating cylindrical case.

A cut point is relatively sharp except at the smaller particle size (because of the liquid accompanying the heavier slurry). Properly operated, the unit can make cuts similar to a decanting centrifuge—2 to 6 microns with barite and 3 to 8 microns with low-gravity solids.

Separation is controlled with a choke in the underflow line. As the choke diameter is reduced,

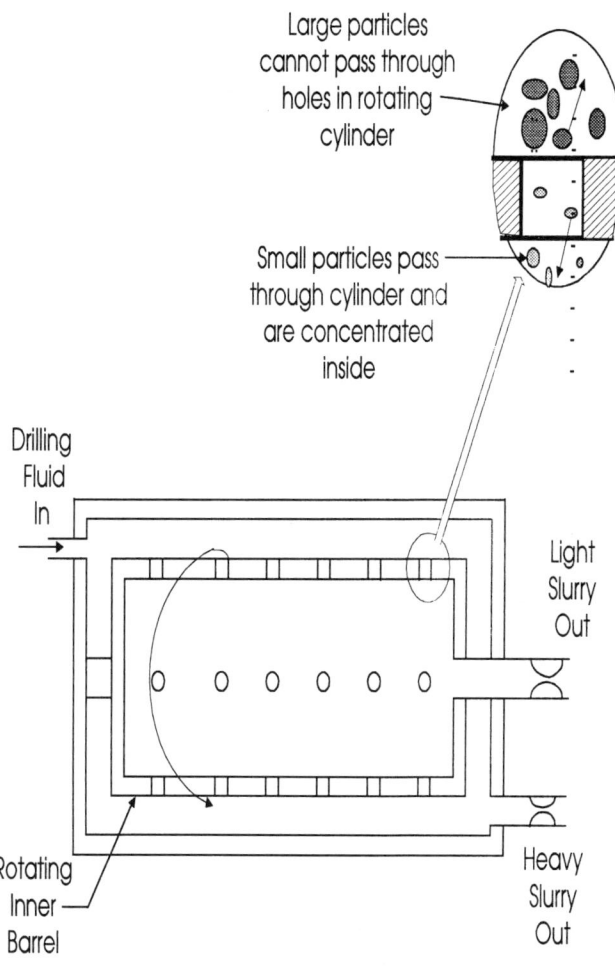

**FIGURE 7-12.** Rotary mud separator.

more of the process slurry is diverted through the performations of the rotor and subjected to the high-gravity field. The smaller mass is carried through the perforated center shaft, while the heavier solids stream continues through the rotor-case annulus to exit as a dense slurry.

The rotary mud separator discharges two streams of liquid slurry capable of being pumped. For this reason, the device may be placed near the pit system instead of being mounted on top of the mud tanks. Both the light and heavy slurries exit with sufficient energy to allow flow into containment or active tanks. Discharge elevation should be less than 20 feet, and discharge slurry density should never exceed 22 pounds per gallon.

The suction hose should be short—under 25 feet—and suction elevation should be limited to 10 feet. Additionally, a strainer should be placed on the suction line intake.

In practice, the rotor mud separator requires approximately seven gallons of water for every gallon of feed mud processed. Unless ample "recovered" water is already on hand, this requirement can significantly increase the total system volume (active + reserve + discards). Any such recovered or recycled water used in the separator must be relatively free of colloidal fines.

The rotary mud separator can process greater volumes of fluid than the decanting centrifuge. This, combined with the fact that it need not be mounted over suction or discharge pits, are its principal advantages over the decanting centrifuge.

The major disadvantage of the unit is that it requires significant quantities of dilution water, which increases total system volume. Drilling fluid treatment material (such as the filtration control additives and polymer additives) requirements are also increased.

Obviously, because of the high quantity of dilution water needed, it cannot be used as a dewatering device. However, the effluent from a rotary mud separator can be effectively processed by a decanting "dewatering" centrifuge. Such dewatered effluent—particularly if the process has been enhanced by coagulation and flocculation—can provide much of the required dilution volume for its operation.

***Unweighted drilling fluid applications.*** In unweighted drilling fluid, a centrifuge may be used to remove particles smaller than a desilter can remove. The $D_{50}$ cut point is around 2 to 10 microns as opposed to 15 to 25 microns for a desilter. The heavy slurry is discarded and the light slurry is retained, identical to that of a desilter.

Centrifuging unweighted drilling fluids removes drilled solids that are too small to be removed by other separtion devices. A dry discharge is desired, which reduces total liquid consumption and disposal volumes, and is an important consideration if the fluid phase contains expensive additives.

Centrifuges can also be used to process hydrocyclone underflow. In this case, the cone apex may be opened to produce a wetter discharge yielding a lower solids concentration to be centrifuged. Frequently, additional fluid from the desilter overflow is added to the underflow to assist centrifuge processing.

Decanting centrifuges are used to process unweighted oil muds, discussed later in this chapter.

***Weighted drilling fluid applications.*** In weighted drilling fluids, the heavy slurry from a centrifuge contains appreciable quantities of barite. Both streams can contain low-gravity solids and barite. Usually, most of the large drilled solids and large barite are contained in the heavy slurry, and most of the colloidal barite and drilled solids are contained in the light slurry. A centrifuge is effective in reducing the colloidal concentration in a weighted drilling fluid.

In weighted mud applications, centrifuges recover barite and "coarse" drilled solids from drilling fluids, while the colloidal-size solids (barite and drilled solids) in the centrate, or light slurry, are discarded. This controls the high shear-rate and low shear-rate viscosity while the recovered solids are returned to the active system.

Viscosity of weighted, water-based drilling fluids can be controlled by feeding a portion of the active circulating system into a decanting centrifuge and discarding the liquid, or colloid, overflow. These colloidals contribute inordinately to viscosity and gel strength problems. The heavy slurry, containing the larger particles (both weighting agent and drilled solids) are returned to the system.

Centrifuge operation on weighted mud is frequently intermittent as needed to control viscosity. If the feed rate into the centrifuge is decreased, so that the centrifuge operates most of the time, drilling fluid properties will be more uniform. The centrifuge performance is enhanced with a lower feed rate because the fluid residence time within the machine is longer.

When centrifuging a weighted mud, bentonite (prehydrated) and other chemicals must be added back into the system. The amount of replacement materials may be calculated from mass balance equations. In general, for gel/lignosulfonate drilling fluids, add one sack of (prehydrated) bentonite per hour of centrifuge operation.

The centrifuge underflow, or heavy slurry, will contain 55% to 60% solids by volume and is too "dry" to flow easily.

Discharge of weighted drilling fluids should be introduced at an angle (45° or greater) into a well-agitated pit to retard settling. If discharged solids are dry and do not move easily down the trough, downstream drilling fluid can be used as a wash fluid. Similarly, solids discarded by two-stage centrifuging, or discards from centrifuging unweighted drilling fluids, can be conveyed by wash fluid supplied from the reserve pits.

## Other Factors of Centrifuge Performance

Pond depth is set by an adjustable weir and controls fluid residence time. Increased residence time increases separation at the expense of flow (and processing) capacity. For fine solids distribution, a deep pond (at lower flow rates) will yield higher separation efficiencies. For coarse solids, less separation efficiency is needed, and a shallower pond (and higher flow rates) may be desirable. This corresponds to shallow pond depths generally associated with upper-hole drilling and deeper pond depths for deeper drilling, and with weighted fluids.

The difference in speed between the bowl and the conveyor controls the velocity with which settled solids are conveyed through the centrifuge. Settled solids are transported through the centrifuge because of the small difference in rotating speed, or rpm differential, which has little influence on settling efficiency, provided that the differential is sufficient to keep pace with the settling rate to avoid undue solids buildup in the beach.

***Discharge dryness.*** As previously discussed, larger particles have less associated surface liquid. Thus, while discharge dryness may seem indicative of satisfactory centrifugal separations, it has been shown to correspond more with median particle size ($D_{50}$).

## Other Applications

***Two-stage centrifuge operations.*** Two-stage centrifuge operations are becoming more common. These units can be operated in series or parallel depending on whether liquid reclaimation is required. In parallel operations, two centrifuges increase the quantity of fluid that can be processed. In series operations, the overflow (or light slurry) from the first machine is fed to the second machine. The underflow (or heavy slurry) from the second machine is discarded and the overflow is returned to the drilling fluid. This discards the majority of the drilled solids larger than the cut point of the second machine and smaller than the first machine. The objective of this process is to save the expensive liquid phase at the cost of returning the small colloidal particles to the drilling fluid.

The most frequent application of two-stage centrifuging is on oil-based muds when cost and/or environmental factors prohibit discard of the liquid phase. The first centrifuge recovers the weighting material and the centrate is directed to the second centrifuge, which opeates at a higher "G"-force.

The efficiency of this operation depends on the efficiency of each stage. The first centrifuge must adequately separate the solids because the discards from the second centrifuge depend on the solids received. The second centrifuge should operate at the highest possible "G"-force with the deepest pond depths.

- *Centrifuge stage one*—The first stage employs a variable speed (1800 to 2800 rpm) conical unit.

- *Centrifuge stage two*—Second-stage units should be a high rpm (2800 or more), contour design. This processes the overflow (normally the discard) of the first-stage separator. This overflow stream contains mostly solids smaller than 3 to 5 microns and the liquid fraction containing drilling fluid treatment chemicals. The high-speed unit separates this overflow into "heavy" and "light" streams. The heavy stream is discarded and the light stream can be used as dilution fluid—thereby reducing the volumes discarded during drilling operations. This removes solids between 1 and 10 microns and recovers most of the liquid.

***Centrifuges in oil drilling fluids.*** Centrifuges have extensive application in oil-based drilling fluids, including:

1. Allowing the fluid phase of hydrocyclone discharge to be returned to the active system
2. Operating in the manner previsouly described, using two centrifuges
3. Cleaning or separating cuttings from a wash fluid

Centrifuging to clean and recondition an oil-based drilling fluid is often performed at a central "mud plant." Eventually, 100% of the drilling fluid is processed. When the ambient temperature is low, the feed fluid will have a high viscosity. This increases the processing time required to remove solids from the drilling fluid. The problem of increased slurry viscosity related to cold temperature is also encountered when drilling wells in ultra-depth waters, where many wells are drilled with synthetic-oil drilling fluids. For example, seawater temperature below 2,000 feet asymptotically approaches 36°F (2°C). Drilling fluid circulating through large diameters risers in deep water undergoes significant cooling and consequent gellation. The cost of these drilling fluids can be on the order of several hundred dollars, so economics dictate that as much liquid as possible be recovered.

Decanting centrifuges are used to process unweighted oil-based drilling fluids when:

1. Drilling fluid is brought from another location and may contain a significant amount of drilled soilids
2. Slow, hard drilling yielding a buildup of ultra-fine solids is anticipated
3. The liquid drilling fluid phase is valuable

***Weighted oil-based mud applications.*** In weighted oil-based mud applications, decanting centrifuges are operated in series. The first unit—a "standard" machine—returns the dense, coarse solids to the system and routes the light, fine solids discharge to a holding tank. A second unit—usually a high-capacity machine—removes and discards the solids with the effluent available for return to the active system.

This process is not as effective as a single unit for viscosity control. A significant portion of the colloidal-size solids are returned to the active system in the overflow of the second unit. The overflow of the first unit is too valuable to discard, especially with synthetic, oil-based drilling fluids. Disposal costs can be significant for many of these drilling fluids. Two decanting centrifuges can decrease the total waste volume from the well.

***Closed-loop systems.*** As part of closed-loop systems, decanting centrifuges process underflow from hydrocyclones to separate the discarded free liquid from fine particles. Chemical enhancement, used to increase effective particle size (see Chapter 13), reduces water and waste volumes. When used for processing hydrocyclone discards or other "waste," centrifuges are operated continuously and the lighter "free liquid" is retained for dilution.

In closed-loop systems, larger high-capacity (75–250 gpm) decanting centrifuges are used, sometimes with "standard" centrifuges, to maximize fine solids separations.

## Operating Reminders

1. Before startup, rotate bowl or cylinder by hand to make sure it rotates freely.
2. Start centrifuge first, before starting drilling fluid feed pump or dilution water.
3. Set drilling fluid mud and dilution rates according to manufacturer's recommendations, which usually vary according to mud weight.
4. Turn drilling fluid feed off before turning dilution water off and prior to turning the machine off.

## DEWATERING UNIT

More and more rigs are incorporating dewatering units in their drilling fluid treatment process. Dewatering units are designed to inhibit free liquid discard. The units can process drilling fluid from the active system or discard material from the solids removal equipment (shakers, desanders, desilters, and centrifuges). Chemical pH adjustments, flocculants, and coagulants are added to the input stream of the dewatering unit. The solids are usually removed with a centrifuge. Colloidal-rich discards can have as little as 10% volume solids and still be treated as a solid. The liquid recovered from the unit is usually returned to the active system. Chemical overtreatment creates problems in the active system, therefore, the liquid recovery stream must be carefully monitored. For more information on Dewatering, see Chapter 13.

## REMOVAL SECTION ARRANGEMENT

All of the compartments, except three, use bottom equalizers. The sand trap should overflow into the next compartment downstream, which provides an opportunity for the solids to settle. Obviously, to increase residence time to allow settling, the drilling fluid should enter the sand trap as close to the upstream compartment wall as possible. Sand traps are very effective when water or seawater are used as drilling fluid. When drilling fluids are treated to provide a large value of the low-shear-rate viscosity, solids will not settle efficiently. This is most noticeable when the shale shaker possum belly is emptied into the sand trap during a trip. Shortly after the startup of the next drill bit, the unsettled solids reach the apex of the desilters and plug them. At this time, all of the drilling fluid processed by the plugged cones retain their drilled solids. This makes it difficult to maintain a low level of drilled solids in the drilling fluid.

The degasser suction compartment tank should equalize overflow from the next compartment downstream. The degasser should process more drilling fluid than is entering its suction compartment from the sand trap. This will cause some drilling fluid to flow from the downstream compartment into the degasser suction compartment.

The equalizing line from the removal section to the additions tank should be through an "L"-shaped pipe that allows the discharge end to be raised or lowered. Normally, the discharge end will be raised so that the removal section maintains a constant level of fluid.

Solids removal equipment is arranged so that the larger solids are removed before the smaller solids. Each piece of equipment should discharge into the next compartment downstream from the suction compartment. Each compartment in the removal section should backflow from the downstream compartment into the upstream compartment, except for the sand trap. A flow analysis should show that all fluid entering the suction compartment of the degasser, desilters, or desanders, from whatever source, is processed through the equipment.

## Compartment Agitation

Each compartment, except the sand trap, should be well stirred to provide uniform solids-loading for the removal equipment. Failure to agitate these compartments can result in short-term overloading of the hydrocyclones. This, in turn, results in bottom orifice plugging that prevents solids removal. Removal efficiency less than 50% is common in unstirred compartments.

## Single-Purpose Centrifugal Pumps

Solids removal equipment should be installed so that operation is straightforward and consistent. There should be a single starter (for an electric motor) or a single diesel engine for each pump. Manifolding of suction or discharge lines rarely results in effective equipment performance. Funds normally used for valves in a "do-anything" manifold, should be used to store an extra pump and motor in case of failure. Pumps and motors should be mounted so that they are easy to install and maintain.

Unfortunately, the above suggestions create a great deal of controversy. Ideally, the fluid system would allow operators to pump from anywhere to anywhere with any pump. However, leaky and improperly aligned valves can destroy removal efficiency. For example, a system operating at 90% drilled solids removal efficiency can quickly deteriorate to 40% efficiency with only two valves improperly set. If the target drilled solids concentration is 8% volume or less, the low drilled solids removal efficiency can easily require two to four times as much drilling fluid to drill a well (see Chapter 8).

# CHAPTER EIGHT

# Dilution

Dilution is the surface addition of base fluid to the drilling fluid system. This addition has the effect of reducing the concentration of all other components or constituents of the drilling fluid.

Although the shale shaker frequently removes a larger fraction of drilled solids, it should be regarded as only one component of solids removal equipment in a drilling fluid processing system. In addition to solids removal equipment, the detrimental effect of drilled solids can be reduced with **chemical treatment** or **dilution**. Both of these methods are expensive but, unlike chemical treatments, dilution is always effective. This chapter will focus on this process, provide some basic guidelines for its use, and demonstrate how expensive dilution can be.

## INTRODUCTION

One method of managing the drilled solids level is to simply empty some of the drilling fluid containing the drilled solids and replace it with clean drilling fluid. One half of the solids can be eliminated if one half of the system is emptied and replaced with clean fluid. Generally, this is an expensive process, therefore, mechanical equipment is used.

With solids control equipment, both solids and liquid are removed. The total volume removed decreases the volume of the circulating system. Consequently, new drilling fluid must be built to keep the pit levels constant. This additional fluid helps dilute the remaining drilled solids in the pit. If the solids-removal equipment is ineffective at removing a substantial quantity of the drilled solids, too much new drilling fluid must be built. In the case where no drilled solids or liquid are removed from the system, no volume addition is required.

Three primary variables—dryness of discarded solids, targeted concentration of drilled solids in the drilling fluid, and removal efficiency—are important in evaluating a solids management system on a drilling rig. These variables play an important role in minimizing drilling fluid and drilling waste costs:

1. The dryness of discarded solids varies significantly with various types of solids control equipment. Shale shakers with coarse mesh screens (such as 12 to 30 mesh) discard very dry solids (60% to 80% volume solids) but also pass a large fraction of solids presented. Shakers with fine mesh screens (such as 100 to 200 mesh) can remove more solids but also discard more liquid drilling fluid. The fluid removed with these fine mesh screens may have 30% to 45% volume solids and contain more or less weighting material than the drilling fluid in the pits. Hydrocyclones settle solids inside the cones in the whole drilling fluid. Spray discharge is desirable but seldom do all cones on a drilling rig discharge solids in a spray discharge. The solids concentration in hydrocyclone discharges can vary from 20% to 40% volume. A centrifuge underflow usually has a solids concentration between 55% to 65% volume and the overflow has a solids concentration between 5% and 15% volume. The mud cleaner screen discharge in an unweighted drilling fluid has approximately 30% to 45% volume solids. A weighted drilling fluid usually resembles the concentration of the centrifuge underflow. Thus, a typical solids concentration of the overall system discard is approximately 35%.

2. The concentration of drilled solids targeted in the drilling fluid is a function of the type of drilling fluid. Polymeric additives that adhere to active solids are attracted to most clay-type drilled solids. These drilling fluids demand low drilled-solids concentrations. The inhibitive fluids also need a low solids concentration, but also aid the removal processes by not allowing clay-type solids to disperse.

3. The removal efficiency of the solids control arrangement can vary significantly depending on the adherence to sound arrangement practices and operating conditions. Many drilling rigs with poor solids management have trouble reducing the mud weight below 9.7 ppg. While others can comfortably create an 8.8 ppg drilling fluid. This is related to the removal efficiency. API RP13C provides a new method of following a tracer to determine the solids removal efficiency. In the following section, a second method is presented. These methods can be used while drilling a reasonably deep hole (such as 1,000 to 2,000 feet). The analysis assumes that no drilling fluid is jetted from the system. The removal efficiency of the system in an unweighted drilling fluid is usually higher than in a weighted drilling fluid. Solids in the same size range as the weighting material cannot be removed with known technology.

## EFFECT OF DRILLED SOLIDS REACHING THE SURFACE

Calculations, summarized in Figures 8-1 through 8-3, illustrate how the effectiveness of the solids removal equipment determines the quantity of drilling fluid that must be built to maintain a targeted drilled-solids concentration and a constant pit volume. These representative calculations describe the fate and effect of 100 barrels of drilled solids arriving at the surface. (Note, this may not correspond to generating 100 bbl of borehole because of lag time and solids storage in the wellbore.) The first set of calculations is based on a solids removal section that is removing 100%, 90%, or 80% by volume of the drilled solids, with the targeted drilled solids level set at 4% volume. For this comparison, assume 100 bbl is approximately the volume of 1,000 feet of a 10-inch diameter hole. This analysis is the basis for calculating the minimum desirable removal efficiency. It also shows the minimum quantity of new drilling fluid and minimizes the volume of discarded fluid. These calculations assume no migration of fluid to or from the wellbore, no rock compressibility, and no gas migration to the surface. These example calculations also assume that the removed drilled solids concentration is 35% by volume.

In the case of 100% removal efficiency, all of the drilled solids reaching the surface are removed along with 65% drilling fluid volume. The pit levels will decrease by the volume of drilled solids and drilling fluid removed from the system (286 bbl as shown in Figure 8-1). Although no

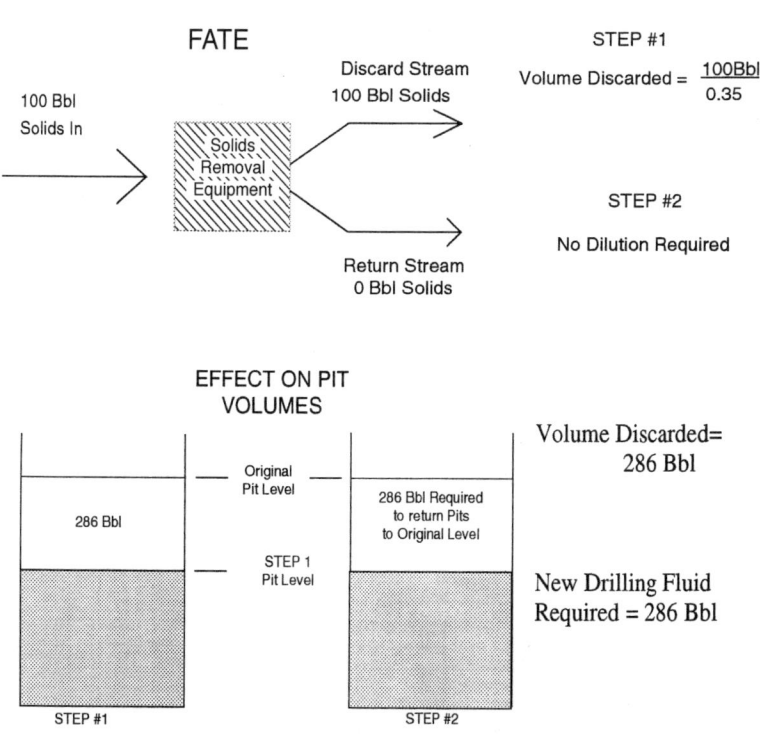

**FIGURE 8-1.** 100% solids removal efficiency.

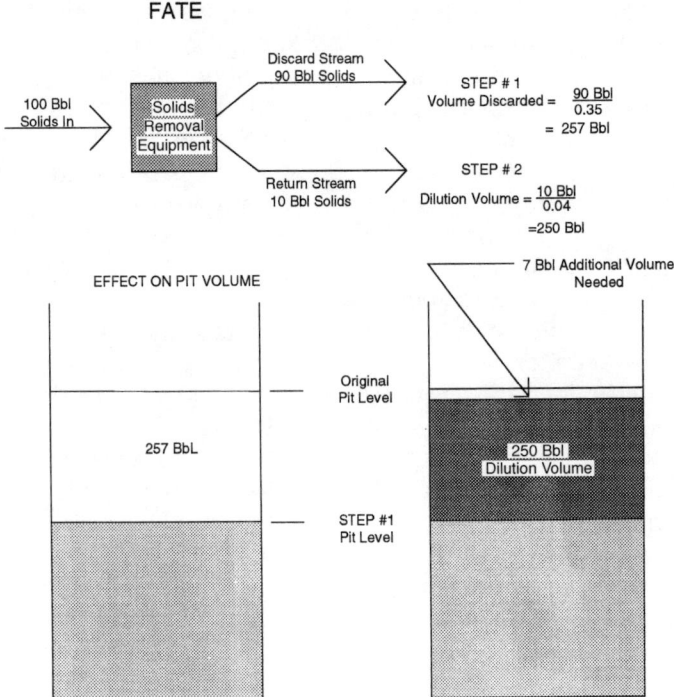

**FIGURE 8-2.** 90% solids removal efficiency.

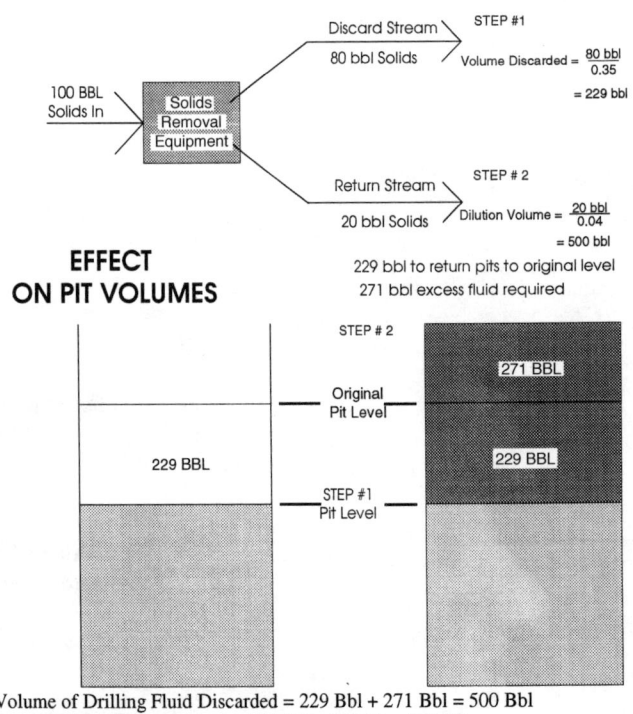

**FIGURE 8-3.** 80% solids removal efficiency.

drilling fluid is needed to dilute the drilled solids returned to the system, 286 bbl of drilling fluid is needed to maintain the pit levels, which will decrease the drilled solids concentration. In other words, a 4% volume drilled solids level cannot be sustained with constant pit volume and 100% removal efficiency.

At 90% removal efficiency, 90 bbl of drilled solids and 167 bbl of drilling fluid are removed from the system. In this case, the pit levels decrease by 257 bbl (Figure 8-2). The solids returned to the system (10 bbl) must be diluted to 4% volume concentration by adding 240 bbl of new drilling fluid. The total addition to the pit system is 10 bbl of drilled solids and 240 bbl of new drilling fluid. This means that 7 bbl of new drilling fluid must be built to maintain the pit levels.

This example illustrates a system that is almost "balanced." If the discarded volume is identical to the volume required for dilution, the minimum quantity of drilling fluid will be built. The optimum removal efficiency for any targeted drilled solids level may be calculated by mathematically equating the removal volume to the required dilution volume.

In the case of 80% removal efficiency, 229 bbl of drilled solids and drilling fluid will be discharged (Figure 8-3). Although this is only 21 bbl less than the 90% removal efficiency, the dilution volumes are significantly higher. The dilution of the 20 bbl of returned drilled solids to a 4% volume level requires the addition of 480 bbl of new drilling fluid to the system. The reconstituted 500 bbl of drilling fluid will contain 20 bbl of drilled solids and 480 bbl of clean drilling fluid. Since only 229 bbl of space is available, 271 bbl of fluid must be discarded to keep the pit levels constant. Therefore, the total discard is the 229 bbl from the solids-removal equipment plus 271 bbl of drilling fluid.

This situation creates a problem: if 271 bbl of drilling fluid is not discarded, the drilled solids concentration will increase significantly above the targeted 4% volume concentration. When large reserve pits were used, drilling fluids were relatively inexpensive and disposal costs were insignificant. Building excess volume was more of an inconvenience than a significant economic burden. If drilling fluid disposal volumes are to be maintained as low as possible (i.e., no drilling fluid jetted from the pits), the removal efficiency must be improved. Otherwise the 4% volume of drilled solids will not be attained.

The minimum discard volume will occur when the system is "balanced" (i.e., no excess drilling fluid is needed to dilute the drilled solids returning to the system). The same solids removal efficiency that provides the minimum quantity of new drilling fluid is also the removal efficiency that generates the minimum discard volume.

These three calculations are summarized in Figure 8-4. The minimum volume required is the lowest point on the curve at the 90% removal efficiency point.

## OPTIMUM REMOVAL EFFICIENCIES

The previous examples, with 35% volume solids in the discard, illustrate that an optimum removal efficiency of around 90% exists for the 4% volume target concentration of drilled solids. These same calculations may be made for other discard and targeted drilled solids concentrations as illustrated in Figure 8-5.

For example, if the removal efficiency is 75%, over 6 bbls of drilling fluid is required for each barrel of drilled solids reporting to the surface when the targeted drilled solids concentration is 4% by volume.

Using the same calculation procedures for 6% volume targeted drilled solids, Figure 8-5 shows that at 60% removal efficiency, over 6.2 bbls of drilling fluid will be required for every barrel of drilled solids reaching the surface. As the efficiency improves, the drilling fluid requirement decreases: at 70% slightly more than 4.5 bbl is required per barrel of solids; at 80% slightly more than 3 bbl is required; and at 85% almost 2.5 bbl is required. For removal efficiencies greater than 85%, the pit level decreases more than the additional new fluid required to dilute the solids to the 6% volume level. This additional fluid will dilute the solids to a value lower than 6% volume.

The volume of drilling fluid required per barrel of drilled solids is similar whether the slurry is wet (25% volume solids) or relatively dry (45% volume solids). When the removal efficiency is below the "balance point," the values are identical. As the concentration of drilled solids decreases, the balance point for any target concentration of drilled solids increases. The volume of new drilling fluid required above the balance point is independent of the dryness of the discarded solids.

## DETERMINING REMOVAL EFFICIENCIES

If a well is spudded with fresh water and the pit levels remain constant during drilling, the drilled solids returned to the system will increase the density. This analysis assumes that no whole drilling fluid is emptied from the system or lost downhole. As the borehole becomes deeper, the mud weight increases at different rates for different re-

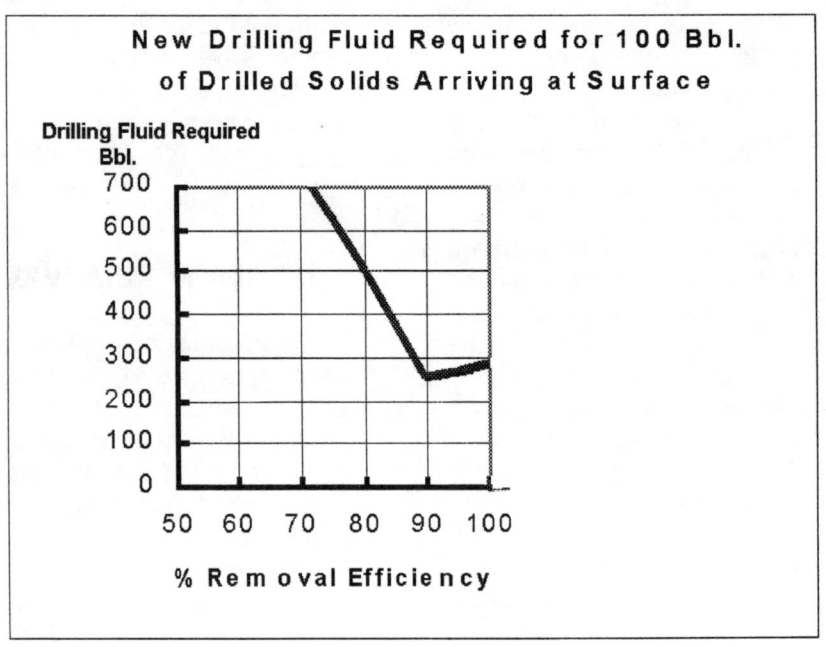

**FIGURE 8-4.** Effect of removal efficiency on drilling fluid requirements for maintaining a 4% volume of drilled solids.

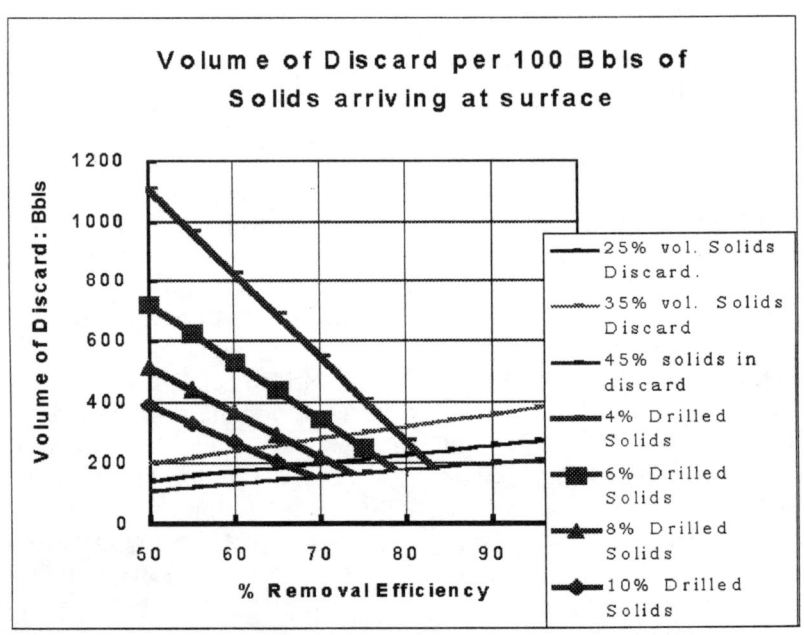

**FIGURE 8-5.** Effect of removal efficiency on drilling fluid requirements for different discard wetness and target drilled solids concentrations.

moval efficiencies (Figure 8-6). For example, assume that the surface volume is 1000 bbl and 2% volume bentonite is added to the new drilling fluid. As shown in Figure 8-6, if the mud weight is 9.3 ppg after drilling 200 bbl of hole, the removal efficiency is 60%. This same calculation can be made in terms of the drilled solids concentration for any density drilling fluid as illustrated in Figure 8-7.

DILUTION 169

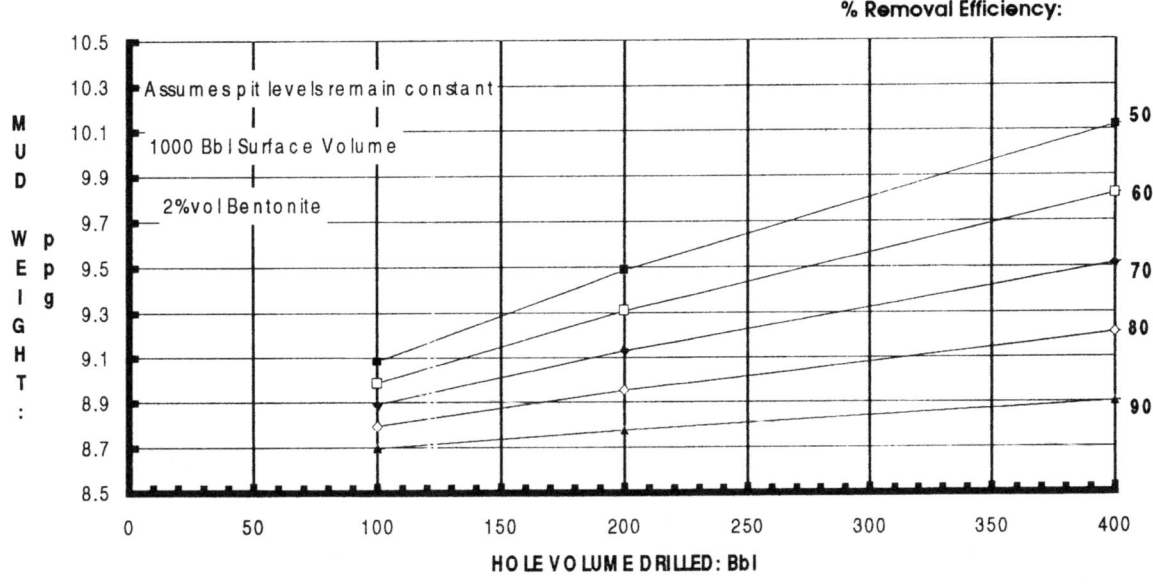

**FIGURE 8-6.** Method to determine removal efficiency.

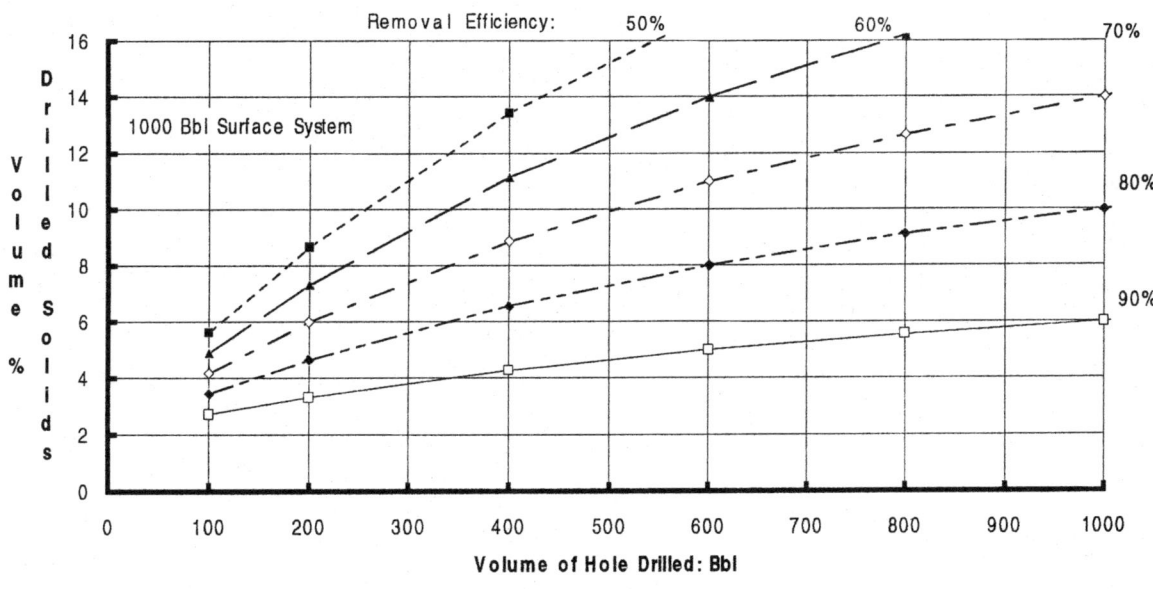

**FIGURE 8-7.** Chart to assist estimating solids removal efficiency.

# CHAPTER NINE

# Cut Points

Cut points are used to indicate the performance of solids control equipment. Cut point curves are determined by examining the mass of a variety of size ranges that are discarded relative to the same mass and size range presented to the removal equipment. Curves can be calculated by determining the ratio of the mass flow rate of solids within each size range in the discard stream, divided by the mass flow rate of solids of the same size range in the input stream.

Cut points for shale shakers may be measured by first determining the particle size distribution of the input stream and the overflow stream. Generally, the distributions may be determined, with a stack of sieves. The flow rate of each stream is then determined and the mass flow rate for each selected size range in both streams is calculated. The mass flow rate of the discard stream for each size range is divided by the mass flow rate of the same size range presented to the shale shaker.

Usually, the sample from the input stream represents a small fraction of the total flow. This creates a problem with material balances. A better method is to sample the discard and the screen throughput streams. Combining these two solids distributions will yield a more accurate cut point curve. This technique is also used on hydrocyclones where the input stream flow rate is much greater than the discard stream.

As an example, assume all of the discard from a shale shaker is collected during a one-minute period while drilling with a 9.6 ppg drilling fluid circulated at 500 gpm. At the same time, the flow stream through the shale shaker (also referred to as the "screen unders") is sampled. The weight of the drilling fluid and the solids content is determined gravimetrically. Since this drilling fluid contains no barite, the fraction of solids in the fluid is specified by its density and the specific gravity of the low-gravity solids. In this case, the solids content is 9.5% volume, indicating a low-gravity solids specific gravity of 2.6.

The screen throughput sample and the discard sample are sieved through a stack of standard ASTM screens. The solids retained on each size screen are dried and weighed. The mass percent of solids in the throughput sample in each size range larger than 400 mesh is then determined. The flow rate through the shaker screen is approximately 4800 lb/min (500 gpm × 9.6 lb/gal).

The 5-gallon bucket used to sample the drilling fluid passing through the screen, samples only about 0.01 minute of flow. The weight of the drilling fluid in the bucket is measured and if the fluid weighs 48.0 pounds, the volume in the bucket is 5.0 gallons. This fluid is then sieved through a stack of standard ASTM screens and the dry weight of the solids on each standard screen is determined. The mass flow rate of the dry solids for each sieve can be determined by dividing the weight of the solids by the time interval of the collection (or 0.01 min.)

All of the shale shaker discard is collected for a specific period of time and weighed. This discard is sieved through the same set of standard ASTM screens. The dry weight of solids on each screen is determined and divided by the time interval required to capture these solids. For example, if the collection period is two minutes, each dry weight would be divided by two minutes, which yields the weight per minute of discard.

The mass flow rate of the discard and throughput streams are combined for each size range. This creates the mass distribution of the input stream. All solids calculations are gravimetric since volumetric measurements are inaccurate.

The mass flow rate onto and off the screen is known, therefore a "cut point" curve may now be established. The ratio of mass flow onto the screen divided by the mass flow off the screen can be calculated for the solids remaining on each sieve.

A typical set of data is shown in Table 9-1. These data, presented in a graph form, create the cut point curve illustrated in Figure 9-1.

CUT POINTS **171**

**TABLE 9-1.** Cut Point of Shaker Screens

| ASTM Standard Screen | | Screen Throughout Sample on ASTM Sieves (gm) | Discard Sample on ASTM Sieves (gm) | Solids Calculated Discard Flow Rate (gm/min) | Solids Calculated Throughput Flow Rate (gm/min) | Calculated Flow Rate of Solids in Flow Line (gm/min) | Percent Discarded by Shaker Screen |
|---|---|---|---|---|---|---|---|
| Mesh Size | Opening (microns) | | | | | | |
| 400 | 37 | 50.0 | 100.0 | 50 | 5,000 | 5,050 | 1.0 |
| 325 | 44 | 45.0 | 180.0 | 90 | 4,500 | 4,590 | 2.0 |
| 250 | 63 | 40.0 | 560.0 | 280 | 4,000 | 4,280 | 6.5 |
| 200 | 74 | 38.0 | 912.0 | 456 | 3,800 | 4,256 | 10.7 |
| 150 | 105 | 25.0 | 2,126.0 | 1,063 | 2,500 | 3,563 | 29.8 |
| 120 | 118 | 10.0 | 1,880.0 | 940 | 1,000 | 1,940 | 48.5 |
| 100 | 140 | 0.5 | 980.0 | 490 | 50 | 540 | 90.7 |
| 80 | 177 | 0.0 | 400.0 | 200 | 2 | 202 | 99.0 |
| 60 | 234 | 0 | 19.8 | 9.9 | 0 | 9.9 | 100.0 |

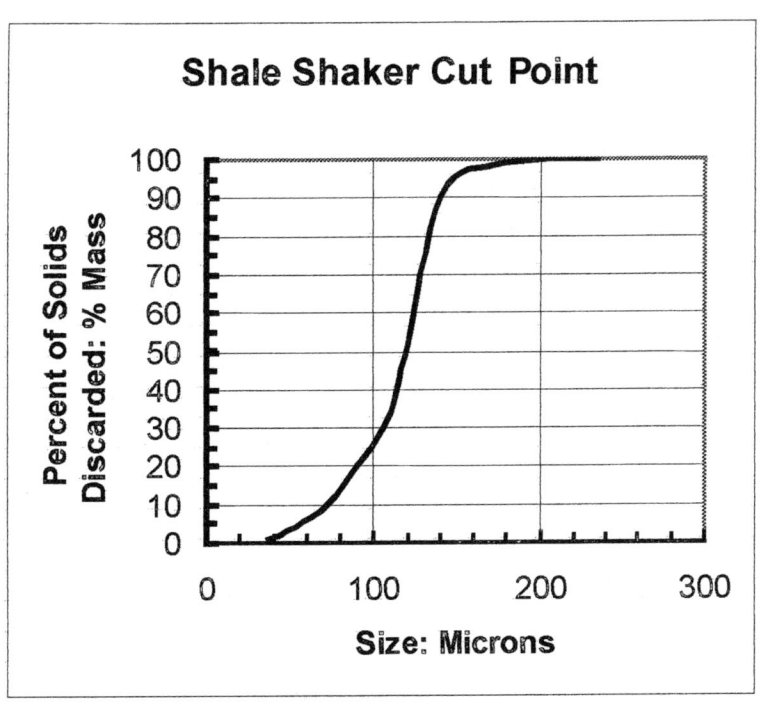

**FIGURE 9-1**

## How to determine cut point curves:

**Feed:**
1. Calibrate the pump circulating drilling fluid in the borehole.
2. Determine the flow rate in well.

**Throughout Screen Sample:**
1. Weigh the bucket.
2. Collect a representative sample of drilling fluid passing through the shale shaker screen.
3. Weigh the bucket and drilling fluid sample.
4. Calculate the weight of drilling fluid in the sample.
5. Calculate the volume of drilling fluid in the bucket (divide weight of drilling fluid in bucket by density of drilling fluid).
6. Calculate the flow rate of the sample passing through the shaker screen (subtract well flow rate from the discard flow rate).
7. Calculate the time (in minutes) during which the volume in the bucket was collected (divide volume of drilling fluid in bucket by sample flow rate).
8. Determine the weight of dry solids captured on each size screen.
9. Divide each weight of dry solids by the time required for these solids to pass through the screen.

**Discard:**
10. Weigh the trough that will be used to collect the shaker discard.
11. Measure the time (in minutes) required to collect all of the shale shaker discard in the trough.
12. Weigh the discard and trough.
13. Determine the weight of discard in the trough (subtract weight of the trough from the combined weight of the trough and discard).
14. Sieve the discard sample through a stack of standard screens using the same screens, or the same sizes, used for the feed sample.
15. Determine the dry weight of the sample on each standard screen.
16. Divide the dry sample weight by the time required to collect the sample (this is the mass discard rate).
17. Sieve the drilling fluid sample through a stack of standard screens.
18. Determine the weight of dry solids captured on each size screen.

**Plotting the Cut Point Curve:**

For each screen size in the standard sieve stack, the mass discard rate is divided by the mass feed rate. This ratio, multiplied by 100, is plotted on the "Y" axis of a graph, and the sieve size range is plotted on the "X" axis.

# CHAPTER TEN

# Calculating Drilled Solids Concentrations

Drilled solids concentration in a slurry (drilling fluid or discharge from solids control equipment) may be determined from the following equation:

$$(\rho_B - \rho_{lg})V_{lg} = 100\rho_f + (\rho_B - \rho_f)V_s - 12MW$$

where $\rho_B$ = Density of high-gravity solids (barite is 4.2 gm/cc)
$\rho_{lg}$ = Density of low-gravity solids
$V_{lg}$ = Volume percent of low-gravity solids
$\rho_f$ = Density of filtrate
$V_s$ = Volume percent of undissolved solids
MW = Mud weight, ppg

For freshwater drilling fluid, assuming the density of the drilled solids is 2.6 gm/cc, the density of the barite is 4.2 gm/cc, and the water density is 1.0 gm/cc, the equation becomes:

$$V_{lg} = 62.5 + 2.0V_s - 7.5MW$$

If the retort solids are 13% volume and the mud weight is 11 ppg, the percent of low-gravity solids is 6% volume. If the fluid contains 2.5% volume of bentonite, the drilled solids concentration is 3.5% volume.

In saline water-based drilling fluid, the salt dissolved in the water phase increases the filtrate density and also remains in the retort cup after the water evaporates from the sample. The density of the filtrate ($\rho_f$) may be calculated from the concentration of the salt:

$$\rho_f = 1.0 + 6.45 \times 10^{-7} \times (NaCl) + 1.67 \times 10^{-3} \times (KCl)$$

$$+ 7.6 \times 10^{-7} \times (CaCl_2) + 0.001(Organic)$$

where NaCl = Concentration of NaCl in filtrate, mg/liter
KCl = Concentration of KCl in filtrate, mg/liter
$CaCl_2$ = Concentration of ($CaCl_2$) in filtrate, mg/liter
Organic = Concentration of organic material in filtrate, lb/bbl

To account for the increase in total solids in a saline drilling fluid, the total volume percent of suspended solids, $V_s$, may be calculated from the equation:

$$V_s = 100 - V_w/\{\rho_f - 10^{-6}([NaCl] + [KCl] + [CaCl_2]$$

$$+ [Organic])\}$$

where $V_w$ = Volume of water (in cc) recovered from the retort

## PROCEDURE FOR DETERMINING ACCURATE, LOW-GRAVITY SOLIDS

The following procedure requires an oven, pycnometer, and an electronic balance to weigh samples. A pycnometer can be made by removing the beam from a pressurized mud balance. Any type of balance may be used to determine weight, however, electronic balances are more convenient.

### Determine Volume of Pycnometer

1. Weigh the pycnometer (assembled)
2. Fill with distilled water
3. Determine the water temperature
4. Reassemble the pycnometer and pressurize
5. Weigh the pycnometer filled with pressurized water
6. Determine the density of water using a density/temperature water table
7. Subtract the pycnometer weight from the weight of the pycnometer filled with water to determine the weight of water in the pycnometer
8. Divide the weight of water in the pycnometer by the density of water to determine the volume of the pycnometer

## Determine the Density of Drilled Solids

1. Select large pieces of drilled solids from the shale shaker and wash with the liquid phase of the drilling fluid (use water for water-based drilling fluid, oil for oil-based drilling fluid, or synthetics for synthetic drilling fluid)
2. Grind the drilled solids and dry in an oven or retort
3. Weigh the assembled, dry pycnometer
4. Add the dry drilled solids to the pycnometer and weigh
5. Add water to the drilled solids in the pycnometer, pressurize, and weigh
6. Determine the water temperature
7. Determine the water density
8. Subtract the weight of the dry pycnometer from the weight of the dry pycnometer containing the dry drilled solids. This is the weight of drilled solids.
9. Subtract the weight of the dry pycnometer containing drilled solids from the weight of the water, drilled solids, and pycnometer. This is the weight of water added to the pycnometer.
10. Divide the weight of the water (determined in step 9) by the water density. This is the volume of water added to the pycnometer.
11. Subtract the volume of the water added to the pycnometer (step 10) from the volume of the pycnometer. This is the volume of drilled solids contained in the pycnometer.
12. Divide the the weight of drilled solids (step 8) by the volume of drilled solids (step 11). This is the density of the drilled solids. This may be used in the previous equations to determine the volume percent of drilled solids in the drilling fluid.

The pycnometer may also be used to more accurately determine drilled solids in the drilling fluid:

1. Determine the density of the drilling fluid with the pycnometer
   a. Weigh the pycnometer filled and pressurized with drilling fluid
   b. Subtract the weight of the dry pycnometer from the weight of the filled, pressurized pycnometer
   c. Divide weight of drilling fluid in pycnometer by the volume of the pycnometer. This is the density of the drilling fluid.
2. Weigh a metal or heat-resistant glass dish
3. Add a quantity of drilling fluid to the dish and weigh
4. Determine the weight of drilling fluid in the dish by subtracting the weight of the dish from combined weight of the dish and drilling fluid.
5. Divide the weight of drilling fluid in the dish by the density of drilling fluid (step 1). This is the volume of drilling fluid.
6. Dry the material in the dish in an oven at 250°F for at least four hours.
7. Weigh the dish containing dry solids
8. Subtract the dish weight from the weight of the dish containing the dry solids. This is the weight of solids.
9. Divide the weight of dry solids (step 8) by the weight of drilling fluid in the dish (step 4). This is the weight fraction of solids in the drilling fluid.
10. Subtract the weight of dry solids (step 8) from the weight of drilling fluid in the dish (step 4). This is the weight of liquid in the drilling fluid.
11. Divide the weight of liquid in the drilling fluid (step 10) by the density of the liquid phase of the drilling fluid. This is the volume of the liquid phase.
12. Subtract the volume of the liquid phase (step 11) from the volume of the drilling fluid (step 4). This is the volume of solids in the drilling fluid.
13. Divide the volume of solids in the drilling fluid (step 12) by the volume of drilling fluid. This is the volume fraction of solids in the drilling fluid.
14. Multiply the volume fraction of solids in the drilling fluid by 100 to obtain the volume percent of solids in the drilling fluid

*Sample calculation.* While drilling a relatively uniform 2,000-foot shale section, a 200 × 200 mesh continuous screen cloth is mounted on a linear shale shaker. An 11.2 ppg, freshwater, gel/lignosulfonate drilling fluid is circulated at 750 gpm while drilling. Given these parameters, the following determinations can be assumed.

Large pieces of shale are removed from the shaker screen and the excess drilling fluid is washed from the surface with distilled water. The shale pieces are ground and dried in a 250°F oven overnight. The shale is placed in a 173.91 cc pycnometer and weighed. Water is added to the pycnometer and pressurized to approximately 350 psi. The increase in weight of the pycnometer determines the volume of water required to fill the pycnometer. (Room and water temperature is 68°F, therefore, the density of water is approximately 1.0 gm/cc.) Subtracting this volume of water from the known volume of the pycnometer calculates the

# CALCULATING DRILLED SOLIDS CONCENTRATIONS

> **Calculation procedure:**
>
> A sample of the discard is placed in the pycnometer and weighed: Pyc. + Sample weight = 898.68 gm
>
> Since the pycnometer weighs 660.61 gm dry and empty, the sample weight is 209.07 gm
>
> The pycnometer with the shaker discard sample is filled with distilled water, pressurized, and weighed:
>
> Pycnometer + Sample + Water = 925.73 gm
>
> The weight of water added is: 925.73 gm – 896.68 gm = 56.05 gm; the volume of water added = 56.05 cc
>
> Since the pycnometer volume is 173.91 cc, the sample volume is: 173.91 cc – 56.05 cc = 117.86 cc
>
> The density of the sample is 209.07 gm/117.86 cc = 1.77 gm/cc; the volume of water added = 56.05 cc
>
> Since the pycnometer volume is 173.91 cc, the sample volume is: 173.91 cc – 56.05 cc = 117.86 cc
>
> The density of the sample is: 209.07 gm/117.86 cc = 1.77 gm/cc

volume of the shale sample. Because the volume of the shale sample and the weight are known, the density may now be calculated as 2.47 gm/cc.

After movement of the solids across the shale shaker screen appears relatively uniform for more than 10 minutes, all of the shaker discard is collected in a bucket. In 16.21 seconds, 3720.7 gms of discard is captured. The discard rate is 13,772 gm/min. and the discard has a density of 1.774 gm/cc or 14.8 ppg.

The objective of the shale shaker is to remove drilled solids, preferably with limited amounts of drilling fluid. The fraction of the discard stream that is water, barite, and low-gravity solids can be determined by the equations presented above. These calculations indicate that the discard stream contains 5.06% volume barite, 38.38% volume low-gravity solids, and 56.56% volume water.

To determine the quantity of drilled solids discarded by the shale shaker, a sample of the discarded material is placed in a metal dish and dried in an oven overnight. The weight percent of dry solids is 68.11% wt. and has a density of 2.78 gm/cc.

The rate of dry solids discarded is calculated from the product of the wet discharge flowrate and the weight fraction of dry solids in the discharge (with the appropriate unit conversion factors):

$$\text{Rate of Dry Solids Discard} = (13{,}772 \text{ gpm})(0.6811) \times (1.0 \text{ lb}/453.59 \text{ gm}) \times (60 \text{ min/hr})$$

$$= 1{,}239 \text{ lb/hr}$$

> **Calculation procedure to determine discarded low-gravity solids:**
>
> The discard from the screen weighs 14.8 ppg and contains 43.44% volume solids. The equation presented in the sample calculation is:
>
> $(\rho_B - \rho_{lg})V_{lg} = 100\rho_f + (\rho_B - \rho_f)V_s - 12MW$
>
> Assuming a barite density of 4.2 gm/cc and 2.47 gm/cc for drilled solids, the equation becomes:
>
> $(4.2 - 2.47)V_{lg} = 100 + (4.2 - 1.0)(43.44) - 12(14.8)$
>
> $V_{lg} = 38.38\%$ volume

> **Experimental and calculation procedure:**
>
> A sample of discard is placed in a 40.10 gm crucible and weighed:
>
>   Crucible + Sample Weight = 114.94 gm
>
> The wet sample weight is 74.84 gm. Since the wet discard density is 1.77 gm/cc, the wet sample has a volume of 74.84 gm/1.77 gm/cc = 42.19 cc.
>
> After heating overnight at 250°F, the crucible and sample weight is 91.08 gm. The dry solids weight in the sample is: 91.8 gm – 40.10 gm = 50.98 gm
>
> The dry solids weight percent in the discard is the weight of dry solids divided by the wet sample weight times 100 or:
>
>   [(50.98 gm)/74.84 gm] × 100 = 68.11 wt%
>
> The volume of the dry sample is calculated by subtracting the volume of water lost from the volume of the wet sample: the 42.19 cc wet sample lost 114.94 cc – 91.08 cc or 23.86 cc of water. The volume of the dry sample is:
>
>   42.19 cc – 23.86 cc = 18.33 cc
>
> The density of the dry solids is the weight of dry solids divided by the volume of dry solids, or 50.98 gm/18.33 cc or 2.78 gm/cc.

## CALCULATION OF BARITE DISCARD

Assuming that all of the drilled and other low-gravity solids in the drilling fluid have a dried density of 2.47 gm/cc and the barite has a density of 4.2 gm/cc, the weight percent of barite in the dry sample may be calculated from the following mass-balance equation:

$$\text{Density of Dry Solids} = \frac{\text{Weight of Solids}}{\text{Volume of Solids}}$$

or

$$\text{Density of Dry Solids} = \frac{\text{Weight of Barite} + \text{Weight of Low Gravity Solids}}{\text{Volume of Barite} + \text{Volume of Low Gravity Solids}}$$

To determine the terms on the right side of the equation:

1. The volume of barite is the density (4.2 gm/cc) divided by the weight of barite

2. The volume of low-gravity solids is the total volume of dry solids minus the volume of barite.
3. The volume of low-gravity solids in one cc of solids = 1 cc minus the volume of barite in 1 cc of solids.

This can be expressed as follows:

$$\begin{bmatrix}\text{Volume of low gravity} \\ \text{solids in one cc of solids}\end{bmatrix} = 1\,\text{cc} - \frac{W_B}{4.2\,\text{gm/cc}}$$

$$\begin{bmatrix}\text{Weight of low gravity} \\ \text{Solids in one cc} \\ \text{of dry solids}\end{bmatrix} = \left[1 - \frac{W_B}{4.2\,\text{gm/cc}}\right]\frac{2.47\,\text{gm}}{\text{cc}}$$

$$\text{Density of Solids (D)} = \frac{W_B + 2.47\,\text{gm/cc}\left[1 - \dfrac{W_B}{4.2\,\text{gm/cc}}\right]1\,\text{cc}}{\dfrac{W_B}{4.2\,\text{gm/cc}} + \left[1 - \dfrac{W_B}{4.2\,\text{gm/cc}}\right]}$$

This equation may be reduced to the expression:

$$D = 0.4119 W_B + 2.47$$

or

$$\text{Weight percent barite} = \frac{D - 2.47}{0.4119}$$

The discard density is 2.78 gm/cc so the weight percent barite is 27.07% weight. The weight of dry discard from the shaker screen is 1239 lb/hr. The quantity of barite discarded is (0.2707) × 1239 lb/hr), or 377 lb/hr. The low-gravity solids discard rate is 1239 lb/hr – 377 lb/hr or 861 lb/hr.

## CALCULATING SOLIDS DISCARD AS WHOLE DRILLING FLUID

The drilling fluid contains 13% volume of solids. This percentage in the liquid phase of the shale shaker discard could be associated with the "whole" drilling fluid.

The weight percent of dry solids discarded from the shaker screen was calculated to be 68.11%; so 31.89% of the discard must be liquid. Assume that this liquid is composed of drilling fluid with the solids distribution of the drilling fluid in the pits. The liquid discard rate is (13,772 gm/min)(0.3189) or 4391.9 gm/min. This liquid should contain 13% volume solids.

Since the drilling fluid contains 13% volume solids, a 100 cc sample contains 87 cc of liquid. In this 100 cc sample the water fraction would weigh 87 gm. With an 11.2 ppg (1.343 gm/cc density) drilling fluid, the 100 cc sample should weigh 134.3 gm. Since the liquid weighs 87 gm, the solids must weigh 47.3 gm. In other words, the drilling fluid contains 47.3 gm of solids for every 87 gm of water. The total liquid discard rate is 4391.9 gm/min. The solids discarded by the screen that are associated with the drilling fluid are: (47.3 gm solids/87 gm water)(4391.9 gm/min) or 2387.8 gm/min or 315.6 lb/hr.

The weight fraction of barite in the drilling fluid is 77.4% wt. and the weight fraction of low-gravity solids in the drilling fluid is 22.4% wt. Of the solids discarded from the screen associated with the "whole" drilling fluid, 244 lb/hr are barite and 71.2 lb/hr are low-gravity solids.

Previously, the dry solids discarded by the shaker screen were calculated to be 377 lb/hr barite and 861 lb/hr low-gravity solids. Subtracting the solids associated with the drilling fluid from the solids removed by the screen yields the discarded solids in excess of those associated with the drilling fluid.

- Barite: 377 lb/hr – 244 lb/hr = 133 lb/hr.
- Low gravity solids: 861 lb/hr – 71.2 lb/hr = 798.8 lb/hr.

This indicates that the 200 mesh screen is removing some excess barite and almost 800 lb/hr of drilled solids.

Note that the technique of using the concentration of barite in the discard does not allow an accurate measurement of the quantity of drilling fluid in the shaker discard. Some measurements even indicate that less barite is in the discard than in the "whole" drilling fluid. Shaker screens can pass much of the small barite and remove it from the liquid before it is discarded by the shaker screen.

# CHAPTER ELEVEN

# Centrifugal Pumps

Many types of pumps move fluids in the oil field. Pumps may lift fluid, force it into pressure vessels, or flow it through pipes. Regardless of pump design, horsepower requirements increase with the specific gravity of any fluid. Although centrifugal pumps are used extensively in drilling fluid systems and will be the primary focus of this chapter, other available pumps are briefly described below.

## COMMONLY USED OIL FIELD PUMPS

A *reciprocating pump* is a positive-acting, displacement pump, which creates flow by displacing liquid from a cylinder or cavity with a moving member, or piston. Each chamber, or cylinder, is filled and emptied by the mechanical motion of a piston that alternately draws-in and then expels liquid. Available horsepower and the strength of the pump's structural parts determine pressure capabilities. Volume, or capacity, delivered per stroke by a reciprocating pump is constant regardless of pressure. The flow rate varies with changes in piston speed, the diameter of the cylinder in which the piston moves, and the stroke length of the piston. Most reciprocating pumps use multiple cylinders (i.e., duplex, triplex, etc.) to regulate the pulsating flow generated by the reciprocating motion. They are used primarily on drilling rigs as mud and cement pumps. Centrifugal or rotary pumps, except for special applications, have replaced small reciprocating pumps, although they are still used where their variable-speed, stroke, and piston cylinder combinations are important considerations. High torque and decreasing efficiencies where solid abrasives are present in the fluid are major disadvantages of this type of pump.

*Reciprocating pumps* offer the following features:

- Positive displacement
- Self-priming
- Pressure limited only by pump strength and motor capacity
- Slow operating speeds (50 to 130 rpm)
- Uses various drives—belt, chain, or geared

A *rotary pump* is simple in design, has few moving parts, and, like a reciprocating pump, also uses positive displacement. A rotary pump consists primarily of two meshed cams, or gears (the idler and the drive gear), in a tight-fitting casing. The drive gear is connected to the power supply and its rotation drives the idler gear. Liquid fills the spaces between the gear teeth and, as they rotate, the liquid is literally "squeezed" out the discharge. As the gear teeth then separate, a partial vacuum is created causing liquid to continuously fill the pump chamber from the suction side.

Rotary pumps including gear, screw, deformed vane, sliding vane, axial-piston, and cam type are used when discharge pressures of 500 to 1000 psi or greater are needed. A rotary pump produces a continuous flow regardless of line backpressure and should not be used to pump erosive slurries. Because of close clearances and metal-to-metal contact by the gear teeth, rotary pumps work best when handling solids-free liquids with adequate lubricating qualities. These pumps are particularly adept at pumping liquids with high viscosity or low-vapor pressures. They can be used to move small capacities at medium discharge heads and where high suction lift is required. Applications include oils or viscous materials such as soaps, molasses, tars, and paints.

Rotary pumps offer the following features:

- Positive displacement
- Self-priming (up to 22 feet suction lift)
- Low speed—spur types—up to 300 rpm
- Medium speed—herringbone type—up to 1750 rpm
- Uses various drives—belt, chain, geared, or directly coupled to motor

*Diaphragm pumps* are classed also as positive displacement pumps. The diaphragm acts as a limited displacement piston. Pumping action is obtained when the diaphragm is forced into reciprocating motion by compressed air, mechanical linkage, or from a pulsating external source. Their construction avoids contact between the liquid being pumped and motive force (motor, shaft, air pressure), largely eliminating leakage and corrosion or erosion of moving parts.

A *jet pump* lifts fluid by a partial vacuum created by a motive stream that is provided by another centrifugal or positive displacement pump. The motive stream is driven through a nozzle into a venturi tube. The partial vacuum, formed as the fluid enters the venturi tube, creates a suction. This vacuum draws additional fluid into the output stream.

The **venturi** tube is a "nozzle in reverse." A nozzle speeds the flow velocity of the fluid, thereby converting pressure into velocity energy, which decreases the ambient pressure around the nozzle discharge. A venturi tube slows the flow, converting kinetic energy back into a pressure head.

These pumps are used to pump large particles, such as shale cuttings, and are frequently used as mud hoppers. Many types of vacuum de-gassers also use this type of pump to remove mud from their vessels.

The *Type II pump*, or progressive cavity pump, is based on the Moineau principle. This principle defines the geometric fit between the rotation element (rotor) and the stationary element (stator) of the pump. The rotor has the shape of a single helix and is normally made of metallic materials. The stator is formed as a double helix and is normally made of an elastomer. The interference (compression) fit between the rotor and the stator creates a series of sealed chambers called "cavities." The rotor, turning eccentrically within the stator, produces the pumping action. Material enters the cavity formed at the inlet and progresses within that cavity to the outlet. The result is a positive, nonpulsing flow that is directly proportional to the pump's speed. These pumps are available in multiple stages (single-stage, two-stage, three-stage etc.); the more stages a pump contains, the more pressure that is produced.

*Submersible pumps* are special vertical centrifugal pumps designed to operate with the entire assembly (both pump and motor) submerged in a fluid. They are used to pump subsurface irrigation water or crude oil from wells. The submersible pump is a multi-stage (several or many impellers) centrifugal pump, directly coupled to a submerged electric motor. The impellers are rotated by a single shaft, each sitting in a bowl so that the flow from one impeller can be directed through a diffuser to the next impeller. The three parts—impeller, bowl, and diffuser—are known as a "stage." The impeller and diffuser widths largely determine the capacity of a multi-stage submersible pump. The diameter, number, and rotational speed of the impellers determine the amount of pressure developed. The more impellers (stages) rotating at higher speeds, the greater the pressure. Submersible pumps are much more efficient than jet pumps.

## THE CENTRIFUGAL PUMP

A *centrifugal pump* is a rotating vane, or **impeller**, within a chamber. This pump increases the pressure of a liquid passing though it by increasing the velocity. Liquid enters the pump at the suction flange and revolves along the rotating impeller vanes until thrown to the inner surface of the chamber. Liquid flows from the tip of the rotating impeller along the inner circumference through the pump discharge. If a long, open-ended pipe is connected to the pump, the impeller causes fluid to move vertically to a height proportional to the velocity of the fluid leaving the tip of the impeller. The height of the fluid is called the **head** created by the centrifugal pump.

A centrifugal pump offers the following features:

- Simple construction
- Low initial cost
- Easy to maintain
- Small space requirements
- No close clearances
- No valves or reciprocating parts
- Impeller and shaft are the only moving parts

Centrifugal pumps can be directly connected or belt driven to various power sources, motors, steam turbines, or engines. They produce a steady flow and have low start-up torque requirements. These pumps are used for a relatively wide range of flow rates and are suited to handle liquid slurries with suspended solid matter. Centrifugal pumps are well adapted to general water supply, irrigation, and sewage service. They are used in mill, chemical and petrochemical process plants, mining operations, and drilling rigs.

Although centrifugal pumps seem simple in design, many incorrect concepts surround the basic fundamentals about these pumps. For example, many believe the flow rate from a centrifugal pump can be increased without changing the head by installing a larger diameter impeller. This is a false assumption because the flow rate from a centrifugal pump depends on the piping system connected to the pump and not on the impeller. The centrifugal pump is a constant head device. The diameter of the impeller at a particular rotation speed controls the head produced by the pump and not the flow rate.

As an example of another common misconception, consider the following scenario: A centrifugal pump is connected to a 200-foot length of casing vertically secured next to a derrick. The top end is open and the lower end is connected to a centrifugal pump. A tank full of water is connected to the pump suction. When the pump is turned on, water stands 120 feet above the water level in the suction tank. Water is drained from the system and replaced with a drilling fluid that is twice as dense as water (16.7 ppg).

The misconception is that the heavy-weight drilling fluid will not rise as high as the level of the water. In reality, the dense fluid will rise to the same height as the water. Head is defined as the height of the fluid column. Constant head means that all fluids, regardless of density, will rise to the same height. Additionally, the horsepower of the pump does not have an impact on the height of the heavy-weight drilling fluid. For example, if a motor is capable of lifting water to a height of 120 feet, it has sufficient horsepower to lift the heavier weight drilling fluid. When a drilling fluid is flowing, the required horsepower increases with fluid density. However, in this case the fluid ceases to flow when it reaches the height of the water. (The last several feet of movement of weighted drilling fluid may be much slower than the water if the horsepower of the pump is very low—but the fluid will eventually reach the same height.)

When liquid leaves the impeller of a centrifugal pump, and thereafter the pump casing, its velocity is roughly the same as that of the impeller vane tips. In addition to providing a containing chamber, the casing acts as a guide, directing the fluid to the discharge. At the discharge of a centrifugal pump, the kinetic energy (velocity energy) creates a hydrostatic pressure, or head. Centrifugal pumps produce a constant head depending on the tip velocity of the impeller.

Centrifugal pumps are generally described by the size of their suction and discharge pipe flange. For example, a 3 × 4 pump has a 3-inch diameter discharge pipe flange and a 4-inch diameter suction pipe flange. Most pumps used in oil fields can be outfitted with different impeller diameters. The head generated by a particular diameter impeller spinning at a particular rotary speed, will be the same no matter what size housing contains the impeller.

Figure 11-1 shows a head curve for several different pump sizes with the same impeller diameter. Each curve shows the head decreasing as the flow rate increases. This is caused by the internal friction

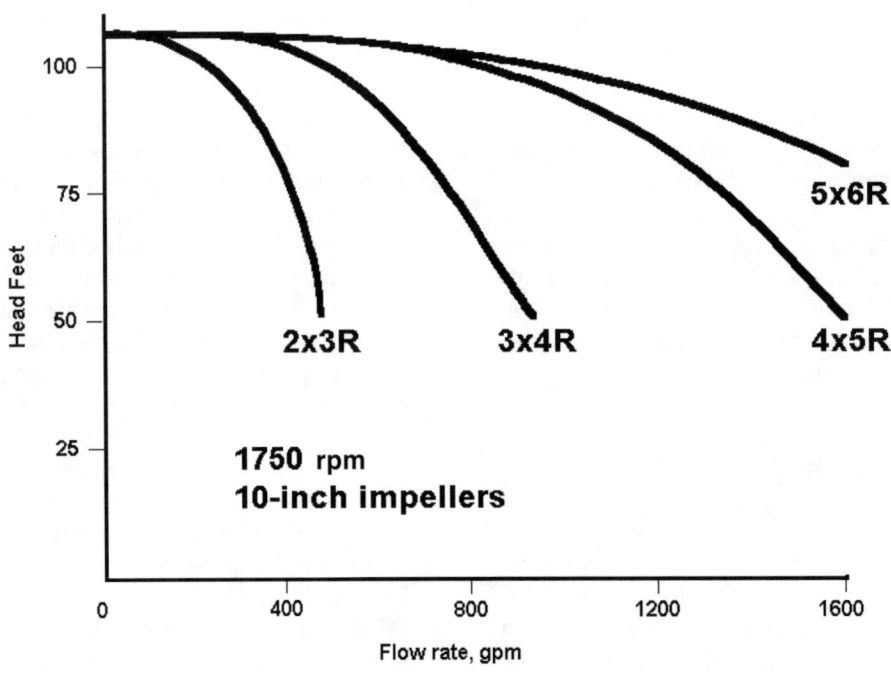

**FIGURE 11-1**

within the pump. Obviously, if fluid is pumped from an external source through a 2 × 3 or a 5 × 6 pump, a greater pressure loss will occur through the smaller diameter pump.

A centrifugal pump acts as a standpipe or water tower that is constantly filled to the same level. The flow rate from such a standpipe depends on the discharge piping connected to it. For example, if a standpipe is filled with fluid to a height of 100 feet, the flow rate of fluid from the standpipe would be much smaller through a 4-inch-diameter pipe than through a 6-inch-diameter pipe.

If the discharge piping is connected at the very bottom of the standpipe, the example would be analogous to a centrifugal pump that generates 100 feet of head.

## Discharge Head

The term "head" is sometimes confused with the term "pressure." In a piping system, head refers to the height of a liquid in a long, open-ended pipe connected to the point at which the head, or pressure, is measured.

A centrifugal pump, with a certain size impeller and rotation speed, pumps fluid to a specific height and maintains the fluid heigth with no further flow. The pressure at the bottom of the column of fluid is a function of the height and density of the fluid. For example, if a well is drilled to 2,000 feet, the head at the bottom of the well is 2,000 feet. The pressure at the bottom of the well depends on the mud weight (or density) of the fluid in the well. The pressure may be determined using the following equation:

Pressure = 0.052(MW)(Depth)

or

Pressure = 0.052(MW)(Head)

where Pressure = Hydrostatic pressure, psig
MW = Mud weight, ppg
Head or Depth = True vertical depth of the well, ft

If the drilling fluid in the 2,000-foot well weighs 10.5 ppg, the bottom-hole pressure would be 1090 psig and the head would still be 2,000 feet. If the mud weight was 15 ppg, the bottom-hole pressure would be 1560 psig and, again, the head would be 2,000 feet.

The pressure at the bottom of the well can also be calculated with the equation:

$$P = \frac{(h)(sp\ gr)}{2.31}$$

where h = Head, in ft of liquid
p = Pressure, lb/in.$^2$
sp gr = Specific gravity of liquid

Pressure read on a gauge attached to a column or tank of fluid is an indication of the pressure created by the static head of liquid in the tank. Centrifugal pumps generate a constant head when the flow rate is low. The velocity of the impeller tip is independent of the fluid density. It generates a constant head because the fluid leaves the impeller tip with a constant velocity. In pump applications, a centrifugal pump can conceptually be considered a constant height stand pipe filled with fluid. The pressure at the bottom of the stand pipe depends on the density, or specific gravity, of the fluid. The head depends only on the height of the liquid in the stand pipe. However, regardless of the source, any pressure can be converted into terms of equivalent head using the previous equation.

The head, or pressure energy, converted into kinetic energy is termed "velocity head." This is not a loss of energy, but a change in energy form. If the velocity were cut in half, half of the kinetic energy converts back to pressure energy, and the gauge reading increases accordingly.

Head lost, or converted due to friction, is termed "friction head." Hydraulically this is a true loss because it is not converted back to pressure energy.

The relationship between liquid flowing past the gauge and its kinetic or velocity head is determined by the equation:

$$h_v = \frac{V^2}{2g}$$

where $h_v$ = Velocity head, ft of liquid
V = Velocity, ft/sec
g = Gravitational acceleration, 32.2 ft/sec$^2$

## Suction Lift

Liquid is forced into the pump's suction when a partial vacuum, or a pressure reduction, is created from the fluid discharge out of the centrifugal pump housing by the impeller. Atmospheric pressure, or line pressure, causes fluid to flow into the pump because of the pressure reduction within the pump. A positive head on the suction side of the centrifugal pump is desirable to prevent the partial vacuum from becoming too low. In reciprocating pumps, this pressure reduction is created by the piston, in rotary pumps by the gear action, and in centrifugal pumps by the liquid forced out of the pump discharge.

One pound per square inch of water pressure is equivalent to 2.31 feet of head in a water column. Atmospheric pressure equals 14.7 pounds per square inch at mean sea level:

14.7 × 2.31 = 33.9 feet (one atmosphere of pressure, on earth, is equivalent to 33.9 feet of head of water)

This is the maximum theoretical suction lift a pump should attain. There are, however, various losses that reduce this theoretical lift including losses through pistons, pump cams, stuffing box leakage, and so forth. Losses in positive-action pumps are less than losses in centrifugal pumps. The maximum practical suction lift in positive acting pumps—based on "water-like" fluid viscosity—is 22 feet and 15 feet with centrifugal pumps.

A low-speed pump will operate safely with a greater suction lift than one with a higher speed. If the suction lift is very high (over 15 feet for water), slower pump impeller speeds and larger pumps are necessary. Increased speeds without proper suction conditions may cause serious problems from **cavitation**, which can ruin a pump very quickly. Head losses from fluid flowing through the suction lines reduce the pressure, or head, at the impeller. This produces the same effect as attempting to lift fluid from a level below the pump.

## Net Positive Suction Head

Net positive suction head (NPSH) is the pressure at pumping temperature (expressed in feet of liquid) available at the pump suction over and above the vapor pressure of the liquid being pumped. In most cases, NPSH is a combination of two parameters: fluid head and pressure existing in the suction line. The pressure in the suction line, if the pump is above the liquid level in the suction tank, may be less than atmospheric. Large friction losses in the suction lines to a centrifugal pump may also decrease the pressure at the pump suction to a value less than atmospheric. Therefore, the net positive suction head may be viewed as the absolute pressure in the suction line at the pump.

NPSH conditions are checked for each pumping application to determine whether or not the fluid will vaporize inside the pump. Vaporization inside a pump is referred to as "cavitation." Cavitation is usually evidenced by excessive noise and vibrations and will sound like gravel being pumped through it.

Pressure on a fluid is reduced as it moves from the pump suction flange to the point where energy is imparted by the impeller. The pressure reduction must be compared to the liquid's vapor pressure, and to the absolute pressure on the fluid as it enters the pump, to determine if the fluid will vaporize.

The pressure reduction inside the pump is referred to as the "required NPSH." The difference between the absolute pressure acting on the fluid and its vapor pressure is called the "available NPSH." When the available NPSH is greater than the required NPSH the pump will not cavitate.

If the available NPSH is not greater than the required NPSH, cavitation, as well as many other serious problems can occur, such as a marked reduction in capacity, or even complete operational failure. Additionally, excessive vibration can occur when some sections of the impeller handle liquid while other sections handle vapor. This will cause pitting and erosion, which drastically reduces pump life. Pitting and erosion result from the collapse of vapor bubbles as they enter regions of higher pressure. As the vapor bubbles collapse, the adjacent walls receive a tremendous shock from the rush of liquid into the cavity left by the collapsed bubble, where small bits of metal actually flake off. Note that the erosion does not occur at the point of lowest pressure, but further upstream where higher pressures cause bubble collapse. Pitting or erosion often occurs on the impeller tips or volute, but are resultant from cavitation caused by insufficient NPSH at, or closer to, the pump suction.

Energy expended to accelerate liquid into voids left by collapsed bubbles causes a drop in the developed head. Consider that there is a volume increase of 50,000 times magnitude when water vaporizes at standard temperature. Even a slight amount of cavitation, therefore, will significantly reduce capacity.

Pumps operating with insufficient NPSH will often pump slugs or spurts of liquid. As the pump is started, liquid accelerates in the suction flange and impeller inlet "eye" until the pump reaches operating capacity. With acceleration of the liquid, friction losses increase and reduce the absolute pressure until the liquid flashes into vapor. When this occurs, pump capacities are reduced and flow decreases, or ceases altogether. In this case, losses are lower with the decreased flow, absolute pressure is higher, liquid ceases to vaporize, and the pump begins to transport fluid again.

The total suction head of the pump can be measured, and vapor pressures for common liquids and brines read from charts. The difference is the available NPSH.

The pump manufacturer determines the required NPSH for each pump by defined testing procedures. NPSH requirements are published on the standard composite performance curves for each pump.

NPSH requirements shown on performance curves are based on boiling water. As the water temperature decreases below the boiling point, less and less head is required to produce the same NPSH due to the difference in vapor pressure. With sufficient reduction in required NPSH, it is then possible to pull a suction lift. This is expressed as:

$$NPSH = P_{atm} + h_{sh} - p_{vp}$$
$$= P_{atm} + h_{ss} - h_{fsuct} - p_{vp}$$

where $P_{atm}$ = Atmospheric pressure in feet of liquid, absolute
$h_{sh}$ = Suction head, feet of liquid
$p_{vp}$ = Liquid's vapor pressure, feet of liquid, absolute
$h_{ss}$ = Static suction head, feet of liquid
$h_{fsuct}$ = Friction head at suction, feet of liquid

## Total Dynamic Suction Lift

Total dynamic suction lift is the sum of two parameters:

1. Vertical distance of pump above liquid (static suction lift)
2. Friction loss in suction pipe

For example, if it is determined that friction losses are equivalent to five feet of head for fluids of "water-like viscosity," the vertical distance of the centrifugal pump suction above the liquid level cannot exceed ten feet. When pumping viscous fluids, the pump suction should be below the "suction tank." In other words, the head that moves fluid from the fluid level in the suction tank to the pump is the "total suction head." The "total discharge head" tends to resist the flow of liquid from the pump to the discharge tank.

These components can be subdivided into three smaller segments:

1. Static head
2. Surface pressure
3. Friction head

***Static head.*** Static suction head is the height of the liquid surface in the suction tank above the centerline of the pump. If the liquid surface is below the centerline, the static suction head assumes a negative value, referred to as "suction lift."

Static discharge head is the height of the highest liquid surface in the discharge section above the centerline of the pump. This can be liquid in the discharge pipe or in the discharge tank, depending on which is highest.

***Surface pressure.*** The suction and/or discharge compartments may be under pressure other than atmospheric. Both suction and discharge surface pressure is converted to feet of liquid.

***Friction head.*** Friction head is the energy necessary per pound of fluid pumped to overcome friction and turbulent losses as fluid flows through the system. The values of suction friction head and discharge friction head depend on pipe lengths, diameters, flow rate, valves and fittings, their configurations, and pipe materials and interior finish.

## Total Dynamic Head

The total dynamic head is the sum of total suction lift plus total discharge head. The sum of the gauge pressure head and the velocity head at the discharge flange, minus the sum of the corresponding heads at the suction flange, equals the energy (in foot-pounds) added per pound of liquid pumped. This is called the "total head" developed by the pump and is expressed as:

$$H = (h_{gd} + h_{vd}) - (h_{gs} + h_{vs})$$

where H = Total head
$h_{gd}$ = Discharge gauge head
$h_{vd}$ = Discharge velocity head
$h_{gs}$ = Suction gauge head
$h_{vs}$ = Suction velocity head

The discharge head of a centrifugal pump is determined by the impeller diameter and its rotating speed, not by the horsepower of the motor.

If the discharge line of a centrifugal pump is closed, the horsepower input decreases and the pressure (head) increases only slightly. As the valve on the discharge is gradually opened, the horsepower input increases rapidly and pressure (head) decreases slightly.

***Pump horsepower.*** Hydraulic horsepower (hhp) is expressed as follows:

$$hhp = \frac{QH (sp\ gr)\ 8.34}{3300} = \frac{(Q)(\Delta P)}{1714}$$

where Q = Flow rate, gpm
H = Head, ft
P = Pressure drop, psi

The actual brake horsepower of a pump is greater than the theoretical hydraulic horsepower by the amount of losses incurred in the pump through friction, leakage, and so forth. The efficiency of a pump is therefore:

$$\text{Pump efficiency} = \frac{\text{Hydraulic horsepower}}{\text{Brake horsepower}}$$

and

$$\text{Horsepower, hp} = \frac{(\text{Flow rate, gpm})(\text{Total head, ft})(\text{Sp Gr})}{3960 (\text{Pump efficiency})}$$

## CENTRIFUGAL PUMP PERFORMANCE CURVES

A pump is a machine that moves a given volume of liquid a given distance in a period of time. Pumps take energy from an external source (motor, turbine, etc.) and develop it in the form of discharge pressure or total dynamic head. A curve can be developed to measure a pump's performance. Tests to establish pump performance curves are run on clear, cold water with a specific gravity of 1.0, at 60°F.

To establish a centrifugal pump performance curve, the pump is run in conjunction with a gauge, a valve, and a flowmeter located on the discharge side. Running the pump with a closed valve on the discharge line, yields the maximum pressure, or feet of head, which the pump will develop at zero flow. The valve is then opened to allow a small flow, and the pressure and flow rate are noted. The required horsepower (or kilowatts of electricity) is also noted. This process is continued over the full operating range of the pump. Plotting the data points reveals the head-capacity and horsepower curves (Figure 11-1).

The flow rate from the pump is determined by the performance and friction curves of the plumbing connected to the pump. The horsepower curves indicated on a performance curve represent the power required when pumping water, or any fluid within the same specific gravity and viscosity as water. Corrections for different specific gravities are, however, quite straightforward—horsepower is directly proportional to the specific gravity of the fluids being pumped. Performance curves are used to determine the flow rate, head (pressure) developed, and horsepower required by a pump having a specific diameter impeller and operating at a given, constant speed.

### Performance Characteristics

The performance characteristics of a centrifugal pump are clearly defined by its performance curve(s). Figure 11-2 shows a typical centrifugal pump performance curve with operations at 1750 rpm.

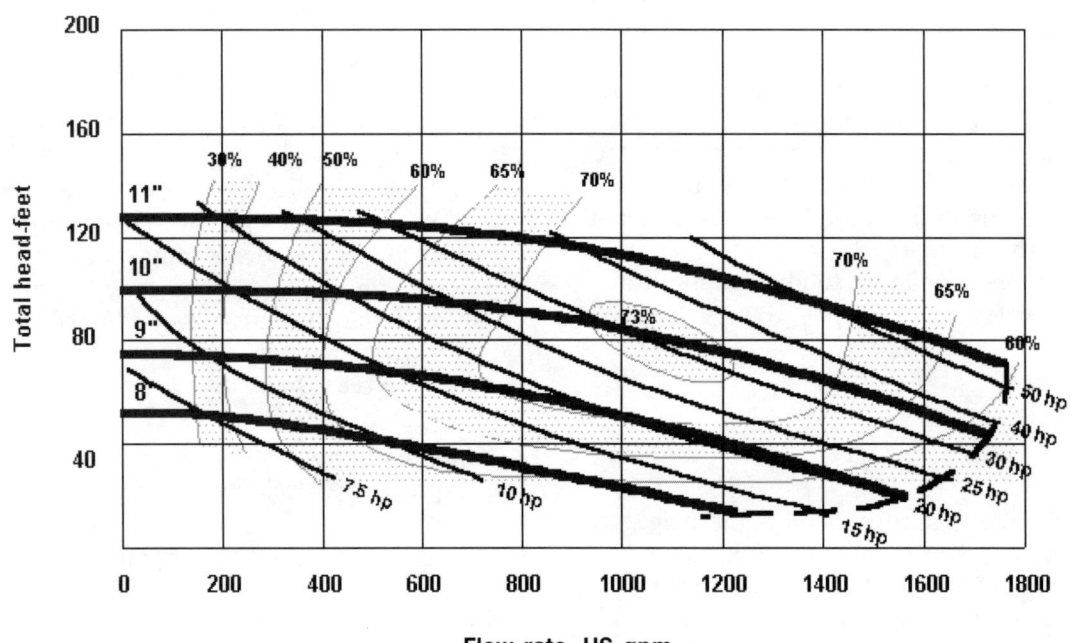

**FIGURE 11-2**

***What flow rate is obtained from a centrifugal pump?*** As stated previously, the flow rate depends on the piping connected to it. Figure 11-3 shows pipe friction curves for 100 feet of common sizes of new steel pipes. This chart shows that 400 gpm flowing through 100 feet of 3-inch diameter pipe will cause 33 feet of head loss. The flow rate from a pump connected to this pipe is determined by the head generated by the pump and the friction loss of the pipe at this flow rate.

The head curve (Figure 11-4) for a particular impeller in a particular pump, would produce a flow rate dependent on the friction curve that represents the piping connected to the pump. This performance curve shows a small lift head required before the friction curves start. If a substantial length of pipe is connected to the pump, the friction curve intersects the head curve at point I, or 400 gpm. This is called the operating point. If less pipe is connected to the same pump, the friction head (or pressure) loss is lower for any flow rate. In this case, the friction curve would intersect the head curve at point II, or 1000 gpm. If much less pipe is connected to the pump, the operating point will move to 1400 gpm at point III.

The flow rate produced by this pump depends on what is connected to it.

Friction curves for a piping system with six hydrocyclones connected to a centrifugal pump is presented in Figure 11-5. The friction curves start with the lift required to move the fluid from the liquid level in the pits to the centerline of the hydrocyclone manifold. This is the physical height, in feet, measured at the rig. It is independent of the flow rate and is, therefore, shown as a horizontal line on the head/flow rate chart. The suction and discharge pipe friction is a function of flow rate and is plotted as a curve above the lift "line." At the flow rate required for the hydrocyclones, the discharge and suction pipe frictions are calculated. For example, if these were 4-inch hydrocyclones, each cone would accept 50 gpm if 75 feet of head is applied. This is a function of the size of the nozzle or opening size on the input side of the hydrocyclone. Since there are six cones in parallel, the friction curves are arranged sequentially. The appropriate head (accounting for lift, suction friction, discharge friction, and hydrocyclones), together with the appropriate flow rate, will determine the impeller size required.

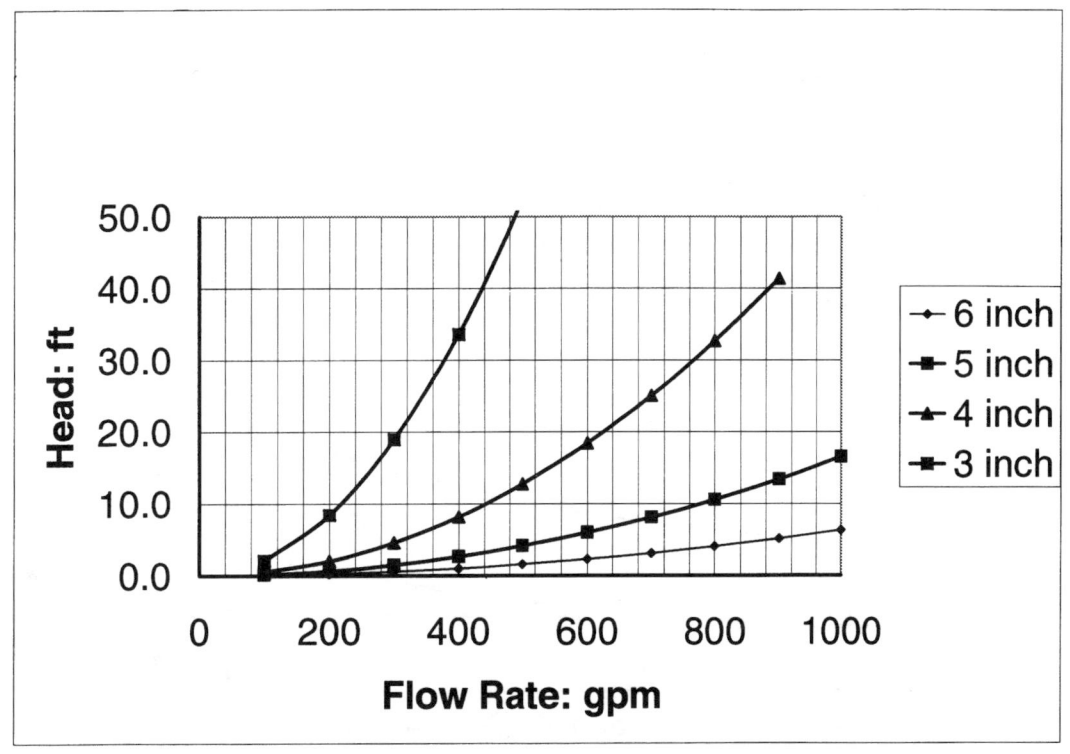

**FIGURE 11-3.** Pipe friction curves for common sizes of 100' new steel pipe.

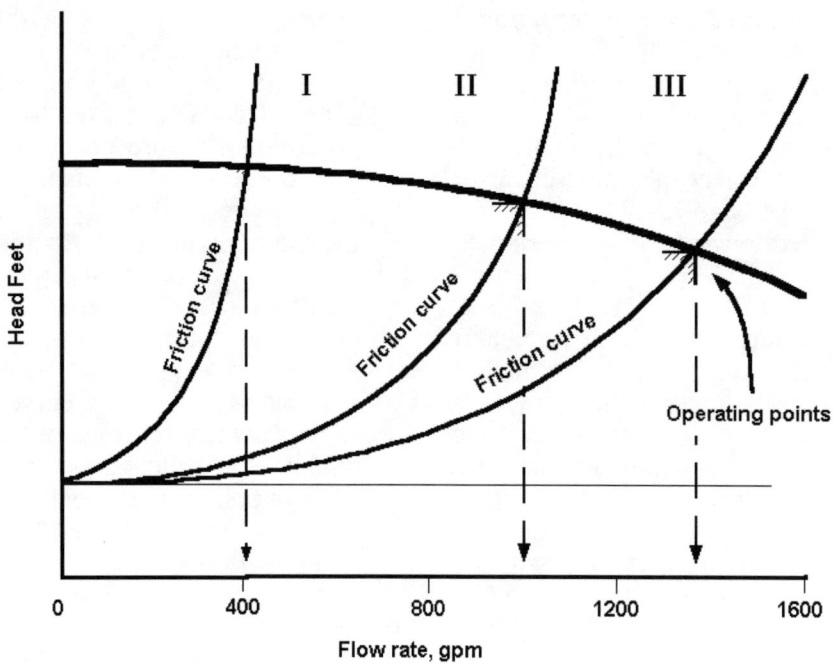

**FIGURE 11-4**

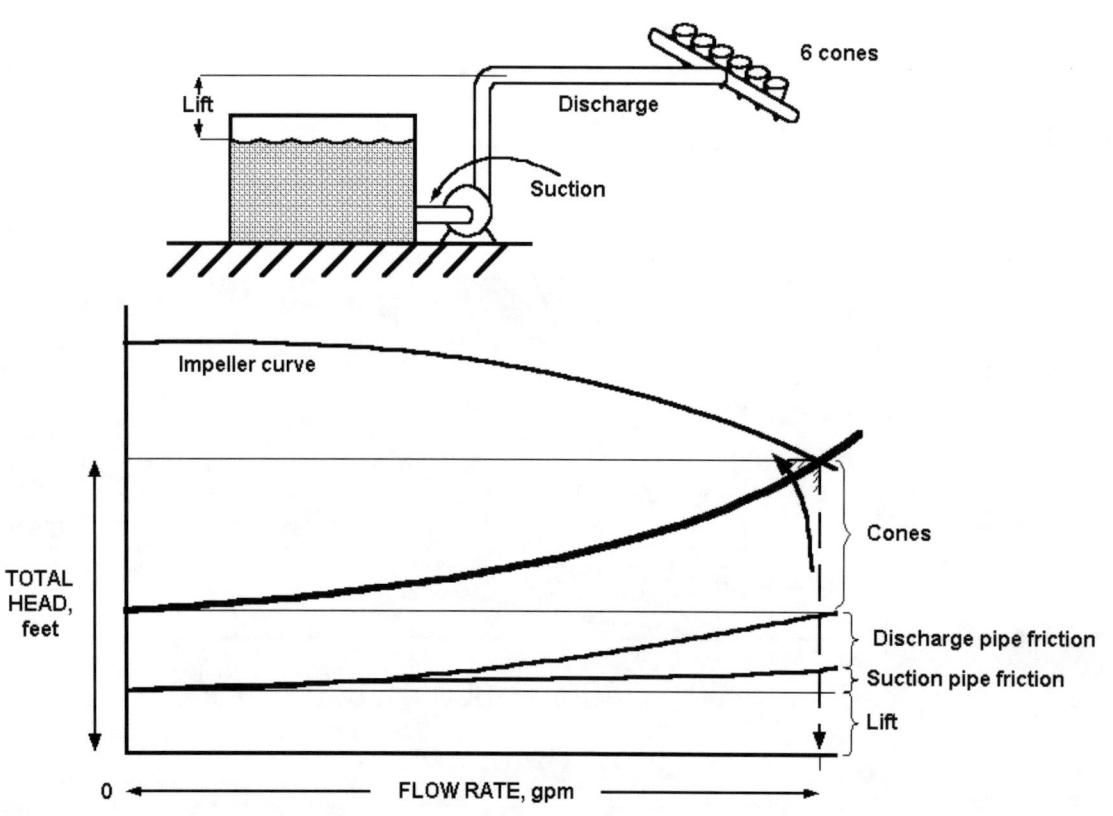

**FIGURE 11-5**

Two characteristics of centrifugal pumps can be observed from these curves:

1. The head produced is limited to that developed at shut-off (no flow). Thus, even if there were a total blockage in the discharge line, the pressure on the equipment between the pump and the blockage would never exceed the shut-off head of the pump.
2. Horsepower decreases as the pump is throttled back to shut-off. Once again, even if there were total blockage in the discharge line, the motor driving the pump would not overload.

Centrifugal pumps are flexible, which allows a casing or volute to be used with various impeller diameters that are reduced from the maximum. Therefore, a composite curve has evolved that details the pump's performance when outfitted to operate at a given speed with various impeller diameters.

*Example #1*

- Capacity (Q) varies in direct proportion to rotating speed.
- Head (h, in feet) varies in direct proportion with the square of the rotating speed.
- Horsepower (hp) varies in direct proportion with the cube of the rotating speed.

At 1750 rpm a centrifugal pump delivers 300 gpm and 50 feet of head, requiring 5 horsepower. What are the ratings at 1450 rpm?

$$\text{Capacity}: \frac{Q_2}{Q_1} = \frac{\text{rpm}_2}{\text{rpm}_1}$$

$$Q_2 = \frac{(1450 \text{ rpm})(300 \text{ gpm})}{1750 \text{ rpm}} = 250 \text{ gpm}$$

$$\text{Head}: \frac{h_2}{h_1} = \frac{(\text{rpm}_2)^2}{(\text{rpm}_1)^2}$$

$$h_2 = \frac{(1450 \text{ rpm})^2 (50 \text{ ft})}{(1750 \text{ rpm})^2} = 34 \text{ ft}$$

$$\text{Horsepower}: \frac{\text{hp}_2}{\text{hp}_1} = \frac{(\text{rpm}_2)^3}{(\text{rpm}_1)^3}$$

$$\text{hp}_2 = \frac{(1450 \text{ rpm})^3 (5 \text{ hp})}{(1750 \text{ rpm})^3} = 2.8 \text{ hp}$$

*Example #2*

- Capacity (Q) varies in direct proportion to impeller diameter.
- Head (h, in feet) varies in direct proportion to the square of the impeller diameter.
- Horsepower (hp) varies in direct proportion to the cube of the impeller diameter.

A pump operating at 650 gpm and 1750 rpm with an 11.5-inch impeller, delivers 75 feet of head and requires 35 horsepower. What are the ratings with a 14-inch impeller?

$$\text{Capacity}: \frac{Q_2}{Q_1} = \frac{D_2}{D_1}$$

$$Q_2 = \frac{(14 \text{ in.})(650 \text{ gpm})}{11.5 \text{ in.}} = 790 \text{ gpm}$$

$$\text{Head}: \frac{h_2}{h_1} = \frac{(D_2)^2}{(D_1)^2}$$

$$h_2 = \frac{(14.0 \text{ in.})^2}{(11.5 \text{ in.})^2} (75 \text{ ft}) = 111 \text{ ft}$$

$$\text{Horsepower}: \frac{\text{hp}_2}{\text{hp}_1} = \frac{(D_1)^3}{(D_2)^3}$$

$$\text{hp}_2 = \frac{(14.0 \text{ in.})^3}{(11.5 \text{ in.})^3} (35 \text{ hp}) = 63 \text{ hp}$$

## HEAD LOSSES THROUGH PIPE

The piping system may consist of suction and discharge tanks, valves, gauges, elbows, tees, piping, process or treating equipment, and the pump itself. As fluid flows through a piping system, friction forces within the fluid and between the fluid and the pipe walls decrease the pressure within the pipe. This pressure loss may be calculated from Darcy's Equation:

$$\frac{h}{L} = \frac{6fv^2}{gd}$$

where h = Head loss, ft
f = Friction factor
L = Pipe length, ft
v = Velocity, ft/sec
g = Acceleration of gravity, ft/sec$^2$
d = Internal diameter of the pipe, in.

> **Sample Calculation:**
>
> For 800 gpm of a 12.0 ppg drilling fluid flowing through a 6-inch diameter, Schedule 80 pipe, the fluid has a PV = 20 cp. and a YP = 18 lb/100 ft². Find the Reynold's number.
>
> The pipe data in Table 11-2 shows that the pipe's inside diameter is 5.76 inches. The velocity of the fluid in the pipe would be:
>
> $$v = \frac{Q}{A} = \frac{(800 \text{ gpm})(231 \text{ in.}^3/\text{gal})(1 \text{ min}/60 \text{ sec})(1 \text{ ft}/12 \text{ in.})}{\pi(5.76 \text{ in.})^2/4} = 9.8 \text{ ft/sec}$$
>
> The 600-rpm Fann reading would be 58 (the 300 rpm reading would be 38). The apparent viscosity would be 29 cp.
>
> $$R_e = \frac{927.8 \, dv(MW)}{\mu} = \frac{927.8(5.76 \text{ in.})(9.8 \text{ ft/sec})(12.0 \text{ ppg})}{29 \text{ cp}} = 22,000$$
>
> This value is larger than 4000, so this flow is turbulent.

This equation is correct for either laminar or turbulent flow.[1] It calculates the head loss of a constant-density fluid flowing in a straight, constant-diameter pipe.

For laminar flow conditions, or Reynold's number ($R_e$) less than 2000, the friction factor (f) is:

$$f_L = \frac{64}{R_e}$$

For turbulent flow conditions, or Reynold's number ($R_e$) greater than 4000, the friction factor is:

$$f_T = \frac{0.184}{R_e^{0.2}}$$

The head loss (H) for fluid with a viscosity, $\mu$, and a flow rate, Q, gpm, in a pipe with a diameter of D, inches, would be:

$$H = f \frac{L_e Q^2}{(32.18) D^5}$$

where $L_e$ = Equivalent length of pipe, ft

Elbows, valves, swedges, and changes in direction of piping can be expressed in equivalent lengths of straight, smooth pipe (Table 11-1). Reynold's number may be calculated from the equation:

$$R_e = \frac{927.8 \, d(v)(MW)}{\mu}$$

where d = Pipe inside diameter, in.
v = Velocity of liquid, ft/sec
MW = Fluid density, pounds per gallon
$\mu$ = Viscosity of the fluid, centipoise

For laminar flow, $R_e < 2000$; and for turbulent flow, $R_e > 4000$.

Viscosity for Newtonian fluids can be easily defined if the pressure and temperatures are known. Drilling fluids, however, are usually shear-thinning, and the viscosity decreases as the shear rate increases. Drilling fluid viscosity also depends on the pressure and temperature. If the flow properties of a drilling fluid are known as a function of shear rate, the viscosity for the appropriate shear rate can be used to calculate head loss in a pipe system. In lieu of this, the apparent viscosity can be used. This value, in centipoise, is calculated by dividing the 600-rpm Fann reading by two. This is the viscosity of the drilling fluid at 1022 sec⁻¹ shear rate.

After determining the Reynold's number and flow regime, the head loss per unit length of pipe may be calculated from the equation:

$$\frac{h}{L} = \frac{6fv^2}{gd}$$

To calculate velocity, v in ft/sec, the following equation can be used instead of using the conversion factors. Here, Q is in gpm and d is the inside pipe diameter in inches:

$$v = 0.409 \frac{Q}{d^2}$$

---
[1] The velocity head is the $v^2/2g$ part of this equation. This should be recognizable from the kinetic energy equation: KE = m($v^2/2g$).

**TABLE 11-1.** Mud Line Sizes and Losses for Pumps with Hydrocyclones (Friction Loss in Fittings: Equivalent Feet of Straight Pipe)

*Note:* Threaded or mitered ells, threaded tees, globe valves, angle valves, swing check valves, and so forth, should not be used and are not included in this table. Suction and discharge pipe usually should have the same diameters. Swage up can be on the pump discharge flange, the pump suction flange, or on the hydrocyclone header. Use the pipe size connected to a swage up to determine the loss in equivalent feet of pipe.

(L) Laminar Flow (T) Turbulent Flow

| FITTING | PIPE SIZE (NOMINAL SCHEDULE 40) | | | | | | | | | | | | | | | | | |
|---|---|---|---|---|---|---|---|---|---|---|---|---|---|---|---|---|---|---|
| | 12" | | 10" | | 8" | | 6" | | 5" | | 4" | | 3" | | 2½" | | 2 | |
| | L | T | L | T | L | T | L | T | L | T | L | T | L | T | L | T | L | T |
| Weld Tee (Branched*) 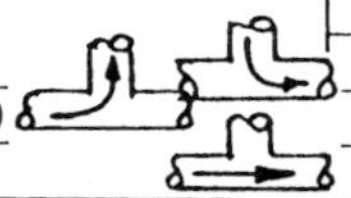 | 73 | 55 | 64 | 48 | 49 | 37 | 37 | 28 | 31 | 23 | 25 | 19 | 19 | 14 | 15 | 11½ | 13 | 9½ |
| Weld Tee (Running*) | 21 | 15½ | 17 | 13 | 13 | 10 | 11 | 8 | 9 | 6½ | 7 | 5 | 5 | 4 | 4 | 3 | 3 | 2½ |
| 90° Weld Ell Long Radius | 125 | 12½ | 105 | 10½ | 85 | 8½ | 65 | 6½ | 55 | 5½ | 40 | 4 | 30 | 3 | 25 | 2½ | 20 | 2 |
| 45° Weld Ell Long Radius | 85 | 8½ | 70 | 7 | 55 | 5½ | 40 | 4 | 35 | 3½ | 30 | 3 | 20 | 2 | 15 | 1½ | 15 | 1½ |
| 45° Lateral or Y (Branched*) 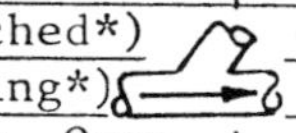 | 41 | 31 | 33 | 25 | 27 | 20 | 21 | 15½ | 17 | 13 | 13 | 10 | 11 | 8 | 8 | 6 | 7 | 5½ |
| 45° Lateral or Y (Running*) | 27 | 20 | 23 | 17 | 17 | 13 | 14 | 10½ | 12 | 9 | 9 | 7 | 7 | 5½ | 5 | 4 | 5 | 3½ |
| Gate or Butterfly Valve, Open | 8½ | 8½ | 7 | 7 | 5½ | 5½ | 4 | 4 | 3½ | 3½ | 3 | 3 | 2 | 2 | 1½ | 1½ | 1½ | 1½ |
| Swage Up: pipe on large end  | | | | | | | | | | | | | | | | | | |
| Ratio 2:4, 3:6, 4:8, 5:10, 6:12 | 412 | 412 | 380 | 380 | 290 | 290 | 215 | 215 | – | – | 140 | 140 | – | – | – | – | | |
| Ratio 3:5, 6:10, Ratio 5:8 | – | – | 123 | 123 | 75 | 75 | – | – | 61 | 61 | – | – | – | – | | | | |
| Ratio 2:3, 4:6, 8:12 | 70 | 70 | – | – | – | – | 38 | 38 | – | – | – | – | 18 | 18 | – | – | – | – |
| Ratio 3:4, 6:8 | – | – | – | – | 14 | 14 | – | – | – | – | 6 | 6 | – | – | – | – | | |
| Ratio 4:5, 8:10 | – | – | 9 | 9 | – | – | – | – | 5 | 5 | – | – | – | – | | | – | – |
| Ratio 5:6, 10:12 | 7 | 7 | – | – | – | – | 4 | 4 | – | – | – | – | | | – | – | – | – |
| Swage Up: pipe on small end  | | | | | | | | | | | | | | | | | | |
| Ratio 5:10, 6:12 | – | – | – | – | – | – | 14 | 14 | 12 | 12 | | | | | | | | |
| Ratio 3:5, 6:10, Ratio 5:8 | | | – | – | – | – | 10 | 10 | 8½ | 8½ | – | – | 5 | 5 | – | – | – | – |
| Ratio 4:6, 8:12 | | | | | 9½ | 9½ | – | – | – | – | 5 | 5 | – | – | – | – | | |
| Ratio 3:4, 6:8 | | | – | – | – | – | 3½ | 3½ | – | – | – | – | 1½ | 1½ | – | – | – | – |
| Ratio 4:5, 8:10 | | | – | – | 3 | 3 | – | – | – | – | 1½ | 1½ | – | – | – | – | | |
| Ratio 5:6, 10:12 | – | – | 3 | 3 | – | – | – | – | 1½ | 1½ | – | – | – | – | | | – | – |
| Reduce Down: pipe on large end 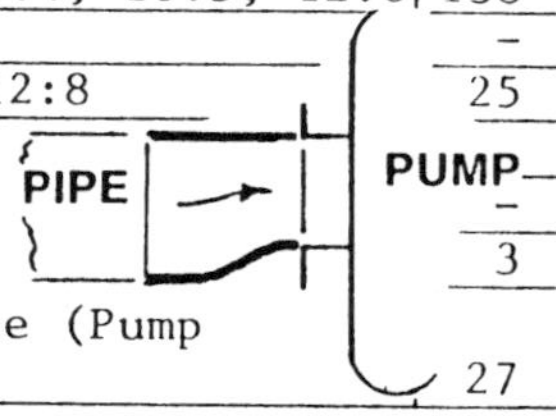 | | | | | | | | | | | | | | | | | | |
| Ratio 4:2, 6:3, 8:4, 10:5, 12:6 | 138 | 138 | 126 | 126 | 83 | 83 | 72 | 72 | | | 38 | 38 | | | – | – | | |
| Ratio 5:3, 10:6 | – | – | 41 | 41 | – | – | – | – | 16 | 16 | – | – | – | – | – | – | – | |
| Ratio 3:2, 6:4, 12:8 | 25 | 25 | – | – | – | – | 10 | 10 | – | – | – | – | 5 | 5 | – | – | – | – |
| Ratio 4:3, 8:6 | | – | – | – | 5 | 5 | – | – | – | – | 3 | 3 | – | – | – | – | | |
| Ratio 5:4, 10:8 | – | – | 4 | 4 | – | – | – | – | 2 | 2 | – | – | – | – | | | – | – |
| Ratio 6:5, 12:10 | 3 | 3 | – | – | – | – | 2 | 2 | – | – | – | – | | | – | – | – | – |
| Submerged Entrance (Pump Suction Only) | 27 | 20 | 23 | 17 | 17 | 13 | 14 | 10½ | 12 | 9 | 9 | 7 | 7 | 5½ | 5 | 4 | 5 | 3½ |

**TABLE 11-2.** Pipe Data for Carbon and Alloy Steel (Stainless Steel)

| Nominal Pipe Size Inches | Outside Diam. Inches | Identification - Steel - Iron Pipe Size | Identification - Steel - Sched. No. | Identification - Stainless Steel Sched. No. | Wall Thickness (t) Inches | Inside Diameter (d) Inches | Area of Metal Square Inches | Transverse Internal Area (a) Square Inches | Transverse Internal Area (A) Square Feet | Moment of Inertia (I) Inches$^4$ | Weight Pipe Pounds per foot |
|---|---|---|---|---|---|---|---|---|---|---|---|
| 1/8 | 0.405 | ... | ... | 10S | .049 | .307 | .0548 | .0740 | .00051 | .00088 | .19 |
|  |  | STD | 40 | 40S | .068 | .269 | .0720 | .0568 | .00040 | .00106 | .24 |
|  |  | XS | 80 | 80S | .095 | .215 | .0925 | .0364 | .00025 | .00122 | .31 |
| 1/4 | 0.540 | ... | ... | 10S | .065 | .410 | .0970 | .1320 | .00091 | .00279 | .33 |
|  |  | STD | 40 | 40S | .088 | .364 | .1250 | .1041 | .00072 | .00331 | .42 |
|  |  | XS | 80 | 80S | .119 | .302 | .1574 | .0716 | .00050 | .00377 | .54 |
| 3/8 | 0.675 | ... | ... | 10S | .065 | .545 | .1246 | .2333 | .00162 | .00586 | .42 |
|  |  | STD | 40 | 40S | .091 | .493 | .1670 | .1910 | .00133 | .00729 | .57 |
|  |  | XS | 80 | 80S | .126 | .423 | .2173 | .1405 | .00098 | .00862 | .74 |
| 1/2 | 0.840 | ... | ... | 5S | .065 | .710 | .1583 | .3959 | .00275 | .01197 | .54 |
|  |  | ... | ... | 10S | .083 | .674 | .1974 | .3568 | .00248 | .01431 | .67 |
|  |  | STD | 40 | 40S | .109 | .622 | .2503 | .3040 | .00211 | .01709 | .85 |
|  |  | XS | 80 | 80S | .147 | .546 | .3200 | .2340 | .00163 | .02008 | 1.09 |
|  |  | ... | 160 | ... | .187 | .466 | .3836 | .1706 | .00118 | .02212 | 1.31 |
|  |  | XXS | ... | ... | .294 | .252 | .5043 | .050 | .00035 | .02424 | 1.71 |
| 3/4 | 1.050 | ... | ... | 5S | .065 | .920 | .2011 | .6648 | .00462 | .02450 | .69 |
|  |  | ... | ... | 10S | .083 | .884 | .2521 | .6138 | .00426 | .02969 | .86 |
|  |  | STD | 40 | 40S | .113 | .824 | .3326 | .5330 | .00371 | .03704 | 1.13 |
|  |  | XS | 80 | 80S | .154 | .742 | .4335 | .4330 | .00300 | .04479 | 1.47 |
|  |  | ... | 160 | ... | .219 | .612 | .5698 | .2961 | .00206 | .05269 | 1.94 |
|  |  | XXS | ... | ... | .308 | .434 | .7180 | .148 | .00103 | .05792 | 2.44 |
| 1 | 1.315 | ... | ... | 5S | .065 | 1.185 | .2553 | 1.1029 | .00766 | .04999 | .87 |
|  |  | ... | ... | 10S | .109 | 1.097 | .4130 | .9452 | .00656 | .07569 | 1.40 |
|  |  | STD | 40 | 40S | .133 | 1.049 | .4939 | .8640 | .00600 | .08734 | 1.68 |
|  |  | XS | 80 | 80S | .179 | .957 | .6388 | .7190 | .00499 | .1056 | 2.17 |
|  |  | ... | 160 | ... | .250 | .815 | .8365 | .5217 | .00362 | .1251 | 2.84 |
|  |  | XXS | ... | ... | .358 | .599 | 1.0760 | .282 | .00196 | .1405 | 3.66 |
| 1¼ | 1.660 | ... | ... | 5S | .065 | 1.530 | .3257 | 1.839 | .01277 | .1038 | 1.11 |
|  |  | ... | ... | 10S | .109 | 1.442 | .4717 | 1.633 | .01134 | .1605 | 1.81 |
|  |  | STD | 40 | 40S | .140 | 1.380 | .6685 | 1.495 | .01040 | .1947 | 2.27 |
|  |  | XS | 80 | 80S | .191 | 1.278 | .8815 | 1.283 | .00891 | .2418 | 3.00 |
|  |  | ... | 160 | ... | .250 | 1.160 | 1.1070 | 1.057 | .00734 | .2839 | 3.76 |
|  |  | XXS | ... | ... | .382 | .896 | 1.534 | .630 | .00438 | .3411 | 5.21 |
| 1½ | 1.900 | ... | ... | 5S | .065 | 1.770 | .3747 | 2.461 | .01709 | .1579 | 1.28 |
|  |  | ... | ... | 10S | .109 | 1.682 | .6133 | 2.222 | .01543 | .2468 | 2.09 |
|  |  | STD | 40 | 40S | .145 | 1.610 | .7995 | 2.036 | .01414 | .3099 | 2.72 |
|  |  | XS | 80 | 80S | .200 | 1.500 | 1.068 | 1.767 | .01225 | .3912 | 3.63 |
|  |  | ... | 160 | ... | .281 | 1.338 | 1.429 | 1.406 | .00976 | .4824 | 4.86 |
|  |  | XXS | ... | ... | .400 | 1.100 | 1.885 | .950 | .00660 | .5678 | 6.41 |
| 2 | 2.375 | ... | ... | 5S | .065 | 2.245 | .4717 | 3.958 | .02749 | .3149 | 1.61 |
|  |  | ... | ... | 10S | .109 | 2.157 | .7760 | 3.654 | .02538 | .4992 | 2.64 |
|  |  | STD | 40 | 40S | .154 | 2.067 | 1.075 | 3.355 | .02330 | .6657 | 3.65 |
|  |  | XS | 80 | 80S | .218 | 1.939 | 1.477 | 2.953 | .02050 | .8679 | 5.02 |
|  |  | ... | 160 | ... | .344 | 1.687 | 2.190 | 2.241 | .01556 | 1.162 | 7.46 |
|  |  | XXS | ... | ... | .436 | 1.503 | 2.656 | 1.774 | .01232 | 1.311 | 9.03 |
| 2½ | 2.875 | ... | ... | 5S | .083 | 2.709 | .7280 | 5.764 | .04002 | .7100 | 2.48 |
|  |  | ... | ... | 10S | .120 | 2.635 | 1.039 | 5.453 | .03787 | .9873 | 3.53 |
|  |  | STD | 40 | 40S | .203 | 2.469 | 1.704 | 4.788 | .03322 | 1.530 | 5.79 |
|  |  | XS | 80 | 80S | .276 | 2.323 | 2.254 | 4.238 | .02942 | 1.924 | 7.66 |
|  |  | ... | 160 | ... | .375 | 2.125 | 2.945 | 3.546 | .02463 | 2.353 | 10.01 |
|  |  | XXS | ... | ... | .552 | 1.771 | 4.028 | 2.464 | .01710 | 2.871 | 13.69 |
| 3 | 3.500 | ... | ... | 5S | .083 | 3.334 | .8910 | 8.730 | .06063 | 1.301 | 3.03 |
|  |  | ... | ... | 10S | .120 | 3.260 | 1.274 | 8.347 | .05796 | 1.822 | 4.33 |
|  |  | STD | 40 | 40S | .216 | 3.068 | 2.228 | 7.393 | .05130 | 3.017 | 7.58 |
|  |  | XS | 80 | 80S | .300 | 2.900 | 3.016 | 6.605 | .04587 | 3.894 | 10.25 |
|  |  | ... | 160 | ... | .438 | 2.624 | 4.205 | 5.408 | .03755 | 5.032 | 14.32 |
|  |  | XXS | ... | ... | .600 | 2.300 | 5.466 | 4.155 | .02885 | 5.993 | 18.58 |

**Identification, wall thickness and weights** are extracted from ANSI B36.10 and B36.19. The notations STD, XS, and XXS indicate Standard, Extra Strong, and Double Extra Strong pipe respectively.

**TABLE 11-2.** (continued)

| Nominal Pipe Size Inches | Outside Diam. Inches | Identification Iron Pipe Size | Identification Steel Sched. No. | Identification Stainless Steel Sched. No. | Wall Thickness (t) Inches | Inside Diameter (d) Inches | Area of Metal Square Inches | Transverse Internal Area (a) Square Inches | Transverse Internal Area (A) Square Feet | Moment of Inertia (I) Inches⁴ | Weight Pipe Pounds per foot |
|---|---|---|---|---|---|---|---|---|---|---|---|
| 3½ | 4.000 | ... | ... | 5S | .083 | 3.834 | 1.021 | 11.545 | .08017 | 1.960 | 3.48 |
| | | ... | ... | 10S | .120 | 3.760 | 1.463 | 11.104 | .07711 | 2.755 | 4.97 |
| | | STD | 40 | 40S | .226 | 3.548 | 2.680 | 9.886 | .06870 | 4.788 | 9.11 |
| | | XS | 80 | 80S | .318 | 3.364 | 3.678 | 8.888 | .06170 | 6.280 | 12.50 |
| 4 | 4.500 | ... | ... | 5S | .083 | 4.334 | 1.152 | 14.75 | .10245 | 2.810 | 3.92 |
| | | ... | ... | 10S | .120 | 4.260 | 1.651 | 14.25 | .09898 | 3.963 | 5.61 |
| | | STD | 40 | 40S | .237 | 4.026 | 3.174 | 12.73 | .08840 | 7.233 | 10.79 |
| | | XS | 80 | 80S | .337 | 3.826 | 4.407 | 11.50 | .07986 | 9.610 | 14.98 |
| | | ... | 120 | ... | .438 | 3.624 | 5.595 | 10.31 | .0716 | 11.65 | 19.00 |
| | | ... | 160 | ... | .531 | 3.438 | 6.621 | 9.28 | .0645 | 13.27 | 22.51 |
| | | XXS | ... | ... | .674 | 3.152 | 8.101 | 7.80 | .0542 | 15.28 | 27.54 |
| 5 | 5.563 | ... | ... | 5S | .109 | 5.345 | 1.868 | 22.44 | .1558 | 6.947 | 6.36 |
| | | ... | ... | 10S | .134 | 5.295 | 2.285 | 22.02 | .1529 | 8.425 | 7.77 |
| | | STD | 40 | 40S | .258 | 5.047 | 4.300 | 20.01 | .1390 | 15.16 | 14.62 |
| | | XS | 80 | 80S | .375 | 4.813 | 6.112 | 18.19 | .1263 | 20.67 | 20.78 |
| | | ... | 120 | ... | .500 | 4.563 | 7.953 | 16.35 | .1136 | 25.73 | 27.04 |
| | | ... | 160 | ... | .625 | 4.313 | 9.696 | 14.61 | .1015 | 30.03 | 32.96 |
| | | XXS | ... | ... | .750 | 4.063 | 11.340 | 12.97 | .0901 | 33.63 | 38.55 |
| 6 | 6.625 | ... | ... | 5S | .109 | 6.407 | 2.231 | 32.24 | .2239 | 11.85 | 7.60 |
| | | ... | ... | 10S | .134 | 6.357 | 2.733 | 31.74 | .2204 | 14.40 | 9.29 |
| | | STD | 40 | 40S | .280 | 6.065 | 5.581 | 28.89 | .2006 | 28.14 | 18.97 |
| | | XS | 80 | 80S | .432 | 5.761 | 8.405 | 26.07 | .1810 | 40.49 | 28.57 |
| | | ... | 120 | ... | .562 | 5.501 | 10.70 | 23.77 | .1650 | 49.61 | 36.39 |
| | | ... | 160 | ... | .719 | 5.187 | 13.32 | 21.15 | .1469 | 58.97 | 45.35 |
| | | XXS | ... | ... | .864 | 4.897 | 15.64 | 18.84 | .1308 | 66.33 | 53.16 |
| 8 | 8.625 | ... | ... | 5S | .109 | 8.407 | 2.916 | 55.51 | .3855 | 26.44 | 9.93 |
| | | ... | ... | 10S | .148 | 8.329 | 3.941 | 54.48 | .3784 | 35.41 | 13.40 |
| | | ... | 20 | ... | .250 | 8.125 | 6.57 | 51.85 | .3601 | 57.72 | 22.36 |
| | | ... | 30 | ... | .277 | 8.071 | 7.26 | 51.16 | .3553 | 63.35 | 24.70 |
| | | STD | 40 | 40S | .322 | 7.981 | 8.40 | 50.03 | .3474 | 72.49 | 28.55 |
| | | ... | 60 | ... | .406 | 7.813 | 10.48 | 47.94 | .3329 | 88.73 | 35.64 |
| | | XS | 80 | 80S | .500 | 7.625 | 12.76 | 45.66 | .3171 | 105.7 | 43.39 |
| | | ... | 100 | ... | .594 | 7.437 | 14.96 | 43.46 | .3018 | 121.3 | 50.95 |
| | | ... | 120 | ... | .719 | 7.187 | 17.84 | 40.59 | .2819 | 140.5 | 60.71 |
| | | ... | 140 | ... | .812 | 7.001 | 19.93 | 38.50 | .2673 | 153.7 | 67.76 |
| | | XXS | ... | ... | .875 | 6.875 | 21.30 | 37.12 | .2578 | 162.0 | 72.42 |
| | | ... | 160 | ... | .906 | 6.813 | 21.97 | 36.46 | .2532 | 165.9 | 74.69 |
| 10 | 10.750 | ... | ... | 5S | .134 | 10.482 | 4.36 | 86.29 | .5992 | 63.0 | 15.19 |
| | | ... | ... | 10S | .165 | 10.420 | 5.49 | 85.28 | .5922 | 76.9 | 18.65 |
| | | ... | 20 | ... | .250 | 10.250 | 8.24 | 82.52 | .5731 | 113.7 | 28.04 |
| | | ... | 30 | ... | .307 | 10.136 | 10.07 | 80.69 | .5603 | 137.4 | 34.24 |
| | | STD | 40 | 40S | .365 | 10.020 | 11.90 | 78.86 | .5475 | 160.7 | 40.48 |
| | | XS | 60 | 80S | .500 | 9.750 | 16.10 | 74.66 | .5185 | 212.0 | 54.74 |
| | | ... | 80 | ... | .594 | 9.562 | 18.92 | 71.84 | .4989 | 244.8 | 64.43 |
| | | ... | 100 | ... | .719 | 9.312 | 22.63 | 68.13 | .4732 | 286.1 | 77.03 |
| | | ... | 120 | ... | .844 | 9.062 | 26.24 | 64.53 | .4481 | 324.2 | 89.29 |
| | | XXS | 140 | ... | 1.000 | 8.750 | 30.63 | 60.13 | .4176 | 367.8 | 104.13 |
| | | ... | 160 | ... | 1.125 | 8.500 | 34.02 | 56.75 | .3941 | 399.3 | 115.64 |
| 12 | 12.75 | ... | ... | 5S | .156 | 12.438 | 6.17 | 121.50 | .8438 | 122.4 | 20.98 |
| | | ... | ... | 10S | .180 | 12.390 | 7.11 | 120.57 | .8373 | 140.4 | 24.17 |
| | | ... | 20 | ... | .250 | 12.250 | 9.82 | 117.86 | .8185 | 191.8 | 33.38 |
| | | ... | 30 | ... | .330 | 12.090 | 12.87 | 114.80 | .7972 | 248.4 | 43.77 |
| | | STD | ... | 40S | .375 | 12.000 | 14.58 | 113.10 | .7854 | 279.3 | 49.56 |
| | | ... | 40 | ... | .406 | 11.938 | 15.77 | 111.93 | .7773 | 300.3 | 53.52 |
| | | XS | ... | 80S | .500 | 11.750 | 19.24 | 108.43 | .7528 | 361.5 | 65.42 |
| | | ... | 60 | ... | .562 | 11.626 | 21.52 | 106.16 | .7372 | 400.4 | 73.15 |
| | | ... | 80 | ... | .688 | 11.374 | 26.03 | 101.64 | .7058 | 475.1 | 88.63 |
| | | ... | 100 | ... | .844 | 11.062 | 31.53 | 96.14 | .6677 | 561.6 | 107.32 |
| | | XXS | 120 | ... | 1.000 | 10.750 | 36.91 | 90.76 | .6303 | 641.6 | 125.49 |
| | | ... | 140 | ... | 1.125 | 10.500 | 41.08 | 86.59 | .6013 | 700.5 | 139.67 |
| | | ... | 160 | ... | 1.312 | 10.126 | 47.14 | 80.53 | .5592 | 781.1 | 160.27 |

Identification, wall thickness and weights are extracted from ANSI B36.10 and B36.19. The notations STD, XS, and XXS indicate Standard, Extra Strong, and Double Extra Strong pipe respectively.

## TABLE 11-2. (continued)

| Nominal Pipe Size Inches | Outside Diam. Inches | Identification Steel - Iron Pipe Size | Identification Steel - Sched. No. | Identification Stainless Steel Sched. No. | Wall Thickness (t) Inches | Inside Diameter (d) Inches | Area of Metal Square Inches | Transverse Internal Area (a) Square Inches | Transverse Internal Area (A) Square Feet | Moment of Inertia (I) Inches[4] | Weight Pipe Pounds per foot |
|---|---|---|---|---|---|---|---|---|---|---|---|
| 14 | 14.00 | ... | ... | 5S | .156 | 13.688 | 6.78 | 147.15 | 1.0219 | 162.6 | 23.07 |
|  |  | ... | ... | 10S | .188 | 13.624 | 8.16 | 145.78 | 1.0124 | 194.6 | 27.73 |
|  |  | ... | 10 | ... | .250 | 13.500 | 10.80 | 143.14 | .9940 | 255.3 | 36.71 |
|  |  | ... | 20 | ... | .312 | 13.376 | 13.42 | 140.52 | .9758 | 314.4 | 45.61 |
|  |  | STD | 30 | ... | .375 | 13.250 | 16.05 | 137.88 | .9575 | 372.8 | 54.57 |
|  |  | ... | 40 | ... | .438 | 13.124 | 18.66 | 135.28 | .9394 | 429.1 | 63.44 |
|  |  | XS | ... | ... | .500 | 13.000 | 21.21 | 132.73 | .9217 | 483.8 | 72.09 |
|  |  | ... | 60 | ... | .594 | 12.812 | 24.98 | 128.96 | .8956 | 562.3 | 85.05 |
|  |  | ... | 80 | ... | .750 | 12.500 | 31.22 | 122.72 | .8522 | 678.3 | 106.13 |
|  |  | ... | 100 | ... | .938 | 12.124 | 38.45 | 115.49 | .8020 | 824.4 | 130.85 |
|  |  | ... | 120 | ... | 1.094 | 11.812 | 44.32 | 109.62 | .7612 | 929.6 | 150.79 |
|  |  | ... | 140 | ... | 1.250 | 11.500 | 50.07 | 103.87 | .7213 | 1027.0 | 170.28 |
|  |  | ... | 160 | ... | 1.406 | 11.188 | 55.63 | 98.31 | .6827 | 1117.0 | 189.11 |
| 16 | 16.00 | ... | ... | 5S | .165 | 15.670 | 8.21 | 192.85 | 1.3393 | 257.3 | 27.90 |
|  |  | ... | ... | 10S | .188 | 15.624 | 9.34 | 191.72 | 1.3314 | 291.9 | 31.75 |
|  |  | ... | 10 | ... | .250 | 15.500 | 12.37 | 188.69 | 1.3103 | 383.7 | 42.05 |
|  |  | ... | 20 | ... | .312 | 15.376 | 15.38 | 185.69 | 1.2895 | 473.2 | 52.27 |
|  |  | STD | 30 | ... | .375 | 15.250 | 18.41 | 182.65 | 1.2684 | 562.1 | 62.58 |
|  |  | XS | 40 | ... | .500 | 15.000 | 24.35 | 176.72 | 1.2272 | 731.9 | 82.77 |
|  |  | ... | 60 | ... | .656 | 14.688 | 31.62 | 169.44 | 1.1766 | 932.4 | 107.50 |
|  |  | ... | 80 | ... | .844 | 14.312 | 40.14 | 160.92 | 1.1175 | 1155.8 | 136.61 |
|  |  | ... | 100 | ... | 1.031 | 13.938 | 48.48 | 152.58 | 1.0596 | 1364.5 | 164.82 |
|  |  | ... | 120 | ... | 1.219 | 13.562 | 56.56 | 144.50 | 1.0035 | 1555.8 | 192.43 |
|  |  | ... | 140 | ... | 1.438 | 13.124 | 65.78 | 135.28 | .9394 | 1760.3 | 223.64 |
|  |  | ... | 160 | ... | 1.594 | 12.812 | 72.10 | 128.96 | .8956 | 1893.5 | 245.25 |
| 18 | 18.00 | ... | ... | 5S | .165 | 17.670 | 9.25 | 245.22 | 1.7029 | 367.6 | 31.43 |
|  |  | ... | ... | 10S | .188 | 17.624 | 10.52 | 243.95 | 1.6941 | 417.3 | 35.76 |
|  |  | ... | 10 | ... | .250 | 17.500 | 13.94 | 240.53 | 1.6703 | 549.1 | 47.39 |
|  |  | ... | 20 | ... | .312 | 17.376 | 17.34 | 237.13 | 1.6467 | 678.2 | 58.94 |
|  |  | STD | ... | ... | .375 | 17.250 | 20.76 | 233.71 | 1.6230 | 806.7 | 70.59 |
|  |  | ... | 30 | ... | .438 | 17.124 | 24.17 | 230.30 | 1.5990 | 930.3 | 82.15 |
|  |  | XS | ... | ... | .500 | 17.000 | 27.49 | 226.98 | 1.5763 | 1053.2 | 93.45 |
|  |  | ... | 40 | ... | .562 | 16.876 | 30.79 | 223.68 | 1.5533 | 1171.5 | 104.67 |
|  |  | ... | 60 | ... | .750 | 16.500 | 40.64 | 213.83 | 1.4849 | 1514.7 | 138.17 |
|  |  | ... | 80 | ... | .938 | 16.124 | 50.23 | 204.24 | 1.4183 | 1833.0 | 170.92 |
|  |  | ... | 100 | ... | 1.156 | 15.688 | 61.17 | 193.30 | 1.3423 | 2180.0 | 207.96 |
|  |  | ... | 120 | ... | 1.375 | 15.250 | 71.81 | 182.66 | 1.2684 | 2498.1 | 244.14 |
|  |  | ... | 140 | ... | 1.562 | 14.876 | 80.66 | 173.80 | 1.2070 | 2749.0 | 274.22 |
|  |  | ... | 160 | ... | 1.781 | 14.438 | 90.75 | 163.72 | 1.1369 | 3020.0 | 308.50 |
| 20 | 20.00 | ... | ... | 5S | .188 | 19.624 | 11.70 | 302.46 | 2.1004 | 574.2 | 39.78 |
|  |  | ... | ... | 10S | .218 | 19.564 | 13.55 | 300.61 | 2.0876 | 662.8 | 46.06 |
|  |  | ... | 10 | ... | .250 | 19.500 | 15.51 | 298.65 | 2.0740 | 765.4 | 52.73 |
|  |  | STD | 20 | ... | .375 | 19.250 | 23.12 | 290.04 | 2.0142 | 1113.0 | 78.60 |
|  |  | XS | 30 | ... | .500 | 19.000 | 30.63 | 283.53 | 1.9690 | 1457.0 | 104.13 |
|  |  | ... | 40 | ... | .594 | 18.812 | 36.15 | 278.00 | 1.9305 | 1703.0 | 123.11 |
|  |  | ... | 60 | ... | .812 | 18.376 | 48.95 | 265.21 | 1.8417 | 2257.0 | 166.40 |
|  |  | ... | 80 | ... | 1.031 | 17.938 | 61.44 | 252.72 | 1.7550 | 2772.0 | 208.87 |
|  |  | ... | 100 | ... | 1.281 | 17.438 | 75.33 | 238.83 | 1.6585 | 3315.2 | 256.10 |
|  |  | ... | 120 | ... | 1.500 | 17.000 | 87.18 | 226.98 | 1.5762 | 3754.0 | 296.37 |
|  |  | ... | 140 | ... | 1.750 | 16.500 | 100.33 | 213.82 | 1.4849 | 4216.0 | 341.09 |
|  |  | ... | 160 | ... | 1.969 | 16.062 | 111.49 | 202.67 | 1.4074 | 4585.5 | 379.17 |
| 22 | 22.00 | ... | ... | 5S | .188 | 21.624 | 12.88 | 367.25 | 2.5503 | 766.2 | 43.80 |
|  |  | ... | ... | 10S | .218 | 21.564 | 14.92 | 365.21 | 2.5362 | 884.8 | 50.71 |
|  |  | ... | 10 | ... | .250 | 21.500 | 17.08 | 363.05 | 2.5212 | 1010.3 | 58.07 |
|  |  | STD | 20 | ... | .375 | 21.250 | 25.48 | 354.66 | 2.4629 | 1489.7 | 86.61 |
|  |  | XS | 30 | ... | .500 | 21.000 | 33.77 | 346.36 | 2.4053 | 1952.5 | 114.81 |
|  |  | ... | 60 | ... | .875 | 20.250 | 58.07 | 322.06 | 2.2365 | 3244.9 | 197.41 |
|  |  | ... | 80 | ... | 1.125 | 19.75 | 73.78 | 306.35 | 2.1275 | 4030.4 | 250.81 |
|  |  | ... | 100 | ... | 1.375 | 19.25 | 89.09 | 291.04 | 2.0211 | 4758.5 | 302.88 |
|  |  | ... | 120 | ... | 1.625 | 18.75 | 104.02 | 276.12 | 1.9175 | 5432.0 | 353.61 |
|  |  | ... | 140 | ... | 1.875 | 18.25 | 118.55 | 261.59 | 1.8166 | 6053.7 | 403.00 |
|  |  | ... | 160 | ... | 2.125 | 17.75 | 132.68 | 247.45 | 1.7184 | 6626.4 | 451.06 |

**Identification, wall thickness and weights are extracted from ANSI B36.10 and B36.19. The notations STD, XS, and XXS indicate Standard, Extra Strong, and Double Extra Strong pipe respectively.

**TABLE 11-2.** (continued)

| Nominal Pipe Size Inches | Outside Diam. Inches | Identification Steel Iron Pipe Size | Identification Steel Sched. No. | Identification Stainless Steel Sched. No. | Wall Thickness (t) Inches | Inside Diameter (d) Inches | Area of Metal Square Inches | Transverse Internal Area (a) Square Inches | Transverse Internal Area (A) Square Feet | Moment of Inertia (I) Inches[4] | Weight Pipe Pounds per foot |
|---|---|---|---|---|---|---|---|---|---|---|---|
| 24 | 24.00 | ... | ... | 5S | .218 | 23.564 | 16.29 | 436.10 | 3.0285 | 1151.6 | 55.37 |
|  |  | ... | 10 | 10S | .250 | 23.500 | 18.65 | 433.74 | 3.0121 | 1315.4 | 63.41 |
|  |  | STD | 20 | ... | .375 | 23.250 | 27.83 | 424.56 | 2.9483 | 1942.0 | 94.62 |
|  |  | XS | ... | ... | .500 | 23.000 | 36.91 | 415.48 | 2.8853 | 2549.5 | 125.49 |
|  |  | ... | 30 | ... | .562 | 22.876 | 41.39 | 411.00 | 2.8542 | 2843.0 | 140.68 |
|  |  | ... | 40 | ... | .688 | 22.624 | 50.31 | 402.07 | 2.7921 | 3421.3 | 171.29 |
|  |  | ... | 60 | ... | .969 | 22.062 | 70.04 | 382.35 | 2.6552 | 4652.8 | 238.35 |
|  |  | ... | 80 | ... | 1.219 | 21.562 | 87.17 | 365.22 | 2.5362 | 5672.0 | 296.58 |
|  |  | ... | 100 | ... | 1.531 | 20.938 | 108.07 | 344.32 | 2.3911 | 6849.9 | 367.39 |
|  |  | ... | 120 | ... | 1.812 | 20.376 | 126.31 | 326.08 | 2.2645 | 7825.0 | 429.39 |
|  |  | ... | 140 | ... | 2.062 | 19.876 | 142.11 | 310.28 | 2.1547 | 8625.0 | 483.12 |
|  |  | ... | 160 | ... | 2.344 | 19.312 | 159.41 | 292.98 | 2.0346 | 9455.9 | 542.13 |
| 26 | 26.00 | ... | 10 | ... | .312 | 25.376 | 25.18 | 505.75 | 3.5122 | 2077.2 | 85.60 |
|  |  | STD | ... | ... | .375 | 25.250 | 30.19 | 500.74 | 3.4774 | 2478.4 | 102.63 |
|  |  | XS | 20 | ... | .500 | 25.000 | 40.06 | 490.87 | 3.4088 | 3257.0 | 136.17 |
| 28 | 28.00 | ... | 10 | ... | .312 | 27.376 | 27.14 | 588.61 | 4.0876 | 2601.0 | 92.26 |
|  |  | STD | ... | ... | .375 | 27.250 | 32.54 | 583.21 | 4.0501 | 3105.1 | 110.64 |
|  |  | XS | 20 | ... | .500 | 27.000 | 43.20 | 572.56 | 3.9761 | 4084.8 | 146.85 |
|  |  | ... | 30 | ... | .625 | 26.750 | 53.75 | 562.00 | 3.9028 | 5037.7 | 182.73 |
| 30 | 30.00 | ... | ... | 5S | .250 | 29.500 | 23.37 | 683.49 | 4.7465 | 2585.2 | 79.43 |
|  |  | ... | 10 | 10S | .312 | 29.376 | 29.10 | 677.76 | 4.7067 | 3206.3 | 98.93 |
|  |  | STD | ... | ... | .375 | 29.250 | 34.90 | 671.96 | 4.6664 | 3829.4 | 118.65 |
|  |  | XS | 20 | ... | .500 | 29.000 | 46.34 | 660.52 | 4.5869 | 5042.2 | 157.53 |
|  |  | ... | 30 | ... | .625 | 28.750 | 57.68 | 649.18 | 4.5082 | 6224.0 | 196.08 |
| 32 | 32.00 | ... | 10 | ... | .312 | 31.376 | 31.06 | 773.19 | 5.3694 | 3898.9 | 105.59 |
|  |  | STD | ... | ... | .375 | 31.250 | 37.26 | 766.99 | 5.3263 | 4658.5 | 126.66 |
|  |  | XS | 20 | ... | .500 | 31.000 | 49.48 | 754.77 | 5.2414 | 6138.6 | 168.21 |
|  |  | ... | 30 | ... | .625 | 30.750 | 61.60 | 742.64 | 5.1572 | 7583.4 | 209.43 |
|  |  | ... | 40 | ... | .688 | 30.624 | 67.68 | 736.57 | 5.1151 | 8298.3 | 230.08 |
| 34 | 34.00 | ... | 10 | ... | .344 | 33.312 | 36.37 | 871.55 | 6.0524 | 5150.5 | 123.65 |
|  |  | STD | ... | ... | .375 | 33.250 | 39.61 | 868.31 | 6.0299 | 5599.3 | 134.67 |
|  |  | XS | 20 | ... | .500 | 33.000 | 52.62 | 855.30 | 5.9396 | 7383.5 | 178.89 |
|  |  | ... | 30 | ... | .625 | 32.750 | 65.53 | 842.39 | 5.8499 | 9127.6 | 222.78 |
|  |  | ... | 40 | ... | .688 | 32.624 | 72.00 | 835.92 | 5.8050 | 9991.6 | 244.77 |
| 36 | 36.00 | ... | 10 | ... | .312 | 35.376 | 34.98 | 982.90 | 6.8257 | 5569.5 | 118.92 |
|  |  | STD | ... | ... | .375 | 35.250 | 41.97 | 975.91 | 6.7771 | 6658.9 | 142.68 |
|  |  | XS | 20 | ... | .500 | 35.000 | 55.76 | 962.11 | 6.6813 | 8786.2 | 189.57 |
|  |  | ... | 30 | ... | .625 | 34.750 | 69.46 | 948.42 | 6.5862 | 10868.4 | 236.13 |
|  |  | ... | 40 | ... | .750 | 34.500 | 83.06 | 934.82 | 6.4918 | 12906.1 | 282.35 |

**Identification, wall thickness and weights** are extracted from ANSI B36.10 and B36.19. The notations STD, XS, and XXS indicate Standard, Extra Strong, and Double Extra Strong pipe respectively.

For a turbulent flow situation, the friction factor (f) may be calculated from the equation:

$$f = \frac{0.184}{(R_e)^{0.2}}$$

Friction curves, presented in Figure 11-6, may be used to simplify the calculations. On the left side of the chart, locate the 6-inch, Schedule 80 value. Follow this value horizontally until the line veers upward; then follow the trend between the lines labeled "inside diameter." This trend line will intersect the vertical line representing a Reynolds number of 22,000. The friction factor may be determined on the left side of the chart. For the example just presented, the friction factor would be 0.025:

$$\frac{h}{L} = \frac{6(0.025)(9.8 \text{ ft/sec}^2)^2}{(32.2 \text{ ft/sec}^2)(5.76 \text{ in.})} = 0.0777 \frac{\text{ft of head}}{\text{ft of pipe}}$$

The length of pipe connected to the pump must now be determined. Valves, elbows, tees, and swages create more pressure loss, or head loss, than a straight pipe. The equivalent lengths of these items are presented in Table 11-1. For

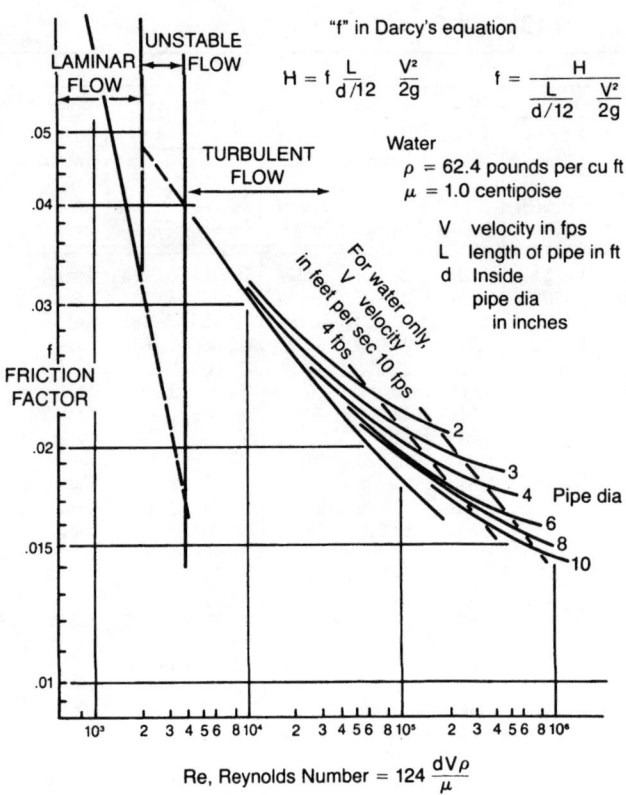

**FIGURE 11-6.** Plot "f" values vs. Reynolds number.

example, in a 6-inch line, two butterfly valves and one welded branched tee (4 ft + 4 ft + 28 ft) would add the equivalent of 36 feet of 6-inch pipe to the length of actual pipe connected to the centrifugal pump.

## PUMPING VISCOUS LIQUIDS

When centrifugal pumps handle viscous fluids, the viscosity must be considered. Viscous, or thick, liquids are not handled as easily as water, particularly by centrifugal pumps. The internal friction losses increase, thereby reducing capacity, head, and efficiency, and requiring even greater horsepower.

A pump operating at a lower speed is less affected by viscosity and lift requirements than if it were operated at higher speeds. Viscous slurries should flow to the pump suction by gravity or under positive pressure to ensure proper filling of the pump.

Liquids that have high specific gravities are not necessarily viscous. For example, mercury has a high specific gravity of 13.7, however, its viscosity is only 0.118 centistokes. Water, at standard temperature, has a viscosity of 1.0 centistokes, and transmission lubricants, which are lighter than water, have viscosities above 500 centistokes. Therefore, specific gravity and viscosity are independent of one another, as are their effect on pump performance and power requirements.

## Measuring Viscosity

Viscosity is expressed in absolute or kinematic units:

$$\text{Kinematic viscosity, centistokes} = \frac{\text{Absolute viscosity, centipoise}}{\text{Specific gravity}}$$

Saybolt viscometers are most common and viscosity is frequently expressed in "Saybolt Seconds." The time for a measured volume of fluid to flow through an orifice of specified dimensions is calculated in seconds and is called the SSU or SSF (Saybolt Seconds Universal or Saybolt Seconds Furol) depending on orifice size.

Other viscometers, such as Redwood, Irany, and Zahn, operate on the same principle. The Ostwald, Bingham, and Ubbelohde viscometers use a capillary tube in place of an orifice to measure fluids of lower viscosity.

A Brookfield viscometer measures the torque on a disc rotating in the fluid. The spring, which transmits the torque to the reading dial, is calibrated so that the reading is measured directly in centipoise. These viscometers are useful in measuring the viscosity of non-Newtonian fluids. By varying the rotating speed, or shear rate, it can determine whether the fluid is plastic, pseudoplastic, dilatant, or thixotropic.

Rotating viscometers yield the most reliable readings in respect to the friction forces encountered in a centrifugal pump. When testing solids-liquid slurries (such as a drilling fluid) solids tend to clog the orifices or capillaries of other viscometers.

Many experiments have been conducted attempting to determine a correlation for pump performance with viscous liquids. Unfortunately, no acceptable general analytical method has been developed because the shear stress-shear rate properties of fluids vary too much. Manufacturers take the position that if exacting viscous performance data are needed, performance tests should be performed using the liquid in question. However, this presumes a liquid of standard composition, density, and flow properties, which is certainly not the case with drilling fluids.

## Errosion-Corrosion

Erosion-corrosion often involves high velocities and/or solids suspended in the fluid. Ideally, pump construction materials should possess corrosion resistance to the media being pumped, strength, ductility, and a high degree of hardness. Unfortunately, few materials combine all these features.

Cavitation is a special form of erosion-corrosion resulting from the collapse or implosion of gas bubbles against the metal surface in high-pressure regions. The stresses created by cavitation are high enough to actually remove metal from the surface and destroy any corrosion-resistant, passive layers. The same qualities for the construction materials mentioned above provide some resistance to cavitation. In most cases adequate suction pressure and reduced pressure drop in suction lines eliminates cavitation problems. This emphasizes the need for proper pump and suction piping designs so that friction losses are minimized.

## Installation Design Considerations

The majority of hydraulic problems encountered in centrifugal pump piping systems originate in the suction piping. Providing the best possible suction piping layout is important, especially in minimizing NPSH problems and optimizing pump performance. In drilling operations, NPSH problems can be minimized by using short, straight suction lines below the liquid level in the mud pits.

As fluid enters the suction, there is a friction pressure loss through both the flange and nozzle. Another pressure reduction occurs, as well as turbulent flow losses, when the liquid is turned to flow radially along the impeller. Any prerotation of the fluid changes the inlet angle and results in more turbulence. It is best to have at least 10 pipe diameters of straight pipe upstream of any pump suction. "Stacks" of flanged elbows should be avoided. Once the liquid is accelerated inside the impeller, pressure begins to increase toward the full discharge head.

The pipe size must be large enough to efficiently carry the fluid volume. Otherwise, the high velocity will increase the friction losses. <u>Fluid velocity should be above 5 ft/sec to minimize solids settling and below 10 ft/sec to reduce friction.</u> If a larger pump replaces a smaller one, consideration must be given to increasing pipe diameters.

Importantly, piping configuration must be such that liquid is properly led into the pump. Again, the ideal piping should have a minimum of 10 pipe diameters of straight pipe between the pump suction and the tank. It is preferred that the suction tank be elevated above the pump.

Piping should be in one plane. If an elbow bolts directly into the pump's suction, the fluid tends to follow the long radius, creating a void—or even reverse flow—along the shorter radius. This starves part of the impeller. Multi-elbow configurations ("stacks") are to be avoided, especially if used in several planes causing many changes in flow directions.

If branching is necessary, a "Y" fitting is preferable to a tee header. Turbulence occurs in the tee header perpendicular to the feed flow direction. If pump suction is close to a tee, turbulence can enter the impeller. In general, however, only one suction line should be connected to a centrifugal pump. Multiple manifolding to provide "flexibility" usually leads to severe problems in drilling fluid systems.

If a reducer is used, it should be an eccentric reducer mounted to eliminate areas where voids or air pockets can form.

Free-falling discharge into the suction pit, which entrains air into the suction fluid, should be avoided. Suction intake should not be close to the tank inlet line and should be surrounded with sufficient fluid so that fluid moving to the intake remains uniform and without turbulence. The suction lines must be deep enough below tank level to avoid whirlpools, which allow entrained air to funnel into the pump intake.

## ANATOMY OF CENTRIFUGAL PUMPS

The typical centrifugal pump consists of the power end, stuffing box, backing plate, impeller, and fluid housing.

***Power end.*** Centrifugal pump power ends consist of bearings, oil or grease seals, bearing housing(s), the shaft-bearing lubrication system, and the pump shaft connecting the drive force motor and driven impeller. The power end assembly is usually made of iron or steel. When venting of the bearing housing is required, the vent must be designed to prevent the entry of water.

***Stuffing box.*** The stuffing box is located in front of the power end behind the fluid housing. It uses either packing or a mechanical seal to prevent excessive liquid loss between the shaft and the stuffing box. There are many problems inherent with either packing or mechanical seals, including:

- Seal mounted on and/or around a rotating shaft, which may or may not be running concentric, or may be worn
- Pressure differentials across the seal or packing may vary
- Abrasive solids may be entrained in the fluid
- Temperature may vary
- Corrosion may occur

In the stuffing box area, the wet shaft surface must be hard enough to resist wearing. Centrifugal pumps often have carbon steel shafts overlain with a replaceable sleeve, which can be made of ceramics, stainless steels, or even glass. Some sleeves are permanently bonded to the shaft, which helps minimize deflection. "Plasma-spray" processes apply ceramic and stainless steel.

***Shaft packing.*** Packing usually consists of a fibrous material woven into a rope, and then cut and molded into packing rings with a rectangular cross-section. Packing materials include braided asbestos, lead, graphite, teflon, and others. The rings must be sufficiently pliable to conform and fill the spaces between the stuffing box and shaft. Pressure is then applied with the use of a packing gland. However, if the shaft has any radial movement caused by run out, deflection, whip, or loose bearings, leakage will occur in the stuffing box.

Packing is usually impregnated with grease that provides lubrication between the shaft and packing rings. The grease also fills the pores between the packing fiber and helps distribute friction caused by heat. Most packing lubricants break down above 250°F. Additional grease can be introduced through a lubrication port in the stuffing box. If pump media is corrosive, the packing is usually made from asbestos, teflon, or graphite fibers.

A small amount of liquid must be allowed to seep through the packing to cool and lubricate the shaft surface contacting the packing. This leakage rate is hard to control, and the usual tendency is to over-tighten the packing gland to stop, or control, the leakage. The consequence of over-tightening can be rapid scouring of the shaft surface. This creates an abrasive surface and makes it difficult, if not impossible, to properly adjust packing compression. The main advantages of a packed stuffing box are:

- Relative inexpensive initial cost
- Slow sealing deterioration allows replacement to be scheduled when pump is off-line
- Simple to adjust and/or replace

With simple water-based fluids, leakage is best tolerated rather than controlled. Expensive, corrosive, or toxic fluids present another problem, usually better approached with mechanical shaft seals.

***Mechanical seals.*** Mechanical seals overcome most of the shortcomings of rope packing. The disadvantages of these seals, however, are their high cost and sudden seal failure, which can result in inconvienent maintenance downtime.

Construction of mechanical seals varies but the basic components are similar. There are two mating seal rings, one stationary mounted inside the power end housing or stuffing box, and one rotating with the shaft. The seal rings are held in contact with each other by compression springs. A flat gasket, or O-ring, provides the seal between the stationary ring and packing gland. O-rings or V-seals are used to seal the rotating seal ring and shaft.

The pumped liquid will form a thin film between the seal faces. This is the key to the success of the mechanical seal. If the liquid has adequate lubricating properties, the seal will provide long service.

***Rotating shafts.*** Bent or out of round pump shafts will erode large holes between the packing and shaft, allowing liquid to escape. Lack of static or dynamic balance in the impeller produces a dynamic bend in the shaft with the same result.

Bent shafts also impair mechanical seal functioning as rotating members of the seal bend with each revolution. If shaft deflection is greater than nominal, the flexible springs do not react sufficiently to keep the seal faces together.

# CENTRIFUGAL PUMPS    197

***Shaft deflection.*** When not in motion, the shaft is straight and centered in the stuffing box. When rotating, any unbalanced radial load on the impeller deflects the shaft rotation from that of a true circle. Under these conditions, the shaft will run in an elliptical or oval pattern. Packing in the stuffing box can be adjusted with the packing gland to allow minimal leakage.

***Shaft whip.*** A cone-shaped pattern of rotation characterizes shaft whip. This creates a hole in the packing larger than the shaft diameter and excessive leakage will occur. A properly balanced impeller will always cause whip because the heavy side of the impeller is always on the same side of the shaft as it rotates. It is also possible to have a combination of shaft deflection and shaft whip at the same time.

***Shaft runout.*** Shaft runout is the amount that the shaft power end section is out of true when measured as the shaft is slowly rotated. Defects such as an out-of-round shaft, eccentricity between the shaft and shaft sleeve, or permanent bend in the shaft produce an oversized hole in the shaft packing, thereby making the stuffing box hard to seal.

***Bearings and bearing life.*** Bearings are one of the most important components of a centrifugal pump. When properly mounted, lubricated, and protected from dirt, a ball bearing will carry axial and radial loads at high rotary speeds for years. For troublefree service, bearings should:

- Allow the shaft to rotate with negligible friction, keeping power requirements to a minimum
- Hold the shaft assembly radially and axially in the proper position relative to the stationary parts of the pump, preventing contact between moving and nonmoving parts
- Absorb forces transmitted to them from impeller loading

Bearings operate for a specific number of revolutions under specified loads before evidence of failure appears. At higher loads, the total number of revolutions decreases, and at lighter loads the number increases. Bearings that control axial movement are designed to limit end movement below 0.002. Generally, end-play above 0.002 has been found to be detrimental to packing and mechanical seal integrity. Bearing temperatures should not exceed 225°F. Excessive noise, heat, or vibration are signs that bearings have reached the end of their service life.

Proper lubrication is essential to bearings. Oil levels must be maintained, contamination prevented, and regreasing performed properly. In general, grease is preferred as a lubricant when:

- Temperatures are moderate—usually not above 200°F
- Rotating speed does not exceed bearing manufacturer's recommendations
- Extra protection from dirt or fumes is needed
- Prolonged periods of operation without qualified maintenance are anticipated

In general, oil is the preferred bearing lubricant in cases where:

- Operating temperatures are consistently high
- Speed exceeds bearing manufacturer's limit for grease
- Dirt and dust are not excessive and oil-tight seals can be used
- Bearing design does not lend itself to grease lubrication

***Bearing assembly checklist:***

1. Does the shaft meet manufacturer's specifications for machined tolerances?
2. Are locating shoulders at 90° angles to the shaft centerline so that the bearing will be squared with the shaft?
3. Are housing bores in straight alignment so that the bearing will not operate in a twisted position?
4. Are grease and/or oil of proper grade and viscosity at correct levels?
5. Do lubricant seals "seal"?

## VIBRATION

Vibration damages the bearings and other parts of the pump. It is measured in the horizontal, vertical, and radial directions and expressed in terms of peak-to-peak displacements.

Vibration may originate within the pump and motor or be transmitted from an outside source. If large enough, it will damage some of the pump parts. The change in vibration magnitude between inspection periods is at least as important as the amount of vibration itself. An increasing vibration level, even at low values, may indicate an impending problem. An unchanged vibration level, even though relatively high in absolute value, gives evidence of a stable operating condition.

**TABLE 11-3.** Maximum Acceptable Vibrations

| Pump speed (rpm) | Displacement (Mils, peak-peak) |
|---|---|
| 3600 | 1.25 |
| 1800 | 2.50 |
| 1200 | 3.75 |

*Note:* Displacement is measured perpendicular to the shaft axis at a point on the bearing housing adjacent to each bearing.

Table 11-3 lists generally accepted vibration limits. If vibration exceeds these limits, corrective action should be taken.

## SUMMARY

1. Centrifugal pumps produce a constant head. When the mud weight changes, the pressure changes but the head remains constant.
2. Head may be converted into pressure using the well control equation:

   $P = 0.052 \, (MW, ppg)(depth, ft)$

   Depth and head are the same. A centrifugal pump will produce a constant head but the pressure will depend on the mud weight.
3. Think of a centrifugal pump as a constantly filled water tower containing drilling fluid. The flow rate from the tower depends on the piping attached to it. For example, the flow rate would be different for 20 feet of 6-inch pipe compared to 2,000 feet of $\frac{1}{4}$-inch pipe. The same head would be applied to both piping systems. The pressure at the bottom of the tower would depend on the mud weight.
4. Centrifugal pumps will produce the same head with 8.34 ppg water as with 18 ppg of drilling fluid. The horsepower requirement to pump the heavier weight mud is higher by a ratio of mud weights.
5. Never operate two centrifugal pumps in parallel supplying the same piece of equipment. In other words, two pumps should not pump through the same pipe at the same time. Operating two centrifugal pumps in parallel is equivalent to connecting two water towers together. Usually, when this occurs in a drilling fluid system, one pump will have fluid flowing in a reverse direction because the heads are not identical.
6. Use a suction strainer to keep rags, rubber chunks, sticks, wrenches, and so forth, out of the pump impeller. Clean the strainer before it clogs.
7. Size the centrifugal-pump piping by flow rate requirements of the equipment, not by the size of a flange on a piece of equipment or on the centrifugal pump.
8. Never adjust a valve in the suction line to regulate flow rate or pressure of the pump. This will cause the pump to cavitate and destroy itself. Cavitation sounds as if gravel is passing through the centrifugal pump.
9. The discharge valve in the piping next to the centrifugal pump may be closed briefly to measure the discharge pressure. This is a diagnostic tool. The no-flow centrifugal pump discharge pressure can be used to monitor the wear on an impeller without dismantling the pump.
10. Never walk away from a discharge valve that you have closed. The fluid in the pump will boil after running for only a few minutes. Centrifugal pumps use less horsepower when the discharge valve is closed. The current to an electric motor will be lower so there is no danger of blowing a circuit breaker by closing the discharge valve.
11. When pumping into a long, empty line with a centrifugal pump, partially close the discharge valve until the line is full. A centrifugal-pump motor requires less current (or horsepower) when nothing is connected to the discharge side of the pump than when the discharge side is completely closed. Always fully open the valve before leaving it.
12. Do not open bypass valves in the piping system to reduce flow rate or pressure on equipment being supplied by a centrifugal pump. This wastes power (or energy) and may overload the motor.
13. A flooded suction is needed to prime a centrifugal pump and prevent cavitation. When a flooded suction is not possible, make certain that the proper NPSH is available.

# Centrifugal pump troubleshooting:

1. **PUMP WILL NOT REACH DESIGN FLOW RATE**
   A. Insufficient NPSH (noise may or may not be present)
      - Redesign suction piping; possibly enlarge diameter
      - Keep elbows, and so forth, at a minimum
   B. System head requirements greater than anticipated
      - Redesign system piping; possibly enlarge diameter
      - Keep elbows, and so forth, at a minimum
      - Increase impeller diameter
   C. Plugged impeller, suction line, or pump casing by fibrous or settled solids
      - Clean and install strainer over suction line
      - For fibrous material, possibly enlarge pump
   D. Entrained air from atmosphere
      - Check suction-line fittings and gaskets for leaks
      - Repack pump stuffing box
      - Check suction pit for vortex
      - Check for minimum suction-line submergence
      - Check degasser
   E. RPM too slow
      - Check motor speed
   F. Wrong rotation direction
      - Reverse any two of the three leads on a three-phase motor
   G. Impeller worn, has improper clearance, etc.
      - Inspect impeller and reset to proper clearance
      - If worn, replace

2. **NO DISCHARGE OR FLOW**
   A. Motor horsepower too small
      - Replace motor with larger size and proper rpm
      - Motor cannot rotate impeller
   B. Not properly primed
      - Repeat priming operations
      - Try to have a flooded suction
   C. Suction lift too high
      - Rearrange piping
      - Increase suction head
      - May need larger impeller
      - Check motor speed
      - May need larger pump for higher suction lift
      - Check suction line and pump for plugging
      - Check direction of rotation; if incorrect, change any two leads to motor
      - Check for entrained air or gas
      - Move pump down to a lower level

3. **PUMP OPERATES BUT CONTINUALLY LOSES PRIME**
   A. Insufficient NPSH
      - Redesign suction piping; possibly enlarge diameter
      - Keep elbows, and so forth, at a minimum
   B. Check for entrained air or gas
   C. If small whirlpools appear on top of liquid level, install baffle plate

4. **EXCESSIVE NOISE—WET END**
   A. Cavitation—insufficient NPSH
      - Redesign suction piping; possibly enlarge diameter
      - Keep elbows, and so forth, to a minimum
      - Abnormal fluid rotation due to complex piping
   B. Impeller rubbing
      - Inspect and reset impeller
      - Inspect bearing assemblies for end play

5. **EXCESSIVE NOISE—POWER END**
   A. Overloaded bearing with races flaking or pitting
      - Suction pressure and/or fluid specific gravity may exceed bearing design
   B. Bearing contamination—raceways scoured, pitted, or rusted
      - Work with clean tools and clean the area around pump; provide cover from wash-down water
      - Clean pump housing before removing
      - Use clean solvents and flushing oil
      - Protect bearings from moisture when not in use (wrap in clean paper or clean cloth)
      - Clean interior of pump housing before replacing bearings
      - Inspect oil lip seals
      - Inspect plugs and tapped openings for tight closure
   C. Worn bearings
      - Inspect and replace bearings
      - Improper lubrication including brown or bluish discoloration on races; stiff, cracked grease; failure of ball retainer; dirty or improper amount of lubricant; improper lubricant grade(s)
      - Check shaft straightness
      - Check bearing alignment
      - Check housing bore for alignments
   D. Check for electric arcing from static electricity, electrical leakage, or short circuiting.

*Note:* When mounting a bearing on a shaft, use a proper size ring and apply pressure against the inner rings. When mounting in a housing, press against the outer ring. Apply pressure slowly and evenly.

## Centrifugal pump troubleshooting (continued):

6. **EXCESSIVE LEAKAGE OF A PACKED PUMP STUFFING BOX**
   A. Loose packing gland
      - Adjust packing gland while pump is running
   B. Packing excessively worn or improperly lubricated
      - Replace packing and inspect for proper lubrication
   C. Bent or scoured shaft
      - Replace shaft and inspect pump for cause

7. **EXCESSIVE VIBRATION**
   A. Steady, regular vibration
      - Inspect shaft balance and straightness
      - Check motor—shaft alignment; inspect coupling
      - Check baseplate bolting and grouting
   B. Unsteady vibration
      - Bad bearings (largest vibration should be near the bad bearing)
   C. Random, extremely high vibration
      - Possible cavitation; improve suction piping layout
   D). Vibration disappears when power is off
      - Check current cps

**TABLE 11-4.** Conversion Factors

PSI = Head, ft × sp. gr. × 0.434

$$\text{Head, ft} = \frac{\text{psi} \times 2.31}{\text{sp. gr.}}$$

Head, feet (sp. gr. = 1.0) = 1.133 × inches mercury

Common barrels:
- 1 bbl beer = 31 U.S. gallons
- 1 bbl whiskey = 45 U.S. gallons
- 1 bbl oil = 42 U.S. gallons

Barrels (oil) per day × 0.02917 = gallons per minute

# CHAPTER TWELVE

# Electric Motors

This chapter discusses electric motors, electric motor standards, motor enclosure and frame designations, hazardous location ratings, and specific motor applications.

Continuous duty electric motors are an integral part of the drilling rigs' solids control and processing systems. Centrifugal pumps that feed hydrocyclones, circulate mud for mixing, transfer mud to and from reserve, and also into the trip tank, are powered by electric motors. Shale shakers, mud cleaners, and pit agitators are also driven by electric motors while they rotate centrifuges.

Continuous duty electric motors meet well-defined performance standards. Motors are designed with conductor, frame, and insulating materials to continuously deliver rated horsepower and not exceed the insulation's temperature limits. A service factor rating defines the ability of the motor to continuously withstand prolonged overload conditions while remaining within the temperature limitations of the insulating material.

The criterion for sizing and selection of any motor is its ability to deliver startup power under the process load, and to then provide power that drives the equipment throughout operation. Adequate torque must be developed to overcome inertia during startup. The **load** must then be accelerated to the desired operating speed and full load power requirements supplied without overheating. These parameters depend on motor design and the full load rating (output horsepower).

Electric motor operating efficiency is the ratio of output power to input power. The power loss is the difference between the power into the motor and the power output of the motor. This power loss is caused by:

- Heat from the electrical resistance of motor windings and rotor
- **Windage losses** from cooling fans or rotor fins
- Magnetic and core losses from currents induced in the laminations of frame and stator
- Friction losses from shaft bearings

The motor's internal heat is a function of load conditions, motor design, and ventilation conditions. Heat produced internally by the motor raises operating temperature and adversely affects insulation used to isolate electrical conductors from each other and from the motor frame. Insulation materials are rated based on thermal capacity, or the ability to withstand heat effects. High-quality insulation systems with high thermal capacity can withstand relatively high temperature increases and deliver a long motor service life at rated performance. Because motors may be operating properly and still be too hot to touch, it is important to check the manufacturer's guidelines.

## VOLTAGE

Motors are rated for operation at specific voltages. Motor performance is affected when the supply voltage varies from the motor's rated voltage. Motors generally operate satisfactorily with voltage variations within ±10%. However, equipment connected to the motor may not always function properly with such variations.

Surge voltage is any higher-than-normal voltage that temporarily exists on one or more of the power lines of a three-phase motor. A surge causes a large voltage rise during an extremely short period of time. Surges are of concern because the higher voltage is impressed on the first few turns of the motor windings. The lead insulation may be destroyed and the motor burned out. Frequent voltage surging can result from line switching of large generators.

Under-voltage at the motor terminals can result when large current demands are placed on the generator, such as starting the top drive motor. Operation below 10% of the marked motor voltage will generally result in excessive overheating and torque reduction. Overheating prematurely deteriorates the insulation system. Torque reduction may result in the motor stalling or, in the case of shale shakers, may result in poor performance.

Figure 12-1 provides general guidelines for the effects on induction motors of voltage variation and the effects of voltage unbalance on motor performance.

## TEMPERATURE CONSIDERATIONS

The National Electric Manufacturer's Association (NEMA) full-load motor ratings are based on an ambient temperature of 40°C (104°F), at a maximum altitude of 1,000 meters (3,300 feet) above sea level. Variations above 1,000 meters may require "derating," or using a larger motor (Table 12-1).

Machines intended for use at altitudes above 1,000 meters (3,300 feet), at an ambient temperature of 40°C (104°F), should have temperature rises at sea level not exceeding the values calculated from the following:

When altitude is expressed in meters:

$$T_{RSL} = T_{RA}[1 - (Alt - 1,000)/10,000]$$

When altitude is expressed in feet:

$$T_{RSL} = T_{RA}[1 - (Alt - 3,300)/33,000]$$

where: $T_{RSL}$ = Test temperature in degrees Celsius at sea level
$T_{RA}$ = Temperature rise in degrees Celsius from tables (See NEMA Standard MG-1, 1993, Section II, Par 14.04.3)
Alt = Altitude above sea level at which machine will be operated

NEMA performance ratings are also based on operation at voltages within 10% of the "nameplate" voltage and a frequency within 5% of the "nameplate" frequency. If both frequency and voltage vary, the combined total variance is not to exceed 10%.

**50 Hz versus 60 Hz operation.** A 50-Hz motor should not be operated at 60 Hz unless it is specifically designed and marked for 60 Hz operation.

A 60-Hz, three-phase induction motor may be operated at 50 Hz if the voltage and horsepower is reduced by 80%. It should be noted that the speed and the slip will also be reduced by 80%. The shaft torque will remain the same.

## MOTOR INSTALLATION AND TROUBLESHOOTING

When replacing a motor, its exact dimensions, as well as speed, horsepower, and torque characteristics should be determined and duplicated if the same performance is desired. When replacing a motor, the entire system should be inspected for internal and external degradation. Neither the motor mounting nor mechanical coupling should exhibit signs of wear. The power supply and connections should not be damaged.

Misalignment between the motor and the driven machine (e.g., centrifugal pump) can cause bearing failures and shaft breakage. Excessive vibrations frequently indicate misalignment. All four motor feet must be fastened to a flat, preferably machined, surface. Otherwise, the frame can bend when the motor is tightened down, which twists the motor frame and causes misalignment. Care should be taken to evenly tension mounting bolts. If torque values are specified, follow the manufacturer's recommendations.

A few thick shims are preferable to many thin shims if it is necessary to align motor and machine (pump) shafts. Misaligned couplings create bearing loading in both the motor and machine, causing high-speed distortions and also increasing power consumption.

The electrical power supply should be frequently checked for proper frequency, voltage, and voltage balance between phases. Poor and broken connections of one of the supply lines are major causes of voltage unbalance. Overload relays for each phase will protect against extreme (greater than 5%) voltage unbalance.

The mounting should also be inspected frequently. If necessary, retighten bolts with the proper torque. If vibration is detected, one or more of the motor feet may have to be shimmed.

Motor bearings should be greased using manufacturer's specified greases in concert with manufacturer's specified lubrication frequency and quantities. Greasing is necessary to maintain a lubricant

**TABLE 12-1.** Effect of Ambient Temperature on Electric Motors

NEMA motor ratings are based on operating at altitudes below 1,000 meters (3,300 feet) and at ambient temperatures of 40°C (104°F). Variance in ambient temperature requires rerating of motor horsepower requirements.

| Ambient Temperature (°C/°F) | Ambient Temperature Factor |
|---|---|
| −20/−4 | 1.27 |
| 0/32 | 1.19 |
| 20/68 | 1.10 |
| 40/104 | 1.00 |
| 60/140 | 0.88 |

# GENERAL SPEED-TORQUE CHARACTERISTICS
## THREE-PHASE INDUCTION MOTORS

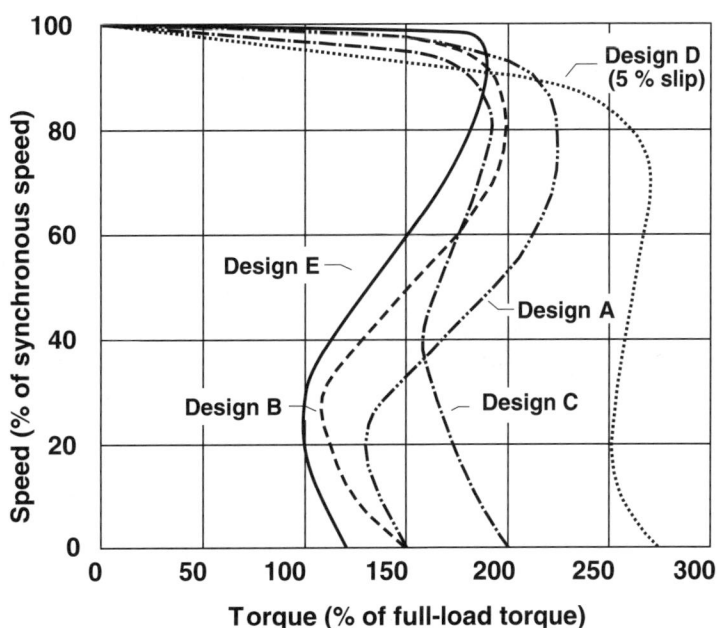

| NEMA DESIGN | LOCKED ROTOR TORQUE | BREAKDOWN TORQUE | LOCKED ROTOR CURRENT | % SLIP | RELATIVE EFFICIENCY |
|---|---|---|---|---|---|
| B | 70 - 275%* | 175 - 300%* | 600 - 700% | 0.5-5% | Medium or High |
| | **Applications:** Fans, blowers, centrifugal pumps and compressors, motor-generator sets, etc., where starting torque requirements are relatively low. | | | | |
| C | 200 - 250%* | 190 - 225%* | 600 - 700% | 1-5% | Medium |
| | **Applications:** Conveyors, crushers, stirring machines, agitators, reciprocating pumps and compressors, etc., where starting under load is required. | | | | |
| D | 275% | 275% | 600 - 700% | 5 - 8%<br>8 - 13%<br>15 - 25% | Medium |
| | **Applications:** High peak loads with or without flywheels, such as punch presses, shears, elevators, extractors, winches, hoists, oil-well pumping, and wire-drawing machines. | | | | |
| E | 75 - 190%* | 160 - 200%* | 800 - 1000% | 0.5-3% | High |
| | **Applications:** Fans, blowers centrifugal pumps and compressors, motor-generator sets, etc., where starting torque requirements are relatively low. | | | | |

Based on NEMA Standards MG 10, Table 2-1. NEMA Design A is a variation of Design B having higher locked-rotor current.

*Higher values are for motors having lower horsepower ratings.

**FIGURE 12-1**

© 1997, Electrical Apparatus Service Association, Inc., St. Louis, MO. Reprinted with permission.

seal where the shaft exits the motor shell. Relubrication is necessary to replenish grease that has broken down by oxidation, or lost by evaporation, and by centrifugal force. Double-shielded and sealed bearings, however, cannot be lubricated after manufacture.

Inspect and keep cooling and ventilation vents clear of obstructions.

If a motor burns out, the windings should be inspected for signs of single-phasing, short-circuiting, overloading, and voltage unbalance. Any cause of winding damage should be identified and corrected. If a motor burns out, the circuit supplying the voltage should also be inspected for broken or shorted wires, burnt contacts, or voltage imbalance.

## ELECTRIC MOTOR STANDARDS

Generally, standards for electric motors are based on the International Electrotechnical Commission (IEC), the National Electrical Manufacturer's Association (NEMA), and the Institute of Electrical and Electronic Engineers (IEEE). Most countries typically have their own standard or a recognized standard.

### U.S. Standards

NEMA designs (A, B, C, D, and E) classify motors according to specific torque characteristics for effective startup and operation of equipment under particular loading and operating situations. Design B motors are commonly used on drilling rigs. These are general purpose motors suitable for normal startup required by pumps, fans, and low-pressure compressors (Figure 12-2).

### European Standards

In Europe and Asia national standards for electric motors are, in general, based on the International Electrotechnical Commission (IEC). The IEC facilitates coordination and unification of motor standards. IEC standards for dimensions, tolerances, and output ratings are contained in IEC Publications 72 and 72A. IEC standards for rating, performance characteristics, and testing of rotating machinery for nonhazardous locations are contained in a series of IEC Publications No. 34, while IEC standards dealing with apparatus for explosive gas atmospheres are contained in a series of Publications No. 79.

The IEC recommendations hold international applicability. The European standards are identical in all countries in regard to their contents and are published as national standards. Before existence of these standards, each country had its own national certifying authority. Today, with certification from recognized "national testing houses," the motors are acceptable in all EEC countries and most other European and Asian countries as well. Electric equipment certified to conform to these standards may be installed and used in any EEC member state. Participating non-EEC states may require additional testing standards.

### Comparison of U.S. and IEC Nomenclature

While there are many similarities and even direct interchangeabilities between U.S. and IEC recognized standards, specific applications must be considered. Motors may be acceptable under all standards but not necessarily certified under all standards.

The IEC "flame-proof" motor is essentially the same as the U.S. "explosion-proof" motor. Each design withstands an internal explosion of a (specified) gas or vapor and prevents ignition of the specified gas or vapor that may surround the motor. However, construction standards are not identical. The U.S. standard is generally more stringent and acceptability can be based on approval of local authorities.

The U.S. totally enclosed "purged and pressurized," or "inert gas filled," motors are manufactured to similar standards as those of IEC pressurized motors. Each operates by first purging the motor enclosure of any flammable vapor and then preventing entry of the surrounding (potentially explosive or corrosive) atmosphere into the motor enclosure by maintaining a positive gas pressure within the enclosure.

IEC Type 'e' (Increased Safety) motors are non-sparking motors with additional features that provide further protection against the possibilities of excess temperature and/or occurrence of arcs or sparks.

NEMA and IEEE standards and testing are more comprehensive than the IEC standards. In general, motors designed to NEMA/IEEE standards should be suitable for application under IEC standards from a rating, performance, and testing viewpoint. Mounting dimensions and tolerances should always be verified.

## ENCLOSURE AND FRAME DESIGNATIONS

### Motor Enclosures

Motors operate best in areas free of airborne particles and corrosives and should have sufficient

## STARTER ENCLOSURES

| TYPE | NEMA ENCLOSURE |
|---|---|
| 1 | General Purpose—Indoor |
| 2 | Driproof—Indoor |
| 3 | Dusttight, Raintight, Sleettight—Outdoor |
| 3R | Raintight, Sleet Resistant—Outdoor |
| 3S | Dusttight, Raintight, Sleettight—Outdoor |
| 4 | Watertight, Dusttight, Sleet Resistant—Indoor & Outdoor |
| 4X | Watertight, Dusttight, Corrosion-Resistant—Indoor & Outdoor |
| 5 | Dusttight, Drip-Proof—Indoor |
| 6 | Occasionally Submersible, Watertight, Sleet Resistant—Indoor & Outdoor |
| 6P | Watertight, Sleet Resistant—Prolonged Submersion—Indoor & Outdoor |
| 12 | Dusttight and Driptight—Indoor |
| 12K | Dusttight and Driptight, with Knockouts—Indoor |
| 13 | Oiltight and Dusttight—Indoor |

**HAZARDOUS LOCATION STARTERS**

| TYPE | |
|---|---|
| 7 | Class I, Group A, B, C or D Hazardous Locations—Indoor |
| 8 | Class I, Group A, B, C or D Hazardous Locations—Indoor & Outdoor |
| 9 | Class II, Group E, F or G Hazardous Locations—Indoor |
| 10 | Requirements of Mine Safety and Health Administration |

## CONVERSION OF NEMA TYPE NUMBERS TO IEC CLASSIFICATION DESIGNATIONS

(Cannot be used to convert IEC Classification Designations to NEMA Type Numbers)

| NEMA ENCLOSURE TYPE NUMBER | IEC ENCLOSURE CLASSIFICATION DESIGNATION |
|---|---|
| 1 | IP10 |
| 2 | IP11 |
| 3 | IP54 |
| 3R | IP14 |
| 3S | IP54 |
| 4 and 4X | IP56 |
| 5 | IP52 |
| 6 and 6P | IP67 |
| 12 and 12K | IP52 |
| 13 | IP54 |

Note: This comparison is based on tests specified in IEC Publication 529.
Reference: Information in the above tables is based on NEMA Standard 250-1991.

**FIGURE 12-2**
© 1997, Electrical Apparatus Service Association, Inc., St. Louis, MO. Reprinted with permission.

cool airflow to dissipate heat developed during operation. Poor ventilation causes many industrial motor failures. Motors should also be protected or shielded from damage by liquids. Rarely, if ever, do all these conditions exist. NEMA has defined various enclosures suitable for different operating environments (Table 12-2).

## NEMA and IEC Frame Dimensions

IEC Publications 72 and 72A provide standards for dimensions, tolerances, and output ratings. Publication 72A, an extension of Publication 72, addresses larger machines. There are small, but significant, differences between the IEC and NEMA frame dimensions (Table 12-3). In most cases, machines built to either series can be adapted by special machining or shimming. For example, the IEC bolt hole is larger than the NEMA bolt hole. Couplings can normally be obtained or machined that accommodate the shafts of either series.

One potential point of confusion between NEMA and IEC dimensional nomenclature is the different letter symbols used to indicate basic mounting dimensions (Table 12-4). For example:

IEC 112M28 = 112 mm foot height, M frame length, 28 mm shaft diameter

IEC 18M1-1/8 = 4.5 foot height, M frame length, $1\frac{1}{8}$-inch shaft diameter

In IEC nomenclature, when a flange exists on the drive end, the flange number is added directly following the shaft diameter. For example:

112M28F215 = 112 mm foot height, M frame length, 28 mm shaft diameter, 215 mm pitch circle diameter flange

## Protection Classes Relating to Enclosures

The ingress protection code (IP), published as European Standard EN 60 529, provides for enclosures that:

- Protect against contact with live or moving parts
- Protect against entry of solid foreign matter
- Protect against water entry

**TABLE 12-2.** Widely Used Electric-motor Enclosures

**Open** motors (IEC class IP 00) have ventilating openings for passage of external air over and around the windings for cooling purposes.

**Drip-proof** motors (IEC class IP 12) are open motors protected from entry of liquids or solids falling on the motor at angles up to 15° from vertical.

**Guarded** motors (IEC class IP 22) are open motors with ventilating openings of such size and shape to prevent fingers or rods from coming in contact with rotating or electrical parts.

**Splash-proof** motors (IEC class IP 46) are open motors protected from entry of liquids or solids falling on the motor or coming in contact with the motor in a straight-line path at angles up to 100° from vertical.

**Totally Enclosed** motors (IEC class IP 44) are constructed to prevent free exchange of air between the inside and outside of the motor case but are not airtight.

**Totally Enclosed Nonventilated (TENV)** motors are totally enclosed but not self-equipped for cooling.

**Totally Enclosed Fan-Cooled (TEFC)** motors are totally enclosed with a shaft-mounted fan that directs air across the external frame (applications include dusty, dirty, corrosive atmospheres).

**Totally Enclosed Blower-Cooled (TEBC)** motors are totally enclosed motors with an independently powered external frame cooling fan.

**Encapsulated** motors have coated windings to protect them from moisture, dirt, and abrasion.

**Dust-Ignition-Proof** motors are totally enclosed motors designed to exclude the entry of combustible dusts into the enclosure or bearing chamber. They are also designed to operate under any normal or abnormal operating condition (including heavily blanketed with dust) such that external surface temperatures of the motor casing do not exceed the motor's maximum operating temperature.

**Explosion-proof** motors (IEC "flame-proof") are totally enclosed motors built to contain the flames and pressures resulting from repeated internal (inside the motor casing) explosions. They are also designed to operate under any normal or abnormal condition such that external surface temperatures of the motor casing do not exceed the motor's maximum operating temperature. Explosion-proof (or flame-proof) motors are used almost exclusively on drilling rigs.

ELECTRIC MOTORS **207**

**TABLE 12-3.** IEC Versus NEMA Mounting Dimensions

For foot-mounted motors, the IEC frame designation consists of the shaft height followed by the shaft extension expressed in millimeters, while the NEMA system uses a specific number for each frame.

| Frame Size IEC/NEMA | H/D IEC/NEMA | A/2E IEC/NEMA | B/2F IEC/NEMA | C/BA IEC/NEMA | K/H IEC/NEMA |
|---|---|---|---|---|---|
| 90S/143 | 90/86.9 | 140/139.7 | 100/101.6 | 56/57.2 | 10/8.6 |
| 90L/145 | 90/88.9 | 140/139.7 | 125/127 | 56/57.2 | 10/8.6 |
| 112S/180 | 112/114.3 | 190/190.5 | 114/114.3 | 70/69.9 | 12/10.4 |
| 112M/184 | 112/114.3 | 190/190.5 | 140/139.7 | 89/88.9 | 12/10.4 |
| 132S/213 | 132/133.4 | 216/215.9 | 140/139.7 | 89/88.9 | 12/10.4 |
| 132M/216 | 132/133.4 | 216/215.9 | 178/177.8 | 89/88.9 | 12/10.4 |
| 160M/254 | 160/158.8 | 254/254 | 210/209.5 | 108/108 | 15/13.5 |
| 160L/256 | 160/158.8 | 254/254 | 254/254 | 108/108 | 15/13.5 |
| 180M/284 | 180/177.8 | 279/279.4 | 241/241.3 | 121/120.6 | 15/13.5 |
| 180L/286 | 180/177.8 | 279/279.4 | 279/279.4 | 121/120.6 | 15/13.5 |
| 200M/324 | 200/203.2 | 318/317.5 | 267/266.7 | 133/133.4 | 19/16.8 |
| 200L/326 | 200/203.2 | 318/317.5 | 305/304.8 | 133/133.4 | 19/16.8 |
| 225S/364 | 225/228.6 | 356/355.6 | 286/285.8 | 149/149.4 | 19/16.8 |
| 225M/365 | 225/228.6 | 356/355.6 | 311/311.1 | 149/149.4 | 19/16.8 |
| 250S/404 | 250/254 | 406/406.4 | 311/311.2 | 168/168.1 | 24/20.6 |
| 250M/405 | 250/254 | 406/406.4 | 349/349.2 | 168/168.1 | 24/20.6 |
| 280S/444 | 280/279.4 | 457/457.2 | 368/368.3 | 190/190.5 | 24/20.6 |
| 280M/445 | 280/279.4 | 457/457.2 | 419/419.1 | 190/190.5 | 24/20.6 |
| 315S/504 | 315/317.5 | 508/508 | 406/406.4 | 216/215.9 | 28/— |
| 315M/505 | 315/317.5 | 508/508 | 457/457.2 | 216/215.9 | 28/— |
| 355S/585 | 355/368.3 | 610/584.2 | 500/508 | 254/254 | 28/— |
| 355M/586 | 355/368.3 | 610/584.2 | 560/558.8 | 254/254 | 28/— |
| 400S/684 | 400/431.8 | 686/685.8 | 560/558.8 | 280/292.1 | 35/— |
| 400M/685 | 400/431.8 | 686/685.8 | 630/635 | 280/292.1 | 35/— |

Note: All dimensions are in millimeters.

**TABLE 12-4.** IEC Versus NEMA Mounting Dimensions Nomenclature

| IEC Letter | NEMA Letter | Dimension |
|---|---|---|
| H | D | Distance from shaft centerline to foot bottom |
| A | 2E | Distance between centerlines of foot mounting holes (end view) |
| B | 2F | Distance between centerlines of foot mounting holes (side view) |
| C | BA | Distance from shoulder on shaft to centerline of mounting holes in the nearest feet |
| K | H | Diameter of holes or width of slots in the feet |
| D | U | Diameter of shaft extension |
| M | AJ | Pitch circle diameter of fixing holes in face, flange, or base |
| N | AK | Diameter of spigot on face, flange, or base |
| S | BF | Diameter of threaded or clearance hole in face, flange, or base |

**208**  SHALE SHAKERS AND DRILLING FLUID SYSTEMS

This classification system uses the letters "IP" followed by two digits (Table 12-5). The first digit of the code indicates the degree that persons are protected against contact with moving parts, and the degree that equipment is protected against solid foreign bodies (tools, wires, etc.) from intruding into an enclosure. The second digit of the code indicates the degree of protection to the equipment from moisture entry by various means such as dripping, spraying, or immersion (Figure 12-3).

**TABLE 12-5.** Scope and Protection According to IP Protection Classes

(If a code character is not necessary, it should be replaced by the letter "X.")

| Digit Protection | First Digit (Physical) | Foreign Body | Second Digit (Water) |
|---|---|---|---|
| 0 | no protection | no protection | no protection |
| 1 | protection against back and body contact | protection against solid foreign bodies, 50 mm (2.08 in) diameter | protection against water-drops falling vertically |
| 2 | finger contact | solid foreign bodies, 12.5 mm 15° from vertical (0.52 in.) diameter | water-drops falling |
| 3 | tool contact | solid foreign bodies 2.5 mm up to 60° (0.1 in.) diameter | water spray at angles |
| 4 | wire contact | solid foreign bodies 1.0 mm directions (0.04 in.) diameter | water spray from all |
| 5 | wire contact | dust | water jets |
| 6 | wire contact | dust-tight | strong water jets |
| 7 | — | — in water | intermittent immersion |
| 8 | — | — | continuous immersion in water |

| Code Letters | First Digit (0–6) Contact and Foreign Body Protection | Second Digit (0–8) Water Protection | Additional Letter (optional) | Supplemental Letter (optional) |
|---|---|---|---|---|
| IP | 5 | 4 | C | S |

Additional letters concern personnel protection against access to dangerous parts as follows:
    by back of hand     "A"
    by finger     "B"
    by tools     "C"
    by wire     "D"

Supplemental letters concern equipment protection as follows:
    for high voltage equipment     "H"
    waterproof during operation     "M"
    waterproof during standstill     "S"
    weather conditions     "W"

If more than one supplemental letter is required, they should be listed in alphabetical order.

**Example: "IP 66/IP 67 CS"**

Personnel are protected against contact with a wire. Equipment enclosure is dust-tight. The equipment is protected against spray from strong water jets and also intermittent immersion in water. Protection is designed for persons using tools 2.5 mm diameter and 100 mm long (0.2 in. × 4 in.). Water testing was performed while equipment was at standstill (not operating).

## EFFECT OF VOLTAGE VARIATION ON INDUCTION MOTOR CHARACTERISTICS

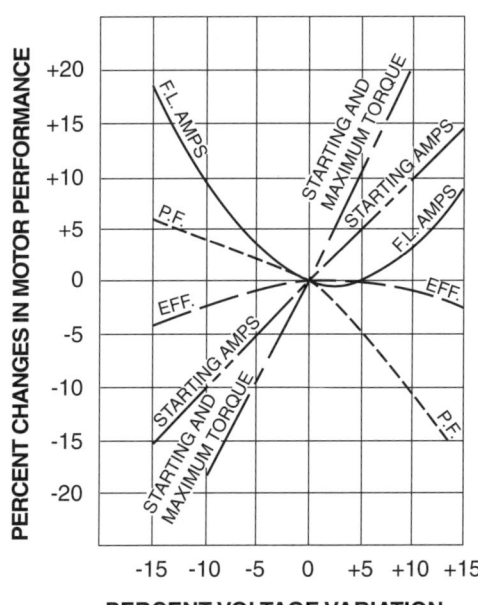

### POWER SUPPLY AND MOTOR VOLTAGES

| NOMINAL POWER SYSTEM VOLTAGE, VOLTS | MOTOR UTILIZATION (NAMEPLATE) VOLTAGE, VOLTS |
|---|---|
| 120 | 115 |
| 208 | 200 |
| 240 | 230 |
| 480 | 460 |
| 600 | 575 |
| 2400 | 2300 |
| 4160 | 4000 |
| 6900 | 6600 |

Reference: NEMA Standards MG-10.

## EFFECT OF VOLTAGE UNBALANCE ON MOTOR PERFORMANCE

When the line voltages applied to a polyphase induction motor are not equal, unbalanced currents in the stator windings will result. A small percentage voltage unbalance will result in a much larger percentage current unbalance. Consequently, the temperature rise of the motor operating at a particular load and percentage voltage unbalance will be greater than for the motor operating under the same conditions with balanced voltages.

Should voltages be unbalanced, the rated horsepower of the motor should be multiplied by the factor shown in the graph below to reduce the possibility of damage to the motor. Operation of the motor at above a 5 percent voltage unbalance condition is not recommended.

Alternating current, polyphase motors normally are designed to operate successfully under running conditions at rated load when the voltage unbalance at the motor terminals does not exceed 1 percent. Performance will not necessarily be the same as when the motor is operating with a balanced voltage at the motor terminals.

### MEDIUM MOTOR DERATING FACTOR DUE TO UNBALANCED VOLTAGE

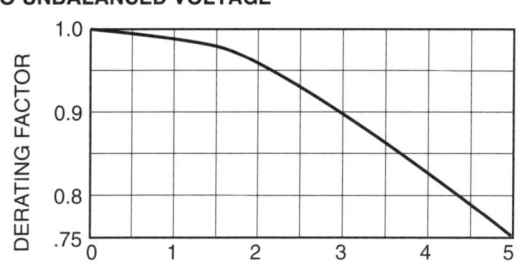

Figure 2

$$\text{Percent Voltage Unbalance} = 100 \times \frac{\text{Max. Volt. Deviation from Avg. Volt.}}{\text{Average Volt.}}$$

**Example:** With voltages of 460, 467, and 450, the average is 459, the maximum deviation from the average is 9, and the

$$\text{Percent Unbalance} = 100 \times \frac{9}{459} = 1.96 \text{ percent}$$

Reference: NEMA Standards MG 1-14.35.

---

**FIGURE 12-3**

© 1997, Electrical Apparatus Service Association, Inc., St. Louis, MO. Reprinted with permission.

# HAZARDOUS LOCATIONS

## Explosion Protection

Construction and installation of all electrical equipment placed in a flammable or potentially explosive location must receive careful consideration. In some drilling and production sites, where the occurrence of explosive mixtures of flammable materials and air cannot be prevented, special construction measures for prevention and/or containment of ignition sources are warranted. Such areas are classified by hazardous ratings, which will be discussed later in this chapter.

**Hazardous locations** are those where potentially explosive atmospheres can occur due to local and/or operational conditions. Leaks inevitably occur during manufacture or movement of volatile, or slightly volatile, liquids. Such leakage may combine with oxygen from the atmosphere to form mixtures of explosive concentrations. Ignition of such mixtures by an electrical spark or by contact with an excessively hot surface may result in an explosion.

An explosive atmosphere occurs when a mixture of air (or oxygen) and flammable substances in the form of gas, vapor, or mist exists in such proportions that the mixture can explode through excessive temperature, arcs, or sparks.

Different techniques are used to minimize the risk of explosion including explosion-proof construction, purging, pressurization, encapsulation, enriching, oil immersion, and intrinsic safety.

## Combustion, Ignition, Flashpoint

When oxygen reacts with other elements or compounds, heat is usually liberated. Because of this, the temperature rises and causes the reaction to proceed at a more rapid rate. Generally, the term "combustion" refers to the vigorous and rapid reaction with oxygen attended by liberation of energy in the form of heat and light.

Reactions other than those involved with oxygen can also liberate heat and light. For example, a jet of acetylene burns brilliantly in chlorine:

$$C_2H_2 + Cl_2 \rightarrow 2HCl + 2C$$

Hydrogen also burns brilliantly in chlorine:

$$H_2 + Cl_2 \rightarrow 2HCl$$

Various substances must be heated to different temperatures before they will ignite and continue to burn in air without suppling additional heat from an outside source. Some substances (sand and clay for example) will not ignite at any temperature because the elements they contain have already combined with as much oxygen as they are capable. In this case further reaction with oxygen is not possible. Some substances ignite at very low temperatures such as white phosphorus, which ignites at 35°C (95°F). Gasoline ignites at a lower temperature than kerosene; kerosene ignites at a lower temperature than motor oil; and ether ignites at a lower temperature than alcohols.

The combustion of some substances is accompanied by the production of flames, which are burning gases. When wood or coal (especially soft coal) is heated to its ignition point or below, combustible gases are released. These combustible gases usually ignite at a lower temperature than the residue of wood or coal, and their combustion produces the effect know as "flame."

Hydrogen burns with an almost colorless flame, as opposed to flames produced by wood, which are generally yellow colored. Flames are usually colorless if solid particles are not present in the burning gases and are produced by the decomposition of substances into the gases as they burn. The hydrogen flame can be yellow if a small quantity of sodium chloride is vaporized and mixed with the burning hydrogen.

**Kindling temperature** is the temperature at which a substance bursts into flames and combustion proceeds without further application of heat. Kindling temperature varies considerably with the state of division of the substance (for instance, the "wood" of a match), its surface area, porosity, and so forth. Finely divided particles offer much more surface area than the same weight of a substance in one large mass. Iron and lead can both be produced in small enough particles (large surface area per unit mass) so that they will ignite without preliminary heating when poured from a container into air.

**Flashpoint** is the lowest temperature at which the vapors above a volatile, combustible substance (such as any petroleum product) ignite momentarily in air due to a spark or small flame applied near the liquid surface. It has also been described as the lowest temperature at which a liquid will give off sufficient vapor to ignite momentarily on application of a flame. The degree of flammability of a substance is mainly expressed by its flashpoint.

An **ignitable mixture** is one within the flammable range (between upper and lower limits) capable of flame propagation away from the source of ignition when ignited. Some evaporation occurs below the flashpoint, but not in quantities sufficient to form an ignitable mixture. This term applies mostly to flammable and combustible liquids, although there are certain solids (such as camphor or naphthalene) that slowly evaporate or volatilize at ordinary room temperatures. Also, liquids, such as benzene, freeze at relatively high temperatures and, therefore, have flashpoints while in the solid state.

The term "propagation of flame" is used to describe the spread of flame from the ignition source through a flammable mixture. A gas or vapor mixed with air in proportions below the lower limit of flammability, may burn at the source of ignition. In other words, they may burn in the zone immediately surrounding the source of ignition, without propagating (spreading away) from the source of ignition. However, if the mixture is within the flammable range, the flame will spread throughout when a source of ignition is supplied. The use of the term "flame propagation," therefore, can be used to distinguish between combustion that takes place only at the source of ignition and that which travels (propagates) through the mixture.

**Ignition temperature** (or auto-ignition temperature) of a substance, whether solid, liquid, or gaseous, is the minimum temperature required to initiate or cause self-sustained combustion in the absence of any ignition source, such as a spark or flame. To avoid the risk of explosion, the temperature of any part or surface must always be below the ignition temperature.

Ignition temperatures observed under one set of conditions may alter significantly with changing conditions, such as:

- Percentage composition of the vapor or gas-air mixture
- Shape or size of the space in which ignition occurs
- Rate and duration of heating
- Reactivity of any other materials present

Thus, ignition temperatures should be viewed as approximations.

There are many differences in ignition temperature test methods including size, shape and composition of ignition chambers, method and rate of heating, residence time, and method of flame detection. Reported ignition temperatures are affected by the test methods employed.

Since ignition temperature is the temperature at which ignition may occur due to contact with a hot surface, it follows that motor selection must be based on the maximum surface temperature that will never exceed the auto-ignition temperature of any potentially explosive mixture likely to exist. The National Electrical Codes (Table 12-6) indicates the maximum external surface temperature for motors in service with combustible materials.

Ignition temperature depends on the type and concentrations of gasses and vapors present. Table 12-7 compares the flashpoint and ignition (or auto-ignition) temperatures for some common materials.

**TABLE 12-6.** NEC Maximum External Surface Temperatures

| Temperature Class | Maximum Surface Temperature | Ignition Temperature of Combustible Material |
|---|---|---|
| T1 | 450°C/842°F | > 450°C/842°F |
| T2 | 300°C/572°F | > 300°C/572°F |
| T2A | 280°C/536°F | > 280°C/536°F |
| T2B | 260°C/500°F | > 260°C/500°F |
| T2C | 230°C/446°F | > 230°C/446°F |
| T2D | 215°C/419°F | > 215°C/419°F |
| T3 | 200°C/392°F | > 200°C/392°F |
| T3A | 180°C/356°F | > 180°C/356°F |
| T3B | 165°C/329°F | > 165°C/329°F |
| T3C | 160°C/320°F | > 160°C/320°F |
| T4 | 135°C/275°F | > 135°C/275°F |
| T4A | 120°C/248°F | > 120°C/248°F |
| T5 | 100°C/212°F | > 100°C/212°F |
| T6 | 85°C/185°F | > 85°C/185°F |

**TABLE 12-7.** Flashpoint and Auto-ignition Temperature for Some Common Materials

| Gas/Vapor/Liquid | Flashpoint | Auto-ignition Temperature | Class | Explosion Group |
|---|---|---|---|---|
| Acetone | −20°C/ −4°F | 465°C/869°F | T1 | IIA |
| Benzyl Alcohol | 93°C/200°F | 436°C/817°F | T2 | IIA |
| Benzene | −11°C/ 12°F | 498°C/928°F | T1 | IIA |
| Gasoline (petrol) | −43°C/−45°F | 280°C/536°F | T2A | IIA |
| No. 2 Diesel Fuel | | | | IIA |
| Hydrogen Sulfide | gas | 260°C/500°F | T2B | IIB |

## Motors for Hazardous Duty

A summary of the hazardous location designations as outlined in the U.S. National Electrical Code with a comparison of the international designation, is outlined in Table 12-8.

IEC standards that address equipment for use in explosive atmospheres are contained in a series of Publications 79-0 through 79-10. Motor classification and applicability differ considerably from U.S. standards and practices. Some of these differences are summarized below.

IEC classifies equipment into two broad categories:

- Group I—Underground mines
- Group II—Use in other industries

This discussion is restricted to motors in the Group II classification, and specifically, to groups IIA, IIB, and IIC, which relate to the gas or vapor involved. A comparative, but not identical, grouping is:

| IEC | U.S. |
|---|---|
| Group II A | Group D |
| Group II B | Group C |
| Group II C | Groups A and B |

The IEC classifies hazardous locations into "zones" according to the probability of a potentially explosive atmosphere occurrence. The degree of danger varies from extreme to rare:

- **Zone 0.** An explosive gas-air mixture is continuously present or present for long periods of time. No electric motors may be used in these areas.
- **Zone 1.** An explosive gas-air mixture is likely to occur in normal operations.
- **Zone 2.** An explosive gas-air mixture is not likely to occur in normal operations, and if it does occur, will only exist for a short time.

The following is a comparison of IEC and U.S. designations:

| IEC Designation | U.S. Designation |
|---|---|
| Group II, Zone 0 | rotating equipment generally not recommended |
| Group II, Zone 1 | Class I, Division 1 |
| Group II, Zone 2 | Class I, Division 2 |

Construction features and test requirements for motors used in hazardous locations are defined by the IEC as:

**TABLE 12-8.** Hazardous (Classified) Location Reference Guide

| | | Specification must include class, division, and group. | |
|---|---|---|---|
| | CLASS I | Areas containing flammable gas or vapor. | |
| | CLASS II | Areas containing combustible dust. | |
| DIVISION 1 | | Explosion hazard may exist under normal operating conditions or due to maintenance, leakage, or breakdown of equipment. | International Equivalent Zone 1 |
| DIVISION 2 | | Explosion hazard may exist under abnormal operating conditions such as rupture of containers or failure of ventilation equipment. | International Equivalent Zone 2 |
| Class I Group | | | International Equivalent |
| | A | Atmospheres containing acetylene | IIC |
| | B | Atmospheres containing hydrogen and the like | IIC |
| | C | Atmospheres containing ethylene and the like | IIB |
| | D | Atmospheres containing acetone, methanol, propane and the like | IIA |
| Class II Group | | | |
| | E | Atmospheres containing combustible metal Dust such as magnesium or aluminum | — |
| | F | Atmospheres containing combustible carbonaceous dust such as coal | — |
| | G | Atmospheres containing combustible dust such as flour, grain, wood, and plastic | — |

For additional information on the properties and group classification of Class I and Class II materials, see *Manual for Classification of Gases, Vapors and Dusts for Electrical Equipment in Hazardous (Classified) Locations*, National Fire Protection Association (NFPA) 497M.

- Flame-proof enclosures
- Pressurized enclosures
- Increased safety protection, "e"

Common IEC symbols are:

- Flame-proof—"d" or (Ex)d
- Pressurized—"p" or (Ex)p
- Increased Safety—"e" or (Ex)e

Table 12-9 provides a detailed description of these terms and symbols.

## SPECIFIC MOTOR APPLICATIONS

With the exception of specialized motors for centrifuge feed and high-speed shale shakers, practically all electric motors encountered in drilling fluid operations are integral-horsepower, **across-the-line start**, horizontal **squirrel-cage motors**. Across-the-line, start-up motors are the simplest and lowest cost. The motor is connected directly to the input power through a starter switch. Full current and torque are realized at startup. This is acceptable with solids control and processing

**TABLE 12-9.** IEC Nomenclature Applied to Motors for Hazardous Areas

| Protection, IEC or European Standard | Basic Principle | Applications |
|---|---|---|
| Flame-proof enclosure, "d" | Parts that can ignite an explosive atmosphere are enclosed to withstand pressure developed during an explosion and prevent transmission of the explosion to explosive atmospheres around the enclosure | Switch gear, control and indicating equipment, control boards, motors, transformers, light fitting, and other spark-producing parts |
| Increased safety, "e" | Additional measures against internal or external arcs or sparks, or excessive temperatures not produced in normal service | Terminal and connection boxes, control boxes and housing, squirrel cage motors, and light fittings |
| Pressurized apparatus, "p" | Entry of surrounding atmosphere is prevented by maintaining a protective gas at pressure higher than the surrounding atmosphere | As above, usually for large equipment and contained rooms |
| Intrinsic safety, "i" | Internal electric circuits are incapable of causing explosions in the surrounding atmosphere | Measurement and control equipment |
| Oil immersion, "o" | Electric apparatus or its parts are immersed in oil to prevent ignition of a surrounding or covering atmosphere | Transformers (rarely used) |
| Powder filling, "q" | Enclosure is filled with a finely granulated material so that an internal arc will not ignite the surrounding atmosphere; also, ignition will not be caused by flame or excessive temperature of enclosure surfaces | Transformers, capacitors, heater strip connection boxes, and electronic assemblies |
| Moulding, "m" | Parts that can ignite an explosion are enclosed (encapsulated) in a resin to prevent ignition of an explosive atmosphere by internal sparking or heating | Only small capacity switch gear, control gear, indicating equipment, and sensors |

equipment, however, it is suggested that centrifugal pumps be started with the discharge valve partially closed to restrict initial pump output and load demand on the motor.

## Electric Motors for Shale Shakers

Shale shaker motors are generally three-phase induction motors that are explosion-proof having NEMA Design B or similar characteristics (Table 12-10).

A shale shaker motor usually has four magnetic poles (1800 rpm synchronous shaft speed at 60 Hz), but may sometimes have six poles (1200 rpm) or two poles (3600 rpm).

The motor should have independent, third-party markings indicating its suitability in explosive or potentially explosive environments. It is recommended that these motors be suitable for Class I, Division 1, Groups C and D, and Group IIB atmospheres. The motor also should have the proper operating temperature or code designation for the anticipated ambient temperature.

A 50-Hz motor driving a shale shaker vibrator should not be operated at 60 Hz since the centrifugal force output will increase by 44%. This will likely damage the bearings and the vibrating screen. A 60-Hz motor driving a shale shaker vibrator can be operated at 50 Hz with the understanding that the centrifugal force output will decrease by 31%. If, at 60 Hz, the centrifugal force is 1000 lbs., the centrifugal force will only be 690 lb at 50 Hz.

For a given frame size, higher speed motors will have high horsepower ratings, low **slip**, high starting torque, and low bearing life. Conversely, lower speed motors will have lower horsepower ratings, high slip, low starting torque, and long bearing life.

Electric industrial vibrators are rated in centrifugal force output, frequency, unbalance (working moment), and horsepower. Centrifugal force is created by a rotating offset mass operating at a given speed. The unbalance is caused by torque resulting from the offset eccentric weight times the moment arm [product of] (distance from the shaft center to the center of gravity of weight). The unbalance provides the amplitude at which the vibrating screen will move (see Chapter 3).

Two counter-rotating shale shaker motors will produce a linear force that should be located through the center of gravity of the shaker basket. The resultant motion is perpendicular to a plane drawn between the rotating shafts directed through the center of gravity of the machine. The shale shaker motor should be selected to meet or exceed the desired stroke of the machine, centrifugal force, and acceleration (G's). Adequate horsepower is required to perform the work and to ensure synchronization. Synchronization results in opposing forces from two counter-rotating vibrators that cancel each other and double directional forces.

Stroke, which is independent of motor speed, is the peak-to-peak displacement imparted to the machine. Dampening may occur in the system affecting the total stroke. Stroke is a function of the unbalance (or torque) of the motor and of the total weight of the shaker basket, including the weight of the motors and the live load. The stroke equals two times the motor unbalance, multiplied by the number of motors, divided by the total weight. Note that the motor unbalance is a function of the eccentric weight setting. For example, the motor unbalance is 50% of the maximum unbalance if the eccentric weights are set at 50%.

The centrifugal force output of the vibrating motor (lbs.) is equal to the shaft speed squared times the unbalance (in.-lb.), divided by 35,211. Once again, the vibrating motor's centrifugal force output is a function of the eccentric weight setting. For example, the centrifugal force is 50% of the maximum centrifugal force if the eccentric weights are set at 50%. Typical acceleration rates for vibrating screens are 4 to 8 G's.

## Electric Motors for Centrifuges

Most centrifuges use the same NEMA Design B explosion-proof motors used for centrifugal pumps: either 1450 rpm at 50 Hz or 1750 rpm at 60 Hz.

**TABLE 12-10.** Electric Motor Specifications for Shale Shakers

| | US DESIGNATION | IEC DESIGNATION |
|---|---|---|
| Terminology | Explosion-proof | Flame-proof |
| Hazardous Location Rating | Class I, Division 1, Group D | Eexd Gas Group IIA |
| Hazardous Location Rating if Hydrogen Sulfide is Encountered | Class I, Division 1, Groups C and D | Eexd Gas Group IIB |

Centrifuges may draw up to seven times the full load current for approximately 15 seconds at startup. NEMA Design B motors are all rated for 7.5 times full load current for 30 seconds at startup. It is, however, considered good practice to limit centrifuge startups to two starts per hour to protect the motor because the current draw closely approaches the range limit.

Oil field centrifuges do not always use a direct drive between the motor and centrifuge. Direct drive requires expensive, variable-speed motors and have restricted availability of replacement parts and repair facilities. Most oil field centrifuges are connected to the motor by a fluid clutch or a hydraulic drive, which uses a system of adding or subtracting motor or hydraulic oil to increase/decrease slippage between the driver and driven coupling halves. Some centrifuges use a variable-speed electric motor startup system that brings the centrifuge slowly up to operational rotating speed. It is important that all personnel understand the manufacturer's recommended startup and shutdown procedures.

Early centrifuges were capable of generating 500 to perhaps 1500 G's of acceleration. Today, machines commonly generate 2000 to 3000 G's. The advent of higher G centrifuges is attributed to improvements in bearing design and manufacturing procedures including hard surfacing with tungsten carbide and precision robot welding.

## Electric Motors for Centrifugal Pumps

The fluid volume necessary to be moved by many centrifugal pumps is related to the rig circulation rate and the specific rig plumbing. Centrifugal pumps should be sized by the particular application and the maximum anticipated flow rate. Piping friction losses—if lines are reasonably short, with few turns or restrictions, and flow velocities between 5 and 10 feet per second—are readily estimated. The pressure or head, which should be delivered to each piece of equipment (for hydrocyclones, typically 75 feet of head), is specified by the manufacturer.

Horsepower requirements for centrifugal pumps, when pumping water or fluids of water-like viscosity, are well established and published with the performance curve for each design (see Chapter 11).

# CHAPTER THIRTEEN

# Solids Dewatering

## INTRODUCTION

The use of "dewatering units," as a regular part of the drilling fluid treatment process on drilling rigs, continues to increase. A dewatering unit, the final phase of a solids control program, is designed to discard no free liquid. A solids control program in which a dewatering unit is used is referred to as a "closed-loop system" (CLS), "sumpless system," or "chemically enhanced centrifugation" (CEC). These units can process drilling fluid from the active mud system or underflow from the solids removal equipment (shakers, desanders, desilters, and centrifuges). Chemical pH adjustments are made to the excess mud; flocculants and coagulants are added to the input stream of the dewatering unit. The flocculated solids are usually removed by a high-speed (high-gravity) centrifuge. Colloidal-rich discards typically contain approximately 50% by volume water but can have as little as 10% by volume solids and still be handled as a solid. These solids can be handled with a front-end loader and hauled on dump trucks, therefore, disposal is typically not a problem. The recovered liquid is usually treated and returned to the active mud system. Treating chemicals, including acids, polymers, strong bases, and clarifying agents, should be carefully selected and additions carefully monitored in order to prevent personnel injury, overtreatment, and excessive treatment cost.

## PROCEDURE

Dewatering of drilling fluids while drilling is the final step when removing solids from excess fluid. All solids are removed from the drilling fluid, related sludge, and waste water. The recovered water is normally reused as treatment water in the active mud system. This process is now relatively common in environmentally sensitive areas. Although the process is expensive, it can be cost effective in areas where disposal costs are excessive.

In many areas, the cost of drilling fluid disposal may greatly exceed the preparation cost. These fluids may be classified as "hazardous" and, therefore, require disposal in hazardous disposal sites. The combined cost of transportation and disposal can run into the hundreds of dollars per barrel. The chemicals in the fluid, such as caustic soda, oil, or salt, are usually major constituents of the drilling fluid. By removing the solids from the fluid and adjusting the chemistry, the remaining water can be reused in the drilling fluid.

A variety of shakers, pumps, centrifuges, tanks, mixers, and chemicals are used in the dewatering process. The solids are removed from the waste fluid to minimize the volume of waste and reduce the cost of disposal. Although the final result may be approximately the same, the solids removal cost and techniques used may vary considerably.

The fluid from which the solids are to be removed may be any type of drilling fluid, including oil muds. The fluid may be excess native mud from the active mud system or may contain concentrated solids from the underflow of mud cleaners, dryer shakers, desilters, or centrifuges. The mud type may be native mud, low solids nondispersed mud, highly treated lignosulfonate mud, salt muds, or even oil muds. Generally, the more highly treated the mud system, the more complicated and expensive it is to dewater and return a clean, usable water.

Water-based muds vary from lightly treated "spud mud" to highly treated, saturated salt mud. The lightly treated muds are simple to dewater and the solids can often be disposed of without restrictions. Polyacrylamide treated muds are usually the simplest to treat because they have a low pH and may contain the same type polyacrylamide that is used as the flocculant. The process becomes more difficult and costly as the chemical content of the mud increases. Saturated salt mud is more expensive to process because the removed solids contain excessive amounts of salt, and the water cannot be disposed of because of the high salt content. The only use for this water is in the mud system.

The first step in the dewatering process is to form units of the aggregated solids that are large enough to be removed by a decanting centrifuge. After these aggregated units are removed, the free water is sent to a holding tank where a "clarifying" chemical is added and the pH increased to neutral (Figure 13-1). In the holding tank the finer solids are allowed to settle by gravity. How clean and clear the water needs to be depends on how the recovered water is to be used.

Adding an acid to the drilling fluids is necessary since these fluids usually have a pH of 8 or higher. The charges on the clay particles are usually negative. For **flocculation** to work properly the charges need to be positive. Adding an acid increases the positive charges on the solids. In order to obtain adequate positive charges on the clays, the pH must be reduced to about 4. The acid used may be hydrochloric, sulfuric, phosphoric, or buffered phosphoric. The use of sulfuric acid is discouraged because it is so dangerous to handle. Although more expensive, buffered phosphoric is the most desirable because it is relatively safe. Next, a flocculant, such as a polyacrylamide or **alum**, is added to increase the size of the flocs. This allows the solids to be brought together in large flocs, which can then be removed by a centrifuge. These additions may be made in batches in a large pit—approximately 150 barrels—or in a continuous flow, computer-controlled dispenser.

After the fluid is flocculated, it is pumped to a high-speed, 3000 to 3400 rpm, decanting centrifuge where the flocked solids are removed. Although the underflow solids from the centrifuge appears very dry and can be handled with a front-end loader, the water content is approximately 50% by volume. It may be possible to use the solids as dike building materials depending on local environmental regulations, the chemical content of the solids, and the water that remains in the solids.

The recovered water is pumped to a holding tank for use as mud treatment water. The pH is increased to between 9 to 10.5, and the turbidity

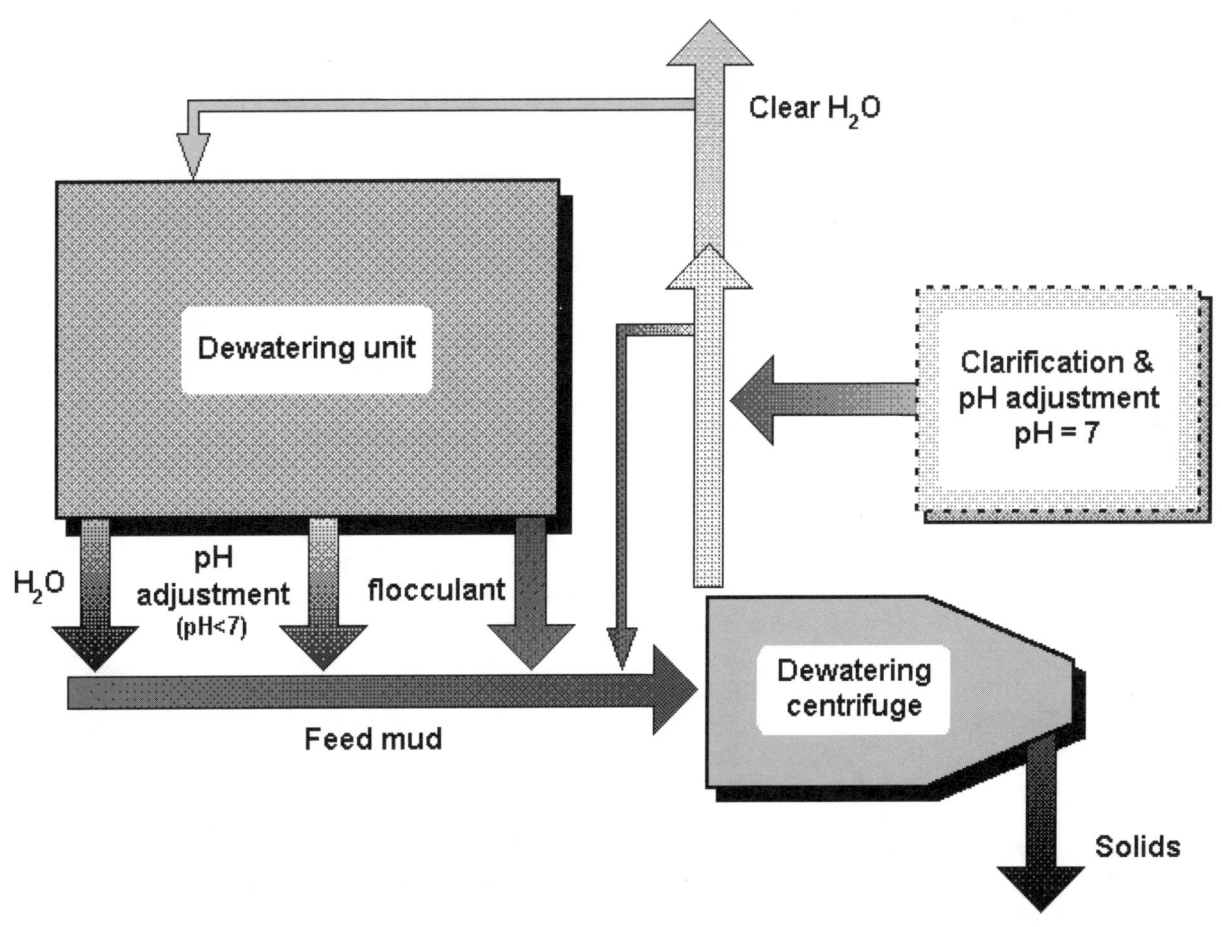

FIGURE 13-1

value is tested prior to adding the water to the mud system. Neutralization of the acid coagulants is commonly accomplished using caustic soda, magnesium hydroxide, or sodium carbonate. Of these neutralizing agents, magnesium hydroxide works at the lowest level of concentration. Using magnesium hydroxide in conjunction with phosphoric acid produces only water and an insoluble phosphate precipitant. Treatments with the other neutralizers, such as caustic soda and soda ash, produce significant amounts of water soluble salts. Continued use of these products will create an unacceptable buildup of the salts and increased turbidity levels. Chlorine, called "everclear," may be added to the recovered water to precipitate the colloidal solids and increase the pH.

Oil muds are more difficult to dewater than water-based muds. The cuttings travel from the shaker to a wash tank containing a detergent and an agitator where the oil is removed from the cuttings. The water/oil mixture is then put into a pit where the oil is removed by a "skimmer," or it is processed through a small ID-high pressure hydrocyclone.

Dewatering of oil-based muds follows approximately the same procedure as that of water-based muds. First, the emulsion must be broken down, which is accomplished by adding an acid and additional water. The oil separates and rises to the surface where it can be removed by a skimmer. The remaining solids are then sent to the processing tank where they are treated in the same manner as water-based mud. Additional washing may be required if the remaining oil content in the sludge exceeds acceptable environmental levels.

Vacuum filtration is another method of removing solids from the liquid. Chemical flocculation may be used to aggregate the solids, or the mud can be processed through a filter depending on the properties of the filter. The solids can then be dried to further reduce any residual water. Chemical flocculation and removal by a centrifuge has proven the least expensive of the two methods, however, the filtration method is continuously being improved and should be given consideration.

In comparison field studies the most effective and economical treatment method has been chemical treatment, settling, and centrifuging of the mud. In these field tests, mud from the active system was treated with anionic polyacrylamide polymers, flocculants, and phosphoric acid coagulants. The flocculated mud was processed through a high-speed (1500 to 2000 times the force of gravity) centrifuge. The "dry" solids removed contained approximately 50% solids by volume. The recovered water, or filtrate, was placed in a holding tank where the pH was neutralized with magnesium hydroxide. After settling, the water was pumped into another holding tank where it was given a final treatment with chlorine prior to returning to the active mud system.

This technique removes dry solids and returns the recovered water to the active mud system, which has no suspended solids, a minimum of dissolved solids, and a pH in the 7 to 10 range.

## POLYMER TECHNOLOGY

A number of additives may be used to coagulate and flocculate solids in waste water, but polymers are the most effective and least expensive. Understanding the dewatering process requires some knowledge of polymers.

### Polymer Chemistry

A polymer is a chemical chain of organic molecules that is produced by the joining of primary chemical units called monomers. The word "polymer" comes from the Greek word *polymeros,* meaning, "many parts."

Depending on their ionic character, polymers used in water treatment applications can be divided into three major types: nonionic, anionic, and cationic. Nonionic polymers possess no ionic charge, anionic polymers possess net negative charges, and cationic polymers possess net positive charges. The basic polymer structure for water treatment polymers is polyacrylamide, which is made from acrylamide monomer. Straight polyacrylamide is nonionic. Charge can be imparted to the polymer by chemically adding functional groups that possess the desired charge type. In this manner, anionic or cationic polymers can be generated.

**Polyelectrolytes**, as these polymers are sometimes called, encompass the entire family of water treatment polymers. Nonionic polymers, although they do not possess net charge, are still classified as polyelectrolytes because they tend to exhibit similar solution properties, as do ionically charged polymers.

Another important characteristic of water treatment polymers is molecular weight, which refers to the polymer chain length. The longer the chain length, the higher the molecular weight. Polymers with molecular weights between 5,000 to 200,000 behave as coagulants in water. These polymers, carrying positive charge, neutralize negatively charged **colloidal matter** in waste water to form small floc particles. Conversely, polymers with molecular weight between 500,000 and several million, act as flocculants in water. These polymers possess long chains that trap and enmesh small floc particles to form larger floc particles.

Nonionic and anionic polymers are classed as flocculants depending on their molecular weight. Coagulants, by principle, must possess low molecular weight and cationic charge. Table 13-1 lists the characteristics of polyelectrolytes used in water and waste water treatment applications.

Current water treatment polymers are available in three physical forms: powder, aqueous solution, and emulsion. The powders are the most common and oldest form for flocculants. Powdered products must be dissolved in water to approximately 2% or less before they can be applied.

Solution polymers have become quite common. These polymers need only to be diluted before applying them. Many flocculant polymers are currently available in solution form, containing 2% to 10% active polymer. These polymers are much easier to handle and apply than the older, powdered forms. Coagulant polymers are always sold in liquid form.

The newest product form for flocculants is the emulsion. This polymer is in the form of small droplets suspended in a hydrocarbon and surfactant carrier. Emulsion polymers must first be activated, or inverted, in water before they can be applied. These require a special polymer makeup and feeding system.

## COAGULATION

The process of chemically neutralizing the negative charge on colloidal matter in waste waters to form small **floc** particles is called "**coagulation**." Most colloidal suspended matter in water carries a negative charge by nature. It can consist of emulsified oil particles, solid particles, dirt, metal fines, or biological particles. These particles will not settle until this charge is neutralized. Water treatment polymers that carry a positive charge and low molecular weight (coagulants) can be added to the water to neutralize the colloidal charges.

Inorganic chemical salts such as aluminum sulfate, ferric sulfate, ferric chloride, and calcium chloride, among others, have historically been used as coagulants. The major disadvantages to these inorganic coagulants are (1) large volumes of resultant sludge, (2) the sludge is stable and difficult to further treat, (3) inorganics require narrow pH ranges for activation, and (4) solids are added to the water that later must be removed.

Polymer coagulants provide excellent coagulation without the drawbacks of inorganic coagulants. The major advantages include (1) minimal dosage requirements, (2) the sludge volume is minimized, (3) they work over wide pH ranges, and (4) a minimum amount of solids are required to be added to the water. Coagulation is usually

**TABLE 13-1.** Characteristics of Water Treatment Polymers

| Class | Charge | Molecular Weight | Form |
|---|---|---|---|
| 1. Cationic Coagulants<br>Polyamines<br>Polyquaternaries<br>Poly DADM<br>DMA-EPI | + | 5,000 to 200,000 | Liquid solutions |
| 2. Cationic Flocculants<br>Copolymers of:<br>Acrylamide and Methacrylate<br>Acrylamide and DADM<br>Mannich Polymers | + | 1,000,000 or more | Powders, solutions, or emulsions |
| 3. Anionic Flocculants<br>Polyacrylates<br>Copolymers<br>Acrylamide and Acrylate | − | 1,000,000 or more | Powders, solutions, or emulsions |
| 4. Nonionic Flocculants<br>Polyacrylamides | 0 | 1,000,000 or more | Powders, solutions, or emulsions |

*Notes:* DADM = Diallydimethyl Ammonium Chloride
DMA = Dimethylamine
EPI = Epichlorohydrin

the first step in water and waste water clarification. This process requires rapid mixing for thorough integration of the chemical and water.

## FLOCCULATION

The process of agglomerating small, neutralized floc particles to form large, settleable flocs is called "**flocculation**." A flocculating agent gathers together small floc particles by means of bridging and enmeshment to form large floc particles that readily settle. Bridging results from the long polymer chains of the flocculant that reach out and gather small floc particles. Enmeshment results from the netting effected on the cross-linking polymer structure. This traps and retains the small floc particles and allows the floc size to grow.

The same inorganic chemical salts used as coagulants have also been used as flocculants. The major drawback to inorganic flocculants is that they are extremely ineffective because they lack the necessary molecular weight. Polymer flocculants, on the other hand, provide excellent water clarity at relatively lower dosages, without voluminous sludge. Effective flocculation requires slow, gentle mixing to allow the floc particles to build and grow. This is one reason why polymer flocculants must be diluted prior to application. Dilution facilitates proper mixing of the polymer and water, which allows the polymer thins to reach out and contact the floc particles.

The major benefit obtained with polymer flocculants is the large increase in floc size, resulting in more rapid sedimentation. It must be remembered that these products function primarily as high molecular weight flocculants by providing more efficient and effective bridging mechanisms. This should be contrasted to cationic polyelectrolytes of lower molecular weight that function as primary coagulants (neutralization). The flocculants do not normally function effectively as primary coagulants except in cases where the charge is of a low order magnitude. The dosage of polymer flocculants for water clarification is normally very low, ranging from 0.05 to 2 ppm (for highly turbid systems).

In practice, flocculants have a considerable advantage over activated silica due to the more predictable control over solution preparation and feeding, as well as lower space requirements for storage. Difficulties in handling high molecular weight flocculants, are primarily due to the lack of understanding in the area of dissolving the pulverized polymer product. Overdosage, particularly when the polymers are being used as a filter aid, can cause plugging and should be avoided.

The mechanism of flocculation by high molecular weight polymers is specific to this type of chemical. Functionally, the polyelectrolyte assumes an elongated shape in the solution because of the electrostatic repulsion between charged groups, which are adjacent to one another along the polymer chain. Generally, nonionic polymers form a random coil due to the lack of charged groups to provide an uncoiling force. The extremely long length and charge density of the synthetic polymers are important characteristics in the polyelectrolyte's behavior. It is generally accepted that a "bridging" mechanism accounts for the flocculation behavior of these compounds. In this process, the polymer molecules attach themselves to the surface of suspended particles at one or more sites with part of the long chain extending into the bulk of the solution. The free end of the molecule is then able, upon contact, to be adsorbed into other suspended particles, thus forming a "bridge," or link, between **turbidity** particles. This dynamic process is one in which adsorption and desorption rapidly take place. It continues until a condition is established in which almost all of the polymer segments become attached, by one or more sites, to two or more turbidity particles. The reactions are fast, with approximately 85% of the polymer being adsorbed within about 10 seconds, and an additional 10% becoming attached after a more prolonged contact. For this reason, extra effort is taken to significantly dilute polymer solutions to allow them to be well distributed into the flocculating medium. The progressive linking of more and more particles results in an ever-increasing sized floc, which is limited only by its ability to withstand the hydraulic shear gradient imposed by the existing turbulence. Theoretical and experimental evidence has shown that the turbidity particle is partially occupied by the adsorbed polymer. Some studies have shown that the maximum flocculation rate is obtained when the turbidity particles have one-half of their surfaces covered with polymer segments. If too many adsorption sites are occupied by polymer segments, bridging is restricted, and if all sites are occupied, bridging is impossible. Both conditions lead to a dispersed rather than flocculated state. Conversely, if too few sites are occupied by polymers, the possibility of bridging between adjacent particles is reduced, and the bonds may be so weak that the floc is unable to withstand shear.

## EMULSION POLYMERS

Emulsion polymers represent the latest technology in the development of water and waste water treatment polymers. These polymers are the fastest growing physical form in the water treatment market.

Years ago, when polyacrylamides were first developed, only dry polymer was available. Dry polymers were difficult to handle due to dusting, as well as slow and difficult to dissolve. Next, liquid solution polymers were developed, which were mainly dilutions of dry polymers. Liquid solution polymers eliminated the dusting problem associated with dry polymers, however, because of their dilute nature, they were extremely freight sensitive.

Emulsion polymers, developed only 12 to 14 years ago, were designed to provide the advantages of both dry and liquid solution polymers, without any of their associated problems. In this respect, emulsion polymers can be viewed as the third generation of water-soluble polymer technology.

Because of their liquid form, emulsion polymers are free from dusting, easy to handle, store, dilute, and feed. Like dry polymers, emulsion polymers are concentrated (normally 25% to 35% active) and, therefore, are not freight sensitive and can be shipped economically long distances from the manufacturing plant.

Another important characteristic of emulsion polymers is that they can be manufactured to possess any ionic character: anionic, cationic, or non-ionic. Also, practically any degree of ionic charge can be exhibited by changing the amount of anionic or cationic monomer used to make the polymer. Thus, an entire series of emulsion polymers is possible. Additionally, certain applications requiring a specific degree of charge can be handled with emulsion polymers.

Competitive cationic emulsion polymers on the market today are quaternized. This means that they are designed to maintain their cationic strength throughout swings in pH. This characteristic broadens the pH range in which cationic emulsion polymers are effective.

Emulsion polymers consist of approximately equal parts of active polymer, water, and solvent carrier. The polymer in an emulsion exists as minute droplets of aqueous, concentrated, polymer solution suspended in a hydrocarbon or other organic continuous fluid. Most commercial emulsion polymers have an opaque, milky appearance, which is due to the refraction of light by these droplets. The polymer's molecular backbone exists as a coiled chain within these suspended aqueous droplets.

Emulsion polymers also contain various surfactants and stabilizing agents, which provide consistency and stability. One type of surfactant helps keep the polymer droplets suspended within the organic solvent carrier. This provides emulsion polymers with long shelf lives. Other types of surfactants help the polymer to "invert" into a continuous aqueous phase upon dilution of the neat product. This makes emulsion polymers easy to dissolve and prepare for application. The surfactants must be balanced to provide this desired effect.

### Inversion

The term "inversion" refers to the change from a water-in-oil emulsion to an oil-in-water emulsion. Inversion allows the coiled polymer chains to uncoil and expand into the water phase. This is accomplished by mixing neat emulsion polymer and water under prescribed mixing and dilution conditions. Optimum inversion of emulsion polymers is the key to achieving maximum performance.

The complete inversion of an emulsion polymer is actually a complex process on a microscopic scale, involving several sequential steps. In order of occurrence, these steps are:

1. Dispersion
2. Swelling
3. Eruption
4. Disentanglement

When an emulsion polymer is added to water, the first thing that occurs is dispersion of the emulsion mass into discreet droplets (Figure 13-2). Microscopic examination reveals a small droplet of oil containing thousands of tiny particles of polymer-in-water. This oil droplet becomes surrounded by the bulk water in the makeup tank.

Immediately upon contact with water, the polymer particles inside the oil droplet start to swell

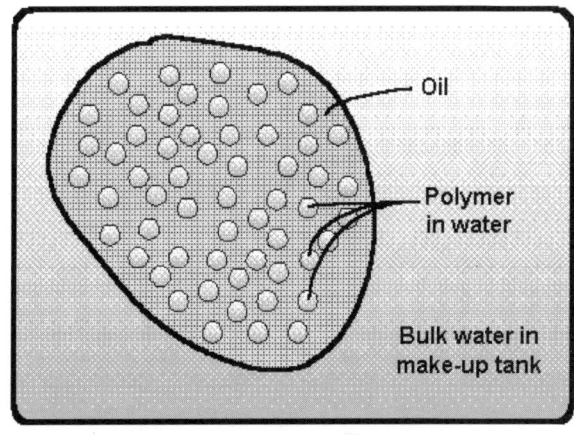

**Dispersion step**

FIGURE 13-2

to approximately five times their original size (Figure 13-3). The particles are absorbing water that migrates across the boundary and through the oil.

Once the small particles-in-water swell, almost all will burst or explode simultaneously. This process is known as microscopic eruption (Figure 13-4).

The last step in this process, disentanglement, allows the polymer to freely dissolve and distribute throughout the water solution (Figure 13-5). In other words, the emulsion has inverted. As a result, the oil is now present as tiny particles dispersed in the continuous water phase. The dissolving polymer makes the solution viscous and stringy.

When carried out to completion, the inversion provides a smooth, consistent, polymer solution free of lumps, fisheyes, or gel particles. The solution is now ready to direct to the point of application, usually along with post-dilution water.

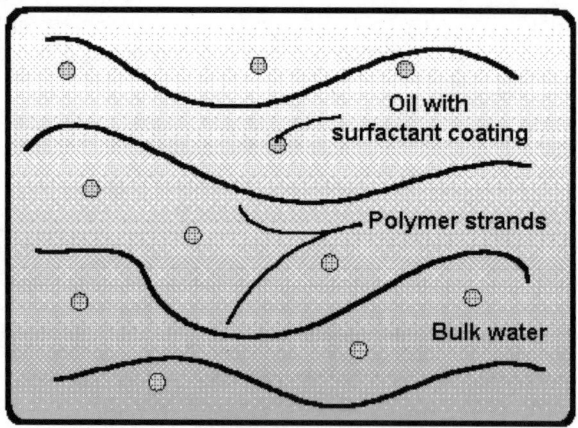

**Disentanglement step**

FIGURE 13-5

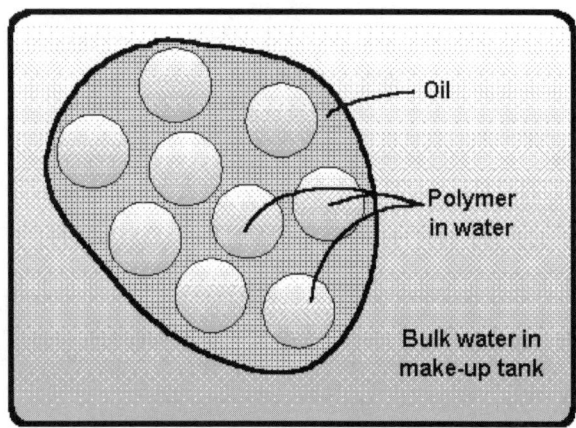

**Swelling step**

FIGURE 13-3

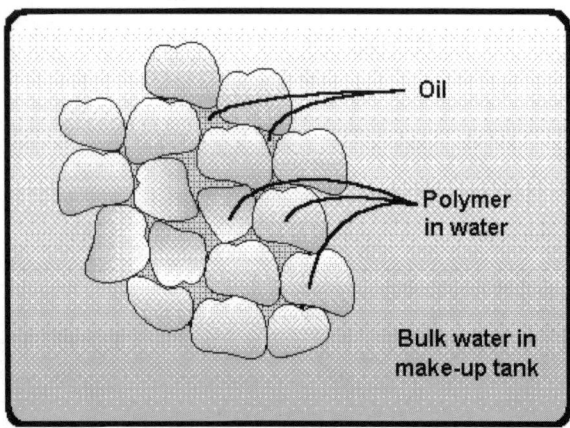

**Eruption step**

FIGURE 13-4

## Stability

Because of their unique physical makeup, emulsion polymers must be specially formulated to remain stable in the drum, and yet be easily inverted and dissolved in water when it is time to use them. There are two aspects of emulsion polymer stability: separation and creaming.

As described previously, an emulsion polymer exists as small polymer/water particles finely dispersed in a hydrocarbon-type fluid. Ideally, these particles would remain evenly dispersed and suspended indefinitely within the hydrocarbon fluid. In reality, these particles begin to settle over time and separation occurs. This process forms a layer of hydrocarbon at the top of the container. It is not uncommon to see a one- to two-inch clear layer at the top of a drum containing emulsion polymer that has been stored undisturbed for a week or longer. This does not indicate a quality problem since all emulsion polymers tend to separate to some extent.

Creaming occurs at the bottom of the emulsion container, where the polymer/water particles begin to settle and concentrate. As storage continues, the polymer/water particles begin to coalesce and form a cream at the bottom of the container. This also can occur, to a smaller degree, after a week or longer of undisturbed storage. Creams can normally be redispersed by mixing, although once formed, tend to resettle more rapidly. Unless the cream formation is redispersed, it can cause plugging of polymer feed pumps and resultant dosage variations.

To minimize any makeup problems caused by separation or creaming, the drum contents should be thoroughly mixed before using. Although horizontal drum rollers can be used for this purpose, it is best to use a side-bung, angle-entry, air driven, or electric drum mixer. A portable drum mixer can be attached to the side bung for easy mixing and redispersion of the emulsion polymer. Center-bung mounted mixers can also be used but the effect is not as great. A drum should be mixed in this fashion for at least 15 minutes before using.

## COAGULATION CONCEPTS AND MECHANISMS

### Colloid Stability

To understand the concept of the mechanisms involved in coagulation reactions, it is important to understand why particles in suspension remain discrete and "stable" as a **colloid**.

Particles of a colloid remain separate and discrete mainly due to electrostatic forces of repulsion that exist between the particles. In any stable colloid, these electrostatic forces must be reduced for particles to contact each other, thereby enabling coagulation to occur.

***Charge neutralization.*** Theoretically, if the like surface charge on particles could be reduced or eliminated, the particles would be able to contact each other and possibly bind together. By adding counter-charged coagulant chemicals, the surface charge on particles is neutralized. Charge neutralization starts the coagulation process.

***Zeta potential.*** The stability of a colloid depends on the electrostatic repulsive forces between particles. The magnitude of these forces are variable and can be measured. The charge sign and magnitude is determined by following the path and speed of particles through an electric field. The "zeta potential" of the particles of a colloid allow for the prediction of an effective coagulant charge selection and dosage.

***Interparticle bridging.*** Once the charge of colloidal particles has been neutralized, particle contact will occur and point flocs will form. The size of these flocs is limited and usually not large enough to provide for an optimal settling velocity. The mechanism primarily responsible for further floc growth is due to the joining, or "bridging," of micro flocs by high-molecular weight, long-chain molecules.

***Van der Waals force of attraction.*** Van der Waals force is a naturally occurring attractive force between two bodies. Although not as powerful as the electrostatic forces of repulsion between colloidal particles, it does enter into the coagulation sequence once charge neutralization has occurred.

In addition to the above, other factors include:

- *PH*—Solution pH is an important influence on particle charge, coagulant effectiveness, and the formation of precipitates of dissolved solids. The optimum pH range for cationic polymers is 4.5 to 10.

- *Temperature*—As water temperature increases, the viscosity decreases and the particles become more active due to the decrease in the shear conditions of the fluid.

- *Mixing Energy*—Refers to both the magnitude of velocity gradients, as well as the duration of mixing. The mixing energy affects the rate of particle collisions and shear conditions.

- *Particle Size*—Affects the floc size and the ultimate settling velocity of flocs.

- *Particle Concentration*—Affects the opportunity for particle collisions.

- *Shear Forces*—Probably the most limiting factor controlling floc growth.

- *Particle Charge*—Dictates the dosage and charge of coagulant needed to attain charge neutralization.

# APPENDIX A

# Derrickman's Pages

This appendix contains comments and guidelines directed specifically toward rig hands that control shale shaker operations. In so far as possible, this section will be written in the language of the rig crew (without some of the more colorful adjectives not used in polite society but commonly used on a rig.) Effective use of shale shakers is greatly dependent upon the treatment that they receive by the hands on the rig.

If some basic rules are not followed, however, all of the theory in the world will not help drill a well. This appendix does not present anything new but it does summarize practical operating guidelines. First, the shale shaker purpose is described; and second, some operating rules are discussed. In other sections this manual describes shale shakers, screens, types of motion, and much other information. In this section, practical rules are discussed.

## PURPOSE

The purpose of a shale shaker is to remove from the mud "dirt" or rock drilled by the bit. The shale shaker is the first piece of solids control equipment to treat or condition the drilling fluid. Good shaker performance is necessary if the entire system is to work properly.

A carpenter's drill bit carries wood cuttings out of a drilled hole with the spirals on the bit itself. In oil or gas drilling the drill string does not have the spirals to move the cuttings out of the hole. In a well, mud brings the cuttings to the surface. These cuttings must be taken out of the mud. Many bad things can happen on a rig if these cuttings are not removed from the drilling fluid.

Cuttings, or drilled solids, left in the mud, create many problems. Some of these problems are:

- Stuck pipe
- Bad cement jobs
- Lost circulation
- Slow drilling rates
- High mud costs
- Pump wear
- Short bit life

Good shaker operation helps prevent these.

## SHALE SHAKER OPERATIONS

Shakers now come in a dazzling assortment of sizes, shapes, and motions. Their performance is controlled by the size and shape of the openings in the screen, the movement of the screen, the mud properties, and the type of cuttings arriving at the shaker. The one selected for your rig may or may not be the best one for the job. Unfortunately, if it is not, it must still be kept operational and with hard work perhaps it can be made to do the job. All commercial shakers, however will remove cuttings—and they do a better job if they are properly treated.

If the mud is not bringing cuttings to the surface, the screen cannot remove them. The solids coming off the end of the shaker screen should have sharp edges. Cuttings that roll around in the borehole on the way to the surface have rounded edges. Rounded edges or round cuttings indicates that the mud is not bringing cuttings to the surface as fast as it should. The mud engineer, or driller, should be told that the cuttings are not reaching the shaker.

If the cuttings are arriving at the shale shaker, several general rules will make certain that the shaker is working properly.

1. The shale shaker should be run continuously while circulating. Cuttings cannot be removed if the shaker is not in motion.
2. The mud should cover most of the screen. If the mud covers only $\frac{1}{4}$ or $\frac{1}{3}$ of the screen, the screen mesh is too coarse.
3. A screen with a hole in it should be replaced immediately. The hole in a panel screen can be plugged. Cuttings are not removed from the mud flowing through the hole.

4. Shaker screen replacements should be made as quickly as possible. Locate and arrange tools and screens before starting.
    - If possible, get help. This will decrease the amount of cuttings being kept in the mud because the shale shaker is not running. If possible, change the screens during a connection.
    - In critical situations, the driller may want to stop the pumps and stop drilling while the screen is replaced.
5. Water should not be added in the possum belly or on the shaker screen. Water should be added downstream.
6. Except for cases of lost circulation, the shale shaker should not be bypassed—even for a short time.
7. The possum belly should not be dumped into the sand trap just before making a trip.
    When the next bit starts drilling, the cuttings will move down the mud system and plug the desilters.

Each of these seven rules is very important to maximize cuttings removal.

---

### Things to Check When Going On Tour

1. Make sure the shale shaker is running properly.
2. Check for mud flowing properly over the shaker screen.
3. Check for holes in the screen.
    - Wash off the screen during a connection for this examination.
    - Sometimes a flashlight is needed to examine the whole surface.
    - Most frequently holes appear where the overflow from the possum belly strikes the screen.
    - If a riser-lift pump is used to help move mud up the riser, divert the flow to one shaker during a connection so that all screens may be examined.
4. Replace the screen, if necessary.
    - On panel screens, the holes may be plugged or blanked.
    - Minimize downtime by planning your work.
5. Make certain water is not being added at the shaker.
6. Check for cuttings, not mud, going off the end of the screens.
7. Check cuttings reaching the shaker to make certain that they have sharp edges.
8. Report findings to the driller and/or rig supervisor.

The above rules will help drill a well at the lowest possible costs. Shaker performance is an important part of a trouble-free, lowest-cost well.

# APPENDIX B

# Equipment Guidelines

When the American Petroleum Industry workgroup met to edit and approve the API 13C equipment guidelines, "Practical Operational Guidelines," each item was discussed in depth. Some of those considerations are included here along with some additional thoughts concerning solids removal equipment. The equipment discussed in this appendix is all-inclusive and may not apply to all drilling applications.

## SURFACE SYSTEMS

1. *The surface system should be divided into three sections each having a distinct function: Removal Section, Addition Section, and Check/Suction Section. Undesirable drilled solids should be removed in the Removal Section. All mud material and liquid additions should be made in the Addition Section. The Check/Suction Section provides volume for blending of new mud materials and verification of desired mud properties.*

    This is a simple concept that requires each surface system to have three easily identifiable sections. If not, changes will quickly pay for themselves.

2. *Minimum recommended "usable" surface mud volume is 100 barrels (less for slim holes) plus enough to fill the hole when the largest drill string the rig can handle is pulled wet and all the mud inside the string is lost. In order to maintain fluid properties in large diameter, soft, fast-drilling holes, the minimum surface volume should be at least five or six times the volume of the hole drilled per day.*

    For safety reasons a rig must have enough drilling fluid to fill the hole at all times. This is the situation described above. A second consideration is not a safety feature but a recognition of practicality. When rapidly drilling a large diameter hole and removing the drilled solids from the system, new fluid must be built quickly. The volume of new fluid is the sum of the solids removed and the drilling fluid clinging to them. Mixing equipment on rigs is not usually geared to rapid additions of drilling fluid products. Drilling operations experience fewer problems if the drilling fluid properties are controlled, which is difficult with large, rapid additions of drilling fluid products.

3. *All removal compartments, except the sand trap, should be well stirred or agitated to ensure even loading of solids removal equipment.*

    Solids control equipment works best when the solids loading remains constant. Slugs of a large quantity of solids tend to plug the lower discharge opening in desilters. When this occurs, drilled solids will not be removed until the plugged cones are cleaned. On a rig, even with diligent crews, some cones will usually remain plugged, which leads to increased drilled solids in the drilling fluid.

4. *The ideal tank depth would be approximately equal to the width, or the diameter, of the tanks. If deeper, special considerations may be necessary for stirring; if shallower, adequate stirring without vortexing will be difficult or impossible.*

    Baffles will help prevent vortexing. When using vertical blade stirrers in circular tanks, baffles are a necessity.

5. *Use top equalization for the sand trap.*

    To take advantage of the maximum settling time, fluid should enter the upstream end of the sand trap. After the fluid moves through the compartment, it exits through an overflow weir into the next compartment. An underflow arrangement would carry settled solids into the next compartment.

    The ability to use 200-mesh screens on shale shakers means most of the "sand-sized" particles are removed and few drilled solids are available to settle. (API defines anything larger than 74 microns, or 200 mesh, as sand.) With the introduction of linear motion shale shakers (using 200-mesh screens) and the emphasis on minimizing rig discharges, fewer rigs are including sand traps in their removal systems. Many rigs place their desander pump suctions in the former sand traps. This requires that these compartments be well agitated.

6. *Use top equalization between the degasser suction and discharge compartments.*

   The degasser should process more fluid than is entering its suction compartment. This will create a backflow from the discharge compartment into the suction compartment. Only processed drilling fluid can flow over the top weir. Any drilling fluid still containing gas will overflow over the top weir back to the degasser suction compartment.

7. *Use bottom equalization between the suction and discharge compartments of desanders, desilters, mud cleaners, and centrifuges.*

   Openings in the partitions between the suction and discharges are primarily needed to allow a small backflow. Since the level can vary in these compartments, an underflow is needed so that no adjustment is necessary while drilling. Also, an overflow tends to entrain air as the fluid cascades into the upstream tank.

8. *Use an adjustable equalizer between the removal and addition sections when cyclones and/or centrifuges are being used. Run with the high position on the downstream side.*

   An adjustable equalizer is usually a curved pipe that can swivel in the equalizing line. The upper end is in the downstream compartment. Fluid exits the bottom of the removal section and flows through the adjustable equalizer. The downstream end of the equalizer is elevated so that the liquid level in the removal section remains constant.

9. *Use bottom equalization in the Addition Section and in the Check/Suction Section.*

   This will allow the maximum use of the fluid contained in these sections.

10. *For removal devices processing flow rates greater than the rig circulating rate, equalizing flows should always be in the reverse (or upstream) direction.*

    A centrifuge usually processes only a small portion of the total rig flow. Backflow should not be expected between the centrifuge suction and discharge compartments. All other removal equipment (degasser, desander, desilter, and mud cleaner) should process all of the fluid entering the suction compartment. This may exceed the rig flow if drilling fluid enters upstream from another process or from mud guns.

11. *Based on experience a rule-of-thumb for the minimum square feet of horizontal area for a compartment is as follows:*

    $$\text{Horizontal Surface Area (sq ft)} = \frac{\text{Max. Circulating Rate (gpm)}}{40}$$

    *It has been found from experience that this rule of thumb provides fluid velocities low enough to allow entrained air bubbles to rise to the surface and break out. Note: This rule-of-thumb was developed by George Ormsby and is included in the IADC Mud Equipment Manual, Handbook 2: "Mud System Arrangements," pp. 2–17.*

    This is strictly an empirical guideline. Following this guideline, usually results in most of the air leaving the drilling fluid, although many types of drilling fluids (polymer, synthetics, relaxed fluid-loss oil muds, mineral oil muds, etc.) were not in existence when this rule was developed. Since the air break-out is a function of the interfacial tension and the low-shear-rate viscosity of the fluid, this formula should be used with caution. This is also the reason that all tanks should be adequately stirred. The entrained air must be brought to the surface so that it can leave the system.

12. *Mechanical Stirrers are preferred for stirring removal compartments.*

    All fluid entering a suction compartment should be processed. Mud guns suctioning from a downstream compartment will increase the quantity of fluid that must be processed. Mechanical stirrers eliminate this problem. Mud guns can be used in removal compartments if each centrifugal pump stirs its own suction.

13. *Mechanical stirrers should be properly sized and installed according to manufacturer's recommendations.*

    Mechanical stirrers are available in two stirring blade shapes. One type is canted to actually pump fluid vertically downward. The second type has vertical blades, similar to the impeller blades in a centrifugal pump, which propel the fluid outward. Both types are designed to move all of the fluid within a certain volume. The turnover rate (TOR) depends on how much fluid is moving. The TOR must be large enough to adequately blend the fluid within the compartment. TOR is calculated as 60 times the tank volume divided by displacement. Displacement or flow rate associated with each type of blade, based on projected area of blade, are presented in Table B-1.

14. *Baffles may be installed around each mechanical stirrer to prevent air vortices and settling in corners. A typical baffle can be 1 inch thick by 12 inches wide and extends from the tank bottom to 6 inches above the top agitator blade. Four baffles are installed around each agitator. They are installed 6 inches past the tips of the agitator blades, along lines connecting the center of the agitator blade with the four actual corners of a square pit or compartment. For a long rectangular pit, with two or more agitators, the tank is divided into imaginary square compartments and a baffle is pointed at each corner (either actual or imaginary).*

    The purpose of installing baffles is to prevent the drilling fluid from swirling in a manner that creates a vortex, which pulls air into the drilling fluid. The baffles can be created in a variety of shapes and positions and still function properly.

15. *Mud guns should not be used in the removal section except where the feed mud to the mud gun(s) comes from the compartment being stirred by the mud gun(s).*

    See item 12.

16. *Mud guns can be used in the Addition and Check/Suction Sections of the surface system and provide the benefits of shear and dividing and re-blending newly added mud materials.*

    Mud guns do an excellent job of shearing new material enabling it to disperse, and are effective at blending new material into the surface system. Agitators may aid and assist mud guns in these capacities. The flow rate through a mud gun can be calculated from the following equation:

    Flow rate, gpm = $19.4(D)^2 \sqrt{H}$

    where D = Nozzle diameter, in.
    H = Head, ft

17. *The sand trap is the only settling compartment in the surface mud system. It should not be stirred nor should any pump take its suction from the sand trap.*

    Sand traps are becoming obsolete except for fast drilling surface holes where seawater is used as a drilling fluid and coarse mesh screens are used on the shale shakers.

18. *If a sand trap is used, the bottom should slope to its outlet at 45 degrees or steeper. The outlet valve should be large, non-plugging, and quick opening and closing.*

    The bottom slope is needed so that the settled solids can be easily removed from the compartment. The quick opening and closing valve is needed to reduce drilling fluid loss. The valve is normally shaped like a plate or rectangular piece of metal.

**TABLE B-1**

| Four 60° Canted Blade Impeller, 57.5 rpm | | Four Flat Blade Impeller, 57.5 rpm | |
|---|---|---|---|
| Impeller Diameter (inches) | Flow Rate (gpm) | Impeller Diameter (inches) | Flow Rate (gpm) |
| 20 | 910 | 20 | 1,100 |
| 24 | 1,600 | 24 | 1,900 |
| 28 | 2,500 | 28 | 2,800 |
| 32 | 3,800 | 32 | 4,400 |
| 36 | 5,400 | 36 | 6,300 |
| 40 | 7,300 | 40 | 8,400 |
| 44 | 9,900 | 44 | 11,300 |
| 48 | 12,500 | 48 | 14,400 |

19. *The degasser (if needed) should be installed immediately downstream of the shaker and upstream of any piece of equipment requiring feed from a centrifugal pump.*

    Degassers are used to remove entrained hydrocarbon gasses from drilling fluid. Another benefit of degassers is to prevent air or gas from entering centrifugal pumps, where even small quantities significantly reduce pump effectiveness. As liquid (or drilling fluid) is thrust to the outside of the impeller chamber, air or gas collects at the center. Eventually, enough air or gas will collect to completely block the suction and no liquid will be able to enter the pump.

20. *The solids removal equipment should be arranged sequentially so that each piece of equipment removes successively finer solids. Although every piece of equipment may not be used or needed, general arrangements are as follows:*

    | **Unweighted Mud** | **Weighted Mud** |
    |---|---|
    | Gumbo Remover | Gumbo Remover |
    | Shale Shaker | Shale Shaker |
    | Degasser | Degasser |
    | Desander | Mud Cleaner |
    | Desilter | Centrifuge |
    | Centrifuge | |
    | Dewatering Units | |

    Including the degasser in the unweighted mud list was not agreed upon by the entire committee. Influx of gas into the drilling fluid normally requires the addition of a weighting agent to the drilling fluid. Very few unweighted drilling fluids are degassed.

21. *The overflow for each piece of solids control equipment should discharge to the compartment downstream from the suction compartment for that piece of equipment. This is termed proper piping, plumbing, or fluid routing.*

    The compartments do not need to be large. Simple partitions in a larger tank can frequently be added to improper systems to improve their performance. Since a backflow is desired between compartments, the partitions do not require a complete seal.

22. *Improper fluid routing always leads to solids-laden fluid bypassing the removal device.*

    Unfortunately, improper routing is all too common on drilling rigs. In the 1980s, approximately 90% of the rigs had flaws in their removal system, and the situation has not improved much in the 1990s.

23. *Two different pieces of solids equipment should not simultaneously operate out of the same suction compartment. Note: Different means, for example, degasser and desander or desander and desilter.*

    If a desander and a desilter take suction from the same compartment and then discharge into the next compartment, some fluid will be desanded and some will be desilted. This is referred to as connecting the equipment in parallel. Desanders are usually used to decrease solids loading in the desilters. If these two pieces of equipment operate in parallel and a significant amount of solids are being processed, the desilters will probably plug.

    In some poorly designed systems, where an insufficient number of compartments are available, the degasser and the desander may be connected in parallel. This assumes that only the degasser, and not the desander, will be used on weighted drilling fluid. Obviously, if they are connected in this manner (parallel), both cannot operate simultaneously.

25. *If two of the same piece of solids control equipment are used simultaneously, the same suction and discharge compartment should be used for both. Example: If two desilter units are used, both should be properly rigged up and have the same suction and discharge compartments.*

    This situation is frequently encountered with desilter banks. In the upper part of a borehole, many hydrocyclones are needed to handle the volume (based on 50 gpm per 4-inch hydrocyclone, 1200 gpm will require 24 cones). While some of these cones can be mounted on mud cleaners (the screen is blanked to discard all underflow), all cones should use the same suction compartment and discharge downstream to the next compartment.

    Another problem can arise when a separate rig pump is used to increase annular velocity in risers. The additional flow rate onto the shale shakers may require adding additional units in parallel. The degasser capacity requirements may demand additional degassers be added to the mud tank system. These degassers should also be connected in parallel.

26. *The degassers, desanders, desilters, and mud cleaners should process 100% of the mud entering their individual suction compartments. In a properly designed system, the processing rate should be at least 10–25% more than the rig circulating rate. (See Rule 5 for Hydrocyclones.)*

The intent of this rule is to provide some guideline so that a backflow will exist. As long as there is more fluid processed than is entering the compartment, except for the backflow, all of the fluid entering the compartment will be processed.

27. *If Rules 21, 25, and 26 are applicable and are followed, equalizing flow between compartments will be in the reverse (or upstream) direction. Backflow confirms that all mud entering the compartment is being processed.*

    In normal drilling operations Rules 21, 25, and 26 are applicable.

28. *Mud should never be pumped from one removal compartment to another except through solids removal equipment.*

    The intent of this rule is to ensure that all drilling fluid is processed in an orderly fashion. Drilling fluid should not bypass any solids removal equipment or be pumped upstream from a suction compartment.

29. *Mud should never enter any removal compartment from outside the removal section to feed mud guns, mixers, or the eductor jet of a vacuum degasser.*

    When a system is designed, the equipment is set up to treat a certain quantity of drilling fluid. Increasing the flow rate demand by returning drilling fluid from downstream usually results in an inefficient removal system.

30. *The power mud for the eductor jet for a vacuum degasser should come from the degasser discharge compartment.*

    The drilling fluid to pull the fluid from the vacuum degasser usually comes from a centrifugal pump. This fluid must be degassed. The fluid in the degasser discharge compartment is pumped through the eductor and returns to the same compartment. This will not interfere with solids removal efficiency.

31. *Single purpose pumps are necessary in the Removal Section to ensure proper fluid routing. One suction and one discharge should be used. Suction and discharges should not be manifolded.*

    Manifolding ruins more good drilling fluid systems than just about any other single design. For this reason, dedicated pumps should be used. Since multiple leaky valves will confuse proper fluid routing, a standby pump be purchased instead of valves. When the desilter pump is on, the system is being desilted and there should be no question concerning routing. All systems, whether dispersed, nondispersed, polymer, water-based, oil-based, synthetic, or salt, will need the same sequential treatment by the removal equipment.

32. *In a properly designed system, solids control devices should not overflow into mud ditches.*

    Many drilling fluid systems have a square channel (approximately 2' × 2') along the top of one side of the tanks. Metal plate openings (ditches) are provided so that drilling fluid can bypass compartments. As drilling fluid drops from a ditch opening into the top of the fluid in the tank, air is entrained. These ditches can be useful for completion fluids and other speciality conditions, however, their use in drilling operations may result in drilling fluid bypassing solids removal equipment.

33. *Exception to rule 32: Based on field experience, mud foaming problems can be reduced by routing the overflow of a desander or desilter into a mud ditch for a horizontal distance of about ten feet before the fluid enters its discharge compartment to allow entrained air to break out. If this is done, ensure that the fluid routing is correct.*

    Flowing drilling fluid down a ditch allows the fluid/air interface to expose more of the entrained air to the surface. Sometimes, however, air does not break-out in this distance and more air is entrained as the fluid drops into the active system. Obviously, experts have many differing opinions concerning ditches.

34. *All mud material additions should be made after the Removal Section. All removal, including all centrifuging, must be finished before the mud material addition begins.*

    This rule is a result of problems which incurred when adding a weighting agent upstream from mud cleaners (or pumping fresh weighting agent upstream through mud guns). API barite allows 3% weight larger than 200 mesh, most of which will be removed by a mud cleaner. All undesirables must be removed before new products are added. One exception to this rule is the addition of flocculents, or other materials, to aid removal of drilled solids.

35. *Mud foaming problems can also be reduced by using a non-air-entraining mud mixing hopper.*

*Jet and venturi hoppers suck air into the mud during mixing.*

Hoppers should be turned off when they are not being used. At the discharge end of the additions line, an inexpensive air removal cylinder can be added without creating much backpressure. This involves a piece of $13\frac{3}{8}$-inch to 20-inch casing approximately $1\frac{1}{2}$-feet tall, vertically welded to the end of the hopper discharge line. A plate with an 8- to 10-inch diameter hole is welded on the top of the casing. Fluid enters tangentially and is swirled as it encounters the piece of casing. This swirling action causes drilling fluid to move to the outside wall, and the air moves to the inside. This acts as a centrifugal separator (see Figure B-1). Air exits through the hole at the top and the drilling fluid drops freely into the pits.

36. *Jet hoppers should include a venturi for better mixing.*

A venturi is needed if the flow line rises to an elevated position. The device converts a velocity head to a pressure head. Without it, fluid does not have enough pressure to rise over the tank wall.

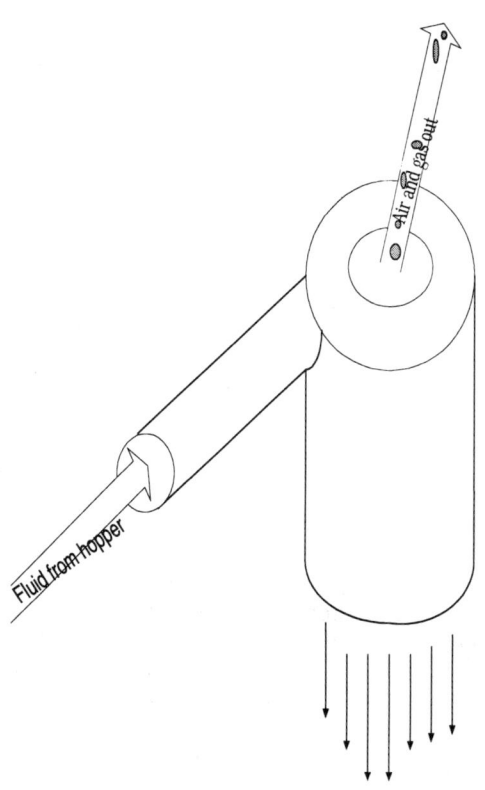

**FIGURE B-1**

37. *The Check/Suction Section of the surface system should contain a 20–50 barrel slugging tank which includes a mud gun system for stirring and mixing.*

An agitator may be used in addition to the mud gun. The mud gun system can be connected to the pump that is used to fill the slug tank. Usually the slug tank is used to prepare a drilling fluid with a higher density. This "slug" is pumped into the drill string. When tripping drill pipe, the fluid level inside the drill pipe will remain below the surface. This prevents spilling drilling fluid when a stand is removed from the drill string. Failure to slug the pipe, or get a good "slug," results in drilling fluid splashing the rig crew as the pipe is pulled and racked.

38. *Mud premix systems should be used on any mud system whose additives require time and shear for proper mixing. Premix systems should especially be used on systems requiring the addition of bentonite, or hard-to-mix polymers, such as CMC, PHPA, XC, etc.*

To be effective, bentonite must be prehydrated and dispersed into platelets as small as possible. It should be added to a well-agitated tank of fresh water. No other additives are required. The addition of lignosulfonate will inhibit dispersion as it thins the slurry. Bentonite should be allowed to hydrate for 24 hours (8 hours minimum). Polymers, such as HP007, require many hours of prehydration and shear before use.

39. *Special shear and mixing devices are recommended for premix systems for mixing polymers (especially PHPA), spotting fluids, specialized coring fluids, and for hydrating bentonite.*

Centrifugal pumps are available that have a modified impeller with holes or nozzles through which the fluid shears. These systems are very effective for shearing polymers.

40. *High-shear devices should not be used on the active system because they will rapidly reduce mud solids to colloidal size.*

Drilled solids are not processed in the same manner as bentonite. The purpose of dispersing bentonite is to take advantage of the very thin clay platelets and their electric charges. Drilled solids usually will not grind as thin as bentonite can disperse. Although they become colloidal, they are still 1,000 times larger than bentonite platelets. Increasing the colloidal content will increase the plastic viscosity, which needs to be as low as possible. Bentonite is the ideal clay because only a small amount is necessary to build yield point or control filtration. The capability of hydrated bentonite to disperse is much greater than drilled solids.

41. *The surface system should include a trip tank.*

    Trip tanks are needed to ensure the wellbore receives the correct amount of fluid as the drill string is pulled. For example, after 10 bbl volume of steel is removed during a trip, the liquid level in the wellbore should drop. This might result in an influx of formation fluid at the bottom of the hole. The trip tank continuously supplies fluid to the bell nipple to keep the well full of drilling fluid. If only 3 bbl of drilling fluid are needed to fill the hole after 10 bbl of steel are removed, the additional fluid must be entering the wellbore from the formations. *A blowout is imminent.* The pipe is run back to the bottom and the situation corrected.

## CENTRIFUGAL PUMPS

1. *Select a pump to handle the highest anticipated flow. Select an impeller size to provide sufficient discharge head to overcome friction in the lines, lift the fluid as required, and have sufficient head remaining to operate the equipment being fed.*

    Initially this guideline suggested that the pump flange size be selected to provide the highest anticipated flow, even though the flange size has nothing to do with the flow rate. Most pump curves are listed in terms of the flange sizes. The size of the pump impeller housing increases as the flange size increases. An impeller rotating at a constant speed will create a constant head independent of the size of the housing or the flanges. An impeller that fits inside a 2 × 3, 3 × 4, 4 × 5, or 5 × 6 pump will produce the same head in each pump if it is rotated at the same speed. Because the housing of a pump with a 2-inch and 3-inch flange is smaller, the internal friction at a high flow rate will be greater than a 5 × 6 pump. This means that the capacity of the various pump sizes will be indicated by their flange sizes. The committee decided to only indicate that the pump should be selected to handle the highest anticipated flow rate, instead of indicating that the flange size is commonly used to specify pump size.

2. *Install the centrifugal pump with a flooded suction which is sumped so that sufficient submergence is available to prevent vortexing or air-locking. Foot valves are not needed or recommended with flooded suctions.*

    A small influx of air into the suction of a centrifugal pump can create cavitation problems and diminished flow. As the air enters the chamber with the impeller, it tends to concentrate in the center of the impeller because of the centripetal acceleration of the drilling fluid. The liquid continues to move through the pump. The air does not always continue to the impeller tip, but tends to remain in the center of the impeller. This bubble of air forms a barrier for the incoming fluid, which diminishes the flow rate into the pump. The air also experiences a significant decrease in pressure—possibly even below atmospheric pressure. This causes implosions of vapor bubbles that can remove metal from the impeller. The pump will sound as if it is pumping gravel. If it continues in this mode for a long period of time, the impeller will be severely damaged.

    Flooded suctions tend to eliminate most of the air influx problems but sometimes a small vortex will form in the mud tank. These small vortexes can entrain a significant amount of air. Increasing agitation in the tanks may prevent a coherent cylinder of air from reaching the suction line. Alternately, a plate can be installed in the tank to interrupt the formation of a vortex.

    In some cases, a centrifugal pump is placed on the ground above a pond or buried tank. Foot valves are needed if the centrifugal pump is operated above the liquid level of the suction tank. Foot valves are check valves that prevent the suction line from draining when the pump is turned off. Care must be taken to eliminate any air leaks in the suction line because the absolute pressure will be below atmospheric pressure. The pump and suction line should be filled with fluid before the pump motor is started. Centrifugal pumps do not move air very well.

    A centrifugal pump suction can only lift fluid a certain height above a liquid level. These heights are determined by observing the NPSH (negative pressure suction head) values listed on the centrifugal pump curves. If the NPSH is exceeded, cavitation can destroy the impeller.

3. *Install a removable screen over the suction to keep out large solids and trash. It can be made out of half-inch expanded metal and should*

*have a total screen area at least five times the cross-sectional area of the suction line so it will not restrict flow. An extended handle arrangement reaching to the tank surface is desirable to allow the screen to be pulled during service and cleaned.*

An expanded metal screen prevents objects (such as gloves, buckets, pieces of clothing, chunks of rubber, etc.) from plugging the suction line or fouling the impeller. A bucket, turned so that the bottom fits into the suction line, can be difficult to diagnose and locate. A box made from expanded metal that covers the suction can prevent these disasters.

If two alignment yokes are welded to the tank walls to hold a one-inch pipe handle, the screen can be removed, cleaned, and easily returned to the suction opening. Without these alignment yokes, reseating the expanded metal box is difficult.

4. *Suction and discharge lines should be properly sized and as short as practical. Flow velocities should be in the range of 5 to 10 ft/sec. Less than 5 ft/sec causes solids to form a tight layer obstructing the bottom of horizontal lines. At velocities at or exceeding 10 ft/sec, pipe-turns tend to erode, headers do not distribute properly, and usually there will be cavitation in the suction lines. To calculate the velocity inside the pipe, use the following equation:*

$$\text{Velocity, ft/sec} = \frac{\text{Flow rate, gpm}}{3.48 \, (\text{Inside diameter, in.})^2}$$

*Suction lines should contain no elbows, swages, or reducers closer than three (3) pipe diameters to the pump suction flange.*

Horizontal pipes will fill with solids until the flowrate reaches 5 ft/sec. Barite in equalizing lines between mud tanks is normally settled until the velocity between the tanks reaches 5 ft/sec. Increasing the diameter of connection lines only causes more barite to settle. Above 10 ft/sec, pressure losses in the pipe become too great. Elbows and swages tend to cause turbulence in the flow stream, which can lead to cavitation.

5. *Eliminate manifolding. One suction and one discharge per pump is most cost effective over time. Do not manifold two pumps on the same suction line. Do not pump into the same discharge line with two or more pumps.*

"Flexibility" of piping so fluid can be pumped from any tank through any equipment to any other tank has created more problems over the years than just about any other concept. A properly plumbed system should only require one suction and one discharge for each piece of solids removal equipment. Ignoring this rule allows rig hands the opportunity to open or close the wrong valves. A leaky or incorrectly opened valve can reduce drilled-solids removal efficiency by up to 50%. This translates to an expensive drilling fluid system. This problem can be eliminated by storing an extra pump and motor. Arrange the centrifugal pumps and motors so that they may be easily replaced. If a pump or motor fails, simply replace the unit. The damaged unit can be replaced during routine maintainence.

Two centrifugal pumps in parallel will not double the head available to equipment because a centrifugal pump is a constant head device. For example, visualize a standpipe that is constantly filled with fluid. If two standpipes of approximately the same height are connected, the flow from both pipes will almost equal the flow from one standpipe. If fluid stands lower in one standpipe than the other, fluid will flow from the highest standpipe to the lowest standpipe. This same flow occurs when two pumps are connected in parallel—fluid will flow backwards through one of the pumps.

6. *Install a pressure gauge between the pump discharge and the first valve. When the valve is closed briefly, the pressure reading may be used for diagnostic evaluation of the pump performance.*

A centrifugal pump uses the smallest amount of power when no fluid is moving through the pump (that is, when the discharge valve is completely closed). If the valve remains closed for longer than five minutes, the fluid within the pump will become hot from the impeller agitation. This hot fluid may damage the seals. Closing the valve for a short time allows a good reading of the no-flow head produced by the pump. This reading should be compared with the pump manufacturer's charts. The diameter of the impeller can then be determined. After the pump has been in service for a period of time, the reading will assess the condition of the impeller. This eliminates the need to dismantle the pump for inspection. If the manifold pressure is incorrect, reading the pump no-flow discharge head will assist in troubleshooting.

7. *Keep air out of the pump by degassing the mud, having adequate suction line submergence, and*

*installing baffles to break mixer vortices. Properly sized, baffled, and agitated compartments will not vortex unless the drilling fluid level becomes extremely low.*

Centrifugal pumps can not pump aerated fluid. The air tends to gravitate toward the center of the impeller while the liquid moves toward the outside. This creates an air bubble at the center of the impeller. When the air bubble becomes as large as the suction line diameter, fluid will no longer enter the pump. This is called "air-locked."

Only a small cylinder of air vortexing into the pump is sufficient to prevent the pump from moving liquid. Since the air accumulates over a period of time, a small vortex the size of a pencil is sufficient to eventually shut down a 6 × 8 pump.

Baffles are inexpensive and easily installed in an empty tank. Any vertical surface that disrupts the swirling motion of the fluid in a compartment is usually sufficient to destroy a vortex. Rig pump efficiency can decrease from 99% to 85% efficiency if the drilling fluid content rises to 6% volume. Air in the drilling fluid may be calculated by measuring the pressurized and unpressurized mud weight.

8. *Do not restrict the flow to the suction side of the pump. Starving the pump suction causes cavitation and this will rapidly damage the pump.*

When a pump begins cavitating, small vacuum bubbles adjacent to the impeller surface start imploding. The pump sounds as if it is pumping gravel. The implosions quickly remove metal from the surface to the impeller blade. In a very short period of time, holes will appear in the metal.

Starving the suction will decrease the output head. If the head, or pressure, produced by the pump is too high, change to a smaller diameter impeller. On a temporary basis, a discharge valve can be partially closed. Even a large centrifugal pump is not damaged if only 10 to 20 gpm is discharged from the pump. In fact, the lower flow rates will require less horsepower to the motor than pumping fluid at a much higher flowrate.

9. *Make sure the impeller rotation is correct.*

Centrifugal pumps will pump fluid even if running in reverse. The head produced by the pump will be lower than it should be. The pressrue gauge installed between the pump and the first valve will assist with this diagnosis. Usually, switching two wires in the lead-in panel box will correct the rotation.

10. *Startup procedure for an electric motor-driven centrifugal pump with a valve on the discharge side between the pump and the equipment being operated, is to start the pump with the valve just slightly open. Once the pump is up to speed, open the valves slowly to full open. This approach will reduce the startup load on the electric motor and will reduce the shock loading on equipment such as pressure gauges and hydrocyclones. An alternate startup procedure is to completely close the valve before startup and then open the valve slowly immediately after startup to prevent overheating and possible damage to the pump seals.*

An electric motor-driven centrifugal pump will immediately try to produce a constant head when it is turned on. If the pump is pumping into an empty line, the flowrate is enormous. Very high flowrates requires very high currents to the electric motor. Circuit breakers can stop the pump and avoid motor burn-out. Lower horsepower is required if the pump is started with the discharge valve closed.

# APPENDIX C

# Pre-Well Checklist

This pre-well checklist contains questions that should be considered before drilling a well. The answers to these questions will assist in the proper selection, sizing, and operation of a solids management system.

## WELL PARAMETERS

Where will the well be drilled?
What is the objective (e.g., oil/gas, geothermal, reentry, etc.)?
What formations and geological features are expected?
What type of well (straight hole, directional, horizontal) will be drilled?
What problems are anticipated?

## DRILLING PROGRAM

What is the expected total depth?
Where are the casing points?
Are there other drilling parameters (hole size, bit type, ROP)?
What type of drilling fluid will be used?
What is the low-gravity-solids tolerance level?
What flowrate is planned?
What is the annular velocity in all sections of the borehole?
What is the hole cleaning capability?
What nozzle selection optimization will ensure immediate cuttings removal from the bottom of the hole?
What are other desired drilling fluid properties (mud weight, plastic viscosity, yield point, ES)?

## EQUIPMENT CAPABILITY

What type and size solids need removal?
What type of solids removal equipment is recommended?
Is this solids removal equipment available? From whom and where?
What are the weights and dimensions of the equipment?
What process rates can be provided?
What is the expected removal efficiency?
What drilling fluid losses are expected (downhole and surface)?
How much power/fuel is required?
What experience does the vendor have in this geographical region (number of units and references)?
What is the expected downtime?
What is the vendor's safety record?
Is a health and safety plan available?
What pit volumes are required?

## RIG DESIGN AND AVAILABILITY

What rig will be used?
What equipment is already installed? Is it installed correctly?
What repairs are needed?
How are removal compartments agitated?
Where are mud guns located?
Where are mud gun suctions located?
Where are additions made to the drilling fluid system?
What size centrifugal pumps are available?
How are addition, suction and check sections blended and agitated?
What is the residence time on the surface for the drilling fluid in each section of the borehole?
What modifications will be required?
Is space available for required modifications?
Is additional power available?

## LOGISTICS

Where is the base of operations?
Where is the stock/service facility?

How many additional people will be required?
Do they need accommodations/meals?
Is additional personal protective equipment required, needed, or available?

## ENVIRONMENTAL ISSUES

Can cuttings be buried or discharged without further treatment?
What treatment and disposal options are available?
What determines when cuttings are "clean"?
What analytical testing is required? Where is it located?
How much time is required to reach required standards for "clean"?
For a specific treatment and disposal option: Is it onsite or offsite? What additional equipment is required? Where will it be located? What are fuel and utility requirements?
Are there weather or site contraints?
What permits are required? Who is responsible for obtaining them?

## ECONOMICS

What is the drilling fluid cost per barrel?
Which is more expensive, commercial solids or the liquid phase?
What does required equipment cost to acquire?
What are installation and modification costs?
What are treatment and disposal costs?
What are expected savings?

# APPENDIX D

# Troubleshooting Guide

When shale shakers cease to operate as expected, a variety of items need to be checked and the problem eliminated. This section presents a general guideline for troubleshooting some common problems observed in shaker operations. Refer to the manufacturer's service manual for specific or special instructions for each design.

Always contact the manufacturer's service department when you have a situation where damage or injury could result, and especially where electrical shock hazards may exist.

| Problem: | Items to check: | Solution: |
| --- | --- | --- |
| Motor will not start/run | Power supply incompatible | Check power and wire motors |
| | Power supply disconnected | Reconnect power |
| | Power cable failure | Replace or repair cable |
| | Motor damage | Replace or repair motor |
| | Overloaded or overheated motor | Replace overload protection with proper type and rating |
| | Over-greasing motors | Remove drain plug before greasing; motor must be rebuilt if damaged |
| | Drive belt too tight | Adjust according to manufacturer's recommendations |
| Motor overheating | Moisture in motor | Open electrical covers and allow to dry; reseal covers properly |
| | Undersized overload heaters | Replace with properly sized heaters |
| | Single phasing | Check that all leads are well connected and voltages are consistent at each leg |
| | Unbalanced terminal voltage | Check for faulty leads, connections, and transformers |
| | Bearing failure | Replace motor |
| | Drive belt problems | Check manufacturer's recommendations and adjust or replace |
| Noisy operation | Worn bearings | Replace motor |
| | Shock mount/float mount problems | Check for obstructions, over/under inflation, or breakage; adjust or replace |
| | Loose or cracked motor mounts | Re-torque and/or replace motor |
| | Shipping brackets not removed | Remove the brackets |
| | Damaged or cracked shaker basket | Repair or replace after checking with manufacturer |
| | Object leaning against or bumping shaker basket | Remove object or reposition to prevent damage |

| Problem: | Items to check: | Solution: |
|---|---|---|
| Cuttings moving erratically on screen | Unit not level | Adjust installation to level position; check tank for level alignment |
| | Motor rotational problem | Check manufacturer's recommendations |
| | Shock mount damage | Replace mounts |
| | Drive belt too tight | Adjust to proper manufacturer's setting |
| | Motor failure | Replace motor |
| Mud overflowing screen | Screens plugged | Replace or clean screens |
| | Wrong screens | Adjust screen size up or down to determine proper selection |
| | Incorrect deck angle | Temporarily adjust screen angle up or down |
| | Water-wet screens in an oil-based drilling fluid | Change screens; add more oil-wetter to drilling fluid |
| Oversized cuttings returned to active system | Torn screens | Replace or repair torn or worn screens |
| | Worn tensioning device | Replace |
| | Leaking bypass valve | Repair or replace |
| | Worn screen supports | Replace or repair |
| | Improper screen tensioning | Adjust screen tension |
| | Drilling fluid not added to system through shale shaker screens | Check all procedures for adding or returning drilling fluid (such as from trip tanks, etc.) |
| No drilling fluid flowing to shale shaker | Valve closed | Open valve |
| | Solids blocking valve or flow line | Locate and remove or clear blockage |
| | Bypass open | Open bypass |
| | | Check bell nipple for overflow |
| Screen failure | Within an hour | Heavy solids loading |
| | | Vibrators operating improperly |
| | | Coarse formations on fine screen cloths |
| | Within a few hours | Loose screen cloth |
| | | Missing tension bolts or locks |
| | | Vibrators operating improperly |
| | | Damaged shaker bed |
| | | Missing channel rubbers |
| | | Improper tension |
| | Within a day | Damaged shaker bed |
| | | Inoperable vibrator |
| | | Extremely high mud weights |
| | | Misalignment of screens |
| | | Failure to maintain screen tension |
| | Within a few days | Hematite used as a weighting agent |
| | | Worn channel rubbers |
| | | Worn side or cross supports |
| | | Bent drawbars |

## Common Causes of Screen Failure:

1. Personnel improperly trained on handling, storage, maintenance, and installation of deck rubber and screen assembly.
2. Improper handling during installation—screens damaged before using.
3. Deck rubber improperly installed.
4. Improper tension during installation. Continuous screens should be retensioned within one-half hour of installation, every four hours during the first 24 hours of operation, and once per hour thereafter.
5. Careless storage allowing grit and sand between layers on multiple screen assembly.
6. Tension strip improperly seated in the screen hook strip.
7. Deck rubber is worn, damaged, or dirty.
8. Tension rails damaged allowing improper tensioning of the screen assembly or damaging of the hook strip.
19. Dried cuttings and drilling fluid left on screen during shutdown of shaker.
10. Cuttings accumulate under edge of screen.
11. Tensioned screen assemblies on shaker used as worktables.
12. Improperly manufactured screen assemblies.

# APPENDIX E

## *Shaker Manufacturers*

### Alfa Laval Oilfield DMVA Double Deck Shale Shaker

**Dimensions & Specifications**

| | |
|---|---|
| Length: | 86.4"/2195mm |
| Width: | 83.1"/2112mm |
| Height: | 69.5"/1765mm |
| Weir height: | 39.2"/995mm |
| Dry weight: | 6248lbs/2840kg |
| G Force: | 4.3 |
| Motion: | Dual (elliptical/linear) |
| Screen Angles: | 6 degree range |

**Screen Areas**

| | |
|---|---|
| Upper Deck: | 12.7ft²/1.18m² |
| Lower Deck: | 23.7ft²/2.2m² |
| Total: | 36.4ft²/3.38m² |

**Key Innovative Features**

- Dual Motion (Patent) - Change between linear and elliptical motion at the flick of a switch
- Aeroflex Screens (Patent) - Maintains optimum lower deck screen tension under all conditions

**Telephone Number:** North America +1 281 580 0715, International +44 1224 424300

## Brandt ATL-1000

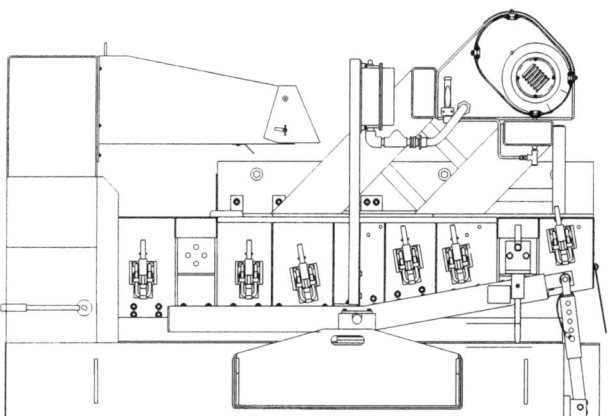

## Technical Specs

**Screens:**
Screen decks: 2
Scalping Screen Area: 10.8 ft² (1 m²)
Fine screen area: 25.0 ft² (2.3 m²)
Scalping Screen Type: 1 (27.5" x 60") Hookstrip style, overslung
Fine screen type: 3 (25" x 49.5") Blue Hex™ (repairable) HCR or Pinnacle™ (three dimensional) All use pre-tensioned rigid-frame panel.

**Screen change space:** Approximately 24 in. in front

**Screens offered:**
Upper deck: 10 mesh through 60 mesh square or B-20 through B-120 oblong
Bottom deck: 24 mesh through 325 mesh and finer upon request.

**Deck angle:** Adjustable while running, 0° to +10°

**Motion:** Linear 4.1 G's @ .095" stroke

**Drive system:** Direct drive, 1 X 5 HP, double shafted motor. Output shafts are direct coupled to gearboxes with counter-rotating eccentric weights attached. 230/460 VAC, 60 Hz Group D EXP (Other VAC and Hz available) Size 1 manual starter is standard.

**Physical Data:**
Overall length: 102" (2591 mm)
Overall width: 79" (2007 mm)
Skid width: 71" (1803 mm)
Overall height: 59" (1500 mm)
Weir height: 43" (1092 mm)
Weight: 4,500 lbs. (2045 kg)

## Brandt ATL-CS (Cascade)

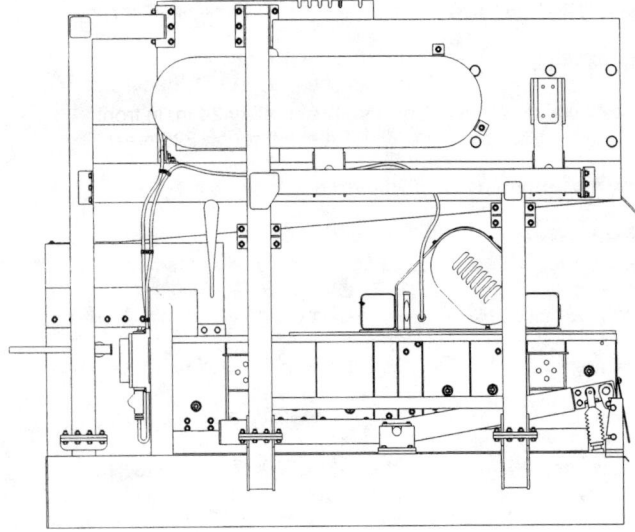

## Technical Specs

**Screens:**
Screen decks: 2
Scalping screen Area: 40 ft² (3.34 m²)
Fine screen area: 25.0 ft² (2.3 m²)
Scalping screen Type: 2 (48" x 60") Hookstrip style overslung top screen/underslung bottom screen.
Fine screen type: 3 (25" x 49.5") Blue Hex™ (repairable) HCR or Pinnacle™ (three dimensional) All use pre-tensioned rigid-frame panel.

**Screen change space:**
Upper deck: Approximately 60 in. in front
Bottom deck: Approximately 24 in. in front

**Screens offered:**
Upper deck: 10 mesh through 120 mesh square or B-20 through B-120 oblong
Bottom deck: 24 mesh through 325 mesh and finer upon request.

**Deck angle:**
Top deck: Fixed 0°
Bottom deck: Adjustable while running, 0° to +10°

**Motion:**
Upper deck: Circular 4.5 G's
Bottom deck: Linear 4.1 G's @ .095" stroke

**Drive system:**
Upper deck: Belt drive, 5 HP motor
Bottom deck: Direct drive, 1 X 5 HP, double shafted motor. Output shafts are direct coupled to gearboxes with counter-rotating eccentric weights attached. 230/460 VAC, 60 Hz Group D EXP (Other VAC and Hz available) Size 1 manual starter is standard.

**Physical Data:**
Overall length: 105" 2667 mm
Overall width: 82" 2083 mm
Overall height: 90" 2286 mm
Weir height: 79.25" 2013 mm

## Brandt ATL-CS (Low-Profile)

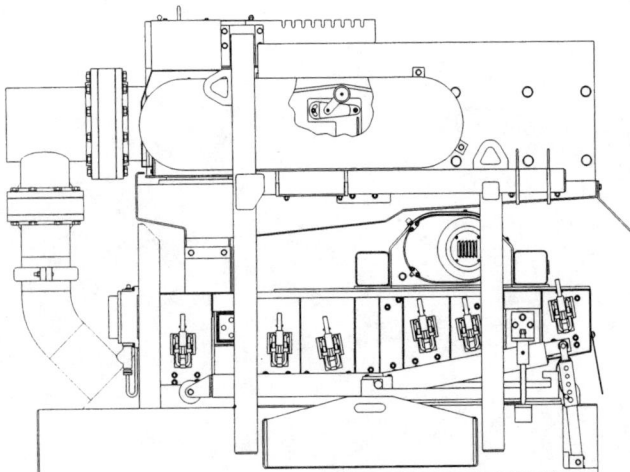

## Low Profile Technical Specs

*Specifications are same as ATL-CS (see above), except as noted.*

**Physical Data:**
Overall length: 100" (2540mm)
Overall width: 77" (1956mm)
Overall height: 74.5" (1892mm)
Weir height: 66.75" (1695mm) "
Weight: 7750 lbs. (3515kg)
Footprint: 100"x70.75" (2540x1797mm)

## Brandt ATL-Mud Conditioners

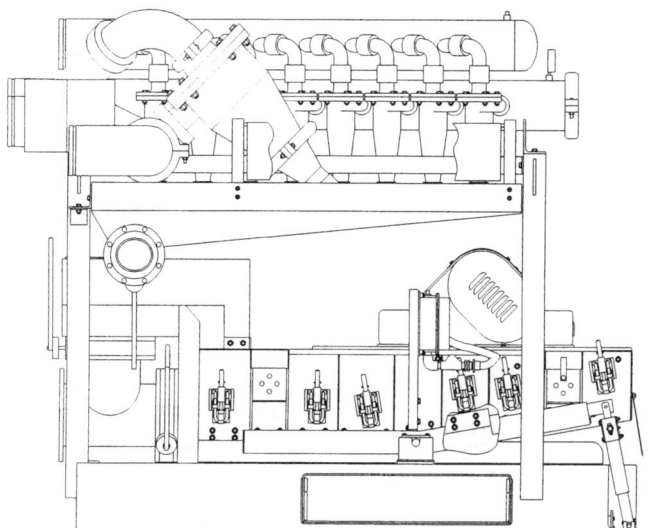

*ATL-16/2 shown*

## Technical Specs

**Screens:**
Screen decks: 1
Screen area: 25.0 ft² (2.3 m²)
Screen type: 3 (25" x 49.5") Blue Hex™ (repairable) HCR or Pinnacle™ (three dimensional) All use pre-tensioned rigid-frame panel.
Screen change space: Approximately 24 in. in front
Screens offered: 24 mesh through 325 mesh and finer upon request.
Deck angle: Adjustable while running, 0° to +10°
Motion: Linear 4.3 G's
Drive system: Direct drive, 1 X 5 HP, double shafted motor. Output shafts are direct coupled to gearboxes with counter-rotating eccentric weights attached. 230/460 VAC, 60 Hz Group D EXP (Other VAC and Hz available) Size 1 manual starter is standard.

**Cones:**
Desander: Brandt 10" or 12" 425 to 500 gpm ea. @ 75 ft. of head
Desilter: Brandt 4" 50 to 65 gpm ea. @ 75 ft. of head

**Physical data:**
Skid width: 71" (1803 mm)
Skid length: 99" (2514 mm)
(See table for specific models)

| Model | Desilter | Desander | Capacity | Length | Width | Height | Weight |
|---|---|---|---|---|---|---|---|
| ATL 1600 | 16 | 0 | 1000 gpm<br>3.785 m³/min | 115"<br>2921 mm | 85"<br>2159 mm | 93"<br>2362 mm | 6050 lb.<br>2750 Kg |
| ATL 2800 | 28 | 0 | 1750 gpm<br>6.624 m³/min | 102"<br>2591 mm | 109"<br>2769 mm | 93"<br>2362 mm | 6600 lb.<br>3000 Kg |
| ATL-16/2 | 16 | 2 | 1000 gpm<br>3.785 m³/min | 115"<br>2921 mm | 92"<br>2337 mm | 93"<br>2362 mm | 6820 lb.<br>3100 Kg |
| ATL-24/3 | 24 | 3 | 1500 GPM<br>5.677 m³/min | 125"<br>3175 mm | 96"<br>2438 mm | 111"<br>2819 mm | 8300 lb.<br>3773 Kg |

## Brandt LCM-2D

*With backtank*

*Without backtank*

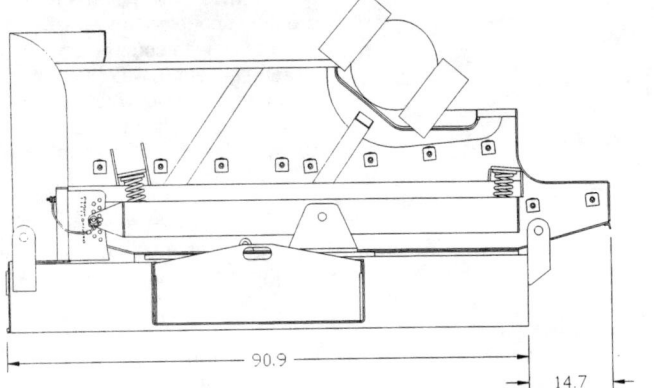

## Technical Specs

**Screens:**
Screen decks: 1
Screen area: 33.7 ft² (3.13 m²)
Screen angle: Contour deck 0 , +5 , +5 dewatering panel)
Screen type: 3 (45.5" x 36")Hook strip PT panel. Diamondback, DX, or HCR flat. Pinnacle™ three dimensional.
Screen mesh: 24 mesh through 325 mesh and finer upon request
Screen support: Overslung
Screen tensioning: Threaded tension bolt or quick-lock

**Deck angle:** Adjustable while running,-5 to +5

**Motion:**
Motion type: Linear
G-force: adjustable, 2.2 to 6.1 G's
Drive system: Direct drive, 2 x 2.5 HP, 230/460 VAC, 60 Hz Explosion Proof (Other VAC and Hz available)

**Physical Data:**
Overall length: 122" (3085 mm)
Skid length: 107.0" (2720 mm)
Overall width: 79.75" (1981 mm)
Skid width: 69.5" (1765 mm)
Overall height: 62.25" (1580 mm)
Weir height: 52.5" (1330 mm)
Weight: 5,200 lbs. (2360 Kg)

## Brandt LCM-2D CS (Cascade)

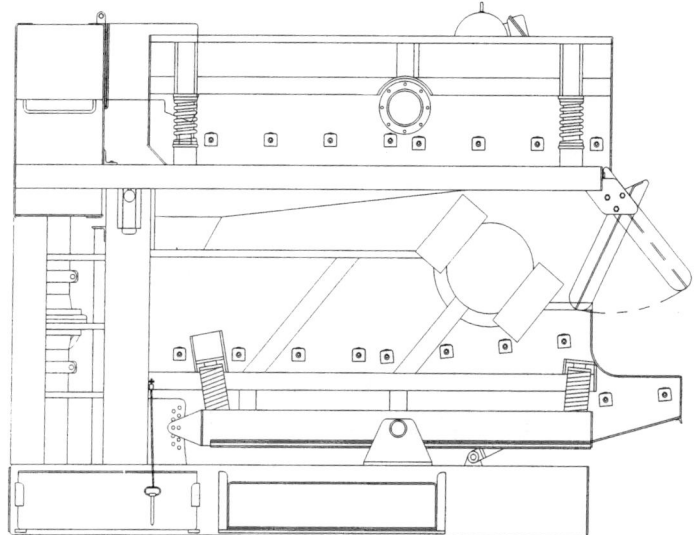

### Technical Specs

**Screens :**
Screen decks: 2
Scalping screen Area: 22.5 ft$^2$ (2.1 m$^2$)
Fine screen area: 33.7 ft$^2$ (3.13 m$^2$)
Scalping screen type: 2 (45.5" x 36") Hookstrip style flat.
Fine screen type: 3 (45.5" x 36") Hookstrip PT panel. Diamondback, DX, or HCR flat. Pinnacle™ three dimensional.
Screen support: Overslung

**Screen change space:**
Upper deck: Approximately 12 in. in front
Bottom deck: Approximately 12 in. in front

**Screens offered:**
Upper deck: 10 mesh through 120 square mesh or B-20 through B-120 oblong
Bottom deck: 24 mesh through 325 mesh and finer upon request.

**Deck angle:**
Top deck: Fixed 0$^0$
Bottom deck: Adjustable while running, -5$^0$ to + 5$^0$

**Motion :**
Upper deck: Circular 4.5 G's
Bottom deck: Linear 2.2 to 6.1 G's

**Drive system:**
Upper deck: Belt drive, 5 HP motor
Bottom deck: Direct drive, 2 x 2.5 HP, 230/460 VAC, 60 Hz Explosion Proof (Other VAC and Hz available)

**Physical Data:**
Overall length: 120" (3048 mm)
Skid length: 102.5 " (2604 mm)
Overall width: 78" (1981 mm)
Skid width: 69.5" (1765 mm)
Overall height: 90.75" (2146 mm)
Weir height: 73.75" (1873 mm)
Weight: 9,100 lbs (4095Kg)

## Brandt LCM-2D Mud Conditioners

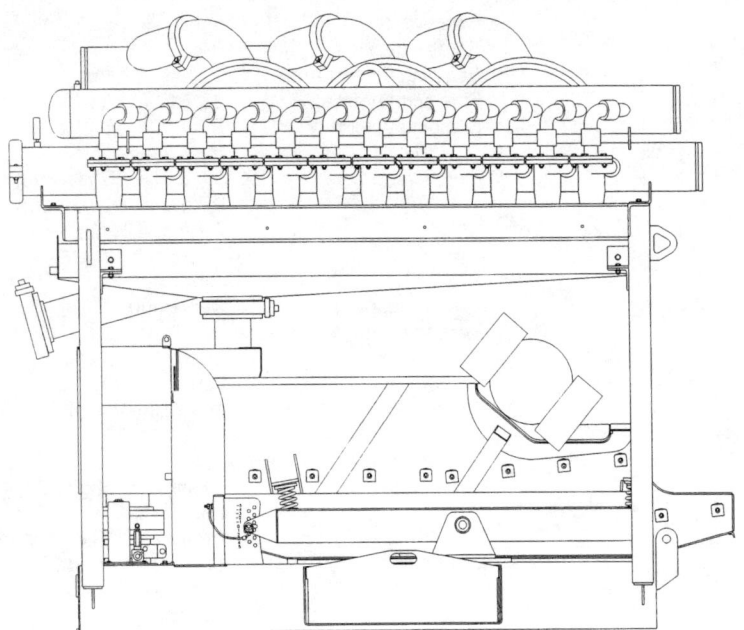

*LCM-2D-24/3 shown*

### Technical Specs

**Screens:**
Screen decks: 1
Screen area: 33.7 ft² (3.13 m²)
Screen angle: Contour deck 0, +5, +5 dewatering panel)

Screen type: 3 (45.5" x 36")Hook strip PT panel. Diamondback, DX, or HCR flat. Pinnacle™ three dimensional.
Screen mesh: 24 mesh through 325 mesh and finer upon request
Screen support: Overslung
Screen tensioning: Threaded tension bolt or quick-lock
**Screen change space:** Approximately 12 in. in front

**Deck angle:** Adjustable while running, -5 to +5

**Motion:**
Motion type: Linear
G-force: adjustable, 2.2 to 6.1 G's

**Drive system:** Direct drive, 2 x 2.5 HP, 230/460 VAC, 60 Hz Explosion Proof (Other VAC and Hz available)

**Cones:**
Desander: Brandt 10" or 12" 425 to 500 gpm ea. @ 75 ft. of head
Desilter: Brandt 4" 50 to 65 gpm ea. @ 75 ft. of head

**Data:**
Skid length: 102.5 " (2604 mm)
Skid width: 69.5" (1765 mm)
Weir height: 73.75" (1873 mm)

*See table for specific models*

| Model | Desilter | Desander | Capacity | Length | Width | Height | Weight |
|---|---|---|---|---|---|---|---|
| LCM-2D 1600 | 16 | 0 | 1000 gpm 3.785 m³/min | 123" 3124 mm | 84" 2134 mm | 106" 2692 mm | 8000 lb. 3600 Kg |
| LCM-2D 3200 | 32 | 0 | 2000 gpm 6.624 m³/min | 140" 3124 mm | 109" 2769 mm | 106" 2692 mm | 8500 lb. 3825 Kg |
| LCM-2D-16/2 | 16 | 2 | 1000 gpm 3.785 m³/min | 140" 3556 mm | 100" 2540 mm | 106" 2692 mm | 8900 lb. 4005 Kg |
| LCM-2D-24/3 | 24 | 3 | 1500 GPM 5.677 m³/min | 142" 3607 mm | 84" 2134 mm | 110" 2794 mm | 8900 lb. 4005Kg |

# Brandt Tandem

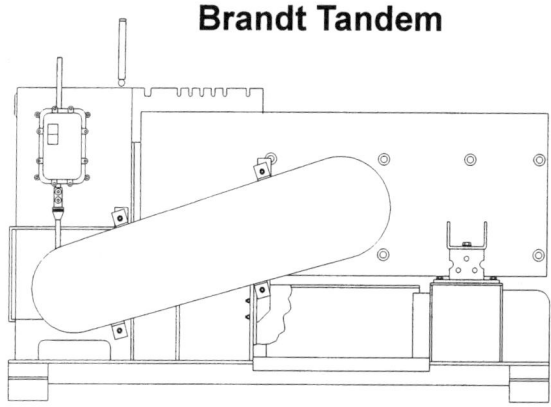

## Technical Specs

**Screens :**
| | |
|---|---|
| Screen decks: | 2 |
| Top Screen Area: | 20 ft$^2$ (1.5 m$^2$) |
| Bottom screen area: | 20 ft$^2$ (1.5 m$^2$) |
| Screen Type: | 2 (48" x 60") Hookstrip style non-pre-tensioned. |
| Screen support: | Underslung |
| Screen change space: | Approximately 60 in. in front |
| Screens offered: | 10 mesh through 120 mesh square or B-20 through B-120 oblong |
| Deck angle: | Fixed 0$^0$ |

**Motion ::**
| | |
|---|---|
| Motion type: | Circular |
| G-force; | 4.9 G's @1400 rpm |

**Drive system:** Belt drive, 5HP motor for each basket. Size 1 manual starter is standard. 230/460 VAC, 60 Hz Explosion Proof (Other VAC and Hz available)

### Physical Data:

**Single Tandem:**
| | |
|---|---|
| Overall length: | 82" (2083mm) |
| Overall width: | 72" (1823mm) |
| Overall height: | 56" (1422mm) |
| Weir height: | 38" (965mm) |
| Weight: | 2865 lbs (1289Kg) |

**Dual Tandem:**
| | |
|---|---|
| Overall length: | 155" (3937mm) |
| Overall width: | 85" (2159mm) |
| Overall height: | 67" (1702mm) |
| Weir height: | 38" (965mm) |
| Weight: | 5000 lbs (250Kg) |

**Triple Tandem:**
| | |
|---|---|
| Overall length: | 226" (5791mm) |
| Overall width: | 86" (2184mm) |
| Overall height: | 65" (1651mm) |
| Weir height: | 38" (965mm) |
| Weight: | 8300 lbs (735Kg) |

## Brandt JTL-500

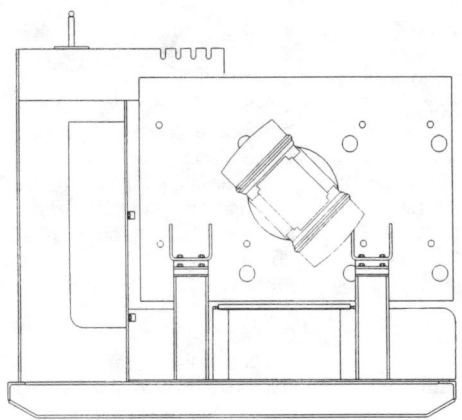

### Technical Specs

**Screens :**
| | |
|---|---|
| Screen decks: | 2 |
| Top Screen Area: | 4.5 ft² (m²) |
| Bottom screen area: | 4.5 ft² (m²) |
| Screen Type: | 2 (" x ") Hookstrip style non-pre-tensioned. |

**Screen support:**
**Screen change space:**
**Screens offered:**
| | |
|---|---|
| Deck angle: | Fixed 0° |

**Motion ::**
| | |
|---|---|
| Motion type: | Linear |
| G-force; | |

**Drive system:**

#### Physical Data:
**Single Tandem**
| | | |
|---|---|---|
| Overall length: | 52" | (1321mm) |
| Overall width: | 63.5" | (1613mm |
| Overall height: | 41" | (1041mm) |
| Weir height: | 31.5" | (800mm) |
| Weight: | 1500 lbs | (675Kg) |

## Brandt Gumbo Chain/Scalper

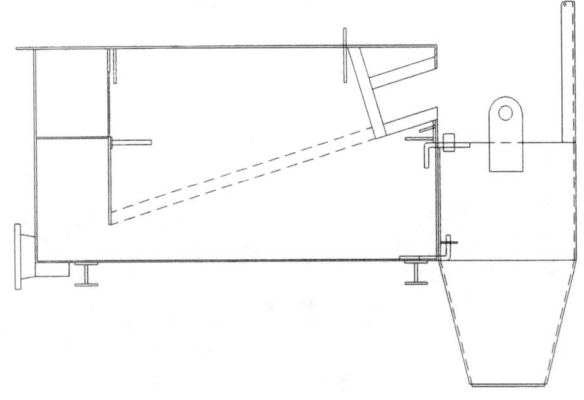

### Technical Specs

**Gumbo Box - Single cartridge - 5 foot**
| | | |
|---|---|---|
| Flow Rate: | 1500 gpm | (5377l/min) |
| Width: | 56.25" | (1429mm) |
| Length: | 98.25" | (2496mm) |
| Height: | 41.625" | (1054mm) |
| Weir Height: | 28" | (711mm) |
| Weight: w/cartridge | 2640 lbs. | (1128Kg) |

**Gumbo Box - Single cartridge - 6 foot**
| | | |
|---|---|---|
| Flow Rate: | 2200 gpm | (8327l/min) |
| Width: | 56.25" | (1429mm) |
| Length: | 110.25" | (2800mm) |
| Height: | 41.625" | (1054mm) |
| Weir Height: | 28" | (711mm) |
| Weight: w/cartridge | 2975 lbs. | (1339Kg) |

**Gumbo Box - Dual cartridge - 6 foot**
| | | |
|---|---|---|
| Flow Rate: | 3800 gpm | (14383l/min) |
| Width: | 100.75" | (2559mm) |
| Length: | 112.5" | (2854mm) |
| Height: | 63.875" | (1620mm) |
| Weir Height: | 50.25" | (1276mm) |
| Weight: w/cartridge | 6350 lbs. | (2858Kg) |

**Drive System - All cartridges – 5 & 6 foot**
Drive system Belt Drive – Kevlar poly-belt 1 HP, 230/460 VAC, 60 Hz Class 1, Group C, Explosion proof (Other VAC and Hz available)

## Cagle Ultra Screen

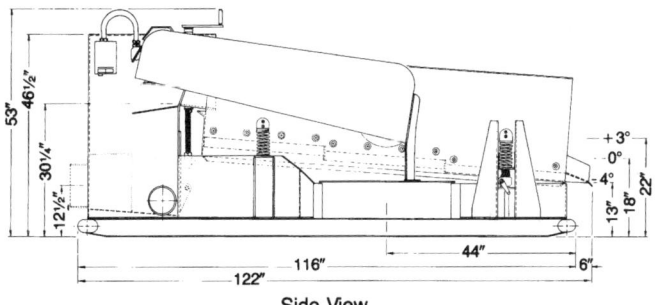

Side View

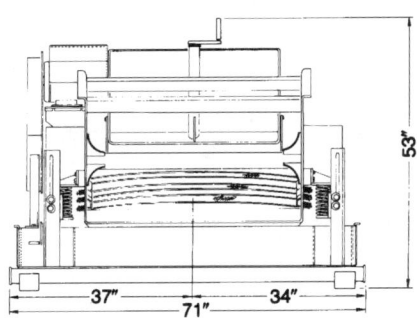

Front View

### Specifications

| Overall dimensions | | |
|---|---|---|
| Height | 53 inches | 135 cm |
| Width | 71 inches | 180 cm |
| Length | 122 inches | 310 cm |
| Mud weir height | 30¼ inches | 77 cm |
| Weight | 4000 lb | 1814 kg |
| Deck angle | +3° to −4° | |
| Vibration speed | 1190 rpm | |
| Motor starter | Manual | |
| Motor hp | 5 | |
| Motor voltage | 230/460 dual voltage | |
| Motor Hz/ph/rpm | 60/3/1750 | |
| Construction | | |
| Skid and deck | Carbon steel | |
| Fasteners | Stainless steel | |

Specifications subject to change without notice.

The Cagle Ultra-Screen, Linear Screen, and Linear Screen-AWD are available in Low Profile, Dual, with the AWD option, or any of these combinations.

CAGLE OILFIELD SERVICES, INC.
4908 S. 79th E. Avenue
Tulsa, OK 74145-6411
Phone: 918-492-7410
Fax: 918-492-6617
E-Mail: cosi@oknet.com

## Cagle Linear Screen

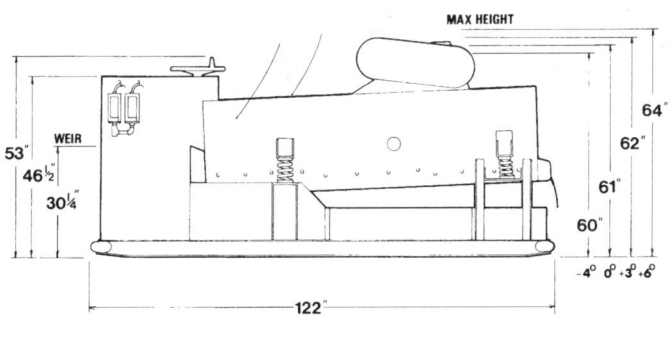

Side View

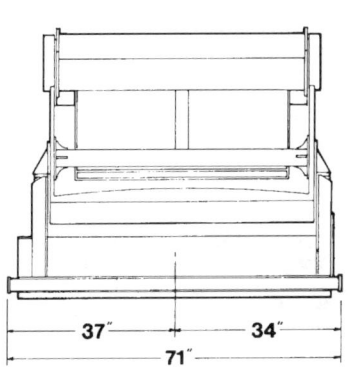

Front View

### Specifications

| Overall dimensions | | |
|---|---|---|
| Height (max) | 64 inches | 163 cm |
| Width | 71 inches | 180 cm |
| Length | 122 inches | 310 cm |
| Mud weir height | 30.25 inches | 77 cm |
| Weight | 4000 lb | 1814 kg |
| Deck angle | +6° to -4° | |
| Vibration speed | 1325 rpm | |
| Motor starters | Manual (2) req. | |
| Motor hp | 3 (2) req. | |
| Motor voltage | 230/460 dual voltage | |
| Motor Hz/ph/rpm | 60/3/1750 | |
| Construction | | |
| Skid and deck | Carbon steel | |
| Fasteners | Stainless steel | |

Specifications subject to change without notice.

The Cagle Ultra-Screen, Linear Screen, and Linear Screen-AWD are available in Low Profile, Dual, with the AWD option, or any of these combinations.

CAGLE OILFIELD SERVICES, INC.
4908 S. 79th E. Avenue
Tulsa, OK 74145-6411
Phone: 918-492-7410
Fax: 918-492-6617
E-Mail: cosi@oknet.com

## Cagle Linear Screen-AWD

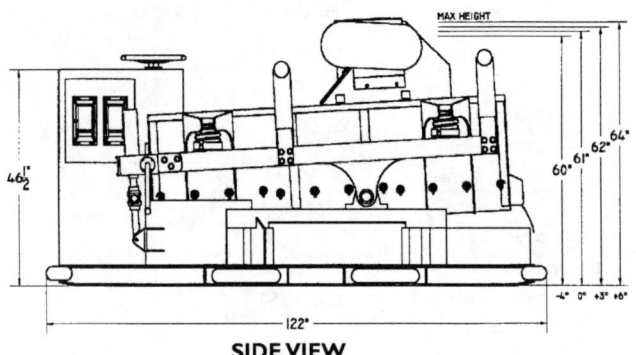

SIDE VIEW

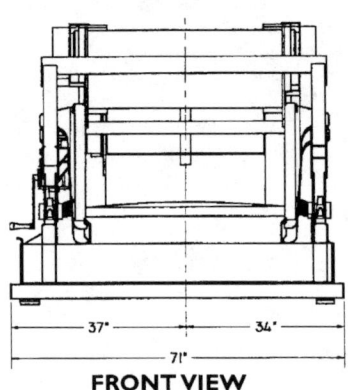

FRONT VIEW

**OVERALL DIMENSIONS**

| | | |
|---|---|---|
| Height (max) | 64" | 163 cm |
| Width | 71" | 180 cm |
| Length | 122" | 310 cm |

| | | |
|---|---|---|
| **Mud Weir Height** | 30.25" | 77 cm |
| **Weight** | 4000 lb | 1814 kg |
| **Deck Angle** | +6° to -4° | |
| **Vibration Speed** | 1325 rpm | |
| **Motor Starters** | Manual (2) req. | |
| **Motor HP** | 3 (2) req. | |
| **Motor Voltage** | 230/460 dual voltage | |
| **Motor Hz/ph/rpm** | 60/3/1750 | |

**Construction**
Skid and Deck    Carbon Steel
Fasteners    Stainless Steel

Specifications subject to change without notice.

The Cagle Ultra-Screen, Linear Screen, and Linear Screen-AWD are available in Low Profile, Dual, with the AWD option, or any of these combinations.

CAGLE OILFIELD SERVICES, INC.
4908 S. 79th E. Avenue
Tulsa, OK 74145-6411
Phone: 918-492-7410
Fax: 918-492-6617
E-Mail: cosi@oknet.com

---

 Flo-Line® Scalper

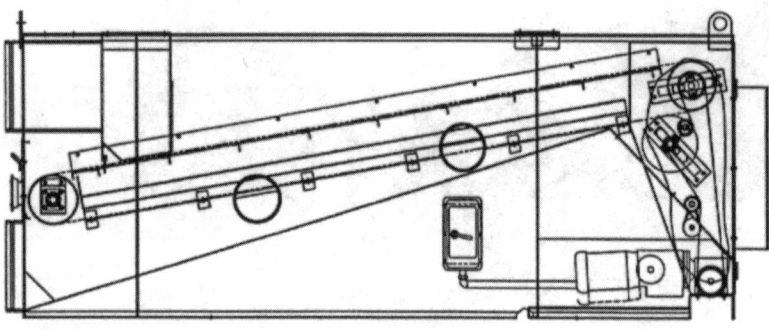

### Features
- Continuous mesh belt
- Variable speed motor
- Counter-rotating cleaning brush
- Stainless steel drive components
- Uphill conveyor bed
- Compact footprint
- Removes gumbo before cuttings reach shakers

**Dimensions**
136 3/8" x 39 1/8" x 50"
(364.4 x 99.4 x 127 cm)
Weir Height: 31 13/16" (99.4 cm)

| | |
|---|---|
| Weight: | 2000 lbs. (908 Kgs.) |
| Screen Bed Type: | Continuous mesh belt |
| # Screen Levels: | 1 |
| API sq.ft. Available for Screening: | N/A |
| # Panels: | N/A |
| Deck Angle: | Fixed at +10° |
| Vibrator: | N/A |
| Avg. G Force: | N/A |
| RPM: | Variable speed |
| Electrical: | 1.5 hp, 230v-60hz, 460v-60Hz, 575v-60Hz |

SHAKER MANUFACTURERS **251**

## Flo-Line® Cleaner 2000
### 2-Panel -- Top Feed

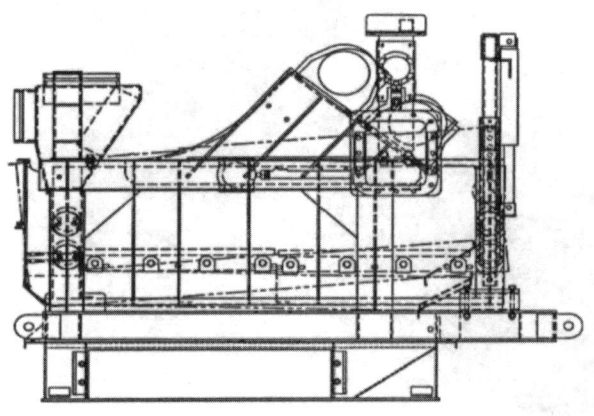

**Dimensions**
81 1/8" x 75 1/4" x 54"
(206.1 x 191.1 x 137.2 cm)
*Weir Height:* 43 1/4" (109.9 cm)

| | |
|---|---|
| Weight: | 3000 lbs. (1361 Kgs.) |
| Screen Bed Type: | Over slung |
| # Screen Levels: | 1 |
| API sq.ft. Available for Screening: | 10.6 (PWP) |
| | 16.6 (Pyramid) |
| | 23.8 (Pyramid+) |
| # Panels: | Two (2) |
| Deck Angle: | Variable, -1° to +5° |
| Vibrator: | Two Super G linear motion vibrating motors, "greased for life" |
| Avg. G Force: | 7.0 Gs |
| RPM: | 1750 |
| Electrical: | 1.5 hp, 230v-60hz, 460v-60Hz, 575v-60Hz |

### Features
- Deck angle is adjustable while running or stopped.
- Can use multiple screen types
- Super G vibrating motors are "greased for life"
- Remote starter panel
- No back tank settling
- Shorter footprint
- Compact size is ideal for production & workover

---

## Flo-Line® Cleaner 2000
### 2-Panel -- Weir Feed

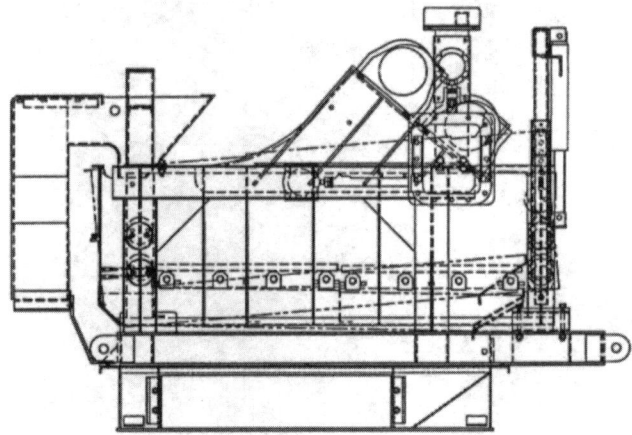

**Dimensions**
94 1/4" x 75 1/4" x 54"
(244.5 x 191.1 x 137.2)
*Weir Height:* 47 1/16" (119.5 cm)

| | |
|---|---|
| Weight: | 3200 lbs. (1452 Kgs.) |
| Screen Bed Type: | Over slung |
| # Screen Levels: | 1 |
| API sq.ft. Available for Screening: | 10.6 (PWP) |
| | 16.6 (Pyramid) |
| | 23.8 (Pyramid+) |
| # Panels: | Two (2) |
| Deck Angle: | Variable, -1° to +5° |
| Vibrator: | Two Super G linear motion vibrating motors, "greased for life" |
| Avg. G Force: | 7.0 Gs |
| RPM: | 1750 |
| Electrical: | 2.5 hp, 230v-60hz, 460v-60Hz, 575v-60Hz |

### Features
- Deck angle is adjustable while running or stopped.
- Can use multiple screen types
- Super G vibrating motors are "greased for life"
- Remote starter panel
- Lower wier height with by-pass

## Flo-Line® Cleaner 2000
### 3-Panel -- Feed Tank w/Bypass

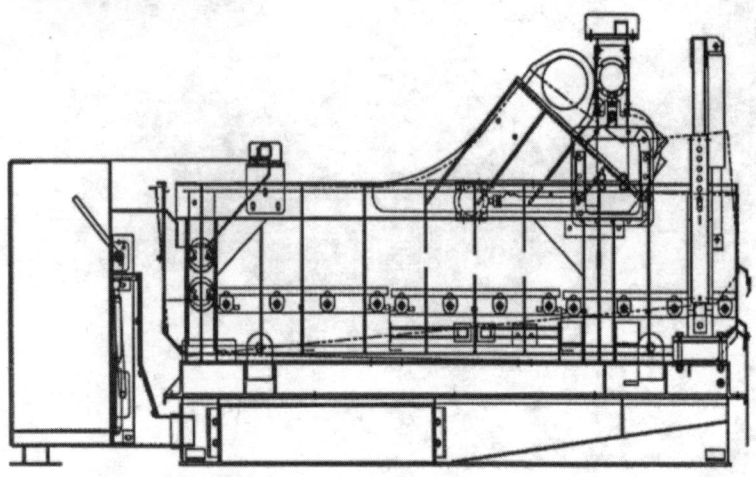

| | |
|---|---|
| **Dimensions** | |
| 118 3/8" x 72 5/16" x 63 3/4" | |
| (300.7 x 183.7 x 161.9) | |
| *Weir Height:* 36" (91.4 cm) | |
| Weight: | 4440 lbs. (2014 Kgs.) |
| Screen Bed Type: | Over slung |
| # Screen Levels: | 1 |
| API sq.ft. Available for Screening: | 15.9 (PWP) |
| | 24.9 (Pyramid) |
| | 35.7 (Pyramid+) |
| # Panels: | Three (3) |
| Deck Angle: | Variable, -1° to +5° |
| Vibrator: | Two Super G linear motion vibrating motors, "greased for life" |
| Avg. G Force: | 7.0 Gs |
| RPM: | 1750 |
| Electrical: | 2.5 hp, 230v-60hz, 460v-60Hz, 575v-60Hz |

**Features**
- Deck angle is adjustable while running or stopped.
- Can use multiple screen types
- Super G vibrating motors are "greased for life"
- Remote starter panel
- Feed tank bypass gate

---

## Flo-Line® Cleaner 2000
### 3-Panel -- Top Feed

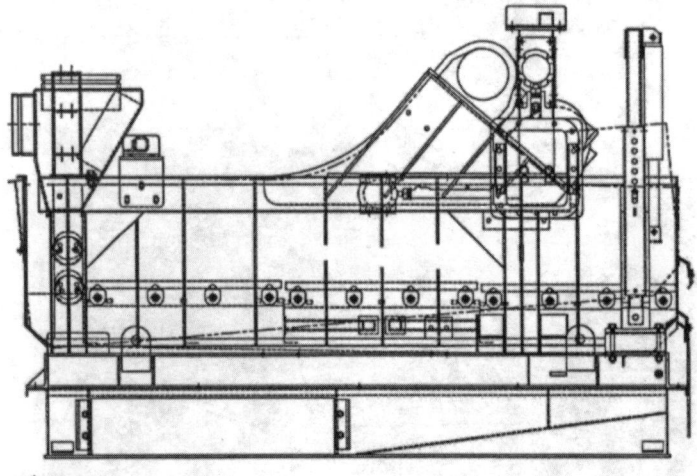

| | |
|---|---|
| **Dimensions** | |
| 93 1/4" x 72 5/16" x 63 3/4" | |
| (236.9 x 183.7 x 161.9 cm) | |
| *Weir Height:* 33 5/16" (84.6 cm) | |
| Weight: | 4005 lbs. (1817 Kgs.) |
| Screen Bed Type: | Over slung |
| # Screen Levels: | 1 |
| API sq.ft. Available for Screening: | 15.9 (PWP) |
| | 24.9 (Pyramid) |
| | 35.7 (Pyramid+) |
| # Panels: | Three (3) |
| Deck Angle: | Variable, -1° to +5° |
| Vibrator: | Two Super G linear motion vibrating motors, "greased for life" |
| Avg. G Force: | 7.0 Gs |
| RPM: | 1750 |
| Electrical: | 2.5 hp, 230v-60hz, 460v-60Hz, 575v-60Hz |

**Features**
- Deck angle is adjustable while running or stopped.
- Can use multiple screen types
- Super G vibrating motors are "greased for life"
- Remote starter panel
- No back tank settling
- Shorter footprint

SHAKER MANUFACTURERS **253**

 ## Flo-Line® Cleaner 2000
3-Panel -- Wier Feed

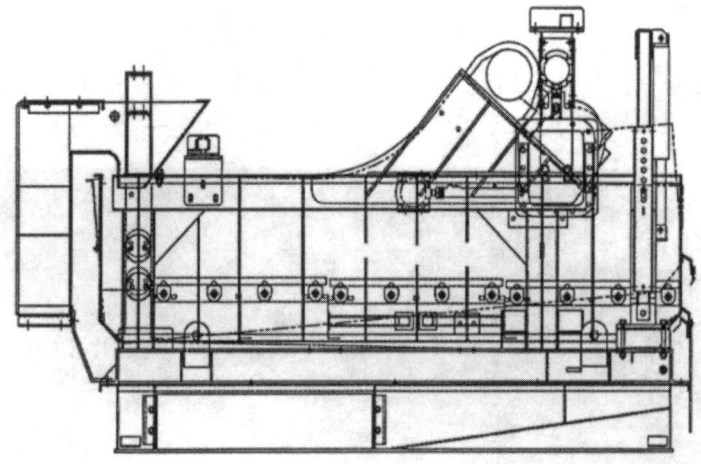

**Dimensions**
107 5/16" x 72 5/16" x 63 3/4"
(272.6 x 183.7 x 161.9 cm)
*Weir Height:* 50 7/16" (128.1 cm)

| | |
|---|---|
| Weight: | 4500 lbs. (2042 Kgs.) |
| Screen Bed Type: | Over slung |
| # Screen Levels: | 1 |
| API sq.ft. Available for Screening: | 15.9 (PWP) |
| | 24.9 (Pyramid) |
| | 35.7 (Pyramid+) |
| # Panels: | Three (3) |
| Deck Angle: | Variable, -1° to +5° |
| Vibrator: | Two Super G linear motion vibrating motors, "greased for life" |
| Avg. G Force: | 7.0 Gs |
| RPM: | 1750 |
| Electrical: | 2.5 hp, 230v-60hz, 460v-60Hz, 575v-60Hz |

### Features
- Deck angle is adjustable while running or stopped.
- Can use multiple screen types
- Super G vibrating motors are "greased for life"
- Remote starter panel
- Lower wier height

 ## Flo-Line® Cleaner 2000
4-Panel -- Feed Tank w/Bypass

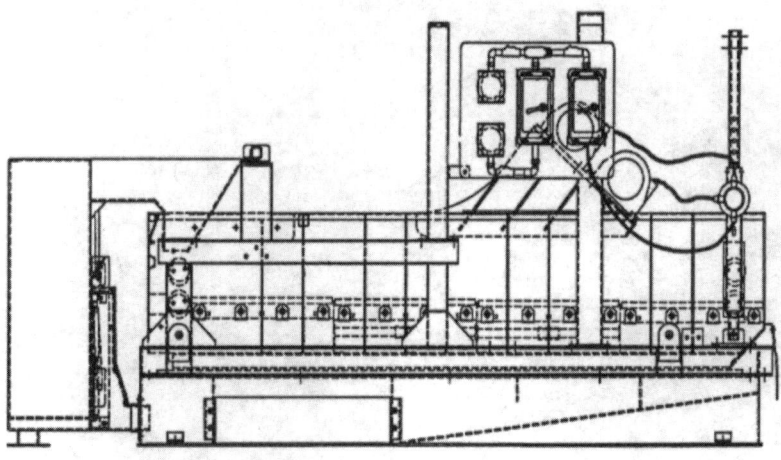

**Dimensions**
148 1/2" x 72 5/16" x 76"
(377.2 x 183.7 x 193 cm)
*Weir Height:* 42 3/16" (107.2 cm)

| | |
|---|---|
| Weight: | 5265 lbs. (2389 Kgs.) |
| Screen Bed Type: | Over slung |
| # Screen Levels: | 1 |
| API sq.ft. Available for Screening: | 21.2 (PWP) |
| | 33.2 (Pyramid) |
| | 47.6 (Pyramid+) |
| # Panels: | Four (4) |
| Deck Angle: | Variable, 0° to +10° |
| Vibrator: | Two Super G linear motion vibrating motors, "greased for life" |
| Avg. G Force: | 7.3 Gs |
| RPM: | 1750 |
| Electrical: | 2.5 hp, 230v-60hz, 460v-60Hz, 575v-60Hz |

### Features
- Deck angle is adjustable while running or stopped.
- Can use multiple screen types
- Super G vibrating motors are "greased for life"
- Remote starter panel
- Feed tank by-pass gate

 ## Flo-Line® Cleaner 2000
### 4-Panel -- Top Feed

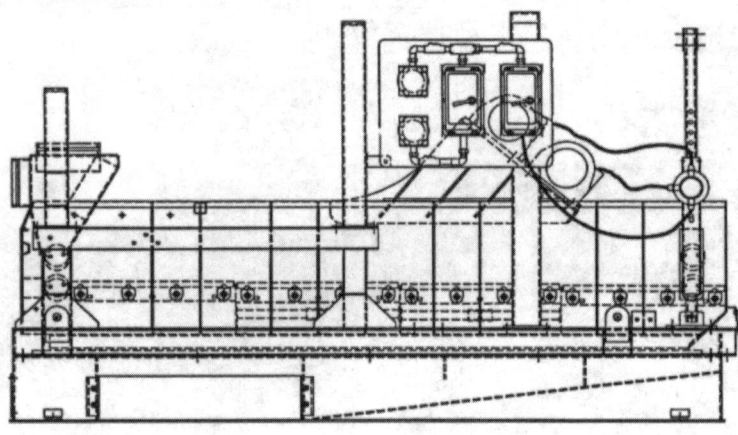

| Dimensions |
|---|
| 123 1/2" x 72 5/16" x 63 3/4" |
| (300.7 x 183.7 x 161.9 cm) |
| *Weir Height:* 44 3/4" (113.7 cm) |

| | |
|---|---|
| Weight: | 4825 lbs. (2189 Kgs.) |
| Screen Bed Type: | Over slung |
| # Screen Levels: | 1 |
| API sq.ft. Available for Screening: | 21.2 (PWP) |
| | 33.2 (Pyramid) |
| | 47.6 (Pyramid+) |
| # Panels: | Four (4) |
| Deck Angle: | Variable, 0° to +10° |
| Vibrator: | Two Super G linear motion vibrating motors, "greased for life" |
| Avg. G Force: | 7.3 Gs |
| RPM: | 1750 |
| Electrical: | 2.5 hp, 230v-60hz, 460v-60Hz, 575v-60Hz |

### Features
- Deck angle is adjustable while running or stopped.
- Can use multiple screen types
- Super G vibrating motors are "greased for life"
- Remote starter panel
- No back tank settling
- Shorter footprint

 ## Cascade System

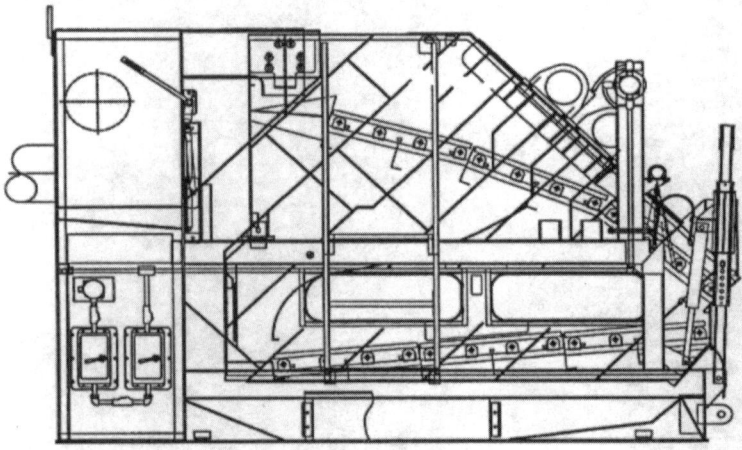

| Dimensions |
|---|
| 129 1/8" x 64 1/2" x 86" |
| (328 x 163.8 x 218.4 cm) |
| *Weir Height:* 71 3/16" (180.8 cm) |

| | |
|---|---|
| Weight: | 8200 lbs. (3720 Kgs.) |
| Screen Bed Type: | Over slung |
| # Screen Levels: | 2 |
| API sq.ft. Available for Screening: | 31.8 each deck (PWP) |
| | 49.8 (Pyramid) |
| | 71.4 (Pyramid+) |
| # Panels: | Three (3) on top |
| | Three (3) on bottom |
| Deck Angle: | Variable: Top; -15°, -22.5°, -40°, w/bot. @ +5°, Bottom; 0° to +5° |
| Vibrator: | Two Super G linear motion vibrating motors, "greased for life" |
| Avg. G Force: | 7.0 Gs |
| RPM: | 1750 |
| Electrical: | 1.5 hp, 230v-60hz, 460v-60Hz, 575v-60Hz |

### Features
- Deck angle is adjustable while running or stopped.
- Can use multiple screen types
- Super G vibrating motors are "greased for life"
- Remote starter panel
- Feed tank bypass gate

SHAKER MANUFACTURERS **255**

# Flo-Line® Cleaner 2000
### 3-Panel --with AWD

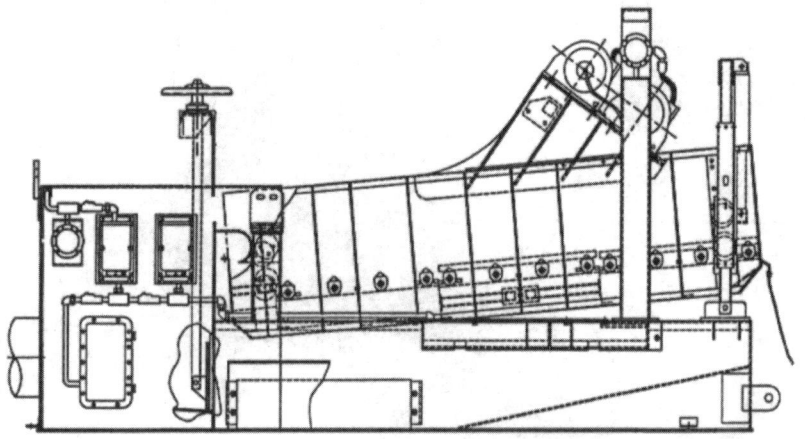

**Dimensions**
128 1/2" x 63" x 57"
(326.4 x 160 x 144.8 cm)
*Weir Height:* 34" (86.4 cm)

| | |
|---|---|
| Weight: | 4950 lbs. (2246 Kgs.) |
| Screen Bed Type: | Over slung |
| # Screen Levels: | 1 |
| API sq.ft. Available for Screening: | 15.9 (PWP) |
| | 24.9 (Pyramid) |
| | 35.7 (Pyramid+) |
| # Panels: | Three (3) |
| Deck Angle: | Variable, -1° to +5° |
| Vibrator: | Two Super G linear motion vibrating motors, "greased for life" |
| Avg. G Force: | 7.0 Gs |
| RPM: | 1750 |
| Electrical: | 2.5 hp, 230v-60hz, 460v-60Hz, 575v-60Hz |

**Features**
- Deck angle is adjustable while running or stopped.
- Can use multiple screen types
- Super G vibrating motors are "greased for life"
- Feed tank bypass gate

---

# Flo-Line® Cleaner +
### Model 48

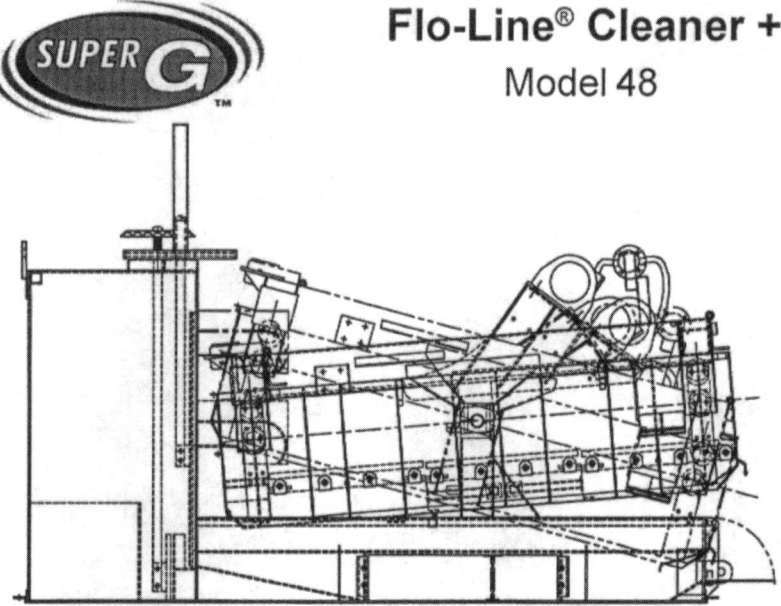

**Dimensions**
129" x 73 3/4" x 83 3/4"
(327.7 x 187.3 x 213.4 cm)
*Weir Height:* 27 1/4" to 43 9/16"
(69.2 to 110.7 cm)

| | |
|---|---|
| Weight: | 6000 lbs. (2722 Kgs.) |
| Screen Bed Type: | Over slung |
| # Screen Levels: | 1 |
| API sq.ft. Available for Screening: | 15.9 (PWP) |
| | 24.9 (Pyramid) |
| | 35.7 (Pyramid+) |
| # Panels: | Three (3) |
| Deck Angle: | Variable, -15° to +5° |
| Vibrator: | Two Super G linear motion vibrating motors, "greased for life" |
| Avg. G Force: | 7.0 Gs |
| RPM: | 1750 |
| Electrical: | 2.5 hp, 230v-60hz, 460v-60Hz, 575v-60Hz |

**Features**
- Deck angle is adjustable while running or stopped.
- Can use multiple screen types
- Super G vibrating motors are "greased for life"
- Remote starter panel
- Widest range of deck angles
- Improves sticky solids conveyance

## DC the double life corp., inc.

## Double Life 4 x 5 Single Tandem Shaker

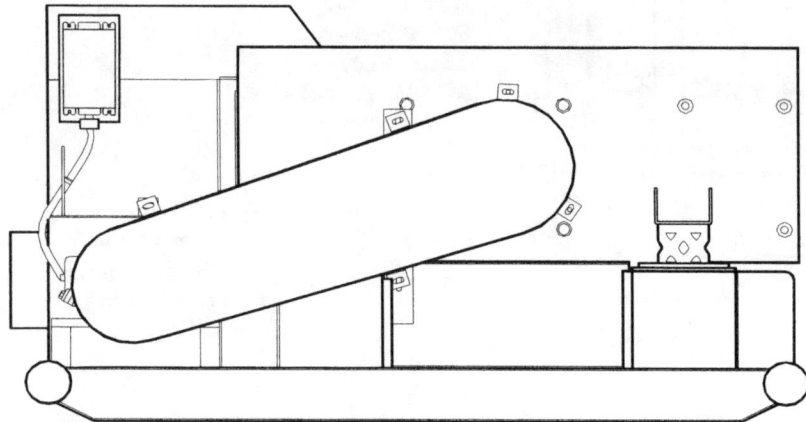

Weir Height: 38-9/16"
L/W/Height: 85-5/8"/72"/46-7/16"
Weight: 2025 lbs.
Screen Area: 40.0 Sq. Ft.
Vibrator Type: Elliptical, Belt Driven Counterweight
Avg. G-Force: 4.5
RPM: 1450
Screens: 4 x 5, Qty. (2) ea.
Deck Angle: 0°
Features: Available in single, double, & triple units

---

## DC the double life corp., inc.

## Double Life 4 x 5 Single Standard Shaker

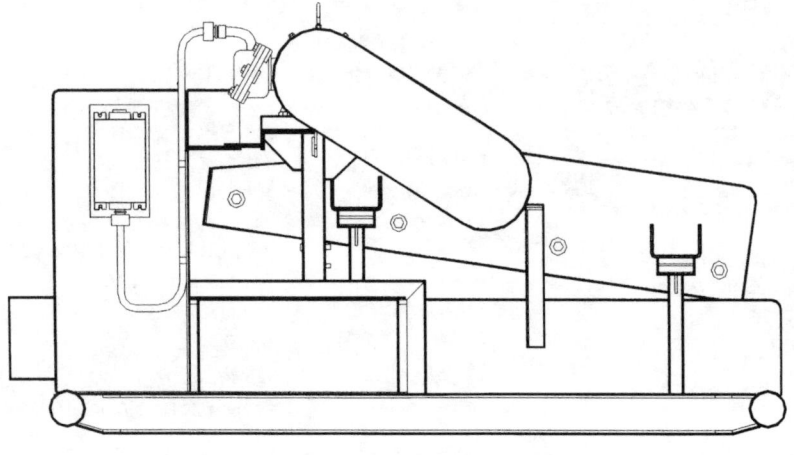

Weir Height: 30.41"
L/W/Height: 85-13/16"/45-1/2"/85-13/16"
Weight: 1930 lbs.
Screen Area: 20.0 Sq. Ft.
Vibrator Type: Elliptical, Belt-Driven Counterweight
Avg. G-Force: 4.0
RPM: 1450
Screens: 4 x 5, Qty. (1) ea.
Deck Angle: -8°
Features: Dual unit available

## DLC the double life corp., inc.

# Double Life 4 x 5 Single Tandem Shaker

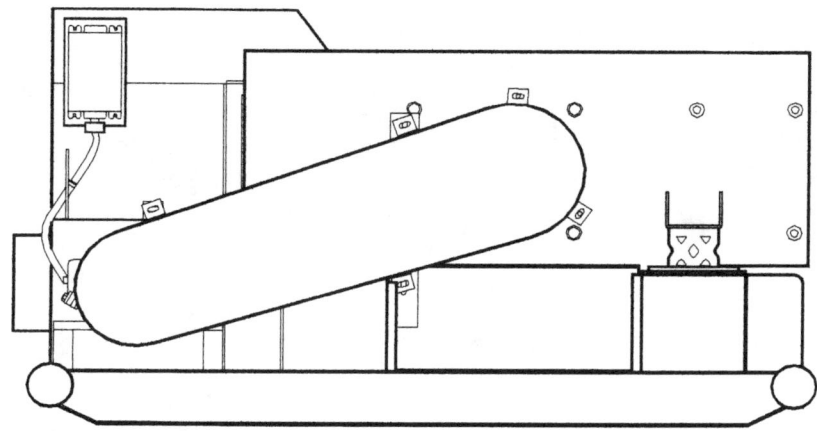

Weir Height: 38-9/16"
L/W/Height: 85-5/8"/72"/46-7/16"
Weight: 2025 lbs.
Screen Area: 40.0 Sq. Ft.
Vibrator Type: Elliptical, Belt Driven Counterweight
Avg. G-Force: 4.5
RPM: 1450
Screens: 4 x 5, Qty. (2) ea.
Deck Angle: 0°
Features: Available in single, double, & triple units

---

## DLC the double life corp., inc.

# Double Life 4 x 5 Single Standard Shaker

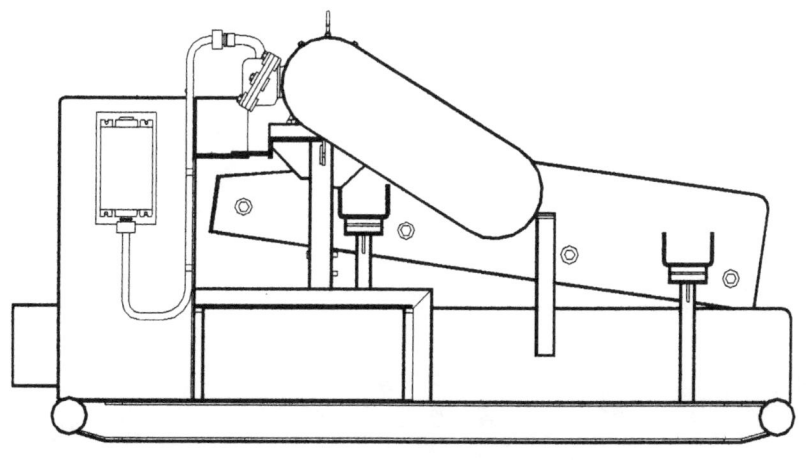

Weir Height: 30.41"
L/W/Height: 85-13/16"/45-1/2"/85-13/16"
Weight: 1930 lbs.
Screen Area: 20.0 Sq. Ft.
Vibrator Type: Elliptical, Belt-Driven Counterweight
Avg. G-Force: 4.0
RPM: 1450
Screens: 4 x 5, Qty. (1) ea.
Deck Angle: -8°
Features: Dual unit available

# the double life corp., inc.

## Double Life – AWD Linear Shaker

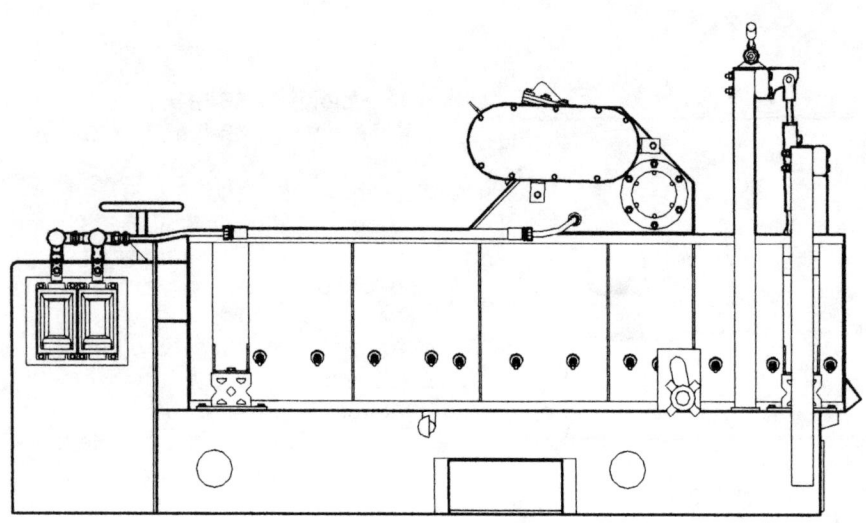

| | |
|---|---|
| Weir Height: | 26" |
| L/W/Height: | 118-5/8"/67"/60-1/2" |
| Weight: | 4000 lbs. |
| Screen Area: | 27.41 Sq. Ft. |
| Vibrator Type: | Linear, Belt Driven |
| Avg. G-Force: | 4.5 to 6.0 (Variable) |
| RPM: | 1750 RPM |
| Screens: | 41-3/8" OCW x27-1/2" (3) ea. |
| Deck Angle: | -1° to +6° (Adjustable While Drilling) |
| Features: | Available, unitized vibrators with adjustable G-Force. Available in Mud Cleaner and Mud Conditioner configurations. |

# Fluid Systems, Inc.

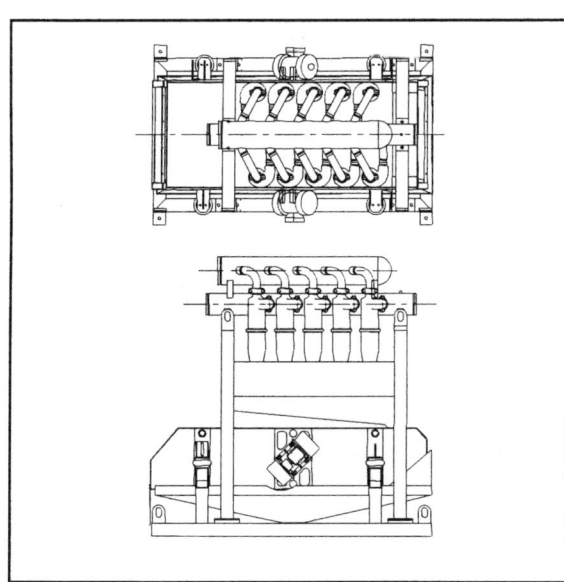

### General Specifications
### 20L VGS™ Power Series Mud Cleaner (Desilting)

| | |
|---|---|
| *Dimensions, in (cm)* | 89 ½ (227.3) L x 53 (134.6) W x 85 ¾ (217.8) H |
| *Weight, lb. (kg)* | 2100 (954) |
| *Average G's* | 4.5 – 10.0 |
| *Electrical* | 3 Phase, 230/460 V, 60 Hz |
| *Motors (HP)* | 2 per unit, 0.56 Ea., 1.12 Total 1200 or 1800 RPM |
| *Screen Type* | Pretensioned Panels, *Wedgelok*™ Fasteners |
| *Screen Panel Size, $ft^2$ ($m^2$)* | 8.45 (.785) |
| *No. Screens Panels* | 2 per unit |
| *Deck Angle* | 0° to 5° |
| *Features* | Quick Screen Change, No Lubrication Required, Modular Construction |

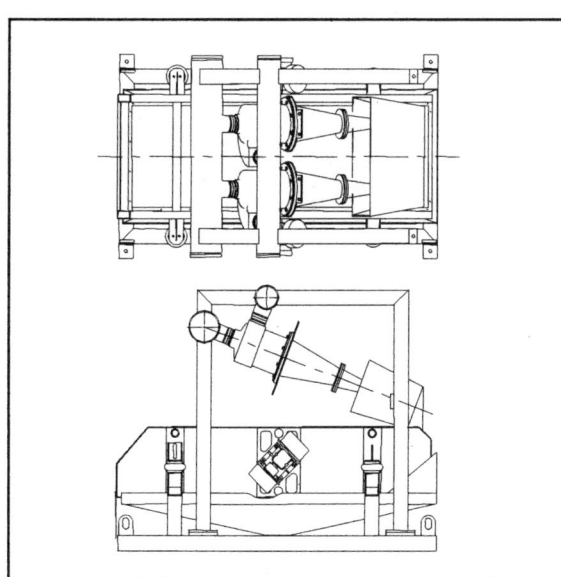

### General Specifications
### 20L VGS™ Power Series Mud Cleaner (Desanding)

| | |
|---|---|
| *Dimensions, in (cm)* | 89 ½ (227.3) L x 53 (134.6) W x 71 ¼ (181.0) H |
| *Weight, lb. (kg)* | 1900 (864) |
| *Average G's* | 4.0 – 10.0 |
| *Electrical* | 3 Phase, 230/460 V, 60 Hz |
| *Motors (HP)* | 2 per unit, 0.56 Ea., 1.12 Total 1200 or 1800 RPM |
| *Screen Type* | Pretensioned Panels, *Wedgelok*™ Fasteners |
| *Screen Panel Size, $ft^2$ ($m^2$)* | 8.45 (.785) |
| *No. Screens Panels* | 2 per unit |
| *Deck Angle* | 0° to 5° |
| *Features* | Quick Screen Change, No Lubrication Required, Modular Construction |

# Fluid Systems, Inc.

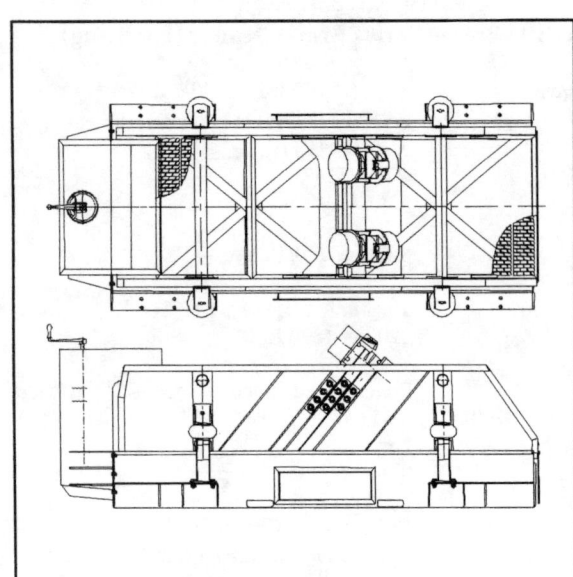

### General Specifications
### 5000 B3X VGS™ Power Series

| | |
|---|---|
| *Weir Height, in (cm)* | 39 (99) |
| *Dimensions, in (cm)* | 146 ¾ (372.7) L x 61 (154.9) W x 47 ¼ (120.0) H |
| *Weight, lb. (kg)* | 3500 (1591) |
| *Average G's* | 4.5 – 10.0 |
| *Electrical* | 3 Phase, 230/460 V, 60 Hz |
| *Motors (HP)* | 2 per unit, 1.29 Ea., 2.58 Total 1200 or 1800 RPM |
| *Screen Type* | Pretensioned Panels, *Wedgelok*™ Fasteners |
| *Screen Panel Size, $ft^2$ ($m^2$)* | 8.45 (.785) |
| *No. Screens Panels* | 4 per unit |
| *Deck Angle* | 0° to 5° |
| *Features* | Quick Screen Change, No Lubrication Required, Modular Construction |

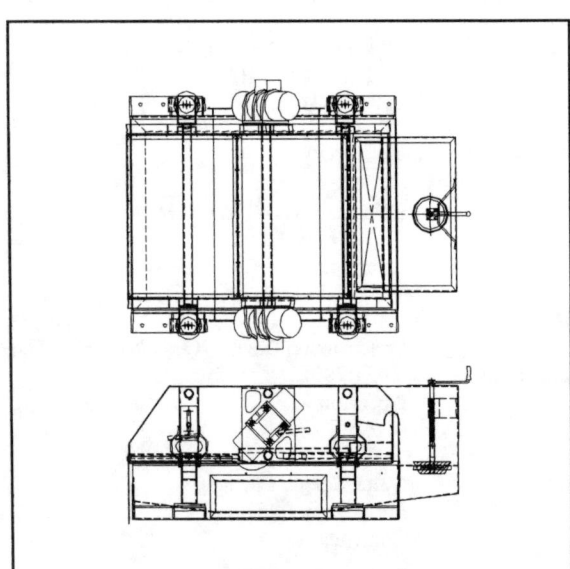

### General Specifications
### 50 B3X VGS™ Power Series

| | |
|---|---|
| *Weir Height, in (cm)* | 28 (64) |
| *Dimensions, in (cm)* | 88 (223.5) L x 69 ¼ (176.0) W x 39 ½ (100.3) H |
| *Weight, lb. (kg)* | 2450 (1114) |
| *Average G's* | 4.5 – 10.0 |
| *Electrical* | 3 Phase, 230/460 V, 60 Hz |
| *Motors (HP)* | 2 per unit, 0.56 Ea., 1.12 Total 1200 or 1800 RPM |
| *Screen Type* | Pretensioned Panels, *Wedgelok*™ Fasteners |
| *Screen Panel Size, $ft^2$ ($m^2$)* | 8.45 (.785) |
| *No. Screens Panels* | 2 per unit |
| *Deck Angle* | 0° to 5° |
| *Features* | Quick Screen Change, No Lubrication Required, Modular Construction |

# Fluid Systems, Inc.

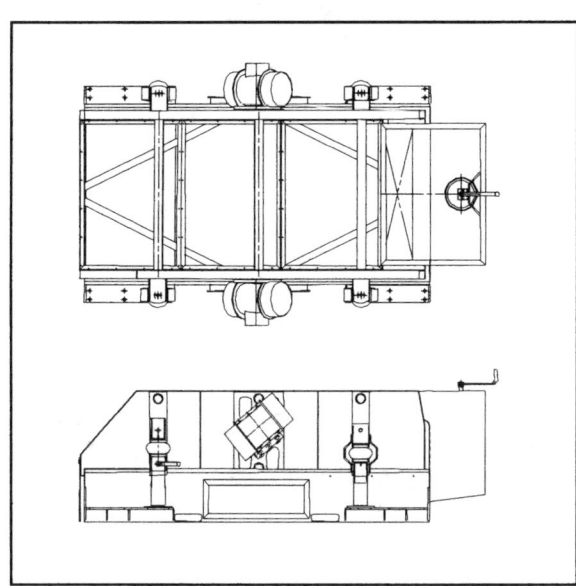

### General Specifications
### 500B3X VGS™ Power Series

| | |
|---|---|
| *Weir Height, in (cm)* | 29 (73.7) |
| *Dimensions, in (cm)* | 117 ½ (298.5) L x 73 ¾ (187.3) W x 43 (109.2) H |
| *Weight, lb. (kg)* | 2750 (1247) |
| *Average G's* | 4.5 - 10.0 |
| *Electrical* | 3 Phase, 230/460 V, 60 Hz |
| *Motors* | 2 Per Unit, 1.29 HP Ea., 2.58 Total 1200 or 1800 RPM |
| *Screen Type* | Pretensioned Panels, *Wedgelok*™ Fasteners |
| *Screen Panel Size, ft² (m²)* | 8.45 (.785) |
| *No. Screens Panels* | 3 per unit |
| *Deck Angle* | 0° to 5° |
| *Features* | Quick Screen Change, No Lubrication Required, Modular Construction |

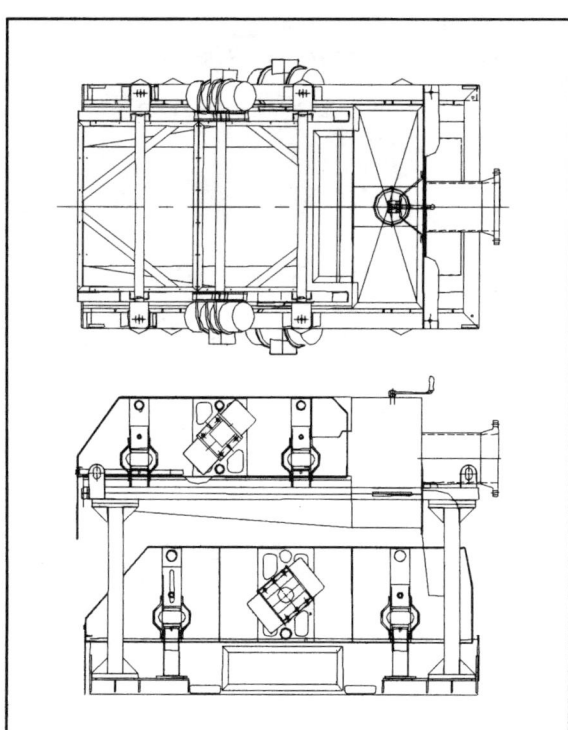

### General Specifications
### 50/500B3X VGS™ Power Series

| | |
|---|---|
| *Weir Height, in (cm)* | 66 (169.6) |
| *Dimensions, in (cm)* | 104 (264.6) L x 73 ¾ (187.3) W x 74 ¼ (188.6) H |
| *Weight, lb. (kg)* | 3500 (1591) |
| *Average G's* | 4.5 – 10 |
| *Electrical* | 3 Phase, 230/460 V, 60 Hz |
| *Motors* | |
| *Upper Unit (HP)* | 2 per unit, 0.68 Ea., 1.36 Total |
| *Lower Unit (HP)* | 2 per unit, 1.29 Ea., 2.58 Total 1200 or 1800 RPM |
| *Screen Type* | Pretensioned Panels, *Wedgelok*™ Fasteners |
| *Screen Panel Size, ft² (m²)* | 8.45 (.785) |
| *No. Screens Panels* | 5 per unit |
| *Deck Angle* | 0° to 5° |
| *Features* | Quick Screen Change, No Lubrication Required, Modular Construction |

# Fluid Systems, Inc.

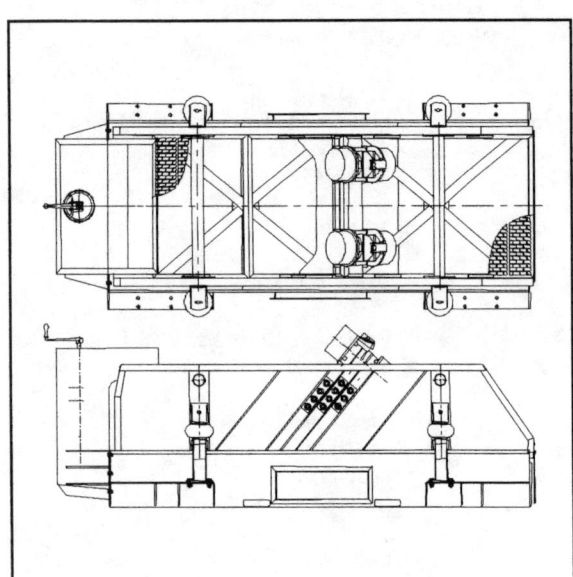

### General Specifications
### 5000 B2X Linear Shaker

| | |
|---|---|
| *Weir Height, in (cm)* | 39 (99) |
| *Dimensions, in (cm)* | 146 ¾ (372.7) L x 61 (154.9) W x 47 ¼ (120.0) H |
| *Weight, lb. (kg)* | 3500 (1591) |
| *Average G's* | 4.5 |
| *Electrical* | 3 Phase, 230/460 V, 60 Hz |
| *Motors (HP)* | 2 per unit, 0.74 Ea., 1.48 Total 1200 RPM |
| *Screen Type* | Pretensioned Panels, *Wedgelok*™ Fasteners |
| *Screen Panel Size, ft² (m²)* | 8.45 (.785) |
| *No. Screens Panels* | 4 per unit |
| *Deck Angle* | 0° to 5° |
| *Features* | Quick Screen Change, No Lubrication Required, Modular Construction |

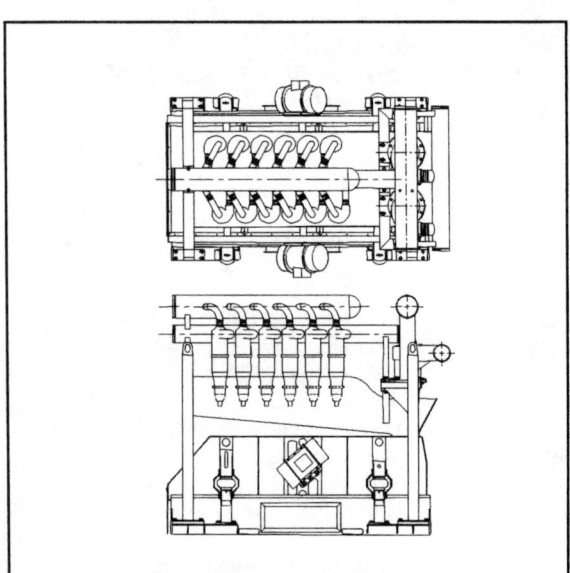

### General Specifications
### 500 B2X DHC Linear Mud Cleaner

| | |
|---|---|
| *Dimensions, in (cm)* | 109 ¾ (278.6) L x 73 ¾ (187.3) W x 90 (228.6) H |
| *Weight, lb. (kg)* | 3500 (1591) |
| *Average G's* | 4.5 |
| *Electrical* | 3 Phase, 230/460 V, 60 Hz |
| *Motors (HP)* | 2 per unit, 0.74 Ea., 1.48 Total 1200 RPM |
| *Screen Type* | Pretensioned Panels, *Wedgelok*™ Fasteners |
| *Screen Panel Size, ft² (m²)* | 8.45 (.785) |
| *No. Screens Panels* | 3 per unit |
| *Deck Angle* | 0° to 5° |
| *Features* | Quick Screen Change, No Lubrication Required, Modular Construction |

# Fluid Systems, Inc.

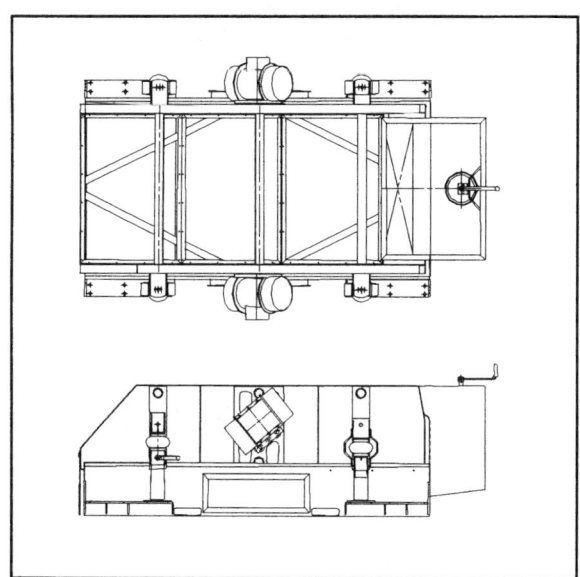

### General Specifications
### 500B2X Linear Shaker

| | |
|---|---|
| *Weir Height, in (cm)* | 29 (73.7) |
| *Dimensions, in (cm)* | 117 ½ (298.5) L x 73 ¾ (187.3) W x 43 (109.2) H |
| *Weight, lb. (kg)* | 2750 (1247) |
| *Average G's* | 4.5 |
| *Electrical* | 3 Phase, 230/460 V, 60 Hz |
| *Motors* | 2 Per Unit, 0.74 HP Ea., 1.48 Total 1200 RPM |
| *Screen Type* | Pretensioned Panels, *Wedgelok*™ Fasteners |
| *Screen Panel Size, $ft^2$ ($m^2$)* | 8.45 (.785) |
| *No. Screens Panels* | 3 per unit |
| *Deck Angle* | 0° to 5° |
| *Features* | Quick Screen Change, No Lubrication Required, Modular Construction |

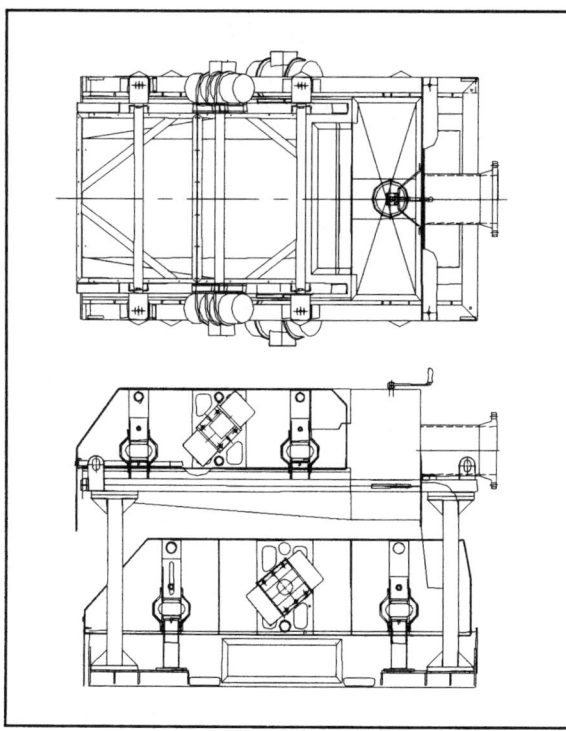

### General Specifications
### 50/500B2X Linear Shakers

| | |
|---|---|
| *Weir Height, in (cm)* | 66 (169.6) |
| *Dimensions, in (cm)* | 104 (264.6) L x 73 ¾ (187.3) W x 74 ¼ (188.6) H |
| *Weight, lb. (kg)* | 3500 (1591) |
| *Average G's* | 4.5 |
| *Electrical* | 3 Phase, 230/460 V, 60 Hz |
| *Motors* | |
|   *Upper Unit (HP)* | 2 per unit, 0.68 Ea., 1.36 Total |
|   *Lower Unit (HP)* | 2 per unit, 0.74 Ea., 1.48 Total 1200 RPM |
| *Screen Type* | Pretensioned Panels, *Wedgelok*™ Fasteners |
| *Screen Panel Size, $ft^2$ ($m^2$)* | 8.45 (.785) |
| *No. Screens Panels* | 5 per unit |
| *Deck Angle* | 0° to 5° |
| *Features* | Quick Screen Change, No Lubrication Required, Modular Construction |

# Fluid Systems, Inc.

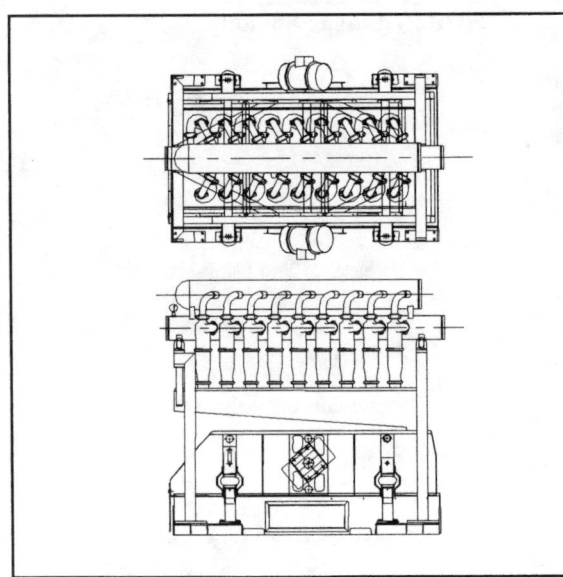

### General Specifications
### 500 B2X Linear Mud Cleaner (Desilting)

| | |
|---|---|
| *Dimensions, in (cm)* | 101 ¾ (258.4) L x 73 ¾ (187.3) W x 85 3/4 (217.8) H |
| *Weight, lb. (kg)* | 3100 (1409) |
| *Average G's* | 4.5 |
| *Electrical* | 3 Phase, 230/460 V, 60 Hz |
| *Motors (HP)* | 2 per unit, 0.74 Ea., 1.48 Total 1200 RPM |
| *Screen Type* | Pretensioned Panels, *Wedgelok* Fasteners |
| *Screen Panel Size, ft² (m²)* | 8.45 (.785) |
| *No. Screens Panels* | 3 per unit |
| *Deck Angle* | 0° to 5° |
| *Features* | Quick Screen Change, No Lubrication Required, Modular Construction |

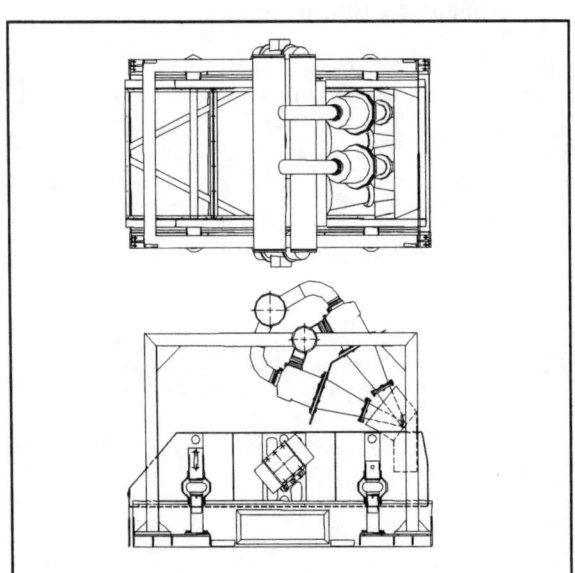

### General Specifications
### 500 B2X Linear Mud Cleaner (Desanding)

| | |
|---|---|
| *Dimensions, in (cm)* | 101 ¾ (258.4) L x 73 ¾ (187.3) W x 93 (236.2) H |
| *Weight, lb. (kg)* | 2850 (1295) |
| *Average G's* | 4.5 |
| *Electrical* | 3 Phase, 230/460 V, 60 Hz |
| *Motors (HP)* | 2 per unit, 0.74 Ea., 1.48 Total 1200 RPM |
| *Screen Type* | Pretensioned Panels, *Wedgelok* Fasteners |
| *Screen Panel Size, ft² (m²)* | 8.45 (.785) |
| *No. Screens Panels* | 3 per unit |
| *Deck Angle* | 0° to 5° |
| *Features* | Quick Screen Change, No Lubrication Required, Modular Construction |

## Harrisburg - Junior Standard Shaker

| | |
|---|---|
| Weir Height: | 24.5" |
| L/W/Height: | 55"/55.5"/43.5" |
| Weight: | 920 lb. |
| Screen Area: | 8.3 sqft |
| Vibrator Type: | Unbalanced Circular |
| G Force: | |
| RPM: | 1750 rpm |
| Screens: | 1 Screen, underslung hookstrip |
| Deck Angle: | Fixed, -8 degrees |
| Electrical: | 230/460 V, 60 Hz or 190/380 V, 50 Hz 3 hp motor. |
| Features: | Principally for a small drilling operation. The Junior Standard Shaker is sandblasted and coated with a zinc primer and a urethane top coat. |

---

## Harrisburg - Single Standard Shaker

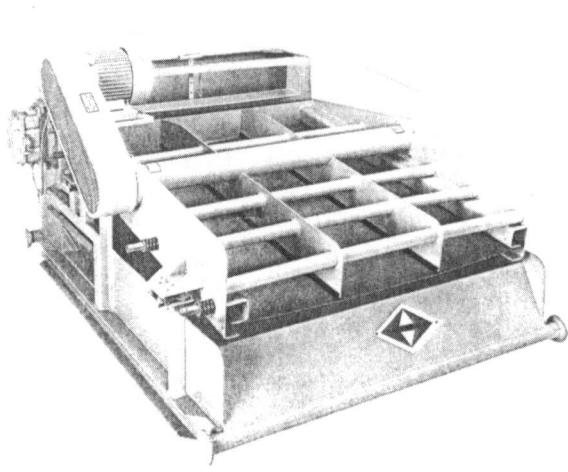

| | |
|---|---|
| Weir Height: | 30" |
| L/W/Height: | 86"/65"/44" |
| Weight: | 1850 lb. |
| Screen Area: | 20 sqft |
| Vibrator Type: | Unbalanced Circular |
| G Force: | |
| RPM: | 1500 rpm |
| Screens: | 1 Screen, underslung hookstrip |
| Deck Angle: | Fixed, -8 degrees |
| Electrical: | 230/460 V, 60 Hz or 190/380 V, 50 Hz 3 hp motor. |
| Features: | Built out of 1/4 inch plate. The Single Standard Shaker is sandblasted and coated with a zinc primer and a urethane top coat. Dual Standard shakers are also available. |

# Harrisburg - Diamond 235 Shaker

| | |
|---|---|
| Weir Height: | 34" |
| L/W/Height: | 119.5/63/63" |
| Weight: | 4200 lb. |
| Screen Area: | 23.5 sq.ft. |
| Vibrator Type: | 2 Adjustable Linear |
| Avg. G Force: | Approximately 5 |
| RPM: | 1850 |
| Screens: | 3 Screens |
| Deck Angle: | +6 -6 |
| Electrical: | 220/460 3 Phase 60 HZ Available in 380V 50HZ |
| Features: | Adjustable Bed Controller, Stainless Steel Bypass |

## Harrisburg - Diamond 270 Shaker

| | |
|---|---|
| Weir Height: | 28" |
| L/W/Height: | 114.5"/78"/43" |
| Weight: | 3600 lb. |
| Screen Area: | 27 sqft |
| Vibrator Type: | Unitized Vibrators, linear motion |
| Avg. G Force: | 4.5 g's in line of motion. |
| RPM: | 1200 rpm for better conveyance |
| Screens: | 3 Screens, pretensioned for easy screen change. Screen installation guides to eliminate screen damage. Improved wedge blocks for screen tiedown. |
| Deck Angle: | Contoured pool design, first screen 0, second +2, third +4, variable +2.5 to -4 |
| Features: | All major componets zinc primer with polyurethane top coat. Stainless steel fasteners. Unique bed vibration isolation. Deck angle adjustment, fast and easy. Improved possum belly design. Quick discharging gate design. |

## Harrisburg - Diamond 180 Shaker

| | |
|---|---|
| Weir Height: | 21.5" |
| L/W/Height: | 73.5"/76"/31" |
| Weight: | 1400 lb. |
| Screen Area: | 18 sqft |
| Vibrator Type: | Unitized Vibrators, linear motion |
| Avg. G Force: | 5.65 g's in the line of motion |
| RPM: | 1800 rpm |
| Screens: | 2 Screens, pretensioned for easy screen change. Improved wedge blocks for screen tiedown. |
| Deck Angle: | Fixed, +4 degrees |
| Features: | Low weir height. Low cost and rugged. Major componets zinc primer with polyurethane top coat. Stainless steel fasteners. Unique bed vibration isolation. |

## Harrisburg - Single Tandem Shaker

| | |
|---|---|
| Weir Height: | 36.75" |
| L/W/Height: | 86"/68.25"/52.5" |
| Weight: | 2800 lb. |
| Screen Area: | 40 sqft on two screen levels |
| Vibrator Type: | Balanced Circular |
| G Force: | |
| RPM: | 1500 rpm |
| Screens: | 2 Screens, underslung hookstrip |
| Deck Angle: | Fixed, 0 degrees |
| Electrical: | 230/460 V, 60 Hz or 190/380 V, 50 Hz 5 hp motor. |
| Features: | Flowback pan to maximize screening surface for final separation. Sandblasted and coated with a zinc primer and a urethane top coat. Also available in dual and triple units. |

## Harrisburg - Tandem Deck Linear Motion

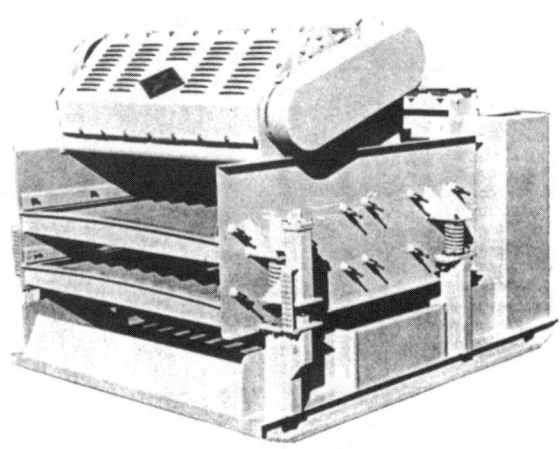

| | |
|---|---|
| Weir Height: | 36.75" |
| L/W/Height: | 86"/65"/43.75" |
| Weight: | 4500 lb. |
| Screen Area: | 38 sqft, on two screen levels |
| Vibrator Type: | Belt driven linear motion |
| G Force: | |
| RPM: | 1750 rpm |
| Screens: | 4 Screens, 2 on each level, overslung hookstrip |
| Deck Angle: | Variable, -3 to +5 degrees |
| Electrical: | 230/460 V, 60 Hz or 190/380 V, 50 Hz using two 5 hp motors. |
| Features: | Sandblasted and coated with a zinc primer and a urethane top coat. Dual Standard shakers are also available. |

# Kem-Tron Technologies, Inc.

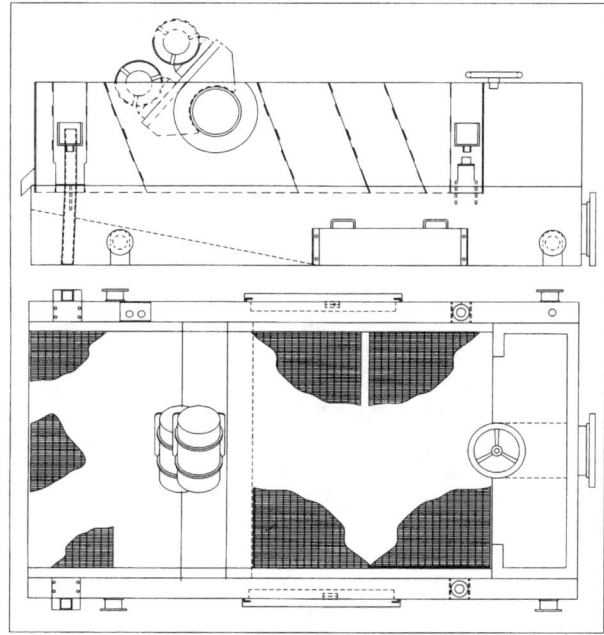

**Linear Motion Shaker
KTL-48 Standard**

## KTL-48 Specifications

- **Length**  120"(305 cm)
- **Width**  70" (178 cm)
- **Height**  59"(150 cm)
- **Weir Height**  31" (79 cm)
- **Weight**  4545 Lbs (2066 Kgs)
- **Screen Area**  28 Sq. ft(2.6 m$^2$)
- **G Force**  3.0 to 6.1
- **Stroke**  0.106
- **Deck Angle**  Adjustable +5$^0$ to -2 1/2$^0$
- **Power**  230/460V, 60Hz or 190/380V, 50 Hz., Two 1.3 HP Electric Vibrators, Two starters., No Belts, No gear box
- **Screen Changing Space Requirement**  Approximately 18" in front of unit
- **Screens Offered**  Hooks strip KDX-24 through KDX-250 multilayer and KDX-10(S) through KDX-120(S), High performance KDX-84R through KDX-250. Option pretension screen with Wedge.
- **Installation**  Mounting location must be firm and level in both direction, fasten by bolting or welding.

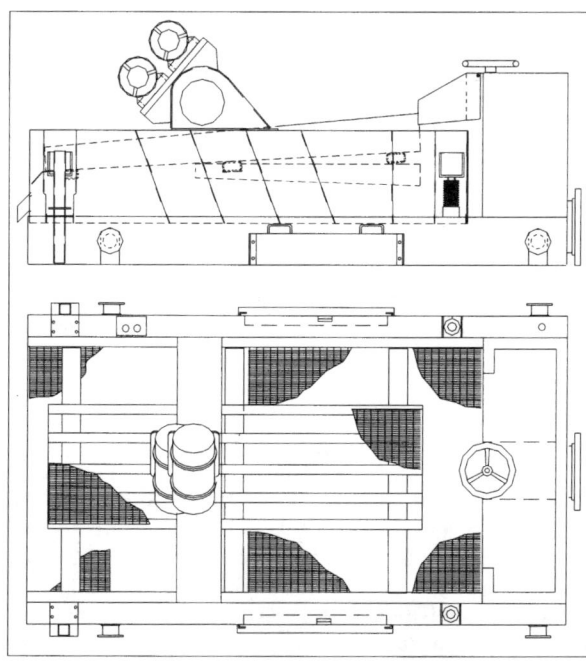

**Linear Motion Shaker
KTL-48D Double Deck**

## KTL-48D Specifications

- **Length**  120"(305 cm)
- **Width**  70" (178 cm)
- **Height**  59"(150 cm)
- **Weir Height**  32" (81.28 cm)
- **Weight**  5170 Lbs (2350 Kgs)
- **Screen Area**  43 Sq. ft. (3.99 m$^2$) Two decks
- **G Force**  3.0 to 6.0
- **Stroke**  0.117
- **Deck Angle**  Adjustable +5$^0$ to -2 1/2$^0$
- **Power**  230/460V, 60Hz or 190/380V, 50 Hz., Two 2.4 HP Electric Vibrators, Two starters., No Belts, No gear box
- **Screen Changing Space Requirement**  Approximately 30" in front of unit
- **Screens Offered**  Hooks strip KDX-24 through KDX-250 multilayer and KDX-10(S) through KDX-120(S), High performance KDX-84R through KDX-250. Option pretension screen with Wedge.
- **Installation**  Mounting location must be firm and level in both direction, fasten by bolting or welding.

# Kem-Tron Technologies, Inc.

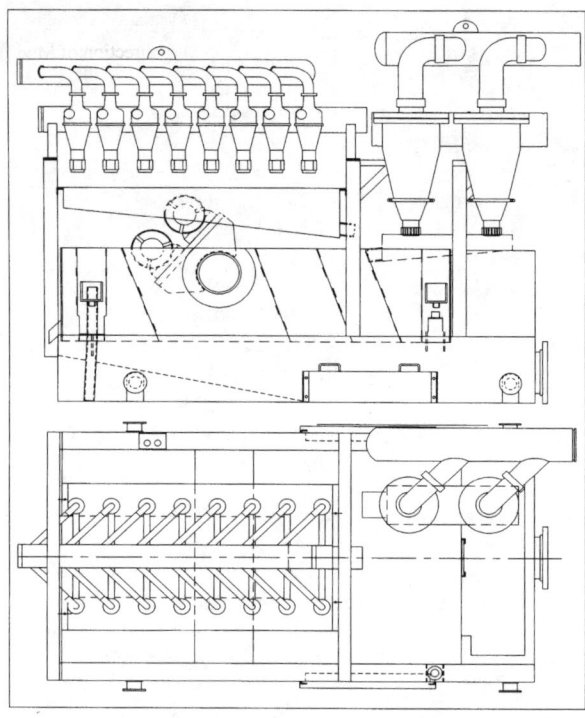

**Linear Motion Mud Cleaner KTL-48A
with Desander Model VSP-212
and Desilter Model KT-16-240**

### KTL-48A Specifications

- **Length** 120"(305 cm)
- **Width** 70" (178 cm)
- **Height** 99"(252 cm) with Hydrocyclone
- **Weight** 6611 Lbs (3005 Kgs)
- **Screen Area** 28 Sq. ft(2.6 m$^2$)
- **G Force** 3.0 to 6.1
- **Stroke** 0.106
- **Deck Angle** Adjustable +5$^0$ to -2$^0$
- **Power** 230/460V, 60Hz or 190/380V, 50 Hz., Two 1.3 HP Electric Vibrators, Two starters., No Belts, No gear box
- **Screen Changing Space Requirement** Approximately 18" in front of unit
- **Screens Offered** Hooks strip KDX-24 through KDX-250 multilayer and KDX-10(S) through KDX-120(S), High performance KDX-84R through KDX-250. Option pretension screen with Wedge.
- **Installation** Mounting location must be firm and level in both direction, fasten by bolting or welding.
- **Others** KTL-48A also available with any combination of Desander VSP-112 or VSP-312 and Desilter KT-8-240 or KT-10-240 or KT-20-240

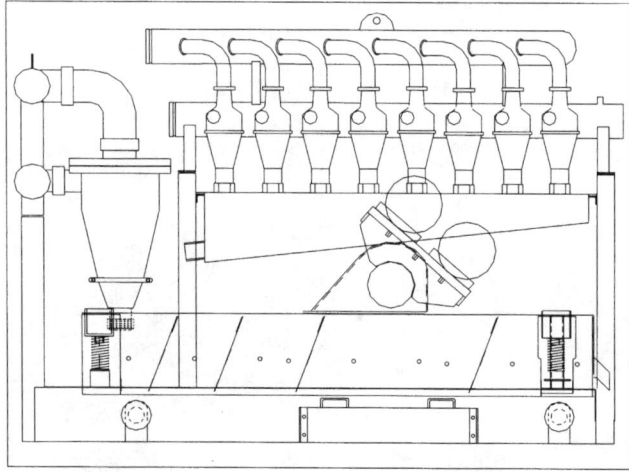

**Linear Motion Mud Cleaner KTL-48B
with Desander Model VSP-212
and Desilter Model KT-16-240**

### KTL-48B Specifications

- **Length** 107"(272 cm)
- **Width** 70" (178 cm)
- **Height** 79"(198 cm) with Hydrocyclone
- **Weight** 5500 Lbs (2500 Kgs)
- **Screen Area** 28 Sq. ft(2.6 m$^2$)
- **G Force** 4 to 8.4
- **Stroke** 0.106
- **Deck Angle** Adjustable +5$^0$ to -2$^0$
- **Power** 230/460V, 60Hz or 190/380V, 50 Hz., Two 2.4 HP Electric Vibrators, Two starters., No Belts, No gear box
- **Screen Changing Space Requirement** Approximately 18" in front of unit
- **Screens Offered** Hooks strip KDX-24 through KDX-250 multilayer and KDX-10(S) through KDX-120(S), High performance KDX-84R through KDX-250. Option pretension screen with Wedge.
- **Installation** Mounting location must be firm and level in both direction, fasten by bolting or welding.
- **Others** KTL-48B also available with any combination of Desander VSP-112 or VSP-312 and Desilter KT-8-240 or KT-10-240 or KT-20-240

# Kem-Tron Technologies, Inc.

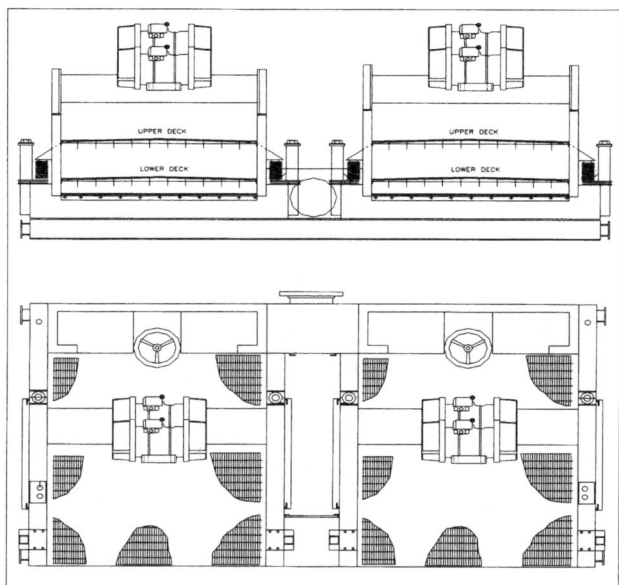

**Linear Motion Shaker
KTL-48DT Dual Tandem Deck**

### KTL-48DT Specifications

- **Length** 80"(203 cm) in direction of Mud flow
- **Width** 148" (375 cm)
- **Height** 66"(168 cm)
- **Weir Height** 38" (97 cm)
- **Weight** 8040 Lbs (3655 Kgs)
- **Screen Area** 74.64 Sq. ft. (6.92 m$^2$) Two decks
- **G Force** 3.0 to 6.25
- **Stroke** 0.132
- **Deck Angle** Adjustable +5$^0$ to -2 1/2$^0$
- **Power** 230/460V, 60Hz or 190/380V, 50 Hz., Two 1.3 HP Electric Vibrators, Two starters., No Belts, No gear box
- **Screen Changing Space Requirement** Approximately 30" in front of unit
- **Screens Offered** Hooks strip KDX-24 through KDX-250 multilayer and KDX-10(S) through KDX-120(S), High performance KDX-84R through KDX-250. Option pretension screen with Wedge.
- **Installation** Mounting location must be firm and level in both direction, fasten by bolting or welding.
- **Others** Replaces earlier Model Elliptical & Circular motion Dual Tandem Shaker

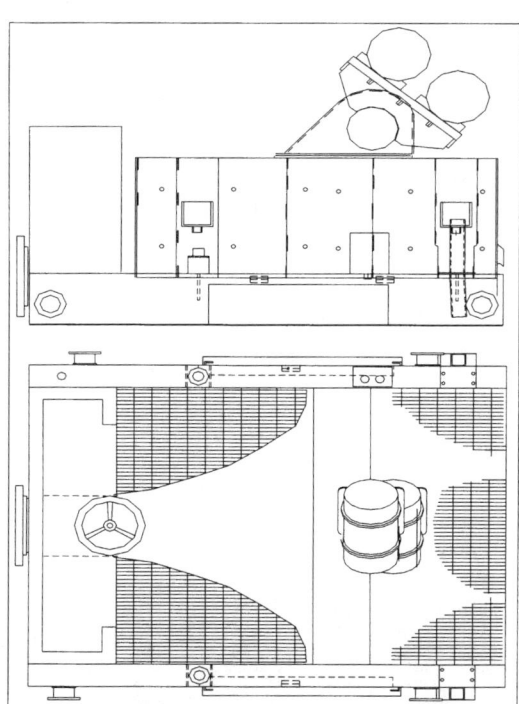

**Linear Motion Shaker
KTL-48T Tandem Deck**

### KTL-48T Specifications

- **Length** 80"(203 cm)
- **Width** 70" (178 cm)
- **Height** 60"(153 cm)
- **Weir Height** 32" (81.28 cm)
- **Weight** 3520 Lbs (1600 Kgs)
- **Screen Area** 37.32 Sq. ft. (3.46 m$^2$) Two decks
- **G Force** 3.0 to 6.25
- **Stroke** 0.132
- **Deck Angle** Adjustable +5$^0$ to -2 1/2$^0$
- **Power** 230/460V, 60Hz or 190/380V, 50 Hz., Two 1.3 HP Electric Vibrators, Two starters., No Belts, No gear box
- **Screen Changing Space Requirement** Approximately 30" in front of unit
- **Screens Offered** Hooks strip KDX-24 through KDX-250 multilayer and KDX-10(S) through KDX-120(S), High performance KDX-84R through KDX-250. Option pretension screen with Wedge.
- **Installation** Mounting location must be firm and level in both direction, fasten by bolting or welding.
- **Others** Replaces earlier Model Elliptical & Circular motion Tandem Shaker

# Kem-Tron Technologies, Inc.

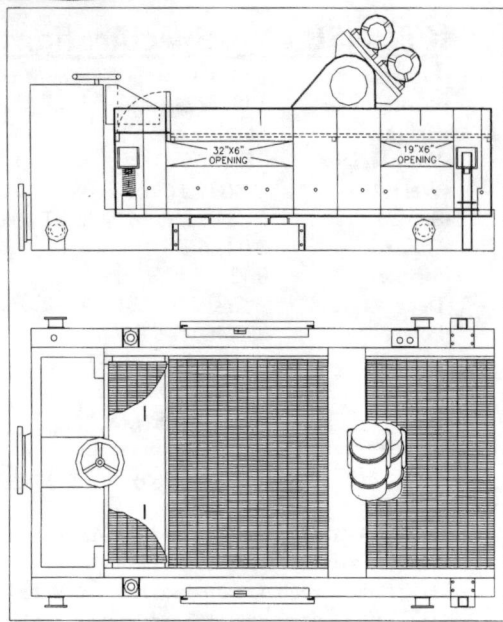

**Linear Motion Shaker
KTL-48SS Double Deck**

### KTL-48SS Specifications
- **Length**         120"(305 cm)
- **Width**          70" (178 cm)
- **Height**         60"(152 cm)
- **Weir Height**    32" (81.28 cm)
- **Weight**         5170 Lbs (2350 Kgs)
- **Screen Area**    54.6 Sq. ft. (5.07 m$^2$) Two decks
- **G Force**        3.0 to 6.0
- **Stroke**         0.129
- **Deck Angle**     Adjustable +5$^0$ to -2 1/2$^0$
- **Power**          230/460V, 60Hz or 190/380V, 50 Hz., Two 2.4 HP Electric Vibrators, Two starters., No Belts, No gear box
- **Screen Changing Space Requirement**    Approximately 36" in front of unit
- **Screens Offered**    Pretension KPT-10(S) through KPT-60(S) &KPT-24 through KPT-250 for top deck. Hooks strip KDX-24 through KDX-250 multilayer and KDX-10(S) through KDX-120(S) for Bottom Deck. High performance KDX-84R through KDX-250 and KPT-84R through KPT-250R.
- **Installation**    Mounting location must be firm and level in both direction, fasten by bolting or welding.

*Patented*

**MULTIFUNCTIONAL
Simultaneous
Linear Motion Shaker and
Mud Cleaner
KTL-48SS MC Double Deck**

### KTL-48SS MC Specifications
- **Length**         120"(305 cm)
- **Width**          70" (178 cm)
- **Height**         88"(224 cm)
- **Weir Height**    32" (97 cm)
- **Weight**         6908 Lbs (3140 Kgs)
- **Screen Area**    54.6 Sq. ft. (5.07 m$^2$) Two decks
- **G Force**        3.0 to 6.0
- **Stroke**         0.129
- **Deck Angle**     Adjustable +5$^0$ to -2 1/2$^0$
- **Power**          230/460V, 60Hz or 190/380V, 50 Hz., Two 2.4 HP Electric Vibrators, Two starters., No Belts, No gear box
- **Screen Changing Space Requirement**    Approximately 36" in front of unit
- **Screens Offered**    Pretension KPT-10(S) through KPT-60(S) &KPT-24 through KPT-250 for top deck. Hooks strip KDX-24 through KDX-250 multilayer and KDX-10(S) through KDX-120(S) for Bottom Deck. High performance KDX-84R through KDX-250 and KPT-84R through KPT-250R.
- **Installation**    Mounting location must be firm and level in both direction, fasten by bolting or welding.
- **Others**    **Multifunctional** Simultaneous independent flows to upper and lower deck permitting use as drying shaker and flowline shaker

# Kem-Tron Technologies, Inc.

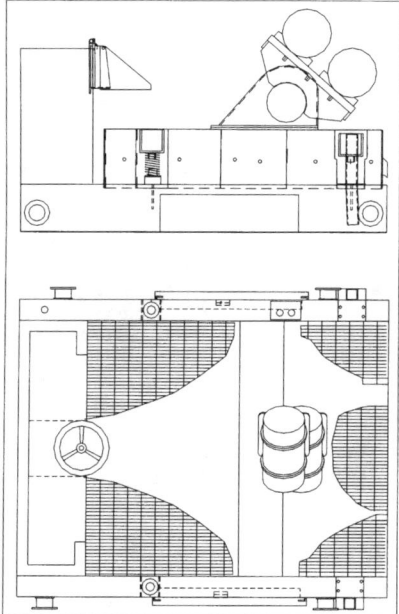

**Linear Motion Shaker
KTL-48TJ JUNIOR**

### KTL-48TJ Junior Specifications

- **Length**     80"(203 cm)
- **Width**     70" (178 cm)
- **Height**     60"(153 cm)
- **Weir Height**     32" (81.28 cm)
- **Weight**     3520 Lbs (1600 Kgs)
- **Screen Area**     37.32 Sq. ft. (3.46 m$^2$) Two decks
- **G Force**     3.0 to 6.25
- **Stroke**     0.132
- **Deck Angle**     Adjustable +5$^0$ to -2 1/2$^0$
- **Power**     230/460V, 60Hz or 190/380V, 50 Hz., Two 1.3 HP Electric Vibrators, Two starters., No Belts, No gear box
- **Screen Changing Space Requirement**
      Approximately 30" in front of unit
- **Screens Offered**    Hooks strip KDX-24 through KDX-250 multilayer and KDX-10(S) through KDX-120(S), High performance KDX-84R through KDX-250. Option pretension screen with Wedge.
- **Installation**     Mounting location must be firm and level in both direction, fasten by bolting or welding.
- **Others**     Replaces earlier Model Elliptical & Circular motion Tandem Shaker

# Rigtech VSM 300

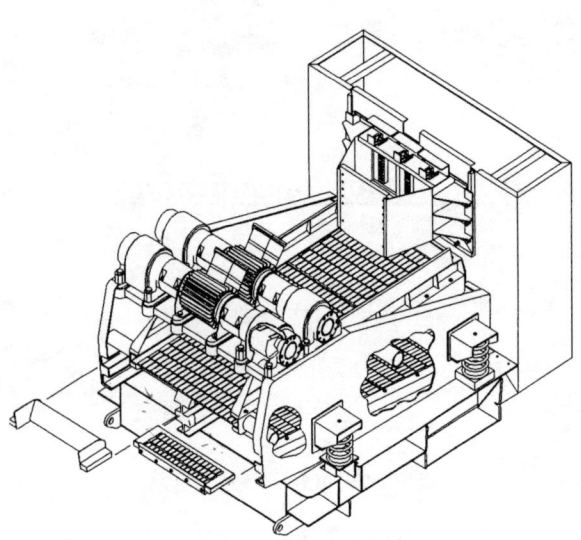

| | | |
|---|---|---|
| **Weir Height** | - | 38.8" |
| | - | 985mm |
| **Dimensions** | - | 108.4 x 73.6 x 59.2" |
| | - | 2754 x 1870 x 1504mm |
| **Weight** | - | 5154lbm |
| | - | 2338kg |
| **No. of Screen Levels** | - | 2 + Secondary Drying Deck |
| **Area per Screen Level** | - | Scalping - 20.56ft² |
| | | - 1.91 m² |
| | | Primary - 26.26ft² |
| | | - 2.44m² |
| | | Secondary - 2.01ft² |
| | | - 0.26m² |
| **No. of panels per Screen Level** | - | 3/4/2 |
| **Deck Angle** | - | Fixed @ - 0° Scalping |
| | | - 7° Primary |
| | | - 7° Secondary |
| | - | or variable to −4.5° |
| **Vibrator No. & Type** | - | Balanced Elliptical – Direct |
| **Average 'G' Force** | - | 6.4 Peak @ 60Hz |
| **RPM** | - | Variable to 2000RPM |
| **Electrical** | - | 6kW, 460V @ 60Hz |
| **Other Features** | - | Quick Release screen clamping system allowing complete screen change-out in < 3 minutes |

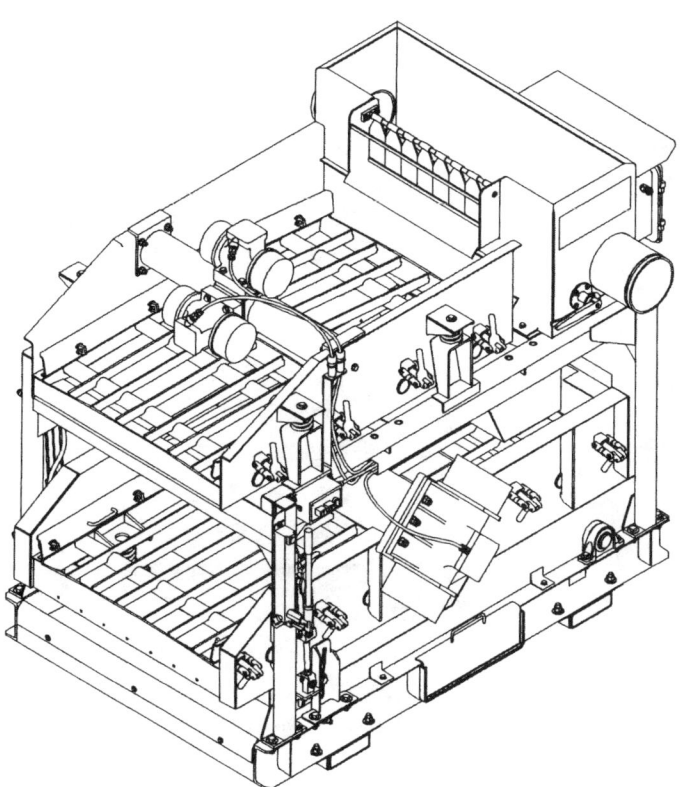

**Swaco ALCS
Adjustable Linear Cascade Shaker**

Length 103"
Width 68.38"
Height 88.00"
Weight 7,660 lbs
Weir Height 72"
Sceen Area 56 ft
Two Top Screens, 3 feet x 4 feet
Two Bottom Screens, 4 feet x 4 feet
Quick-Release Tension Bolts
Bottom Deck Adjustable +5° to -3°

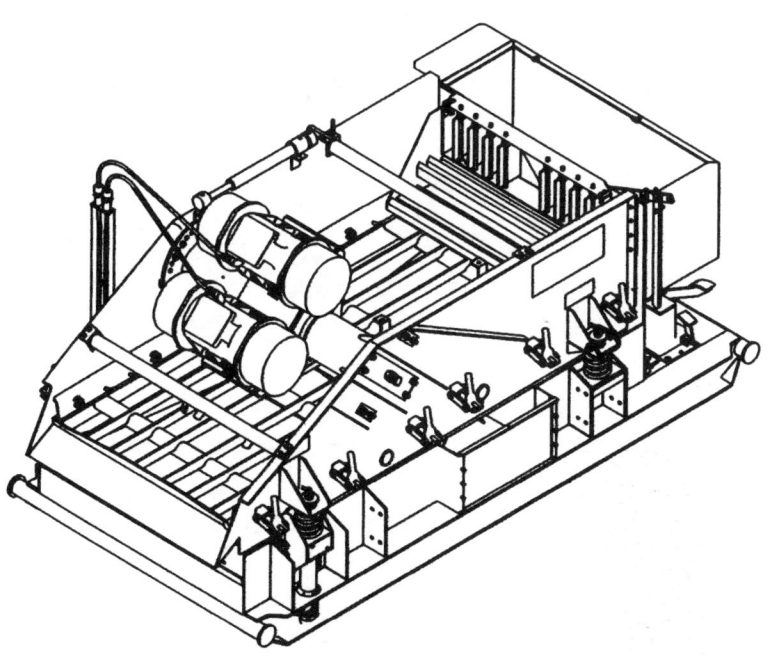

**Swaco ALS-II™
Adjustable Linear Shaker**

Length 128.75"
Width 63.00"
Height 61.00"
Weight 3,500 lbs
Weir Height 33.6"
Sceen Area 32 ft$^2$
Two Screens, 4 feet x 4 feet
Quick-Release Tension Bolts
Deck Angle -3° to +3°

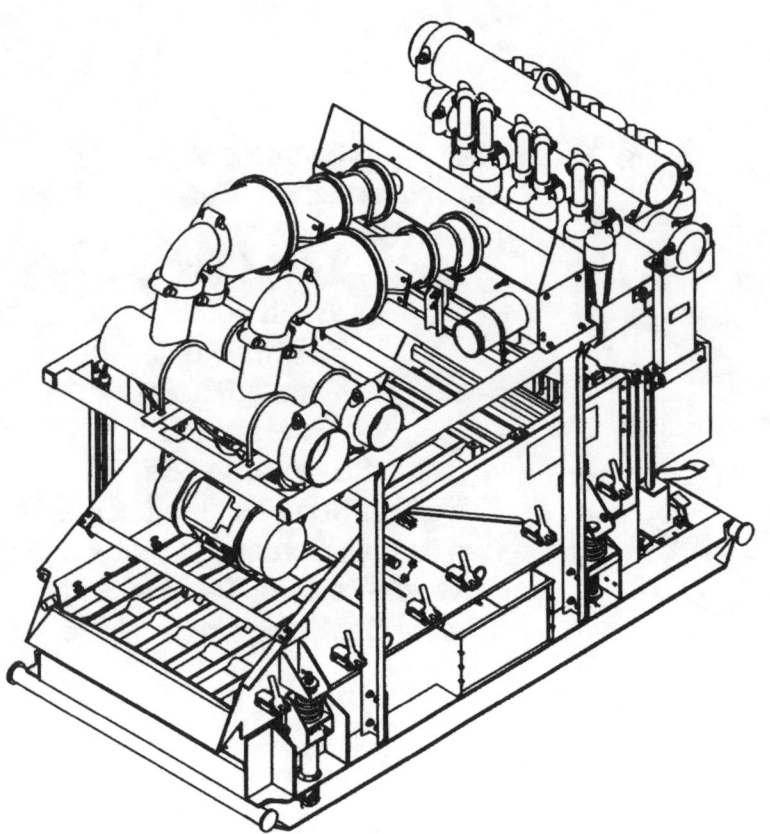

Swaco 212/6T54 ALS-II™
High Volume Mud Cleaner

Length 128"
Width 73"
Height 97"
Weight 7,500 lbs
Weir Height 33.6"
Sceen Area 32 ft²
Two Screens, 4 feet x 4 feet
Quick-Release Tension Bolts
Capacity 1000 gpm Desander
Capacity 900 gpm Desilter
Deck Angle -3° to +3°

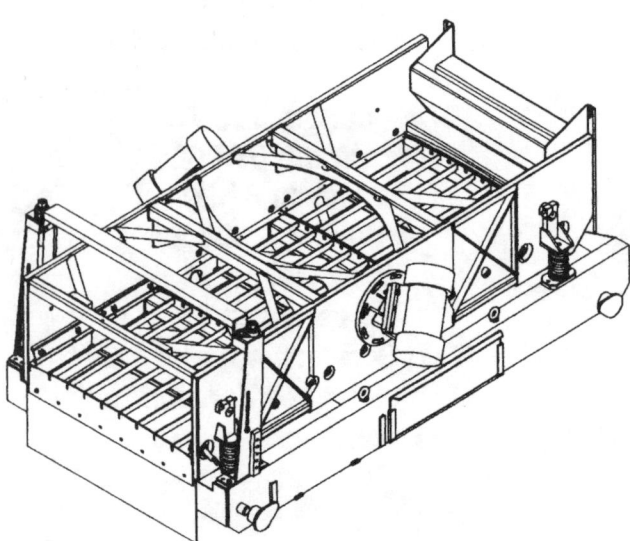

Swaco BEM™
Balanced Elliptical Shaker

Length 121"
Width 83.5"
Height 47.5"
Weight 5,100 lbs
Weir Height 32"
Sceen Area 33.7 ft²
Three Bottom Screens, 3 feet x 4 feet
Quick-Release Tension Bolts
Deck Adjustable +5° to 0°

# Tri-Flo International, Inc.

P.O. BOX 2626 • CONROE, TEXAS 77305

Business (409) 856-8551 • Fax (409) 856-5668 • Houston (281) 350-9190

**Solids Control**

### Tri-Flo 123 Single Shale Shaker

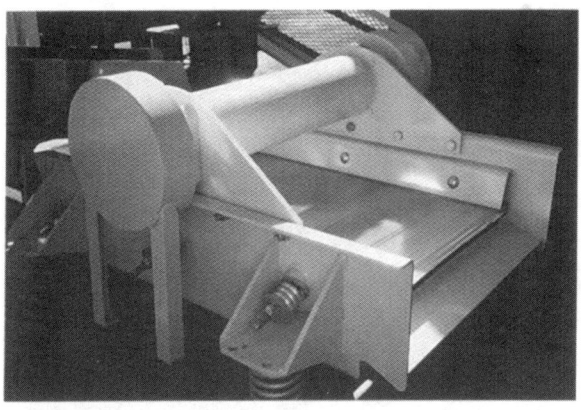

### Tri-Flo 146 Single Shale Shaker

**Dimensions:**
Width.......... 5'-5½"     Weight...... 2575 lbs.
Length......... 8'-2"      Volume...... 520 gpm
Height......... 3'-2"

### Tri-Flo 126 Single Shale Shaker

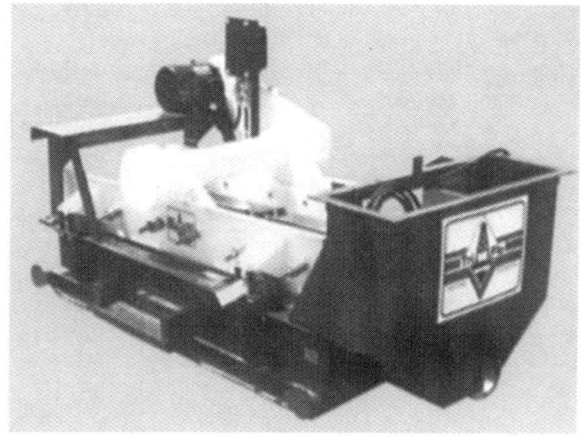

**Dimensions:**
Width.......... 3'-8"      Weight...... 1420 lbs.
Length......... 7'-9"      Volume...... 300 gpm
Height......... 3'-5"

**Dimensions:**
Width.......... 13'-0"     Weight...... 7280 lbs.
Length......... 7'-8"      Volume...... 1040 gpm
Height......... 4'-1½"

### Tri-Flo 246 Dual Shale Shaker

The Tri-Flo 246 represents a culmination of efficient engineering ideas in shale shaker technology. Two Tri-Flo 146 fine screen shale shakers are unitized on a single low profile skid to avoid delicate alignment jobs. The oversized entry flume allows a greater volume of mud to enter the possum belly and individually adjusted diverters maintain an even distribution of mud onto the primary screening surfaces.

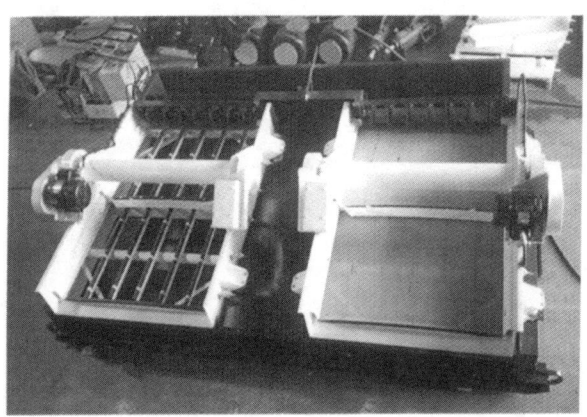

# Solids Control

# Tri-Flo International, Inc.
P.O. BOX 2626 • CONROE, TEXAS 77305
Business (409) 856-8551 • Fax (409) 856-5668 • Houston (281) 350-9190

### Tri-Flo 148 Linear Shaker

The unit consists of a linear motion vibrating screen single deck 4'0" wide by 8'0" long complete with screens in three sections, mounted on a single heavy-duty oilfield skid complete with:

10" feed and bypass manifold in possum belly and gates on each side for direct discharge to mud tanks.

Oil bath lubrication, tank assembly with mud box, lifting mechanism to change slope from 0 to 5 degree uphill. Neoprene splash seal at feed end around flume. Unit operates at 1750 RPM having 1/8" straight line motion with a manually adjustable 45 degree line of force.

### Tri-Flo 148 Linear Shaker

**Dimensions:**
Width . . . . . . . . . . 6'-8"  Weight . . . . . . 4300 lbs.
Length . . . . . . . . . 9'-9"  Volume . . . . . . 750 gpm
Height . . . . . . . . . 4'-2"

**Specifications of the standard motor and drive:**
Timing belt drive and 3 HP 230/460 V, 3 Phase, 60HZ, 1750 RPM, UL/CSA approved, explosion proof motor and manual starter with overload protection.

### Tri-Flo 148 Linear Scalping Shaker

The unit consist of a single deck Tri-Flo 148 Linear Shaker with a 4' x 8' deck and a top mounted scalper deck 8' long and 2' wide. The entire assembly is mounted on a single heavy-duty oilfield skid. The standard motor is a 5 HP, 230/460 V, 3 Phase, 60 HZ, 1750 RPM UL/CSA approved explosion proof motor with manual starter and overload protection.

**Dimensions:**
Width . . . . . . . . . . 6'-8"   Height . . . . . . . . . . 5'-0"
Length . . . . . . . 10'-10"   Weight . . . . . . 4700 lbs.

# Glossary

**Abnormal Pressure**  A formation pore pressure that is higher than that resulting from a water gradient.

**Absolute Temperature**  Temperature related to absolute zero; the temperature where all molecular activity ceases. Calculated by adding 460°F to the temperature in Fahrenheit to obtain the absolute temperature in degrees Rankine or by adding 273°C to the temperature in degrees Celsius to obtain the absolute temperaure in degrees Kelvin.

**Absorb**  To take in and make part of an existing whole. *See:* Absorption, Adsorption, Adsorb, Adsorbed Liquid, Bound Liquid.

**Absorption**  The penetration or apparent disappearance of molecules or ions of one or more substances into the interior of a solid or liquid. For example, in hydrated bentonite, the planar water that is held between the mica-like layers is the result of absorption. *See:* Absorb, Adsorption, Adsorb, Adsorbed Liquid, Bound Liquid.

**Acid**  Any chemical compound containing hydrogen capable of being replaced by positive elements or radicals to form salts. In terms of the dissociation theory, it is a compound, which, on dissociation in solution, yields excess hydrogen ions. Acids lower the pH. Examples of acids or acidic substances are: hydrochloric acid, HCl, sodium acid pyrophosphate, SAPP, sulfuric acid, and $H_2SO_4$. *See:* pH, Acidity.

**Acidity**  The relative acid strength of liquid as measured by pH. A pH value below 7. *See:* pH, Acid.

**Across-the-line-start**  A motor starting method that provides full line voltage to the motor windings during starting.

**Active System**  The volume of drilling fluid being circulated to drill a hole. It consists of the volume of drilling fluid in the hole plus the volume of drilling fluid in the surface tanks through which the fluid circulates.

**Addition Section**  The compartment(s) in a drilling fluid system between the removal section and the suction section, which provides a well-agitated location within the fluid circulation system for the addition of commercial chemicals, liquids, and bulk products.

**Adhesion**  The force that holds unlike molecules together.

**Adsorb**  (1) The liquid on the surface of a solid particle that cannot be removed by draining or centrifugal force. (2) To hold a liquid on the surface of a solid particle that cannot be removed by draining or centrifugal force. *See:* Absorption, Adsorption, Adsorb, Adsorbed Liquid, Bound Liquid.

**Adsorbed Liquid**  The liquid film adhering to the surfaces of solid particles, which cannot be moved by draining, even with centrifugal force. *See:* Absorb, Absorption, Adsorption, Adsorb, Bound Liquid.

**Adsorption**  A surface phenomenon exhibited by a solid (adsorbent) to hold or concentrate gases, liquids, or dissolved substances (adsorptive) upon its surface, a property due to adhesion. For example, water, held to the outside surface of hydrated bentonite, is adsorbed water. Adsorption refers to liquid that is on the outside of some material, and absorbed refers to the liquid that becomes part of the material. *See:* Absorb, Absorption, Adsorb, Adsorbed Liquid, Bound Liquid.

**Aeration**  (1) The technique of injecting air or gas in varying amounts into a drilling fluid for the purpose of reducing hydrostatic head. (2) The inadvertent mechanical incorporation and dispersion of air or gas into a drilling fluid. If not selectively controlled, it can be very harmful. *See:* Air Cutting, Gas Cut.

**Agglomerate**  The larger groups of individual particles usually originating in sieving or drying operations.

**Agglomeration**  A group of two or more individual particles held together by strong forces. Agglomerates are stable to normal stirring, shaking, or handling as powder or a suspension. They may be broken by drastic treatment such as the ball milling of a powder or the shearing of a suspension.

**Aggregate**  To gather or "clump" together. A flocculated drilling fluid will aggregate if more flocculant is added.

**Aggregation**  (1) Formation of aggregates. (2) In drilling fluids, aggregation results in the stacking of the clay platelets face-to-face. As a consequence, the viscosity and gel strength of the fluid decreases.

**Agitation**  The process of rapidly moving a slurry within a tank to obtain and maintain a uniform mixture.

**Agitator**  A mechanically driven mixer that stirs the drilling fluid by turning an impeller in the bottom of a drilling fluid compartment to assist in blending liquids, suspending solids, and maintaining an even consistency in the drilling fluid. *See:* Mechanical Agitator, Mechanical Stirrer, Mud Gun.

**Air Cutting**  The inadvertent mechanical incorporation and dispersion of air into a drilling fluid system. *See:* Aeration, Gas Cut.

**Air Lock**  A condition causing a centrifugal pump to stop pumping because of a large bubble of air or gas in the center of the pump impeller. This prevents the liquid from entering the pump suction. *See:* Air Locking.

**Air Locking**  *See:* Air Lock.

**Alkali**  Any compound having pH properties higher than the neutral state. *See:* Base.

**Alkalinity**  The combining power of a base measured by the maximum number of equivalents of an acid with which it can react to form a salt. In water analyses, its represents the carbonates, bicarbonates, hydroxides, and occasionally the borate, silicates, and phosphates in the water. It is determined by titration with standard acid to certain datum points. See API Bulletin RP 13B for specific directions for determining phenolphthalein (Pf) and methyl orange (Mf) alkalinities of the filtrate in drilling fluids and the (Pm) alkalinity of the drilling fluid itself. *See:* Alkali, Base, Pf, Mf, Pm.

| | |
|---|---|
| **Alum** | Aluminum Sulfate, $Al_2(SO_4)_3$, a common inorganic coagulant. |
| **Aluminum Stearate** | An aluminum salt of stearic acid used as a defoamer. *See:* Stearate. |
| **Amorphous** | The property of a solid substance that does not crystallize and is without any definite characteristic shape. |
| **Ampere** | The measurement of electric flow per second. |
| **Amplitude** | The distance from the mean position to the point of maximum displacement. In the case of a vibrating screen with circular motion, amplitude would be the radius of the circle. In the case of straight-line motion or elliptical motion, amplitude would be one-half of the total movement of the major axis of the ellipse; thus one-half stroke. *See:* Stroke. |
| **Anhydrite** | A mineral compound, $CaSO_4$, that is often encountered while drilling. It may occur as thin stringers or massive formations. *See:* Calcium Sulfate, Gypsum. |
| **Anhydrous** | Without water. |
| **Aniline Point** | The lowest temperature at which equal volumes of freshly distilled aniline and an oil sample that is being tested, are completely miscible. This test indicates the characteristics (paraffinic, naphthenic, asphaltic, aromatic, etc.) of the oil. The aniline point of diesels or crude oils used in drilling fluid is also an indication of the deteriorating effect that these materials may have on natural or synthetic rubber. The lower the aniline point of a particular oil, the greater the propensity for damaging rubber parts. |
| **Anion** | A negatively charged atom or radical, such as $Cl^-$, $OH^-$, $SO^{-4}$, and so forth, in the solution of an electrolyte. Anions move toward the anode (positive electrode) under the influence of an electrical potential. |
| **Annular Pressure Loss** | The pressure on the annulus required to pump the drilling fluid from the bottom of the hole to the top of the hole in the annular space. *See:* ECD. |
| **Annular Space** | *See:* Annulus. |
| **Annular Velocity** | The velocity of a fluid moving in the annulus usually expressed in ft/min or m/min. |
| **Annulus** | The space between the drill string and the wall of the hole or the inside surface of the casing. |
| **Antifoam** | A substance used to prevent foam by increasing the surface tension of a liquid. *See:* Defoamer. |
| **Aperture** | (1) An opening in a screen surface. (2) The opening between the wires in a screen cloth. *See:* Mesh. |
| **Apex** | The lower end of a hydrocyclone. *See:* Underflow Opening. |

| | |
|---|---|
| **Apex Valve** | *See:* Apex, Underflow Opening. |
| **API Bulletin RP 10B** | Recommended Practice for Testing Well Cement. Published by the American Petroleum Institute. |
| **API Bulletin RP 13B** | Recommended Practice for Standard Procedure for Testing Drilling Fluids at the rig. Published by the American Petroleum Institute. |
| **API Bulletin RP 13C** | Recommended Practice for Drilling Fluid Systems Process Evaluation. Published by the American Petroleum Institute. |
| **API Bulletin RP 13E** | Recommended Practice for Shaker Screen Cloth Design. Published by the American Petroleum Institute. |
| **API Filter Press** | A device used to measure the fluid loss under API conditions. *See:* API Fluid Loss. |
| **API Fluid Loss** | This fluid loss is measured under ambient conditions. Usually these are room temperature and 100 psi differential pressure. *See:* API Filter Press. |
| **API Gravity** | The gravity (weight per unit volume) of crude oil or other related fluids as measured by a system recommended by the American Petroleum Institute. It is related to specific gravity by the following formula: Degree API = [141.5/Specific Gravity] − 131.5. |
| **API Sand** | Solids particles that are too large to pass through a U.S. Standard No. 200 Screen (74 micron openings). *See:* API Bulletin RP 13B, Sand, Sand Content. |
| **Apparent Viscosity** | The apparent viscosity in centipoise, as determined by the direct indicating viscometer, is equal to one-half the 600 RPM reading. It is the viscosity of a fluid at a shear rate of 1,022 sec$^{-1}$. *See:* Viscosity, Plastic Viscosity, Yield Point, API RP 13B. |
| **Aromatic Hydrocarbons** | Hydrocarbons that include compounds containing aliphatic or aromatic groups attached to aromatic rings. Benzene is the simplest example. *See:* Live Oil. |
| **Asphalt** | A natural or mixed blend of solid or viscous bitumen found in natural beds or obtained as a residue from petroleum distillation. Asphalt, blends containing asphalt, and altered asphaltic materials (e.g., air-blown, chemically modified, etc.) have been added to drilling fluids for purposes such as lost circulation, emulsification, fluid loss control, lubrication, seepage loss, shale stability, and so forth. |
| **Atom** | The smallest quantity of an element that is capable of entering into chemical combination or that can exist alone. |
| **Atomic Weight** | The relative weight of an atom of any element as compared with the weight of one atom of oxygen. The atomic weight of oxygen is 16. |
| **Attapulgite Clay** | A colloidal, viscosity building clay used principally in saltwater drilling fluids to increase the low shear viscosity. Attapulgite, a special fuller's earth, is a hydrous magnesium aluminum silicate, which has long, needle-like platelets as opposed to the broader, more symmetrical platelets of bentonite. |

| | |
|---|---|
| **Axial Flow** | Flow from a mechanical agitator in which the fluid first moves along the axis of the impeller shaft (usually down toward the bottom of a tank) and then away from the impeller. *See:* Radial Flow. |
| **Back Pressure** | The frictional or blocking pressure opposing fluid flow in a conduit. *See:* Differential Pressure. |
| **Back Tank** | The compartment on a shale shaker that receives drilling fluid from the flow line. *See:* Possum Belly, Mud Box. |
| **Backing Plate** | The plate attached to the back of screen cloth(s) for support. |
| **Backup Screen** | *See:* Support Screen. |
| **Baffles** | Plates or obstructions built into a compartment to change the direction of fluid flow. |
| **Balanced Design Hydrocyclone** | A hydrocyclone that has the lower apex adjusted to the diameter of the cylinder of air formed within the cone by the cyclonic forces of drilling fluid spinning within the cone. This tends to minimize liquid discharge when there are no separable solids. |
| **Balanced Elliptical Motion** | An elliptical motion of a shale shaker screen such that all ellipse axes are tilted at the same angle toward the discharge end of the shale shaker. |
| **Ball Valve** | A valve that uses a spherical closure with a hole through its center, which rotates 90° to open and close. |
| **Barite** | Natural barium sulfate, $BaSO_4$, is used for increasing the density of drilling fluids. The API standard requires a minimum of 4.20 specific gravity. Commercial barium sulfate ore can be produced from a single ore or a blend or ores, and may be a straight-mined product or processed by flotation methods. It may contain accessory minerals other than the barium sulfate mineral. Because of mineral impurities, commercial barite may vary in color from off-white or gray to red or brown. Common accessory minerals are silicates such as quartz and chert, carbonate compounds such as siderite and dolomite, and metallic oxide and sulfide compounds. |
| **Barite Recovery Efficiency** | Barite recovery efficiency is the ratio of the mass flowrate of barite returning to a drilling fluid from a solids control device, divided by the mass flowrate of barite in the feed to the solids control device. |
| **Barium Sulfate** | $BaSO_4$. *See:* Barite. |
| **Barrel** | A volumetric unit of measure used in the petroleum industry consisting of 42 U.S. gallons. |
| **Barrel Equivalent** | One gram of material in 350 ml of fluid is equivalent to a concentration of 1 lb of that material in an oil field barrel of fluid. *See:* Barrel, Pound Equivalent. |
| **Base** | A compound of a metal, or a metal-like group, with hydrogen and oxygen in proportions that form an $OH^-$ radical when ionized in an aqueous solution, yielding excess hydroxyl ions. Bases are formed when metallic oxides react with water. Bases increase the pH. Examples of bases are caustic soda (NaOH) and lime ($Ca(OH)_2$). |

| | |
|---|---|
| **Base Exchange** | The replacement of the cations associated with the surface of a clay particle by another species of cation (e.g., the substitution of sodium cations by calcium cations on the surface of a clay particle). See: Methylene Blue Titration, Methylene Blue Test, MBT, Cation Exchange Capacity, CEC. |
| **Basicity** | pH value above 7. Ability to neutralize or accept protons from acids. *See:* pH. |
| **Basket** | That portion of a shale shaker containing the deck upon which the screen(s) is mounted; supported by vibration isolation members connected to the bed. |
| **Beach** | Area between the liquid pool and the solids discharge ports in a decanting centrifuge or hydrocyclone. |
| **Bed** | Shale shaker support member consisting of mounting skid, or frame, with or without bottom, flow diverters to direct screen underflow to either side of the skid, and mountings for vibration isolation members. |
| **Bentonite** | A colloidal clay largely made up of the mineral sodium montmorillonite, a hydrated aluminum silicate. Used for developing a low-shear-rate viscosity and/or good filtration characteristics in water-based drilling fluids. The generic term "bentonite" is neither an exact mineralogical name, nor is the clay of definite mineralogical composition. See: Gel, Montmorillonite. |
| **Bicarb** | *See:* Sodium Bicarbonate. |
| **Bingham Model** | A mathematical description that relates shear stress to shear rate in a linear manner. This model requires only two constants (plastic viscosity and yield point) and is the simplest rheological model possible to describe a non-Newtonian liquid. It is very useful for analyzing drilling fluid problems and treatment. *See:* Viscosity, Pseudoplastic Fluid, Plastic Viscosity, Yield Point, Gel Strength. |
| **Blade** | *See:* Flight, Flute. |
| **Blinding** | A reduction of open area in a screening surface caused by coating or plugging. *See:* Coating, Plugging. |
| **Blooie Line** | The flow line for air or gas drilling. |
| **Blowout** | An uncontrolled escape of drilling fluid, gas, oil, or water from the well caused by the formation pressure being greater than the hydrostatic head of the fluid being circulated in the wellbore. *See:* Kick, Kill Fluid. |
| **Bonded Screens** | Screens bonded to each other or screens bonded together with plastic to a metal plate. |
| **Bonding Material** | Material used to secure screen cloth to a backing plate or support screen. |
| **Bottom Flooding** | The behavior of a hydrocyclone when the underflow discharges whole drilling fluid rather than separated solids. |
| **Bound Liquid** | Adsorbed liquid. *See:* Absorb, Absorption, Adsorb, Adsorption, Adsorbed Liquid. |
| **Bow** | *See:* Crown. |

| | |
|---|---|
| **Bowl** | The outer rotating chamber of a decanting centrifuge. |
| **Brackish Water** | Water containing low concentrations of any soluble salts. |
| **Break Circulation** | To start movement of the drilling fluid after it has been quiescent in a borehole. |
| **Bridge** | An obstruction in a well formed by the intrusion of subsurface formations and/or cuttings or material that prevents a tubular string from moving down a borehole. |
| **Brine** | Water containing a high concentration of common salts such as sodium chloride, calcium chloride, calcium bromide, zinc bromide, and so forth. |
| **Bromine Value** | The number of centigrams of bromine that are absorbed by 1 gram of oil under certain conditions. The Bromine Check is is a test for the degree of unsaturation of a given oil. |
| **Brownian Movement** | Continuous, irregular motion exhibited by particles suspended in a liquid or gaseous medium, usually as a colloidal dispersion. |
| **BS & W** | Base sediment and water. |
| **Buffer** | Any substance or combination of substances which, when dissolved in water, produces a solution that resists a change in its hydrogen ion concentration upon the addition of acid or base. |
| **Cable Tool Drilling** | A method of drilling a well by allowing a weighted bit (or chisel) at the bottom of a cable to fall against the formation being penetrated. The cuttings are then bailed from the bottom of the wellbore using a bailer. *See:* Rotary Drilling. |
| **Cake Consistency** | According to API Bulletin RP 13B, terms such as "hard," "soft," "tough," "rubbery," "firm," and the like, may be used to convey some idea of cake consistency. |
| **Cake Thickness** | (1) A measurement of the filter cake thickness deposited by a drilling fluid against a porous medium, usually filter paper, according to the standard API filtration test. Cake thickness is usually reported in 32nds of an inch or millimeters. (2) The filter cake thickness deposited on the wall of a borehole. *See:* Filter Cake, Wall Cake. |
| **Calcium** | One of the alkaline earth elements with a valence of 2 and an atomic weight of about 40. Calcium compounds are a common cause of water hardness. Calcium is also a component of lime, gypsum, limestone, and so forth. |
| **Calcium Carbonate** | (1) $CaCO_3$. An acid soluble calcium salt sometimes used as a weighting material (limestone, oyster shell, etc.) in specialized drilling fluids. (2) A term used to denote a unit and/or standard to report hardness. *See:* Limestone. |
| **Calcium Chloride** | $CaCl_2$. A very soluble calcium salt sometimes added to drilling fluids to impart special inhibitive properties, but primarily used to increase the density of the liquid phase (water) in completion fluids and as an inhibitor to the water phase of invert oil emulsion drilling fluids. |
| **Calcium Contamination** | Dissolved calcium ions in sufficient concentration to impart undesirable properties in a drilling fluid such as flocculation, reduction in bentonite |

yield, increase in fluid loss, and so forth. *See:* Calcium Sulfate, Gyp, Anhydrite, Lime, Calcium Carbonate.

**Calcium Hydroxide**  $Ca(OH)_2$. The active ingredient of slaked lime. It is also the main constituent in cement (when wet) and is referred to as "lime" in field terminology. *See:* Lime.

**Calcium Sulfate**  Anhydrite ($CaSO_4$), plaster of Paris ($CaSO_4 \cdot \frac{1}{2}H_2O$), and gypsum ($CaSO_4 \cdot 2H_2O$). Calcium sulfate occurs in drilling fluids as a contaminant or may be added as a commercial product to certain drilling fluids to impart special inhibitive properties. *See:* Gypsum, Anhydrite.

**Calcium Treated Drilling Fluids**  Drilling fluids to which quantities of soluble calcium compounds have been added or allowed to remain from the formation drilled in order to impart special inhibitive properties to the drilling fluid.

**Calendered Wire Cloth**  Wire cloth that has been passed through a pair of heavy rollers to reduce the thickness of the cloth, or to flatten the intersections of the wire, and produce a smooth surface. This process is usually done to the coarser backing clothes. *See:* Market Grade Cloth, Mill Grade Cloth.

**Capacity**  The maximum volume flowrate at which a solids control device is designed to operate without detriment to separation. *See:* Feed Capacity, Solids Discharge Capacity.

**Cascade**  Gravity-induced flow of fluid from one unit to another.

**Cascade Shaker Arrangement**  A system that processes the drilling fluid through two or more shakers arranged in series.

**Cation**  The positively charged particle in the solution of an electrolyte that, under the influence of an electrical potential, moves toward the cathode (negative electrode). Examples are $Na^+$, $H^+$, $NH_4^+$, $Ca^+$, $Mg^{++}$, and $Al^{+++}$.

**Cation Exchange Capacity (CEC)**  The total amount of cations adsorbed on the basal surfaces or broken bond edges of a clay sample, expressed in milli-equivalents per 100 grams of dry clay. *See:* Base Exchange, Methylene Blue Titration, Methylene Blue Test, MBT, CEC.

**Caustic**  *See:* Sodium Hydroxide.

**Caustic Soda**  *See:* Sodium Hydroxide.

**Cave-In**  A severe form of sloughing. *See:* Sloughing.

**Cavernous Formation**  A formation having voluminous voids, usually the result of dissolution by formation waters that may or may not be still present.

**Caving**  A severe form of sloughing. *See:* Sloughing, Heaving.

**Cavitation**  The formation and collapse of low pressure bubbles in a liquid. Cavitation in centrifugal pumps occurs when the pressure within the impeller chamber decreases below the vapor pressure of the liquid. As these vapor bubbles move to the impeller tip and into a higher pressure region, they implode or collapse. The pressure at the suction entry may be considerably below atmospheric pressure if the pressure loss in the suction line is too large, the flowrate from the pump is too large for the inlet

| | |
|---|---|
| | size, or the fluid must be lifted to excessive heights. As the bubbles move out to the impeller tips, they implode releasing a large amount of energy that can actually chip metal pieces from the impeller blade. Cavitation frequently sounds like the centrifugal pump is pumping gravel. *See:* Centrifugal Pump. |
| **CEC** | *See:* Cation Exchange Capacity. |
| **Cement** | A mixture of calcium aluminates and silicates made by combining lime and clay while heating. Slaked cement contains about 62.5% calcium hydroxide, which can cause a major problem if cement contaminates the drilling fluid. |
| **Centipoise (cp)** | Unit of viscosity equal to 0.01 poise. Poise equals 1 dyne-second per square centimeter. The viscosity of water at 20°C is 1.005 cp (1 cp = 0.000672 lb/ft-sec). |
| **Centrifugal Force** | The force that tends to impel matter outward from the center of rotation. *See:* "G"-Force. |
| **Centrifugal Pump** | A machine for moving fluid by spinning it using a rotating impeller in a pump casing with a central inlet and a tangential outlet. The fluid flows in an increasing spiral from the inlet at the center to the outlet, tangent to the annulus. In the annular space between the impeller vane tips and the casing wall, the fluid velocity is roughly that of the impeller vane tips. The pump is effective when some of the spinning fluid flows out of the casing tangential outlet into the pipe system. Power from the motor is used to accelerate the fluid entering the inlet up to the speed of the fluid in the annulus. (Some of the motor power is expended as friction of the fluid in the casing and impeller.) |
| **Centrifugal Separator** | A general term applicable to any device using centrifugal force to shorten and/or to control the settling time required to separate a heavier mass from a lighter mass. |
| **Centrifuge** | A centrifugal separator, specifically a device rotated by an external force for the purpose of separating materials of different masses. This device is used for the mechanical separation of solids from a drilling fluid. Usually in a weighted drilling fluid, it is used to eliminate colloidal solids. In an unweighted drilling fluid it is used to remove solids larger than the collodials. The centrifuge uses high-speed, mechanical rotation to achieve this separation as distinguished from the cyclone-type separator in which the fluid energy alone provides the separating force. *See:* Hydrocyclone, Desander, Desilter. |
| **Ceramics** | A general term for heat hardened clay products that resist abrasion; used to extend the useful life of wear parts in pumps and hydrocyclones. |
| **Check/Suction Section** | The last active section in the surface system. It provides a location for rig pump and drilling fluid hopper suction. This section should be large enough to check and adjust drilling fluid properties before the drilling fluid is pumped downhole. |
| **Chemical Barrel** | A container in which soluble chemicals can be mixed with a limited amount of fluid prior to addition to the circulating system. |
| **Chemical Treatment** | The addition of chemicals (such as caustic, thinners, or viscosifiers) to the drilling fluid to adjust the drilling fluid properties. |

| | |
|---|---|
| **Chemicals** | In drilling fluid terminology, a chemical is any material that produces changes in the low-shear-rate viscosity, yield point, gel strength, fluid loss, pH, or surface tension. |
| **Choke** | An opening, aperture, or orifice used to restrict a rate of flow or discharge. |
| **Chromate** | A compound in which chromium has a valence of 6 (e.g., sodium dichromate). Chromate may be added to drilling fluids either directly or as a constituent of chrome lignites or chrome lignosulfonates to assist with rheology stabilization. In certain areas, chromate is widely used as an anodic corrosion inhibitor, often in conjunction with lime. |
| **Chrome Lignite** | Mined lignite, usually leonardite, to which chromate has been added and/or reacted. The lignite can also be causticized with either sodium or potassium hydroxide. The chrome lignite is used for rheology stabilization and filtration control of the drilling fluid. |
| **Circular Motion** | A shale shaker screen moves in a uniform circular motion when the vibrator is located at the center of gravity of the vibrating basket. |
| **Circulation** | The movement of drilling fluid through the flow system on a drilling or work-over rig. This circulation starts at the suction pit and travels through the mud pump, drill pipe, bit, annular space in the hole, flow line, fluid pits, and back again to the suction pit. The time involved is usually referred to as circulation time. *See:* Reverse Circulation. |
| **Circulating Sytem** | All of the drilling fluid moving through the surface system and downhole. |
| **Circulation Rate** | The volume flow rate of the circulating drilling fluid, usually expressed in gallons per minute or barrels per minute. *See:* Flow Rate. |
| **Clabbered** | A slang term commonly used to describe moderate to severe flocculation of drilling fluid due to various contaminants. *See:* Gelled Up. |
| **Clarification** | Any process or combination of processes, the primary purpose of which is to reduce the concentration of suspended matter in liquid. |
| **Clay** | (1) A soft, variously colored earth, commonly hydrous silicates of alumina, formed by the decomposition of feldspar and other aluminum silicates. Clay minerals are essentially insoluble in water but disperse under hydration, grinding, or velocity effects. Shearing forces break down the clay particles to sizes varying from sub-micron to 100 microns or larger. (2) Physical Description: Solid particles of less than two micrometer equivalent spherical diameter. *See:* Attapulgite Clay, Bentonite, High-Yield Clay, Low-Yield Clay and Natural Clays. |
| **Clay Extender** | Substances, usually high molecular weight organic compounds, which when added in low concentrations to bentonite or other specific clay slurries, will increase the low-shear-rate viscosity of the system. An example would be polyvinyl acetate-maleic anhydride copolymer. *See:* Low Solids Drilling Fluids. |
| **Clay-Size Particles** | *See:* Clay. |
| **Close Loop Mud Systems** | A drilling fluid processing system that minimizes the liquid discard. Usually as much as possible of the liquid phase normally separated with drilled solids is returned to the active system. |

| | |
|---|---|
| **CMC** | *See:* Sodium Carboxymethylcelluose. |
| **Coagulation** | The destabilization and initial aggregation of colloidal and finely divided suspended matter by the addition of a floc-forming agent. |
| **Coalescence** | (1) The change from a liquid to a thickened curd-like state by chemical reaction. (2) The combination of globules in an emulsion caused by molecular attraction of the surfaces. |
| **Coarse Solids** | Solids larger than 2000 microns in diameter. |
| **Coating** | (1) A material adhering to a surface to change the properties of the surface. (2) A condition where material forms a film that covers the apertures of the screening surface. *See:* Blinding, Plugging. |
| **Cohesion** | The attractive forces between molecules of the same kind (i.e., the force), which holds the molecules of a substance together. |
| **Colloid** | A particle smaller than 2 microns is called a colloid. The size and electrical charge of the particles determine the different phenomena observed with colloids (e.g., Brownian movement). *See:* Clay, Colloidal Solids. |
| **Colloidal Composition** | A colloidal suspension containing one or more colloidal constituents. |
| **Colloidal Matter** | Finely divided solids that will not settle but may be removed by coagulation. |
| **Colloidal Solids** | Particles smaller than 2 microns. These particles are so small that they do not settle when suspended in a drilling fluid. Commonly used as a synonym for "clay." *See:* Clay, Colloid. |
| **Colloidal Suspension** | Finely divided particles that are so small they remain suspended in a liquid by Brownian movement. |
| **Combining Weight** | *See:* Equivalent Weight. |
| **Conductance** | The permeability of a shaker screen per unit thickness of the screen, measured in units of kilodarcies/millimeter, while the screen is stationary. |
| **Conductivity** | Measure of the quantity of electricity transferred across unit area per unit potential gradient per unit time. It is the reciprocal of resistivity. Electrolytes may be added to a fluid to alter its conductivity. *See:* Resistivity. |
| **Cone** | *See:* Hydrocyclone, Hydroclone. |
| **Connate Water** | Water trapped within sedimentary deposits, particularly as hydrocarbons displace most of the water from a reservoir. |
| **Consistometer** | A thickening time tester having a stirring apparatus to measure the relative thickening time for drilling fluid or cement slurries under predetermined temperatures and pressures. *See:* API Bulletin RP 10B. |
| **Contamination** | In a drilling fluid, the presence of any material that may tend to harm the desired properties of the drilling fluid. |
| **Continuous Phase** | (1) The fluid phase that completely surrounds the dispersed phase. (2) The fluid phase of a drilling fluid: either water, oil, or synthetic oil. The dispersed phase (noncontinuous phase) may be solids or liquid. |

**Controlled Aggregation** The condition in which the clay platelets are maintained stacked by a polyvalent cation, such as calcium.

**Conventional Drilling Fluid** A drilling fluid containing essentially clay and water.

**Conventional Mud** *See:* Conventional Drilling Fluid.

**Conventional Shale Shakers** Usually refers to shale shakers that vibrate screens with a circular or unbalanced elliptical motion. These shakers are usually limited to processing drilling fluid through shale shaker screens up to 100 mesh.

**Conveyance** Movement of solids toward the discharge end of a shaker screen.

**Conveyor** A mechanical device for moving material from one place to another. In a decanting centrifuge, this is a hollow hub fitted with flights rotating in the same direction but at a different speed than the centrifuge bowl. These flights are designed to move the coarse solids out of the bowl and are part of the conveyor.

**Co-polymer** A substance formed when two or more substances polymerize at the same time to yield a product, which is not a mixture of separate polymers, but a complex substance having properties different from either of the base polymers. Examples are polyvinyl acetate-maleic anyhdride copolymer (clay extender and selective flocculant), acrylamide-carboxylic acid copolymer (total flocculant). *See:* Polymer.

**Corrosion** A chemical degradation of a metal by oxygen in the presence of moisture. An oxide is the by-product of corrosion.

**Corrosion Inhibitor** An agent which, when added to a system, slows down or prevents a chemical reaction or corrosion. Corrosion inhibitors are widely used in drilling and producing operations to prevent corrosion of metal equipment exposed to hydrogen sulfide, carbon dioxide, oxygen, salt water, and so forth. Common inhibitors added to drilling fluids are filming amines, chromates, and oxygen scavengers.

**Crater** The formation of a large, funnel-shaped cavity at the top of a hole resulting from either a blowout or from caving.

**Creaming of Emulsions** The settling or rising of particles from the dispersed phase of an emulsion as observed by a difference in color shading of the layers formed. This separation can be either upward or downward, depending on the relative densities of the continuous and dispersed phases.

**Created Fractures** Induced fractures by means of hydraulic or mechanical pressure exerted in a formation by the drill string and/or circulating fluid.

**Critical Velocity** That velocity at the transitional point between laminar and turbulent types of fluid flow. This point occurs in the transitional range of Reynolds numbers between approximately 2,000 to 3,000.

**Crown** The curvature of a screen deck or the difference in elevation between its high and low points. *See:* Bow.

**Cubic Centimeter (cc)** A metric system unit for the measure of volume. A cube measuring 1 cm on each side would have a volume of 1 cubic centimeter (cc, $cm^3$). It is

| | |
|---|---|
| | essentially equal to the milliliter and commonly used interchangeably. One cubic centimeter of water at room temperature weighs approximately 1 gram. |
| **Cut Point** | Cut point curves are developed by dividing the mass of solids in a certain size range removed by the total mass of solids in that size range that enters the separation device. A cut point usually refers to the size particle that has a 50% chance of being discarded. *See:* Median Cut. |
| **Cutt Points** | Pronounced "Koot." The equivalent spherical diameters corresponding to the ellipsoidal volume distribution of a screen's opening sizes, as determined by image analysis. *See:* API Bulletin RP 13E. |
| **Cuttings** | The pieces of formation dislodged by the bit and brought to the surface in the drilling fluid. Field practice is to refer to all solids removed by the shaker screen as "cuttings," although some can be sloughed material from the wall of the borehole. *See:* Drilled Solids, Low-Gravity Solids, Samples. |
| **Cyclone** | A device for the separation of solid particles from a drilling fluid. The most common cyclones used for solids separation are a desander or desilter. In a cyclone, fluid is pumped tangentially into a cone, and the fluid rotation provides enough centrifugal force to separate particles by mass weight. *See:* Desander, Desilter, Hydrocyclone, Hydroclone. |
| **Cyclone Bottom** | *See:* Apex, Apex Valve, Underflow Opening. |
| **Darcy** | A unit of permeability. A porous medium has a permeability of 1 darcy when a pressure of 1 atm on a sample 1 cm long and 1 sq cm in cross-section will force a liquid of 1 cp viscosity through the sample at the rate of 1 cc per sec. *See:* Millidarcy, Permeability. |
| **Decanter** | *See:* Decanting Centrifuge. |
| **Decanting Centrifuge** | A centrifuge that removes solids from the feed slurry and discharges them as damp underflow. Ultrafine colloidal solids are discharged with the liquid overflow. The decanting centrifuge has an internal auger that moves the solids, which have settled to the bowl walls, out of a pool of liquid and to the underflow. *See:* Centrifuge. |
| **Deck** | The screening surface in a shale shaker basket. |
| **Deflocculant** | Chemical that promotes defloccuation. *See:* Thinner. |
| **Deflocculation** | (1) The process of thinning the drilling fluid by bonding with (neutralizing or covering) the positive electrical charges on drilling fluid additives to prevent one particle of drilling fluid to be attracted to another particle. (2) Breakup of flocs of gel structures by use of a thinner. |
| **Defoamer** | Any substance used to reduce or eliminate foam by reducing the surface tension of a liquid. *See:* Antifoam. |
| **Degasser** | A device that removes entrained gas from a drilling fluid, especially the very small bubbles that do not float readily in viscous drilling fluid. |
| **Dehydration** | Removal of free or combined water from a compound. |
| **Deliquescence** | The liquification of a solid substance due to the solution of the solid by absorption of moisture from the air (e.g., calcium chloride deliquesces in humid air). |

**Density**  
Density is the mass per unit volume expressed in pounds per gallon (ppg), grams per cubic centimeter (gm/cc), and pounds per cubic ft (lb/cu.ft). Drilling fluid density is commonly referred to as "mud weight."

**Desand**  
To remove most API sand (>74 microns) from drilling fluid.

**Desander**  
A hydrocyclone with an inside diameter of 6 inches or larger that can remove a high proportion of solids larger than 74 micrometer. Generally, desanders are used on unweighted muds. *See:* Cyclone, Hydrocyclone, Hydroclone, Desilter.

**Desilt**  
To remove most silt particles greater than 15–20 microns from an unweighted drilling fluid. The desilter is not normally used on weighted drilling fluids because it can remove large amounts of barite.

**Desilter**  
A hydrocyclone with an inside diameter less than 6 inches. They can remove a large fraction of solids larger than 15–20 microns. *See:* Cyclone, Hydrocyclone, Hydroclone, Desander.

**Destabliziation**  
A condition where colloidal particles no longer remain separate and discrete, but instead contact and agglomerate with other particles.

**Diatomaceous Earth**  
A natural earth compound composed of siliceous skeletons of diatoms, which is very porous. Sometimes used for controlling lost circulation, seepage losses, and as an additive to cement.

**Diesel Oil Plug**  
*See:* Gunk Plug.

**Differential Angle Deck**  
A screen deck in which successive screening surfaces of the same deck are at different angles.

**Differential Pressure**  
The difference in pressure between two points. Usually, differential pressure refers to the difference in pressure at a given point in the wellbore between the hydrostatic pressure of the drilling fluid column and the formation pressure. Differential pressure can be positive, zero, or negative with respect to the formation pressure. *See:* Back Pressure.

**Differential Pressure Sticking**  
Sticking which occurs when a portion of the drill string (usually the drill collars) becomes embedded in the filter cake resulting in a nonuniform distribution of pressure around the circumference of the pipe. The conditions essential for sticking require a permeable formation and a positive pressure (from wellbore to formation) differential across a drill string imbedded in a poor filter cake. *See:* Stuck.

**Differential Sticking**  
*See:* Differential Pressure Sticking.

**Diffusion**  
The spreading, scattering, or mixing of material (gas, liquid, or solid).

**Dilatant Fluid**  
Opposite of shear thinning. A dilatant or inverted plastic fluid is usually made up of a high concentration of well-dispersed solids, which exhibit a nonlinear consistency curve passing through the origin. The apparent viscosity increases instantaneously with increasing shear rate. The yield point, as determined by conventional calculations from the direct indicating viscometer readings, is negative. *See:* Apparent Viscosity, Viscosity, Bingham Model, Plastic Viscosity, Yield Point, Gel Strength.

**Diluent**  
Liquid added to dilute or thin a solution or suspension.

| | |
|---|---|
| **Dilution** | (1) Decreasing the percent of drilled solids concentration by addition of a liquid phase. (2) Increasing the liquid content of a drilling fluid by adding water or oil. |
| **Dilution Factor** | The ratio of the actual volume of drilling fluid required to drill a specified interval of footage using a solids removal system, versus a calculated volume of drilling fluid required to maintain the same drilled solids fraction over the same specified interval of footage with no solids removal system. |
| **Dilution Rate** | The rate in gallons per minute or barrels per hour that fluids and/or premix is added to the circulating system for the purpose of solids management. |
| **Dilution Ratio** | Ratio of volume of dilution liquid to the volume of raw drilling fluid in the feed prior to entering a liquid-solids separator. |
| **Dilution Water** | Water used for dilution of water-based drilling fluid. |
| **Direct Indicating Viscometer** | Commonly called a "V-G meter." The direct-indicating viscometer shears fluid between a rotating outer cylinder and a stationary cylindrical bob in the center of the rotating cylinder. The bob is constrained from rotating by a spring. The spring reads the drag force on the bob that is related to the shear stress. The rotational speed of the outer cylinder and the spacing between the bob and the cylinder determine the shear rate. Viscosity is the ratio of shear stress to shear rate, so this instrument may be used to determine viscosity of a fluid at a variety of shear rates. Gel strengths may also be determined after a quiescent period of a drilling fluid between the bob and the cylinder. *See:* API Bulletin RP 13B. |
| **Disassociation** | The splitting of a compound or element into two or more simple molecules, atoms, or ions. Usually applied to the effect of the action of heat or solvents on dissolved substances. The reaction is reversible and not as permanent as decomposition (i.e., when the solvent is removed, the ions recombine). |
| **Discharge** | Material removed from a system. *See:* Effluent. |
| **Discharge Spout or Lip** | Extension at the discharge area of a screen. It may be vibrating or stationary. |
| **Dispersant** | (1) Any chemical that promotes the subdivision of a material phase. (2) Any chemical that promotes dispersion of particles in a fluid. Frequently, a deflocculant is inaccurately called a dispersant (e.g., caustic soda is a dispersant, not a deflocculant). |
| **Disperse** | To separate into component parts. Bentonite disperses by hydration into many smaller pieces. |
| **Dispersed Phase** | The scattered phase (solid, liquid, or gas) of a dispersion. The particles are finely divided and completely surrounded by the continuous phase. |
| **Dispersion** | (1) To break-up and to scatter (as a reduction of particle size) and cause to spread apart. (2) Subdivision of aggregates. Dispersion increases the specific surface of the particle, which results in an increase in viscosity and gel strength. |
| **Dispersoid** | A colloid or finely divided substance. |

| | |
|---|---|
| **Distillation** | Process of first vaporizing a liquid and then condensing the vapor into a liquid (the distillate), leaving behind nonvolatile solid substances of a drilling fluid. The distillate is the water and/or oil content of a fluid. |
| **Divided Deck** | A deck having a screening surface longitudinally divided by partition(s). |
| **Dog Leg** | The "elbow" caused by a sharp change in drilling direction in the wellbore. |
| **Double Flute** | The flutes or leads advancing simultaneously at the same angle and 180° apart. *See:* Flute, Flight, Blade. |
| **Drill Stem Test (DST)** | A post-drilling and pre-production test, which allows formation fluids to flow into the drill pipe under controlled conditions, to determine whether oil and/or gas in commercial quantities have been encountered in the penetrated formations. |
| **Drilled Solids** | Formation solids that enter the drilling fluid system, whether produced by a bit or from the side of the borehole. *See:* Low-Gravity Solids, Cuttings. |
| **Drilled Solids Fraction** | The average volume fraction of drilled solids maintained in the drilling fluid over a specified interval of footage. |
| **Drilled Solids Removal System** | All equipment and processes used while drilling a well that remove the solids generated from the hole and carried by the drilling fluid (i.e., settling, screening, desanding, desilting, centrifuging, and dumping). |
| **Drilled Solids Removal System Performance** | A measure of the performance of a system to remove drilled solids from the drilling fluid. |
| **Drilling Fluid** | The term applied to any liquid or slurry pumped down the drill string and up the annulus of a hole to facilitate drilling. *See:* Drilling Mud, Mud. |
| **Drilling Fluid Additive** | Any material added to a drilling fluid to achieve a particular effect. |
| **Drilling Fluid Analysis** | Examination and testing of the drilling fluid to determine its physical and chemical properties and functional ability. *See:* API Bulletin RP 13B. |
| **Drilling Fluid Cycle Time** | The time necessary to move a fluid from the kelly bushing to the flow line in a borehole. The cycle in minutes equals the barrels of drilling fluid in the hole, minus pipe displacement, divided by barrels per minute of circulation rate: $(Hole_{bbl} - Pipe\ Volume_{bbl})/Circulation\ Rate_{bbl/min}$. |
| **Drilling Fluid Engineer** | One versed in drilling fluids, rig operations and solids and waste management whose duties are to manage and maintain the drilling fluid program at the well site. Also called a mud engineer. |
| **Drilling Fluid Program** | A proposed plan or procedure for application procedure and properties of drilling fluid(s) used in drilling a well with respect to depth. Some factors that influence the drilling fluid program are the casing program and formation characteristics such as type, competence, solubility, temperature, pressure, and so forth. |
| **Drilling In** | The drilling operation starting at the point of drilling into the producing formation. |
| **Drilling Mud** | Drilling Fluid is the preferred term. *See:* Drilling Fluid. |

| | |
|---|---|
| **Drilling Out** | The operation of drilling the casing shoe after the cementing a casing or liner in place. Drilling out of the casing is required before a bore hole is deepened. |
| **Drilling Rate** | The rate at which hole depth progresses, expressed in linear units per unit of time (including connections) as feet/minute or feet/hour. *See:* ROP, Rate of Penetration, Penetration Rate. |
| **Dry Bottom** | An adjustment to the underflow opening of a hydrocyclone that causes a dry beach, usually resulting in severe plugging. *See:* Dry Plug. |
| **Dry Plug** | The plugging of the underflow opening of a hydrocyclone caused by operating with a dry bottom. *See:* Dry Bottom. |
| **Dryer** | A shale shaker with a fine mesh screen that removes excess fluid and fine solids from discarded material from other shale shakers and hydrocyclones. Typically, this is used to decrease the liquid waste from a drilling fluid to reduce discarded volumes. *See:* Mud Cleaner. |
| **Dryer Shaker** | *See:* Dryer. |
| **Dual Wound Motors** | Motors that may be connected to either of two voltages and starter configurations. |
| **Dynamic** | The state of being active, or in motion, as opposed to static. |
| **ECD** | *See:* Equivalent Circulating Density, Annular Pressure Loss. |
| **Eductor** | (1) A device consisting of a fluid stream discharging under high pressure from a jet through an annular space to create a vacuum. When properly arranged, it can evacuate degassed drilling fluid from a vacuum-type degasser. (2) A device using a high velocity jet to create a vacuum that draws in liquid or dry material to be blended with the drilling fluid. |
| **Effective Screening Area** | The portion of a screen surface available for solids separation. |
| **Effluent** | A discharge of liquid. Generally, used to describe a stream of liquid after some attempt at separation or purification has been made. *See:* Discharge. |
| **Elastomer** | Any rubber or rubber-like material (such as polyurethane). |
| **Electric Logging** | Logs run on a wire line to obtain information concerning the porosity, permeability density, and/or fluid content of the formations drilled. The drilling fluid characteristics may need to be altered to obtain satisfactory logs. |
| **Electrolyte** | A substance that dissociates into charged positive and negative ions when in solution or a fused state. This electrolyte will then conduct an electric current. Acids, bases, and salts are common electrolytes. |
| **Elevation Head** | The pressure created by a given height of fluid. *See:* Hydrostatic Pressure Head. |
| **Emulsifier** | A substance used to combine two liquids that do not solubilize in each other or maintain a stable mixture when agitated in the presence of each other. Emulsifiers may be divided, according to their behavior, into ionic and nonionic agents. The ionic types may be further divided into anionic, cationic, and amphoteric, depending on the nature of the ionic groups. |

| | |
|---|---|
| **Emulsion** | A substantially permanent heterogeneous mixture of two or more liquids that do not normally dissolve in each other, but which are held in a dispersed state, one within the other. This dispersion is accomplished by the combination of mechanical agitation, fine solids, and/or emulsifiers. Emulsions may be mechanical, chemical, or a combination of both. Emulsions may be either oil-in-water or water-in-oil. *See:* Interfacial Tension, Surface Tension. |
| **Emulsoid** | Colloidal particle that takes up water. |
| **Encapsulation** | A process of enclosing a material in a covering. Used to describe the process of totally enclosing electrical parts or circuits with a polymeric material (usually epoxy); or coating drilled solids with a polymer. |
| **End Point** | Indicates the end of a chemical testing operation when a clear and definite change is observed in the test sample. In titration, this change is frequently in color of an indicator or marker, which has been added to the solution or the disappearance of a colored reactant. |
| **Enriching** | The process of increasing the concentration of a flammable gas or vapor to a point where the atmosphere has a concentration of that flammable gas or vapor above its upper flammable or explosive limit. |
| **EP Additive** | *See:* Extreme Pressure Lubricant. |
| **EPL** | *See:* Extreme Pressure Lubricant. |
| **Epm** | *See:* Equivalents Per Million, Parts Per Million. |
| **Equalizer** | An opening for flow between compartments in a surface fluid holding system. |
| **Equivalent Circulating Density** | The effective drilling fluid weight at any point in the annulus of the wellbore during fluid circulation. ECD includes drilling fluid density, cuttings in the annulus, and annular pressure loss. *See:* Annular Pressure Loss, ECD. |
| **Equivalent Spherical Diameter** | The theoretical dimension usually referred to when the sizes of irregularly shaped, small particles are discussed. These dimensions can be determined by several methods such as settling velocity, electrical resistance, and light reflection. *See:* ESD, Particle Size. |
| **Equivalent Weight** | The atomic weight or formula weight of an element, compound, or ion divided by its valence. Elements entering into combination always do so in quantities proportional to their equivalent weights. *See:* Combining Weight. |
| **Equivalents Per Million** | Unit chemical weight of solute per million unit weights of solution. The epm of a solute in solution is equal to the ppm (parts per million) divided by the equivalent weight. *See:* Parts Per Million. |
| **ESD** | *See:* Equivalent Spherical Diameter, Particle Size. |
| **Extreme Pressure Lubricant** | Additives which, when added to the drilling fluid, impart lubrication to bearing surfaces when subjected to extreme pressure conditions. *See:* EPL. |
| **Fault** | Geological term denoting a formation break across the trend of a subsurface strata. Faults can significantly affect the drilling fluid and casing programs due to possibilities for lost circulation, sloughing hole, or kicks. |

## GLOSSARY

**Feed** — A mixture of solids and liquid (including dilution liquid) entering a liquid-solids separation device.

**Feed Capacity** — The maximum volume flowrate at which a solids control device is designed to operate without detriment to separation efficiency. This capacity depends on particle size, particle concentration, viscosity, and other variables of the feed. *See:* Capacity, Flow Capacity, Solids Discharge Capacity.

**Feed Chamber** — That part of a device, that receives the mixture of diluents, drilling fluid, and solids to be separated.

**Feed Head** — The equivalent height, in feet or meters, of a column of fluid at the cyclone feed header.

**Feed Header** — A pipe, tube, or conduit to which two or more hydrocyclones are connected and from which they receive their feed slurry.

**Feed Inlet** — The opening through which the feed fluid enters a solids separation device. *See:* Feed Opening.

**Feed Mud** — *See:* Feed.

**Feed Opening** — *See:* Feed Inlet.

**Feed Pressure** — The actual gauge pressure measured as near as possible to, and upstream of, the feed inlet of a device.

**Feed Slurry** — *See:* Feed.

**Fermentation** — Decomposition process of certain substances (e.g., starch) in which a chemical change is brought about by enzymes, bacteria, or other microorganisms. Often referred to as "souring."

**Fibrous Materials** — Any tough, stringy material used to prevent circulation loss or to restore circulation. In field use, fiber generally refers to the larger fibers of plant origin.

**Filter Cake** — The suspended solids that are deposited on a porous medium during the process of filtration. *See:* Wall Cake.

**Filter Cake Texture** — The physical properties of a cake as measured by toughness, slickness, and brittleness. *See:* Cake Consistency.

**Filter Paper** — Porous paper without surface sizing for filtering solids from liquids. The API filtration test specifies 9-cm diameter filter paper (Whatman No. 50, S&S No. 576, or equivalent).

**Filtrate** — The liquid that is forced through a porous medium during the filtration process. *See:* Fluid Loss.

**Filtration Rate** — *See:* Fluid Loss.

**Fill-Up Line** — The line through which fluid is added to the annulus to maintain the fluid level in the wellbore during the extraction of the drilling assembly.

**Filter Cake** — (1) The solid residue deposited by a drilling fluid against a porous medium, usually filter paper, according to the standard API filtration test.

(2) The solid residue deposited on the wall of a borehole during the drilling of permeable formations. *See:* Wall Cake.

**Filter Cake Thickness** — *See:* Cake Thickness.

**Filter Press** — A device for determining the fluid loss of a drilling fluid having specifications in accordance with API Bulletin RP 13B. *See:* API Bulletin RP 13B.

**Filter Run** — The interval between two successive backwashing operations of a filter.

**Filterability** — The characteristic of a clear fluid that denotes both the ease of filtration and the ability to remove solids while filtering.

**Filtrate Loss** — *See:* Fluid Loss.

**Filtration** — (1) The separation process of suspended solids from liquid by forcing the liquid through a porous medium while screening back the solids. Two types of fluid filtration occur in a well: dynamic filtration while circulating, and static filtration when the fluid is at rest. (2) The process of drilling fluid losing a portion of the liquid phase to the surrounding formation. *See:* Water Loss.

**Filtration Rate** — *See:* Fluid Loss.

**Fine Screen Shaker** — A vibrating screening device designed for screening drilling fluids through screen cloth finer than 80 mesh.

**Fine Screen Shale Shakers** — Usually refers to shale shakers that vibrate screens with a balanced elliptical or linear motion. These are usually capable of processing large flow rates of drilling fluid through 120- to 250-mesh screens.

**Fine Solids** — Solids 44–74 microns in diameter or sieve size 325–200 mesh. *See:* API Bulletin RP 13C.

**Fishing** — Rig operations for the purpose of retrieving sections of pipe, collars, or other obstructive items that are in the hole and would interfere with drilling or logging operations.

**Flat Decked** — Shaker screens that do not have a crowned, or bowed, surface.

**Flat Gel** — A condition wherein the gel strength does not increase appreciably with time and is essentially equal to the initial gel strength. Opposite of progressive gel. *See:* Progressive Gel, Zero-Zero Gel.

**Flight** — On a decanting centrifuge, one full turn of a spiral helix such as a flute or blade of a screw-type conveyor. *See:* Blade, Flute.

**Flipped** — A slang term for an extreme imbalance in a drilling fluid. In a water-in-oil emulsion, the emulsion is identified as "flipped" when the continuous and dispersed phases separate and the solids begin to settle.

**Floc** — Small gelatinous masses of solids formed in a liquid.

**Flocculates** — A group of aggregates or particles in a suspension formed by electrostatic attraction forces between negative and positive charges. Bentonite clay particles have negatively charged surfaces that will attract positive charges such as those of other bentonite positive-edge charges.

| | |
|---|---|
| **Flocculating Agent** | Substances (e.g., most electrolytes, a few polysaccharides, certain natural or synthetic polymers) that cause the thickening of a drilling fluid. In Bingham Plastic fluids, the yield point and gel strength increase with flocculation. |
| **Flocculation** | (1) Loose association of particles in lightly bonded groups, sometimes called "flocs," with non-parallel association of clay platelets. In concentrated suspensions, such as drilling fluids, flocculation results in gelation. In some drilling fluids, flocculation may be followed by irreversible precipitation of colloids and certain other substances from the fluid (e.g., red beds, polymer flocculation). (2) A process in which dissimilar electrical charges on clay platelets are attracted to each other. This increases the yield point and gel strength of a slurry. |
| **Flooding** | (1) The effect created when a screen, hydrocyclone, or centrifuge is fed beyond its capacity. (2) Flooding may also occur on a screen as a result of blinding. |
| **Flow Capacity** | The rate at which a shaker can process drilling fluid and solids. This depends on many variables including shaker configuration, design and motion, drilling fluid rheology, solids loading, and blinding by near-size particles. *See:* Feed Capacity. |
| **Flow Line** | The pipe (usually) or trough that conveys drilling fluid from the rotary nipple to the solids separation section of the drilling fluid tanks on a drilling rig. |
| **Flow Rate** | The volume of liquid or slurry moved through a pipe in one unit of time (i.e., gallons per minute, barrels per minute, etc.). *See:* Circulation Rate. |
| **Flow Streams** | With respect to centrifugal separators, all liquids and slurries entering and leaving a machine, such as feed drilling fluid stream plus dilution stream equals overflow stream plus underflow stream. |
| **Flow-Back Pan** | A pan or surface below a screen that causes fluid passing through one screen to flow back to the feed end of a lower screen. |
| **Fluid** | Any substance that will readily assume the shape of the container in which it is placed. The term includes both liquids and gases. It is a substance in which the application of every system of stresses (other than hydrostatic pressure) will produce a continuously increasing deformation without any relation between time rate of deformation at any instant and the magnitude of stresses at the instant. |
| **Fluid Flow** | The state of dynamics of a fluid in motion as determined by the type of fluid (e.g., Newtonian plastic, pseudo-plastic, dilatant), the properties of the fluid such as viscosity and density, the geometry of the system, and the velocity. Thus, under a given set of conditions and fluid properties, the fluid flow can be described as plug flow, laminar (also called Newtonian, streamline, parallel, or viscous) flow, or turbulent flow. *See:* Reynolds Number. |
| **Fluid Loss** | Measure of the relative amount of fluid loss (filtrate), through permeable formations or membranes, when the drilling fluid is subjected to a pressure differential. *See:* Filtrate Loss, API Bulletin RP 13B. |
| **Fluidity** | The reciprocal of viscosity. The measure of rate with which a fluid is continuously deformed by a shearing stress. Ease of flowing. |

| | |
|---|---|
| **Fluorescence** | Instantaneous re-emission of light of a greater wavelength than that of the light originally absorbed. |
| **Flute** | A curved, metal blade wrapped around a shaft, as on a screw conveyor in a cetrifuge. *See:* Blade, Flight. |
| **Foam** | (1) A two-phase system, similar to an emulsion, where the dispersed phase is a gas or air. (2) Bubbles floating on the surface of the drilling fluid. The bubbles are usually air but can be formation gas. |
| **Foaming Agent** | A substance that produces fairly stable bubbles at the air-liquid interface due to agitation, aeration, or ebullition. In air or gas drilling, foaming agents are added to turn water influx into aerated foam. This is commonly called "mist drilling." |
| **Foot** | Unit of length in British (foot-pound-second) system. |
| **Foot Pound** | Unit of work or of mechanical energy—which is the capacity to do work. One foot-pound is the work performed by a force of one pound acting through a distance of one foot; or the work required to lift a one pound weight a vertical distance of one foot. |
| **Foot Valve** | A check valve installed at the suction end of a suction line. |
| **Formation Damage** | Damage to the productivity of a well as a result of invasion of the formation by drilling fluid particles, drilling fluid filtrates, and/or cement filtrates. Formation damage can also result from changes in pH and a variety of other conditions. Asphalt from crude oil will also damage some formations. *See:* Mudding Off. |
| **Formation Sensitivity** | The tendency of certain producing formations to adversely react with the drilling and completion process. |
| **Founder Point** | The bit loading value (weight and rotary speed) that causes the drill bit to redrill cuttings already broken from the formation. |
| **Free Liquid** | The liquid film that can be removed by gravity draining or centrifugal force. *See:* Absorb, Absorption, Adsorption, Adsorb, Bound Liquid. |
| **Freshwater Drilling Fuid** | A drilling fluid in which the liquid phase is freshwater. |
| **Freshwater Mud** | *See:* Freshwater Drilling Fluid. |
| **Friction Loss** | *See:* Pressure Drop, Pressure Loss. |
| **Functions of Drilling Fluids** | Drilling fluids in rotary drilling must remove cuttings from the bottom of the hole, bring those cuttings and any material from the side of the hole to the surface, control subsurface formation pressures, cool the drill bit, lubricate the drill string, create an impermeable filter cake, refrain from invading the formations with excessive quantities of drilling fluid filtrate, and provide a wellbore that can be evaluated and produce hydrocarbons. |
| **Funnel Viscosity** | *See:* Kinematic Viscosity, Marsh Funnel Viscosity. |
| **Galena** | Lead sulfide (PbS). Technical grades (specific gravity about 7.0) are used for increasing the density of drilling fluids to points impractical or |

impossible with barite. Almost entirely used in preparation of "kill fluids." *See:* Kill Fluid.

**Gas Buster** *See:* Poor Boy Degasser, Mud/Gas Separator.

**Gas Cut** Gas entrained by a drilling fluid. *See:* Air Cutting, Aeration.

**Gel** (1) A state of a colloidal suspension in which shearing stresses below a certain finite value fail to produce permanent deformation. The minimum shearing stress that will produce permanent deformation is known as the shear or gel strength of the gel. Gels commonly occur when the dispersed colloidal particles have a great affinity for the dispersing medium (i.e., are lyophilic). Thus, gels commonly occur with bentonite in water. (2) A term used to designate highly colloidal, high yielding, viscosity building, commercial clays, such as bentonite and attapulgite. *See:* Gel Strength.

**Gel Cement** Cement having a small to moderate percentage of bentonite added as a filler and/or reducer of the slurry weight. The bentonite may be dry blended into the mixture or added as a prehydrated slurry.

**Gel Strength** (1) The ability or measure of the ability of a colloid to form gels. Gel strength is a pressure unit usually reported in lb/100 sq ft. It is a measure of the same inter-particle forces of a fluid as determined by the yield point, except that gel strength is measured under static conditions and the yield point is measured under dynamic conditions. The common gel strength measurements are initial, 10-minute, and 30-minute gels. (2) The measured initial gel strength of a fluid is the maximum reading (deflection) taken from a direct reading viscometer after the fluid has been quiescent for 10 seconds. It is reported in lb/100 sq ft. *See:* API Bulletin RP 13B, Shear Rate, Shear Stress, Thixotropy.

**Gelation** Association of particles forming continuous structures at low shear rates.

**Gelled Up** Oil field slang usually referring to any fluid with a high gel strength and/or highly viscous properties. Often a state of severe flocculation. *See:* Clabbered.

**"G"-Factor** The acceleration of an object relative to the acceleration of gravity.

**"G"-Force** Refers to the centrifugal force exerted on a mass moving in a circular path. *See:* "G"-Factor.

**Glossary** A collection of glosses, such as a vocabulary of specialized terms with accompanying definitions.

**Glosses** Explanations or comments to elucidate some difficulty or obscurity in the text; or annotations.

**Grains per Gallon (gpg)** Ppm equals gpg × 17.1.

**Greasing Out** In some cases, certain organic substances, usually fatty acid derivatives, which are added to drilling fluids as emulsifiers (e.g., lubricants, etc.), may react with ions such as calcium and magnesium to form a water-insoluble, greasy material that separates out from the drilling fluid. This separation process is called greasing out.

**Guar Gum** A naturally occurring hydrophilic polysaccharide derived from the seed of the guar plant. The gum is chemically classified as a galactomannan.

|  |  |
|---|---|
|  | Guar gum slurries developed in clear fresh or brine water possess pseudoplastic flow properties. |
| **Gum** | Any hydrophilic plant polysaccharides or their derivatives that, when dispersed in water, swell to produce a viscous dispersion or solution. Unlike resins, they are soluble in water and insoluble in alcohol. |
| **Gumbo** | Small, sticky drilled solids that hydrate as they move up an annulus forming large agglomerations of cuttings. Usually, gumbo is characteristically observed with water-based drilling fluids when drilling shales containing large quantities of smectite clay. |
| **Gumbo Buster** | *See:* Gumbo Remover. |
| **Gumbo Remover** | A device that removes gumbo from a drilling fluid; usually mounted in the flowline between the well and shakers. |
| **Gunk Plug** | A volume of bentonite in oil that are pumped in a well to combat lost circulation. When the bentonite encounters water, it expands and creates a gunk plug with a high viscosity and gel structure. The plug may or may not be squeezed. *See:* Diesel Oil Plug. |
| **Gunning the Pits** | Mechanical agitation of the drilling fluid in a pit by means of a mud gun. *See:* Mud Gun. |
| **Gyp** | *See:* Gypsum. |
| **Gypsum** | Gypsum, $CaSO_4 \cdot 2H_2O$, is calcium sulfate and is frequently encountered while drilling. It may occur as thin stringers or in massive formations. *See:* Anhydrite, Calcium Sulfate. |
| **Hardness (Water)** | The hardness of water is due principally to the calcium and magnesium ions. The total hardness is measured in terms of parts per million of calcium carbonate or calcium and sometimes equivalents per million of calcium. *See:* API Bulletin RP 13B. |
| **Head** | The height a column of fluid would stand in an open-ended pipe if it was attached to the point of interest. The head at the bottom of a 1,000 foot well would be 1,000 feet, but the pressure would depend on the density of the drilling fluid in the well. |
| **Heaving** | The partial or complete collapse of the walls of a hole resulting from internal pressures due primarily to swelling from hydration or formation pressures or internal stresses. *See:* Sloughing. |
| **Heavy Solids** | *See:* High-Gravity Solids (HGS). |
| **Hertz** | A unit of frequency: Cycles per second. |
| **Heterogeneous** | A substance that consists of more than one phase and is not uniform, such as colloids, emulsions, and the like. It has different properties in different parts. |
| **High-Gravity Solids (HGS)** | Solids purchased and added to a drilling fluid specifically and solely to increase drilling fluid density. Barite (4.2 specific gravity) and hematite (5.05 specific gravity) are the most common. *See:* Low-Gravity Solids (LGS). |

| | |
|---|---|
| **High-pH Drilling Fluid** | A drilling fluid with a pH range above 10.5. A high-alkalinity drilling fluid. *See:* pH. |
| **High-Yield Clay** | A classification given to a group of commercial drilling clay preparations having a yield of 35 to 50 bbl/ton, an intermediate rating between bentonite and low-yield clays. High-yield drilling clays are usually prepared by peptizing low-yield calcium montmorillonite clays or, in a few cases, by blending some bentonite with the peptized low-yield clay. *See:* Low-Yield Clay, Bentonite. |
| **HLB** | *See:* Hydrophilic-Lipophilic Balance. |
| **Homogeneous** | Of uniform or similar nature throughout; or a substance or fluid that has the same property or composition at all points. |
| **Hookstrips** | The hooks on the edges of a screen section of a shale shaker that accept the tension member for screen mounting. |
| **Hookstrip Panel** | One of the two main screen panel-types, which consists of one to three layers of screen bordered by metal strips running parallel to the loom. The metal strips have a "U"-shaped cross-section that allow them to be secured and stretched by the shaker tensioning drawbars. These screens are nonpretensioned. *See:* Rigid Frame Panel. |
| **Hopper** | A large, funnel-shaped or cone-shaped device for mixing desirable dry solids or liquids into a drilling fluid stream to uniformly mix these desirable components into the slurry. The solids are wetted prior to entry into the drilling fluid system. The system usually consists of a jet nozzle, an open top hopper, and downstream venturi. *See:* Mud Hopper. |
| **Horsepower** | The rate of doing work or of expending mechanical energy (i.e., horsepower) is work performed per unit of time.<br>• 1 hp = 550 ft-lb/sec = .7067 Btu/sec. = 0.7457 kilowatt<br>• Rated horsepower, converted to kilowatts—horsepower × 0.746 = kilowatts<br>• Motor nameplate horsepower is the maximum steady load that the motor can pull without damage. |
| **Horsepower-Hour** | Horsepower-Hour (hp-hr) and Kilowatt-Hour (kw-hr) are units of work.<br>• 1 hp-hour = 1,980,000 ft-lb = 2545 Btu<br>• 1 hp-hour = 0.7457 kilowatt hours (kw-hr)<br>• 1 kw-hour = 1.341 hp-hr = 3413 Btu = 2,655,000 ft-lb |
| **Horseshoe Effect** | The "U" shape formed by the leading edge of drilling fluid moving down a shale shaker screen. The drilling fluid usually tends to pass through the center of a crowned screen faster than it passes through the edges, thereby creating the "U" shape. |
| **HTHP** | High Temperature High Pressure. |
| **HTHP Filter Press** | A device used to measure the fluid loss under HTHP conditions. *See:* HTHP Fluid Loss. |
| **HTHP Fluid Loss** | The fluid loss measured under HTHP conditions, usually 300°F and 500 psi differential pressure. *See:* HTHP Filter Press. |

| | |
|---|---|
| **Humic Acid** | Organic acids of indefinite composition found in naturally occurring leonardite lignite. The humic acids are the active constituents that assist in the positive adjustment of drilling fluid properties. *See:* Lignin. |
| **Hydrate** | A substance containing water combined in molecular form (such as $CaSO_4 \bullet 2H_2O$). A crystalline substance containing water of crystallization. |
| **Hydration** | The act of a substance to take up water by means of absorption and/or adsorption; usually results in swelling, dispersion, and disintegration into colloidal particles. *See:* Absorb, Absorption, Adsorb, Adsorbed Liquid. |
| **Hydroclone** | *See:* Cyclone, Hydrocyclone. |
| **Hydrocyclone** | A liquid-solids separation device that uses centrifugal force for settling. Fluid enters tangentially and spins inside the cone. The heavier solids settle to the walls of the cone and move downward until they are discharged at the cone bottom (cone apex). The spinning fluid travels partway down the cone and then back up to exit out the top of the cone through the vortex finder. |
| **Hydrocyclone Balance Point** | (1) That adjustment of the apex that creates an opening approximately the same diameter as the air cylinder inside of the hydrocyclone. (2) In the field, to adjust a balanced design hydrocyclone during the setup of the solids control system so that it discharges only a slight drip of water at the underflow opening. |
| **Hydrocyclone Size** | The maximum inside working diameter of the cone part of a hydrocyclone. |
| **Hydrocyclone Underflow** | The discharge stream from a hydroclone that contains a higher percentage of solids than does the feed. *See:* Solids Discharge. |
| **Hydrogen Ion Concentration** | A measure of either the acidity or alkalinity of a solution, normally expressed as pH. *See:* pH. |
| **Hydrolysis** | The reaction of a salt with water to form an acid or base. For example, soda ash ($Na_2CO_3$) hydrolyzes basically, and hydrolysis is responsible for the increase in the pH of water when soda ash is added. |
| **Hydrometer** | A floating instrument for determining the specific gravity or density of liquids, solutions, and slurries. |
| **Hydrophile** | Any substance, usually in the colloidal state or an emulsion, which is wetted by water (i.e., it attracts water or water adheres to it). *See:* Lipophile. |
| **Hydrophilic** | A property of a substance having an affinity for water or one that is wetted by water. *See:* Lipophilic. |
| **Hydrophilic-Lipophilic Balance** | The relative attraction of an emulsifier for water and for oil. It is determined largely by the chemical composition and ionization characteristics of a given emulsifier. The HLB of an emulsifier is indirectly related to its solubility, but it determines the type of an emulsion that tends to be formed. It is an indication of the behavioral characteristics and not an indication of emulsifier efficiency. |
| **Hydrophobe** | Any substance, usually in the colloidal state, that is not wetted by water. |
| **Hydrophobic** | Any substance, usually in the colloidal state or an emulsion, which is not wetted by water (i.e., it repels water or water does not adheres to it). |

| | |
|---|---|
| **Hydrostatic Pressure Head** | The pressure exerted by a column of fluid, usually expressed in pounds per square inch. To determine the hydrostatic head in psi at a given depth, multiply the depth in feet by the density in pounds per gallon by the conversion factor, 0.052. |
| **Hydroxide** | A designation that is given for basic compounds containing the OH- radical. When these substances are dissolved in water, the pH of the solution is increased. *See:* Base, pH. |
| **Hygroscopic** | The property of a substance enabling it to absorb water from the air. |
| **ID** | Inside diameter of a pipe. |
| **Impeller** | A spinning disc in a centrifugal pump with protruding vanes used to accelerate the fluid in the pump casing. |
| **Indicator** | Substances in acid-base titrations that, in solution, change color or become colorless as the hydrogen ion concentration reaches a definite value. These values vary with the indicator. In other titrations, such as chloride, hardness, and other determinations, these substances change color at the end of the reaction. Common indicators are phenolphthalein, methyl orange, potassium chromate, and so forth. |
| **Inertia** | The force that makes a moving particle tend to maintain its same direction, or a particle at rest to remain at rest. |
| **Inhibited Drilling Fluid** | A drilling fluid, having an aqueous phase with a chemical composition, which tends to retard and even prevent (inhibit) appreciable hydration (swelling) or dispersion of formation clays and shales through chemical and/or physical reactions. *See:* Calcium Treated Drilling Fluids, Saltwater Drilling Fluid. |
| **Inhibited Mud** | *See:* Inhibited Drilling Fluid. |
| **Initial Gel** | *See:* Gel Strength. |
| **Inlet** | The opening through which the feed mud enters a solids control device. *See:* Feed Inlet, Feed Opening. |
| **Interfacial Tension** | The force required to break the surface definition between two immiscible liquids. The lower the interfacial tension between the two phases of an emulsion, the greater the ease of emulsification. When the values approach zero, emulsion formation is spontaneous. *See:* Emulsion, Surface Tension. |
| **Intermediate (Solids)** | Particles whose diameter is between 250–2000 microns. |
| **Interstitial Water** | Water contained in the interstices or voids of formations. |
| **Intrinsic Safety** | A feature of an electrical device or circuit in which any spark or thermal effect from the electrical device or circuit is incapable of causing ignition of a mixture of flammable or combustible material in air. |
| **Invert Drilling Fluid** | *See:* Invert Oil Emulsion Drilling Fluid. |
| **Invert Oil Emulsion Drilling** | A water-in-oil emulsion where water (sometimes containing sodium or calcium chloride) is the dispersed phase, and diesel oil, crude oil, or some |

other oil is the continuous phase. Water addition increases the emulsion viscosity and oil reduces the emulsion viscosity. The water content exceeds 5% by volume. *See:* Oil-Based Drilling Fluid.

**Iodine Number**  The number indicating the amount of iodine absorbed by oils, fats, and waxes, which yields a measure of the unsaturated linkages present. Generally, the higher the iodine number, the more severe the destructive action of the oil on rubber.

**Ions**  Acids, bases, and salts (electrolytes), when dissolved in certain solvents, especially water, are more or less dissociated into electrically charged ions or parts of the molecules. This condition is due to loss or gain of electrons. Loss of electrons results in positive charges producing a cation. A gain of electrons results in the formation of an anion with negative charges. The valence of an ion is equal to the number of charges borne by the ion. *See:* Anion, Cation.

**Irreducible Fraction**  *See:* Adsorbed Liquid, Bound Liquid.

**Jet Hopper**  A device, which has a jet that facilitates the addition of drilling fluid additives to the system. *See:* Hopper, Mud Hopper.

**Jetting**  The process of periodically removing a portion of the water, drilling fluid, and/or solids from the pits, usually by means of pumping through a jet nozzle to agitate the drilling fluid while simultaneously removing it from the pit.

**Jones Effect**  The net surface tension of all salt solutions first decreases with an increase in concentration, passes through a minimum, and then increases as the concentration is raised. The initial decrease is called the Jones Effect.

**Kelly**  A heavy, square or hexagonal pipe that passes through rollers in the kelly bushing on the drill floor to transmit rotational torque to the drill string.

**Key Seat**  A section of a hole, usually of abnormal deviation and relatively soft formation, which has been eroded or worn by drill pipe to a size smaller than the tool joints or collars of the drill string. This keyhole type configuration resists passage of the shoulders of these pipe upset (box) configurations when pulling out of the hole.

**Kick**  The term used to express the situation caused when the annular hydrostatic pressure in a drilling well temporarily (and, usually, relatively suddenly) becomes less than the formation, or pore, pressure in a permeable downhole section. A kick occurs before control of the fluid intrusion is totally lost. A blow out is an uncontrolled influx of formation fluid into the wellbore. *See:* Blow Out, Kill Fluid.

**Kill Fluid**  A fluid built with a specific density aimed at controlling a kick or blowout. *See:* Galena.

**Kill Line**  A line connected to the annulus below the blowout preventers for the purpose of pumping into the annulus while the preventers are closed.

**Killing A Well**  (1) Bringing a well kick under control. (2) The procedure of circulating a fluid in a well to overbalance formation fluid pressure after the bottom hole pressure has been less than the formation fluid pressure. *See:* Kick, Blowout, Kill Fluid.

| | |
|---|---|
| **Kilowatt-Hour** | Horsepower-Hour (hp-hr) and Kilowatt-Hour (kw-hr) are units of work.<br>• 1 hp-hour = 1,980,000 ft-lb = 2545 Btu<br>• 1 hp-hour = 0.7457 kilowatt hours (kw-hr)<br>• 1 kw-hour = 1.341 hp-hr = 3413 Btu = 2,655,000 ft-lb |
| **Kinematic Viscosity** | The ratio of the viscosity (e.g., cp in g/cm-sec) to the density (e.g., g/cc) using consistent units. In several common commercial viscometers the kinematic viscosity is measured in terms of the time of efflux (in seconds) of a fixed volume of liquid through a standard capillary tube or orifice. *See:* Marsh Funnel Viscosity. |
| **Laminar Flow** | The movement of fluid in plates or sections with a differential velocity across the front of the flow profile that varies from zero at the wall to a maximum toward the center for flow. These fluid elements flow along fixed streamlines, which are parallel to the walls of the flow channel. Laminar flow is the first stage in a Newtonian fluid and the second stage in a Bingham plastic fluid. This type of motion is also called parallel, streamline, or viscous flow. *See:* Plug Flow, Parallel Flow, Turbulent Flow. |
| **LCM** | *See:* Lost Circulation Materials. |
| **Lead** | In a decanting centrifuge, the slurry conducting channel formed by the adjacent walls of the flutes or blades of the screw conveyor. |
| **Leonardite** | A naturally occurring oxidized lignite. *See:* Humic Acid, Lignin. |
| **Light Solids** | *See:* Low-Gravity Solids. |
| **Lignin** | Mined lignin is a naturally occurring special lignite (e.g., leonardite) produced by strip mining from special lignite deposits. The active ingredients are the humic acids. Mined lignins are used primarily as thinners, which may or may not be chemically modified. *See:* Leonardite, Humic Acid. |
| **Lignosulfonates** | Organic drilling fluid additives derived from by-products of the sulfite paper manufacturing process from coniferous woods. Some of the common salts, such as ferro-chrome, chrome, calcium, and sodium, are used as deflocculants, while other lignosulfonates are used selectively for calcium treated systems. In large quantities, the "heavy metal" ferro-chrome and chrome salts are used for fluid loss control and shale inhibition. |
| **Lime** | $Ca(OH)_2$. Commercial form of calcium hydroxide. |
| **Lime Treated Drilling Fluids** | Commonly referred to as "lime-based" muds. These high pH systems contain most of the conventional freshwater drilling fluid additives to which slaked lime has been added to impart special inhibition properties. The alkalinities and lime contents of the fluids may vary from low to high. *See:* Calcium Treated Drilling Fluids. |
| **Limestone** | $Ca(CO)_3$. *See:* Calcium Carbonate. |
| **Line Sizing** | Ensuring that the fluid velocity through all piping within the surface system has the proper flow and pipe diameter combination to prevent solids from settling and pipe from eroding. Generally, fluid flow should be between 5 and 9 ft/sec as determined by the following:<br>$V_{fps} = [Q_{gpm} \times 0.4087]/d^2$ |

where $V_{fps}$ = Velocity of flow, ft/sec
$Q_{gpm}$ = Flow rate, gal/min
d = Inside diameter of the pipe, in.

| | |
|---|---|
| **Linear Motion** | The linear motion of a shale shaker screen is produced by two counter-rotational motors located above the shaker basket in such a way that a line connecting the two motor axes is perpendicular to a line passing through the center of gravity of the basket. Because the acceleration is applied directly through the center of gravity of the basket, the basket is dynamically balanced; the same pattern of motion will exist at all points along the shaker screen. The resultant screen motion is linear and the angle of this uniform motion is usually 45° to 60° relative to the shaker screen deck. |
| **Lipophile** | Any substance, usually in the colloidal state or an emulsion, which is wetted by oil (i.e., it attracts oil or oil adheres to it). *See:* Hydrophile. |
| **Lipophilic** | A property of a substance having an affinity for oil or one that is wetted by oil. *See:* Hydrophilic. |
| **Liquid** | Fluid that will flow freely and takes the shape of its container. |
| **Liquid Discharge** | *See:* Underflow. |
| **Liquid Film** | The liquid surrounding each particle discharging from the solids discharge of cyclones and screens. *See:* Bound Liquid, Free Liquid. |
| **Liquid-Clay Phase** | *See:* Overflow. |
| **Live Oil** | Crude oil that contains gas and distillates and has not been stabilized or weathered. This oil can cause gas cutting when added to drilling fluid and is a potential fire hazard. *See:* Aromatic Hydrocarbons. |
| **Load** | A device connected to a motor that is receiving output mechanical power from the motor. |
| **Logging** | *See:* Mud Logging, Electric Logging. |
| **Loom** | *See:* Warp. |
| **Loss of Circulation** | *See:* Lost Circulation. |
| **Lost Circulation** | The result of drilling fluid escaping into a formation, usually in fractures, cavernous, fissured, or coarsely permeable beds, evidenced by the complete or partial failure of the drilling fluid to return to the surface as it is circulated in the hole. |
| **Lost Circulation Additives** | Materials added to the drilling fluid to gain control of or prevent the loss of circulation. These materials are added in varying amounts and are classified as fibrous, flake, or granular. |
| **Lost Circulation Materials** | *See:* LCM, Lost Circulation Additives. |
| **Lost Returns** | *See:* Lost Circulation. |
| **Low-Gravity Solids** | With the exception of barite or other commercial weighting materials, all solids in drilling fluid including salts, drilled solids of every size, |

commercial colloids, and lost circulation materials. Salt is considered a low specific gravity solid. *See:* Heavy Solids, High-Gravity Solids.

**Low-Silt Drilling Fluid** — An unweighted drilling fluid that has all the sand and most of the silts removed and contains a substantial content of bentonite or other water-loss reducing clays.

**Low-Silt Mud** — *See:* Low-Silt Drilling Mud.

**Low Solids Drilling Fluids** — A drilling fluid that has polymers, such as CMC or XC Polymer, partially or wholly substituted for commercial or natural clays. For comparable viscosity and densities, a low-solids drilling fluid will have a lower volume percent solids content. In general, the lower the solids content in a mud, the faster a bit can drill.

**Low-Solids, Non-Dispersed (LSND) Drilling Fluids** — A drilling fluid to which polymers have been added to simultaneously extend and flocculate bentonite and drilled solids. These fluids contain low concentrations of dispersed bentonite and do not contain deflocculants such as lignites, lignosulfonates, and the like.

**Low-Solids Muds** — *See:* Low-Solids Drilling Fluids.

**Low-Yield Clay** — Commercial clay, specifically the calcium montmorillonite type, having a yield of approximately 15 to 30 barrels per ton. *See:* High-Yield Clay, Bentonite.

**Lyophilic** — Having an affinity for the suspending medium, such as bentonite in water.

**Lyophlic Colloid** — A colloid that is not easily precipitated from a solution and is readily dispersible after precipitation by an addition of a solvent.

**Lyophobic Colloid** — A colloid that is readily precipitated from a solution and cannot be re-dispersed by addition of the solution.

**Main Shaker** — The shale shaker that processes drilling fluid from the flow line through the finest mesh screen.

**Manifold** — (1) A length of pipe with multiple connections for collecting or distributing fluid. (2) A piping arrangement through which liquids, solids, or slurries from one or more sources can be fed to or discharged from a solids separation device.

**Market Grade Cloth** — A group of industrial wire cloth specifications selected for general purpose work, made of high strength, square mesh cloth in several types of metals. The common metal for oil field use is 304 or 316 stainless steel. The wire diameters are marginally larger than mill grade cloth, resulting in a lower percentage of open area. Market grade and mill grade clothes are used mostly as support screens for fine mesh screens. *See:* Mill Grade Cloth, Tensile Bolting Cloth, Ultra Fine Wire Cloth, Support Screen, Calendered.

**Marsh Funnel** — An instrument used in determining the Marsh funnel viscosity. The Marsh funnel is a container with a fixed orifice at the bottom so that when filled with 1,500 cc freshwater, 1 qt (946 ml) will flow out in 26 ± 0.5 sec. For 1,000 cc out, the efflux time for water is 27.5 ± sec. It is used for comparison values only and not to diagnose drilling fluid problems. *See:* API Bulletin RP 13B, Funnel Viscosity, Marsh Funnel Viscosity, Kinematic Viscosity.

**Marsh Funnel Viscosity**  Commonly called funnel viscosity. The Marsh funnel viscosity is reported as the time in seconds required for 1 qt of fluid to flow through a API standardized funnel. In some areas, the efflux quantity is 1,000 cc. *See:* API Bulletin RP 13B, Funnel Viscosity, Kinematic Viscosity, Marsh Funnel.

**Martin's Radii**  The distance from the centroid of an object to its outer boundary. The direction of this measurement is specified by the azimuth orientation of the line (the radii in the 0°, 90°, 180°, and 270° angle from horizontal).

**Mass**  The inertial resistance of a body to acceleration, considered in classical physics to be a conserved quantity independent of speed. The weight of a body is the product of the mass of the body and the acceleration of gravity for the specific location. In space, the mass would stay constant but the weight would disappear as the gravitational acceleration approaches zero.

**MBT**  *See:* Methylene Blue Test.

**Mechanical Agitator**  A device used to mix, blend, or stir fluids by means of a rotating impeller blade. *See:* Agitator, Mechanical Stirrer.

**Mechanical Stirrer**  *See:* Agitator, Mechanical Agitator.

**Median Cut**  The median cut is the particle size that reports 50% of the weight to the overflow and 50% of the weight to the underflow. Frequently identified as the $D_{50}$ point. *See:* Cut Point.

**Medium (Solids)**  Particles whose diameter is between 74 and 250 microns.

**Meniscus**  The curved upper surface of a liquid column, concave when the containing walls are wetted by the liquid, and convex when they are not wetted.

**Mesh**  (1) The number of openings (and fraction thereof) per linear inch in a screen, counted in both directions from the center of a wire. (2) An indication of the weave of a woven material, screen, or sieve. A 200-mesh sieve has 200 openings per linear inch. A 200-mesh screen with a wire diameter of 0.0021 inch (0.0533 mm) has an opening of 0.074 mm and will pass a spherical particle of 74 microns in diameter. *See:* Micron.

**Mesh Count**  A mesh count, such as 30 × 30 or often 30 mesh, indicates the number of openings per linear inch of screen and is a square mesh. A designation of 70 × 30 mesh indicates a rectangular mesh with 70 openings per inch in one direction and 30 openings per inch in a perpendicular direction.

**Mesh Equivalent**  The U.S. Sieve number as used in oil field drilling applications, which has the same size opening as the minimum opening of the screen in use.

**Methylene Blue Test (MBT)**  A test that serves to indicate the amount of active clay in a fluid system, clay sample, or shale sample. Methylene blue is titrated into a slurry until all of the negative charge sites are covered with the methylene blue. This indicates the number of active charge sites present in the slurry. *See:* Base Exchange, Methylene Blue Titration, MBT, Cation Exchange Capacity, CEC.

**Methylene Blue Titration (MBT)**  Methylene blue is a cation that seeks all negative charges on a clay sur face after the surface has been properly prepared (*see* API RP13B). By titrating with a known concentration, this test provides an indication of the amount of clay present in the drilling fluid. *See:* Methylene Blue Test, MBT, Cation Exchange Capacity, CEC.

| | |
|---|---|
| **Mf** | The methyl orange alkalinity of the filtrate, reported as the number of milliliters of 0.02 normal sulfuric acid required per milliliter of filtrate to decrease the pH to reach the methyl orange end point (pH 4.3). |
| **Mica** | Naturally occurring mineral flake material of various sizes used in controlling lost circulation. An alkali aluminum silicate. |
| **Micelles** | Organic and inorganic molecular aggregates occurring in colloidal solutions. Long chains of individual structural units chemically joined to one another and deposited side by side to form bundles. When bentonite hydrates and certain sodium or other metallic ions go into the solution, the clay particle plus its complement of ions is technically known as a micelle. |
| **Micron** | A unit of length equal to one thousandth of a millimeter. Used to specify particle sizes in drilling fluids and solids control discussions (25,400 microns = 1 inch). |
| **Mil** | A unit of length equal to 1/1000 inch. |
| **Milk Emulsion** | *See:* Oil-In-Water Emulsion Drilling Fluid. |
| **Mill Grade Cloth** | A group of industrial wire cloth specifications with lighter wire than market grade cloth. The standard wire diameter of this grade produces a median percentage of open area. Market grade and mill grade clothes are used mostly as support screens for fine mesh screens. *See:* Market Grade Cloth, Tensile Bolting Cloth, Ultra Fine Wire Cloth, Support Screen, Calendered. |
| **Millidarcy** | 1/1000 Darcy. *See:* Darcy. |
| **Milliliter** | A metric system unit for the measurement of volume. Literally 1/1000th of a liter. In drilling fluid analyses, this term is used interchangeably with cubic centimeter (cc). One quart is equal to approximately 946 ml. |
| **Mini Still** | An instrument used to distill oil, water, and any other volatile material in a drilling fluid to determine oil, water, and total solids content in volume percent. *See:* Distillation, Mini Still, Mud Still. |
| **Mist Drilling** | A method of rotary drilling whereby water and/or oil is dispersed in air and/or gas as the drilling fluid. *See:* Foam. |
| **Ml** | *See:* Milliliter. |
| **Molecule** | Atoms combine to form molecules. For elements or compounds, a molecule is the smallest unit that chemically still retains the properties of the substance in mass. |
| **Monovalent** | *See:* Valence. |
| **Montmorillonite** | A clay mineral commonly used as an additive to drilling muds. Sodium montmorillonite is the main constituent of bentonite. Each platelet of the crystalline structure of montmorillonite has two layers of silicon tetrahedra attached to a center layer of alumina octhahedra. The platelets are thin and have a broad surface. Exchangeable cations are located on the clay surfaces between the platelets. Calcium montmorillonite is the main constituent in low-yield clays. *See:* Gel, Bentonite. |
| **Mud** | Drilling Fluid is the preferred term. *See:* Drilling Fluid. |

| | |
|---|---|
| **Mud Analysis** | *See:* Drilling Fluid Analysis, API Bulletin RP 13B. |
| **Mud Balance** | A beam-type balance used in determining drilling fluid density (mud weight). It consists primarily of a base, graduated beam with constant volume cup, lid, rider, knife-edge, and counterweight. *See:* API Bulletin RP 13B. |
| **Mud Box** | *See:* Back Tank, Possum Belly. |
| **Mud Cleaner** | A device that places a screen in series with the underflow of hydrocyclones. The hydrocyclone overflow returns to the mud system, and the underflow reports to a vibrating screen. Solids discharged from the screen are discarded and the screen throughput returns to the system. |
| **Mud Compartment** | A subdivision of the removal, addition, or check/suction sections of a surface system. *See:* Mud Pits, Mud Tanks. |
| **Mud Ditch** | A trough built along the upper edge of many surface systems that is used to direct flow to selected compartments of the surface system. *See:* Mud Pits, Mud Compartment. |
| **Mud Engineer** | *See:* Drilling Fluid Engineer. |
| **Mud Gun** | A submerged nozzle used to stir the drilling fluid with a high-velocity stream. *See:* Gunning the Pits. |
| **Mud Hopper** | *See:* Hopper. |
| **Mud House** | A structure at the rig to store and shelter materials used in drilling fluids. |
| **Mud Inhibitor** | Additives such as salt, lime, lignosulfonate, and calcium sulfate, which prevent clay dispersion. |
| **Mud Logging** | A process that helps determine the presence or absence of oil or gas in various formations penetrated by the drill bit, and provides a variety of indicators that assist drilling operations. Drilling fluid and cuttings are continuously tested on their return to the surface, and the results of these tests are correlated with the drilling depth for depth of origin. |
| **Mud Mixing Devices** | The most common device for adding solids to the drilling fluid is by means of the jet hopper. Some other devices to assist mixing are eductors, mechanical agitators, paddle mixers, electric stirrers, mud guns, and chemical barrels. |
| **Mud Pit** | *See:* Mud Compartments, Mud Tanks. |
| **Mud Pump** | Pumps at the rig used to circulate drilling fluids. |
| **Mud Scales** | *See:* Mud Balance. |
| **Mud Still** | *See:* Distillation, Mini Still. |
| **Mud Tanks** | (1) Drilling fluid system compartments constructed of metal and mounted so they can be moved from location to location, either as a part of the rig (such as on a semi-submersible rig) or separately on unitized skids (as on most land rigs). (2) Earthen or steel storage facilities for the surface system. Mud pits are of two types: circulating and reserve. Drilling fluid testing and conditioning is normally performed in the circulating pit system. |

## GLOSSARY 313

**Mud Weight** — A measurement of slurry density usually reported in lb/gal, lb/cu ft, psi/1000 ft., or specific gravity. *See:* Density.

**Mud/Gas Separator** — A vessel into which the choke line discharges when a "kick" is being taken. Gas is separated in the vessel as the drilling fluid flows over baffle plates. The gas flows through a line to a flare. The liquid mud discharges into the shale shaker back tank. *See:* Gas Buster, Poor Boy Degasser.

**Mudding Off** — A condition promoting reduced production caused by the penetrating, sealing, or plastering effect of a drilling fluid. *See:* Formation Damage.

**Mudding Up** — Process of mixing drilling fluid additives to a clay/water slurry to achieve some properties not possible with the previous fluid.

**MW** — Abbreviation for mud weight. *See:* Density, Mud Weight.

**Natural Clays** — Natural clays, as opposed to commercial clays, are encountered when drilling various formations. The yield of these clays varies greatly, and they may or may not be purposely incorporated into the drilling fluid system. *See:* Attapulgite Clay, Bentonite, High-Yield Clay, Low-Yield Clay, Clay.

**Near-Size Particles** — Solids approximately the same size as a screen opening.

**Near-Size Plugging** — A term used in describing screen plugging and refers to particles with a dimension only slightly larger than the screen opening. *See:* Blinding, Plugging.

**Neat Cement** — A slurry composed only of Portland cement and water.

**Negative Deck Angle** — The angle of adjustment to a screen deck that causes the screened solids to travel "downhill" to reach the discharge end of the screen surface. This "downhill" travel decreases the fluid throughput of a screen but usually lengthens the life of a screen. *See:* Positive Deck Angle.

**Neutralization** — A reaction in which the hydrogen ion of an acid and the hydroxyl ion of a base unite to form water, the other ionic product being a salt.

**Newtonian Flow** — *See:* Newtonian Fluid.

**Newtonian Fluid** — The most basic fluids from the standpoint of viscosity in which the shear force is directly proportional to the shear rate. These fluids will immediately begin to move when a pressure or shear stress in excess of zero is applied. Examples of Newtonian fluids are water, diesel oil, and glycerine. The yield point as determined by a direct indicating viscometer is zero. *See:* Newtonian Flow.

**Non-Conductive Drilling Fluid** — Any drilling fluid, usually oil based or invert emulsion drilling fluid, whose continuous phase does not conduct electricity. The spontaneous potential (SP) and normal resistivity cannot be logged, although other logs such as the gamma rays, induction, acoustic velocity, and so forth, can be run.

**Non-Dispersed** — A condition in which the clays do not separate into individual platelets. Dispersion is inhibited.

**Normal Solution** — A solution of such a concentration that it contains 1 gram equivalent of a substance per liter of solution.

**Oblong Mesh**  A screen cloth that has more openings per inch in one direction than in the perpendicular direction. For example, a 70 × 30 mesh has 70 openings per inch in one direction and 30 openings per inch in the perpendicular direction, creating a rectangular opening. The smaller opening dimension controls the sizing of spherical material. *See:* Rectangular Screen.

**Oblong Weave**  *See:* Oblong Mesh.

**OD**  Outside diameter of a pipe.

**Ohm**  The measurement of resistance or electrical friction.

**Oil-Based Drilling Fluid**  The term "oil-based mud" is applied to a special drilling fluid where oil is the continuous phase and water is the dispersed phase. Oil-based drilling fluid contains from 1% to 5% water emulsified into the system with lime and emulsifiers. Oil-based muds are differentiated from invert emulsion muds (both water-in-oil emulsions) by the amounts of water used, the method of controlling viscosity, the thixotropic properties, wall building materials, and fluid loss. *See:* Invert Oil Emulsion Drilling Fluid.

**Oil Breakout**  Oil that has risen to the surface of a drilling fluid. This oil had been previously emulsified in the drilling fluid or may derive from oil-bearing formations that have been penetrated.

**Oil Content**  The oil content of any drilling fluid is the amount of oil in volume percent.

**Oil Immersion**  An oil-filled construction where an electrical device has no electrical connections, joints, terminals, or arcing parts at or above the normal oil level.

**Oil Wet**  A surface on which oil easily spreads. If the contact angle of an oil droplet on a surface is less than 90°, the surface is oil wet. *See:* Lipophilic, Water Wet.

**Oil-In-Water Emulsion Drilling Fluid**  Any conventional or special water-based drilling fluid to which oil has been added. A drilling fluid in which the oil content is usually kept between 3% to 7% and seldom over 10% (it can be considerably higher). Commonly called "emulsion mud." The oil becomes the dispersed phase and may be emulsified into the mud either mechanically or chemically. The oil is emulsified into fresh- or saltwater with a chemical emulsifier.

**Overflow**  The discharge stream from a centrifugal separation that normally contains a higher percentage of liquids than does the feed.

**Overflow Header**  A pipe into which two or more hydrocyclones discharge their overflow.

**Overslung**  Field terminology denoting that the support ribs for the shaker screen are located below the screen surface. *See:* Underslung.

**Packer Fluid**  A fluid placed in the annulus between the tubing and casing above a packer. The hydrostatic pressure of the packer fluid is used to reduce the pressure differentials between the formation and the inside of the casing and across the packer.

**Panel Mounted Units**  Shale shaker screens mounted to a rigid frame.

**Parallel Flow**  *See:* Laminar Flow.

| | |
|---|---|
| **Particle** | A discrete unit of solid material that may consist of a single grain or of any number of grains stuck together. |
| **Particle Size** | Particle diameter expressed in microns. *See:* ESD, Equivalent Spherical Diameter. |
| **Particle Size Distribution** | The volume classification of solid particles into each of the various size ranges as a percentage of the total solids of all sizes in a fluid sample. |
| **Parts Per Million** | The unit weight of solute per million unit weights of solution (solute plus solvent), corresponding to weight percent. For example, the results of standard API titration of chloride hardness are correctly expressed in milligrams (mg) per liter, not in ppm. At low concentrations, mg/l is about numerically equal to ppm. A correction for the solution specific gravity or density in g/ml must be made as follows: |

Ppm = [milligrams/liter] / Solution Density (grams/liter)

Weight % = [milligrams/liter] / [10,000 × Solution Density (grams/liter)]

Weight % = [Ppm] / [10,000]

Thus, 316,000 mg/l salt is commonly called 316,000 ppm or 31.6%, which corrected should be 264,000 ppm and 26.4%, respectively. *See:* Ppm.

| | |
|---|---|
| **Pay Zone** | A formation that contains oil and/or gas in commercial quantities. |
| **Penetration Rate** | The rate at which the drill bit penetrates the formation, usually expressed in feet per hour or meters per hour. *See:* Rate of Penetration, ROP. |
| **Peptization** | An increased flocculation of clays caused by the addition of electrolytes or other chemical substances. *See:* Deflocculation Dispersion, High-Yield Clay. |
| **Peptized Clay** | A clay to which an agent has been added to increase its initial yield. For example, soda ash is frequently added to calcium montmorillonite clay to increase the yield. *See:* High-Yield Clay. |
| **Percent Open Area** | Ratio of the area of the screen openings to the total area of the screen surface. |
| **Percent Separated Curve** | A plot of mass distributions of solid sizes discarded from a solids separation device divided by the mass distributions of each size of solids fed to the device. |
| **Perforated Cylinder Centrifuge** | A mechanical centrifugal separator in which the rotating element is a perforated cylinder (the rotor) inside of and concentric with an outer stationary cylindrical case. |
| **Perforated Panel Screen** | A screen in which the backing plate used to provide support to the screen cloths is a metal sheet with openings. |
| **Perforated Plate Screen** | Shale shaker screens mounted on metal plates that have holes punched through them. |
| **Perforated Rotor** | The rotating inner cylinder of the perforated cylinder centrifuge. *See:* Perforated Cylinder Centrifuge. |

| | |
|---|---|
| **Permeability** | A measure of the ability of a formation to allow passage of a fluid. Unit of permeability is the darcy. *See:* Darcy, Porosity. |
| **Pf** | The phenolphthalein alkalinity of the filtrate is reported as the number of milliliters of 0.02 normal sulfuric acid required per milliliter of filtrate for the pH to reach the phenolphthalein end point, which is a pH of 8.3. |
| **pH** | The negative logarithm of the hydrogen ion concentration in gram ionic weights per liter. The pH range is numbered from 0 to 14, with 7 being neutral, and is an index of the acidity (below 7) or alkalinity (above 7) of the fluid. At a temperature of 70°F a neutral pH is 7 or a hydrogen ion concentration of $10^{-7}$. The neutral pH is a function of temperature. At higher elevated temperatures the neutral pH is lower. The pH of a solution offers valuable information as to the immediate acidity or alkalinity, in contrast to the total acidity or alkalinity, which may be determined by titration. |
| **Phosphate** | Certain complex phosphates, commonly sodium tetraphosphate ($Na_6P_4O_{13}$) and sodium acid pyrophosphate (SAPP, $Na_2H_2P_2O_4$), used as either drilling fluid thinners or for treatment of various forms of calcium and magnesium contamination. |
| **Piggy Back** | "Piggybacking" is the attachment of fine solids particles to the surface of larger solids particles due to surface attraction, fluid consistency, and particle concentration. This attachment phenomena causes fine solids to be discharged from the screen that would normally pass through the screen. |
| **Pill** | A small volume of a special fluid slurry pumped through the drill string and normally placed in the annulus. *See:* Slug. |
| **Pilot Testing** | A method of predicting behavior of drilling fluid systems by adding various chemicals to a small quantity of drilling fluid (usually 350 cc) and then examining the results. One gram of an additive in 350 cc is equivalent to 1 lb/bbl. |
| **Plastic Flow** | *See:* Plastic Fluid. |
| **Plastic Fluid** | A complex, non-Newtonian fluid in which shear force is not proportional to shear rate. A definite pressure is required to start and maintain fluid movement. Plug flow is the initial flow type and only occurs in plastic fluids. Most drilling fluids are plastic fluids. The yield point, as determined by a direct indicating viscometer, is in excess of zero. *See:* Plastic Flow. |
| **Plastic Viscosity** | A measure of the internal resistance to fluid flow attributable to the concentration, type, and size of solids present in a given fluid and the viscosity of the continuous phase. This value, expressed in centipoise, is proportional to the slope of the shear stress/shear rate curve determined in the region of laminar flow for materials whose properties are described by Bingham's Law of Plastic Flow. When using the direct indicating viscometer, plastic viscosity is determined by subtracting the 300 RPM reading from the 600 RPM reading. *See:* Viscosity, Yield Point, API Bulletin RP 13B. |
| **Plasticity** | The property possessed by some solids, particularly clays and clay slurries, of changing shape or flowing under applied stress without developing shear planes or fractures (i.e., it deforms without breaking). Such bodies have yield points, and stress must be applied before movement begins. Beyond the yield point, the rate of movement is proportional to the stress applied, but movement ceases when the stress is removed. *See:* Fluid. |

**Plug Flow**  
The movement of material as a unit without shearing within the mass. Plug flow is the first type of flow exhibited by a plastic fluid after overcoming the initial force required to produce flow. *See:* Bingham Model, Newtonian Fluid, Laminar Flow, Turbulent Flow.

**Plugging**  
The wedging or jamming of openings in a screening surface by near-size particles, preventing passage of undersize particles and leading to blinding of the screen. *See:* Blinding, Coating.

**Pm**  
The phenolphthalein alkalinity of drilling fluid is reported as the number of milliliters of 0.02 normal (N/50) sulfuric acid required per milliliter of drilling fluid for the pH to reach the phenolphthalein end point of 8.3.

**Polyelectrolytes**  
(Polymers) Long chain organic molecules possessing ionizable sites that, when dissolved in water, become charges.

**Polymer**  
A substance formed when two or more molecules of the same kind are linked end to end to another compound having the same elements in the same proportion, but higher molecular weight and different physical properties (e.g., paraformaldehyde). Polymers are used in drilling fluids to maintain viscosity and control fluid loss. *See:* Co-polymer.

**Polymer Drilling Fluid**  
A drilling fluid to which polymers have been added to increase the low-shear-rate viscosity.

**Polyurethane**  
A high-performance, elastomer polymer used in construction of hydrocyclones for its unique combination of physical properties, especially abrasion, toughness, and resiliency.

**Pool**  
(1) The reservoir or pond of fluid, or slurry, formed inside the wall of hydrocyclones and centrifuges and where classification or separation of solids occurs due to the settling effect of centrifugal force. (2) The reservoir or pond of fluid which can form on the feed end of an uphill shaker basket (a shaker basket with a positive deck angle).

**Poor Boy Degasser**  
*See:* Gas Buster, Mud/Gas Separator.

**Porosity**  
The volume of void space in a formation rock, usually expressed as percent void volume per bulk volume.

**Ports**  
The openings in a centrifuge for entry or exit of materials. Usually applied in connection with a descriptive term (i.e., feed ports, overflow ports, etc.).

**Positive Deck Angle**  
The angle of adjustment to a screen deck, which causes the screened solids to travel "uphill" to reach the discharge end of the screen surface. This "uphill" travel increases the fluid throughput of a screen but also shortens the life of a screen. *See:* Negative Deck Angle.

**Possum Belly**  
The compartment on a shale shaker into which the flow line discharges, and from which the drilling fluid is either fed to the screens or to a succeeding tank. *See:* Back Tank, Mud Box.

**Potassium**  
One of the alkali metal elements with a valence of 1 and an atomic weight of approximately 39. Potassium compounds, most commonly potassium hydroxide (KOH), are sometimes added to drilling fluids to impart special properties, usually inhibition.

**Potential Separation Curve**  A distribution curve of sizes determined by the optical image analysis for separation potential.

**Pound Equivalent**  A laboratory unit used in pilot testing. One gram of a material added to 350 ml of fluid is equivalent to 1 lb of material added to one barrel. *See:* Barrel, Barrel Equivalent.

**Ppm**  *See:* Parts Per Million.

**Precipitate**  Material that separates out of solution or slurry as a solid. Precipitation of solids in a drilling fluid may follow flocculation or coagulation.

**Pre-Hydration Tank**  A tank used to hydrate materials (such as bentonite, polymers, etc.) that require a long time (hours to days) to fully hydrate and disperse before being added to the drilling fluid. *See:* Premix System.

**Premix System**  A compartment used to mix materials (such as bentonite, polymers, etc.) that require time to fully hydrate or disperse before they are added to the drilling fluid. *See:* Pre-Hydration Tank.

**Preservative**  Any material used to prevent starch or any other organic substance from fermenting through bacterial action. A common preservative is paraformaldehyde. *See:* Fermentation.

**Pressure Drop**  *See:* Friction Loss, Pressure Loss.

**Pressure Head**  Pressure within a system equal to the pressure exerted by an equivalent height of fluid (expressed in feet or meters). *See:* Head, Hydrostatic Head, Centrifugal Pump.

**Pressure Loss**  The pressure lost in a pipeline or annulus due to the liquid velocity in the pipeline, the properties of the fluid, the condition of the pipe wall, and the configuration of the pipe. *See:* Friction Loss, Pressure Drop.

**Pressure Surge**  A sudden, usually brief, increase in pressure. When pipe or casing is run into a hole too rapidly or the drill string is set in the slips too quickly, an increase in the hydrostatic pressure results due to a pressure surge that may be great enough to create lost circulation. *See:* ECD, Annular Pressure Loss.

**Pressurization**  The process of supplying an enclosure with a protective gas, with or without continuous flow, at sufficient pressure to prevent the entrance of a flammable gas or vapor, a combustible dust, or an ignitable fiber.

**Pretensioned Screen**  A screen cloth that is bonded to a frame or backing plate with proper tension applied prior to its installation on a shaker. *See:* Backing Plate, Perforated Panel Screen.

**Progressive Gel**  A condition wherein the 10 min gel strength is greater than to the initial gel strength. Opposite of Flat Gel. *See:* Flat Gel, Zero-Zero Gel.

**Pseudoplastic Fluid**  A complex, non-Newtonian fluid that does not possess thixotropy. A pressure or force in excess of zero will start fluid flow. The apparent viscosity or consistency decreases instantaneously with increasing shear rate until at a given point the viscosity becomes constant. The yield point, as determined by a direct indicating viscometer, is positive, as in Bingham plastic fluids. However, the true yield point is zero. An example of a pseudoplastic fluid is guar gum in fresh or saltwater. *See:* Viscosity, Bingham Model, Plastic Viscosity, Yield Point, Gel Strength.

**Purging** The process of supplying an enclosure with a protective gas at a sufficient flow and positive pressure to reduce the concentration of any flammable gas or vapor initially present to an acceptable level.

**Quebracho** An additive used extensively for thinning/dispersing to control low-shear-rate viscosity and thixotropy. It is a crystalline extract of the quebracho tree consisting mainly of tannic acid. *See:* Thinner.

**Quicklime** Calcium oxide, CaO. Used in certain oil based drilling fluids to neutralize the organic acid.

**Quiescence** The state of being quiet, at rest, or being still. *See:* Static.

**Radial Flow** Flow of a fluid outwardly in a 360° pattern. This describes the flow from a mechanical agitator in which fluid moves away from the axis of the impeller shaft (usually horizontally toward a mud tank wall). *See:* Axial Flow.

**Radical** Two or more atoms behaving as a single chemical unit (e.g., sulfate, phosphate, and nitrate are radicals).

**Rate of Penetration** The rate at which the drill bit penetrates the formation expressed in lineal units of feet/minute. *See:* Penetration Rate, ROP.

**Rate of Shear** The change in velocity between two parallel layers divided by the distance between the layers. Shear rate has the units of reciprocal seconds ($sec^{-1}$). *See:* Shear Rate.

**Raw Drilling Fluid** Drilling fluid, before dilution, that is to be processed by solids removal equipment.

**Rectangular Screen** *See:* Oblong Mesh.

**Reduced Port** A valve whose bore size is less than the area of the pipe to which it is attached.

**Removal Section** The first section in the drilling fluid system consisting of a series of compartments and solids removal equipment to remove gas and undesirable solids.

**Reserve Pit** (1) An earthen pit used to store drilling waste in land drilling operations. (2) A section of a surface system used to store drilling fluid.

**Residence Time** Time a slurry or solids remain in a given location or region.

**Resin** A semi-solid, or solid complex, or amorphous mixture of organic compounds having no definite melting point or tendency to crystalize. Resin may be a component of compounded materials that can be added to drilling fluids to impart special properties to the system (i.e., wall cake, fluid loss, etc.).

**Resistivity** A characteristic electrical property of a material that is equal to the electrical resistance of one meter cube of the material to passage of one ampere electric current perpendicular to two parallel faces. The electrical resistance offered to the passage of a current, expressed in ohm-meters. The reciprocal of conductivity. Freshwater muds are usually characterized by high resistivity and saltwater muds by a low resistivity. *See:* Conductivity.

**Resistivity Meter** — An instrument for measuring the electrical resistivity of drilling fluids.

**Retention Time** — The time any given particle of material is retained in a region; for example, the time a particle is actually on a screening surface, within a hydroclone or the bowl of a centrifuge.

**Retort** — An instrument used to distill oil, water, and other volatile material in a drilling fluid to determine oil, water, salt and total solids contents in volume-percent. *See:* Mud Still, Mini Still, API Bulletin RP 13B.

**Reverse Circulation** — The method by which the normal flow of a drilling fluid is reversed by circulating down the annulus, then up and out the drill string. *See:* Circulation.

**Reynolds Number** — A dimensionless number, $R_e$, that occurs in the theory of fluid dynamics. Reynolds number for a fluid flowing through a cylindrical conductor is determined by the equation: $R_e = DV\rho/\mu$; where diameter = D; velocity = V; density = $\rho$, and viscosity = $\mu$. The number is important in fluid hydraulics calculations for determining the type of fluid flow (i.e., whether laminar or turbulent). The transitional range occurs approximately from 2,000 to 3,000. Below 2,000 the flow is laminar and above 3,000 the flow is turbulent. *See:* Fluid Flow.

**Rheology** — The science that deals with deformation and flow of matter. *See:* Viscosity, Bingham Model, Plastic Viscosity, Yield Point, Gel Strength.

**Rig Pump** — The reciprocating, positive displacement, high-pressure pump on a drilling rig used to circulate the hole. *See:* Mud Pump.

**Rig Shaker** — A slang term for a shale shaker.

**Rigid Frame Panel** — One of the two main screen panel types that consists of a rigid panel to which the screen or layers of screen are attached. The screen panel fastening device can be designed for fast panel replacement. *See:* Hookstrip Panel.

**ROP** — *See:* Rate of Penetration, Penetration Rate.

**Rope Discharge** — The characteristic underflow of a hydrocyclone so overloaded with separable solids that not all separated solids can travel out the underflow opening (apex). This causes the solids that can exit to form a slow moving, heavy, rope-like stream. Also referred to as "rope" or "rope underflow."

**Rotary Drilling** — The method of drilling wells in which a drill bit, attached to a drill string, is rotated on the formation to be drilled. A fluid is circulated through the drill pipe to remove cuttings from the bottom of the hole, bring cuttings to the suface, and perform other functions. *See:* Cable Tool Drilling.

**Rotary Mud Separator** — A centrifuge consisting of a perforated cylinder rotating inside an outer cylinder housing. As drilling fluid flows outside of the perforated cylinder, only the very small particles pass through the perforations.

**RMS** — *See:* Rotary Mud Separator.

| | |
|---|---|
| **Round Trip** | *See:* Trip. |
| **RPM** | Revolutions per minute. |
| **Salt** | A class of similar compounds formed when the acid hydrogen of an acid is partially or wholly replaced by a metal or a metallic radical. Salts are formed by the action of acids on metals, or oxides and hydroxides, directly with ammonia, as well as by other methods. *See:* Sodium Chloride. |
| **Saltwater Drilling Fluid** | A water-based drilling fluid fluid whose external liquid phase contains sodium chloride or calcium chloride. |
| **Saltwater Mud** | *See:* Saltwater Drilling Fluid. |
| **Samples** | Cuttings obtained for geological information from the drilling fluid as it emerges from the hole. They are washed, dried, and their depth is recorded. |
| **Sand** | (1) Particle-size classification for solids larger than 74 microns. (2) A loose, granular material resulting from the disintegration of rocks with a high silica content. *See:* API Bulletin RP 13B, API Sand. |
| **Sand Content** | The solids particles retained on a U.S. Standard No. 200 test screen, expressed as the bulk percent by volume of the drilling fluid slurry sample. The opening in the test screen is 74 microns. The retained solids may be of any mineral or chemical composition and characteristic. For example, barite, shale, mica, silica, steel, or chert, larger than 74 microns are called API sand. *See:* API Sand. |
| **Sand Trap** | The first compartment and the only unstirred compartment in a well-designed drilling fluid system intended as a settling compartment. |
| **Scalping Shakers** | The first set of shale shakers after the flow line in a cascade shaker arrangement. These are usually circular or elliptical motion shakers with coarse mesh screens that are used to remove the bulk of large-diameter drilled solids or gumbo. This initial fluid preparation allow the second set of fine screen shale shakers in the series to operate more efficiently with less possibility of flooding. *See:* Fine Screen Shale Shakers, Flooding, Blinding. |
| **Screen** | *See:* Screen Cloth. |
| **Screen Cloth** | A type of screening surface, woven in square, rectangular, or slotted openings. *See:* Wire Cloth. |
| **Screen Support Rubbers** | Elastomers that cushion the contact between screens and shale shaker frames. |
| **Screen Underflow** | The discharge stream from a screening device that contains a greater percentage of liquids than does the feed. *See:* Liquid Discharge. |
| **Screening** | A mechanical process resulting in a division of particles on the basis of size by their acceptance or rejection by a screening surface. |
| **Screening Surface** | The medium containing the openings for passage of undersize material. |
| **Scroll** | *See:* Flute. |

| | |
|---|---|
| **Self-Lubricating** | Units that provide their own means of lubrication. |
| **Separation Potential** | The size distribution of equivalent spherical volumes calculated by determining the equivalent ellipsoidal volumes of at least 1,500 openings in a screen as determined by image analysis. Also called the "cutt" point distribution. *See:* Cutt Point. |
| **Settling Velocity** | The velocity a particle achieves in a given fluid when gravity forces equal friction forces of the moving particle (i.e., when the particle achieves its maximum velocity). |
| **Shale** | Stone of widely varying hardness, color, and compaction that is formed of clay-sized grains (less than two microns). *See:* Natural Clay. |
| **Shale Shaker** | Any of several mechanical devices for removing cuttings and other large solids from drilling fluid. Common examples are the vibrating screen, rotating cylindrical screen, and the like. |
| **Sharpness Of Cut or "Separation"** | The slope of a straight line drawn between the solids separated at the 84% point and the 16% point on a graph of the percent solids separated versus particle size. The more vertical the slope, the sharper the cut. |
| **Shear Rate** | The change of velocity with respect to the distance perpendicular to the velocity changes. *See:* Rate of Shear. |
| **Shear Stress** | The force per unit of an area parallel to the force that tends to slide one surface past another. *See:* Viscosity, Bingham Model, Plastic Viscosity, Yield Point, Gel Strength. |
| **Shear Thinning** | Opposite of dilatant. The apparent viscosity decreases instantaneously with increasing shear rate. *See:* Apparent Viscosity, Viscosity, Bingham Model, Plastic Viscosity, Yield Point, Gel Strength. |
| **Short-Circuiting** | A hydraulic condition existing in parts of the tank basin, reservoir, or hydrocyclone where the time of travel of liquid/solids is less than the normal flow-through time. For instance, if the surface tanks contain very viscous fluid but the returns from the flow line have a very low viscosity; the flow line returns might tend to channel across the top of the surface system toward the pump suction. In this case, the flow line returns would be "short-circuiting" or bypassing the solids separation equipment. In hydroclones, separable solids that pass directly from the feed inlet to the vortex finder, without passing through the cone section, have "short-circuited" the hydrocyclone processing system. |
| **Shute** | In a woven cloth, the direction of the wires running perpendicular to the loom or running across the roll of cloth. In wire cloth production, these are the short or transverse wires. *See:* Weft. |
| **Sieve** | *See:* Testing Sieve. |
| **Sieve Analysis** | The mass classification of solid particles passing through, or retained on, a sequence of screens of increasing mesh count. Analysis may be by wet or dry methods. *See:* Partical Size Distribution. |
| **Silt** | Materials whose particle size generally falls between 2 and 74 microns. A certain portion of dispersed clays and barite falls into this particle size range, as well as drilled solids. |

| | |
|---|---|
| **Size Distribution** | *See:* Particle Size Distribution. |
| **Slip** | The difference between synchronous speed and operating speed compared to synchronous speed, expressed as a percentage. If expressed in RPM, slip is the difference between synchronous speed and operating speed. |
| **Sloughed Solid** | A solid entering the wellbore from the exposed formation; not a drilled solid. |
| **Sloughing** | A situation in which portions of a formation fall away from the walls of a hole as a result of incompetent unconsolidated formations, tectonic stresses, high angle of repose, wetting along internal bedding planes, or swelling of formations. *See:* Caving, Cuttings, Heaving. |
| **Slug** | A small volume of weighted fluid pumped into the drill string to keep the drilling fluid liquid level below the rig floor while pulling drill pipe during a trip. This prevents drilling fluid from spilling on the rig floor as the pipe is pulled. *See:* Pill. |
| **Slug Tank** | A small compartment (normally adjacent to the suction compartment) used to mix special fluids to pump downhole. The most common use is to prepare a slug or a small volume of weighted mud before a trip. *See:* Pill Tank. |
| **Slurry** | A mixture or suspension of solid particles in one or more liquids. |
| **Sodium Bicarbonate** | $NaHCO_3$. A material used extensively for treating cement contamination and, occasionally, other calcium contamination of drilling fluids. It is the half neutralized salt of carbonic acid. *See:* Bicarb. |
| **Sodium Carboxymethylcelluose** | An organic polymer, available in various grades of purity, used to control filtration, suspend weight material, and build low-shear-rate viscosity in drilling fluids. It can be used in conjunction with bentonite where low-solids drilling fluids (muds) are desired. *See:* CMC, Low-Solids Drilling Fluids. |
| **Sodium Chloride** | $NaCl$. Commonly known as salt. Salt may be present in the drilling fluid as a contaminant or may be added purposely for inhibition. *See:* Salt. |
| **Sodium Chromate** | $Na_2CRO_4$. *See:* Chromate. |
| **Sodium Hydroxide** | $NaOH$. Commonly referred to as "caustic" or "caustic soda." A chemical used primarily to raise the pH. |
| **Sodium Polyacrylate** | A synthetic, high-molecular weight polymer of acrylo-nitrile used primarily for fluid loss control. |
| **Sodium Silicate Drilling Fluids** | Special class of inhibited chemical drilling fluid using sodium silicate, salt water, and clay. |
| **Solid** | A firm substance that holds its form; not gaseous or liquid. |
| **Solids** | All particles of matter in the drilling fluid (i.e., drilled formation cuttings, barite, bentonite, etc.). |
| **Solids Content** | The total amount of solids in a drilling fluid. This is usually determined by distillation that measures the volume fraction of both the dissolved and the suspended, or undissolved, solids. The suspended solids content |

may be a combination of high and low specific gravity solids and native or commercial solids. Examples of dissolved solids are the soluble salts of sodium, calcium and magnesium. Suspended solids make up the wall cake; dissolved solids remain in the filtrate. The total suspended and dissolved solids contents are commonly expressed as percent by volume and, less commonly, as percent by weight. *See:* Retort.

**Solids Discharge**  The stream from a liquid-solids separator containing a higher percentage of solids than does the feed.

**Solids Discharge Capacity**  The maximum rate at which a liquid-solids separation device can discharge solids without overloading.

**Solids Separation Equipment**  Any and all of the devices used to remove solids from liquids in drilling (i.e., shale shaker, desander, desilter, mud cleaner, and centrifuge).

**Solubility**  The degree to which a substance will dissolve in a specific solvent.

**Solute**  A substance that is dissolved in another (the solvent).

**Solution**  A mixture of two or more components that form a homogeneous single phase. An example of a solution is salt dissolved in water.

**Solvent**  Liquid used to dissolve a substance (the solute).

**Souring**  A term commonly used to describe fermentation.

**Specific Gravity**  The weight of a specific volume of a liquid, solid, or slurry in reference to the weight of an equal volume of water at a reference temperature of 3.89°C (water has a density of 1.0 gm/cc at this temperature).

**Specific Heat Capacity**  The number of calories required to raise one gram of a substance 1°C.

**Spray Bar**  A pipe located over the bed of a shale shaker through which dilution fluid is sprayed onto the screen surface during separation of the drilled solids. In practice, spray bars may supply a mist or small amount of liquid—not a hard spray—to prevent washing fine solids through the screen panels and back into the circulating system.

**Spray Discharge**  *See:* Spray Underflow.

**Spray Underflow**  The characteristic underflow of certain balanced hydrocyclones discharging to the atmosphere and not overloaded with separable solids.

**Spud Mud**  The drilling fluid used when drilling starts at the surface, often a thick bentonite-lime slurry.

**Spudding In**  The initiating of the drilling operations in the first top hole section of a new well.

**Spurt Loss**  The flux of fluids and solids that occurs in the initial stages of any filtration before pore openings are bridged and a filter cake is formed. *See:* Surge Loss.

**Square Mesh**  Screen cloth with the same mesh count in both directions.

**Square Weave**  *See:* Square Mesh.

| | |
|---|---|
| **Squeeze** | A procedure whereby slurries of cement, drilling fluid, gunk plug, and so forth, are forced into the formation by pumping into the hole while maintaining a backpressure. This is usually achieved by closing the blowout preventers or by using a retrievable downhole packer. |
| **Squirrel Cage Motor** | An induction motor that gets its name from the rotor assembly that resembles a squirrel cage. The cage consists of rotor bars secured at each end to the shorting rings. An induction motor is one in which there is no physical electrical connection to the rotor. Current in the rotor is induced by the magnetic field of the stator. |
| **Stability Meter** | An instrument to measure the breakdown voltage of oil-based drilling fluids. This gives an indication of the emulsion stability. |
| **Stacking a Rig** | Storing a drilling rig once a job is completed and the rig is to be withdrawn from service for a period of time. |
| **Starch** | A group of carbohydrates occurring in plant cells. Starch is specially processed (pre-gelatinized) for use in drilling fluids to reduce the filtration rate and, occasionally, to increase the viscosity. Without the proper preservative, starch can ferment. |
| **Static** | Not moving, or at rest. Opposite of dynamic. *See:* Quiescence. |
| **Stearate** | Salt of stearic acid, which is a saturated, 18-carbon fatty acid. Certain compounds, such as aluminum stearate, calcium stearate, and zinc stearate, have been used in drilling fluids for one or more purposes such as a defoamer, lubrication, air drilling in which a small amount of water is encountered, and so forth. |
| **Stirrer** | *See:* Agitator, Mechanical Agitator. |
| **Stokes Law** | Stokes Law states that the terminal settling velocity of a spherical particle is proportional to the square of the particle diameter, the acceleration of gravity, and the density difference between the density of the particle and the density of the liquid medium; and the terminal settling velocity is inversely proportional to the viscosity of the liquid medium. |

$$V_T = [gD_P^2(\rho_S - \rho_L)(10^{-6})] / 116\mu$$

where $V_T$ = Terminal settling velocity, in inches/second
$D_P$ = Particle diameter, in microns
$\rho_S$ = Density of the solids, in grams/cm³
$\rho_L$ = Density of the liquid, in grams/cm³
$\mu$ = Viscosity of the feed slurry, in centipoise

| | |
|---|---|
| **Stormer Viscometer** | A rotational shear viscometer used for measuring the viscosity and gel strength of drilling fluids. This instrument has been largely replaced by the direct indicating viscometer. |
| **Streaming Potential** | The electrokinetic portion of the spontaneous potential (SP) electric-log curve, which can be significantly influenced by the characteristics of the filtrate and filter cake of the drilling fluid. |
| **Streamline Flow** | *See:* Laminar Flow. |

**Stroke**  
The distance between the extremities of motion or total displacement normal to the screen (i.e., the diameter of a circular motion or twice the amplitude). *See:* Amplitude.

**Stuck**  
A condition whereby the drill pipe, casing, or any other device inserted into the wellbore inadvertently becomes lodged in the hole. Sticking may occur while drilling is in progress, while casing is being run in the hole, or while the drill pipe is being tripped.

**Stuck Pipe**  
*See:* Differential Pressure Sticking, Stuck.

**Suction Compartment**  
(1) The area of the check/suction section from which drilling fluid is picked up by the suction of the mud pumps. (2) Any compartment from which a pump moves fluids.

**Sump**  
(1) A disposal compartment or earthen pit for holding discarded liquids and solids. (2)The pan or compartment below the lowest shale shaker screen.

**Super-Saturation**  
If a solution contains a higher concentration of a solute in a solvent than would normally correspond to its solubility at a given temperature, a state of super-saturation exists. This is an unstable condition because the excess solute separates when the solution is seeded by introducing a crystal of the solute. The term "super-saturation" is frequently used erroneously for hot salt drilling fluids.

**Support Screen**  
A heavy, wire mesh—either plain or calendered—that supports a finer mesh(s) screen for use in filtering or screen separation. *See:* Back-up Screen.

**Surface Active Materials**  
*See:* Surfactant.

**Surface Tension**  
Generally, the cohesive forces acting on surface molecules at the interface between a liquid and its own vapor. This force appears as a tensile force per unit length along the interface surface and is usually expressed in units of dynes per centimeter. Since the surface tension is between the liquid and air, values measured against air are commonly referred to as "surface tension," and measurements at an interface between two liquids or a liquid and a solid are termed "interfacial tension." *See:* Interfacial Tension, Emulsion.

**Surfactant**  
Material that tends to concentrate at an interface of an emulsion or a solid liquid. Used in drilling fluids to control the degree of emulsification, aggregation, dispersion, interfacial tension, foaming, defoaming, wetting and the like.

**Surfactant Drilling Fluid**  
A drilling fluid that contains a surfactant. Usually refers to a drilling fluid containing surfactant material to effect control over the degree of aggregation and dispersion or emulsification.

**Surge**  
The pressure increase in a wellbore caused by lowering tubulars. Viscous drilling fluid flowing up the annulus, displaced by drill pipe, tubing, or casing, creates the pressure surge.

**Surge Loss**  
This is a colloquial term used to describe a spurt of filtrate and solids, which occurs in the initial stages of any filtration before pore openings are bridged and a filter cake is formed. The preferred term is "spurt loss." *See:* Spurt Loss.

| | |
|---|---|
| **Suspensoid** | A mixture that consists of finely divided colloidal particles floating in a liquid. The particles are kept in motion by the moving molecules of the liquid (Brownian movement) and, therefore, do not settle. |
| **Swabbing** | When pipe is withdrawn from the hole in a viscous drilling fluid or if the bit is balled, a decrease in pressure in the wellbore can cause formation fluid to flow into the well. |
| **Swelling** | *See:* Hydration. |
| **Synergism** | Term describing the effect obtained when two or more products are used simultaneously to obtain a certain result. Rather than the result of each product being additive to the other, the result is a multiple of the effects. |
| **Synergistic Properties** | *See:* Synergism. |
| **Tannic Acid** | Tannic acid is the active ingredient of quebracho and other quebracho substitutes such as mangrove bark. chestnut extract, hemlock, and so forth. |
| **Temperature Survey** | An operation to determine temperatures at various depths in the wellbore. This survey is used to find the location of inflows of water into the borehole or where proper cementing of the casing has occurred. |
| **Ten Minute Gel** | *See:* Gel Strength. |
| **Tensile Bolting Cloth** | A group of industrial wire cloth specifications woven of extremely smooth and durable stainless steel in a square mesh pattern. The wire diameter is lighter than mill grade cloth producing a higher percentage of open area. *See:* Market Grade Cloth, Mill Grade Cloth, Ultra Fine Wire Cloth and Calendered. |
| **Tensioning** | The stretching of a screening surface of a shale shaker, within the vibrating frame, to the proper tension. |
| **Testing Sieve** | A cylindrical or tray-like container with a screening surface bottom of standardized apertures. *See:* Sieve. |
| **Thermal Decomposition** | Chemical breakdown of a compound or substance by temperature into simple substances or into its constituent elements. For example, starch thermally decomposes in drilling fluids as the temperature approaches 300°F. |
| **Thinner** | Any of the various organic agents (eg. tannins, lignins,lignosulfonates, etc.) and inorganic agents (pyrophosphates, tetraphosphates, etc.) that are added to a water-based drilling fluid to reduce the low-shear-rate viscosity and/or thixotropic properties by deflocculation. |
| **Thixotropy** | The ability of a fluid to develop gel strength with time. That property of a fluid at rest that causes it to build-up a rigid or semi-rigid gel structure if allowed to remain at rest. The fluid can be returned to a liquid state by mechanical agitation. This change is reversible. *See:* Gel Strength. |
| **Thrust** | A force that pushes; for example as solids experience a thrust on a shale shaker screen. |
| **Tighten Up Emulsion or Mud** | Jargon describing the condition in oil-based drilling fluids where either chemicals or shear, or both, are used to emulsify water in oil into smaller droplets to prevent the emulsion from breaking, or coming apart. |

| | |
|---|---|
| **Titration** | The process of using a standard solution to determine of the amount of some substance in another solution. The known solution is usually added in a definite quantity to the unknown solution until a reaction is complete. |
| **Tool Joint** | A drill-pipe coupler consisting of a threaded pin and a box of various designs and sizes. |
| **Torque** | (1) The turning effort caused by a force acting normal to the radius at a specified distance from the axis of rotation. Torque is expressed in pound-feet (pounds at a radius of one foot). Torque, lb-ft = Force, lbs × lever arm, ft. (2) Drill string connections require a specific torque to be properly tightened. The drill string in a borehole experiences a frictional force as it is rotated, which causes a torque in the drill string. Torque reduction can usually be accomplished by the addition of various drilling fluid additives. |
| **Total Depth (TD)** | The greatest depth reached by the drill bit in a particular well. |
| **Total Dilution** | The volume of drilling fluid that is built to maintain a specified fraction of drilled solids over a specified interval of footage if no solids are removed from the system. |
| **Total Hardness** | *See:* Hardness (Water). |
| **Total Head** | The sum of all the heads within a system (Total Head = velocity head + pressure head + elevation head). |
| **Total Non-Blanked Area** | The net unblocked area, in square feet, that will permit the passage of fluid through a screen. Some screen designs can eliminate as much as 40% of the gross screen panel area from fluid flow due to backing plate and bonding material blockage. |
| **Tour** | A person's turn in an orderly schedule. The word, which designates the shift of a drilling crew, is pronounced as if it were spelled t-o-w-e-r. |
| **Trenchless Drilling** | Excavating material near the surface of tunnels, cables, pipelines, and so forth, by drilling instead of digging ditches. |
| **Trip** | The process of pulling the drill string from the hole and running it back to the bottom again. One way (either in or out) is referred to as a "half-trip." *See:* Round Trip. |
| **Trip Tank** | A gauged and calibrated vessel used to account for fill and displacement volumes as pipe is pulled from and run into the hole. Close observation allows early detection of formation fluid entering the wellbore and of drilling fluid loss to a formation. |
| **Turbidity** | A condition in a clear fluid that causes a lack of clarity caused by the presence of suspended matter resulting in the scattering and absorption of light rays. |
| **Turbulent Flow** | Fluid flow in which the velocity at a given point that constantly changes in magnitude and flow direction; pursues erratic and continually varying courses. *See:* Critical Velocity, Reynolds Number. |
| **Twist-Off** | The severing or failure of a drill pipe joint caused by excessive torque. |

| | |
|---|---|
| **Ultra Fine Wire Cloth** | A group of industrial wire cloth specifications with lighter than normal wire. The wire diameter of this grade produces the highest percentage of open area of all other grades for any specific mesh size. This cloth is used in multiple-layer screens. *See:* Market Grade Cloth, Mill Grade Cloth, Tensile Bolting Cloth, Calendered. |
| **Ultra Fine Solids** | Particles whose diameter is between 2–44 microns. |
| Ultraviolet Light | Light waves shorter than the visible blue and violet waves of the spectrum. Crude oil, colored distillates, residium, a few drilling-fluid additives, and certain minerals and chemicals fluoresce in the presence of ultraviolet light. These substances, when present in drilling fluid, may cause the drilling fluid to fluoresce. |
| **Unbalanced Elliptical Motion** | An elliptical motion of a shale shaker screen such that the ellipse axes at the feed end are tilted toward the discharge end of the screen, and the ellipse axes at the discharge end are tilted toward the feed end. Usually, these screens are tilted downward to assist solids removal from the end of the screen. The vibrator is usually located above the center of gravity of the shaker basket. |
| **Underflow** | (1) Centrifugal Separators: The discharge stream that contains a higher percentage of solids than does the feed. (2) Screen Separators: The discharge stream that contains a lower percentage of solids than does the feed. |
| **Underflow Header** | A pipe, tube, or conduit into which two or more hydrocyclones discharge their underflow. |
| **Underflow Opening** | *See:* Apex, Apex Valve. |
| **Undersize Solids Particles** | (1) Particles, in a given situation, that will pass through the mesh of the screen in use. (2) Particles, in a given situation, that will remain with the liquid phase when subjected to centrifugal force. |
| **Underslung** | Field terminology denoting that the support ribs for the shaker screen are located above the screen surface. *See:* Overslung. |
| **Unoccluded Area** | Unobstructed area of a screen opening. |
| **Unweighted Drilling Fluid** | A drilling fluid that does not contain commercial suspended solids added for the purpose of increasing the drilling fluid density. |
| **V.G. Meter** | *See:* Direct Indicating Viscometer. |
| **VAC** | Alternating Current Voltage. |
| **Valence** | A number representing the combining power of an atom (i.e., the number of electrons lost, gained, or shared by an atom in a compound). It is also a measure of the number of hydrogen atoms with which an atom will combine or replace (e.g., an oxygen atom combines with two hydrogens, hence has a valence of 2). Thus, there are mono-, di-, tri-, and so forth, valent ions. |
| **Valence Effect** | Generally, the higher the valence of an ion, the greater the loss of stability to emulsions, colloidal suspensions, and the like, these polyvalent ions will impart. |

**Velocity**
Time rate of motion in a given direction and sense. It is used as a measure of the fluid flow, and may be expressed in terms of linear velocity, mass velocity, volumetric velocity, and so forth. Velocity is one of the factors that contributes to the carrying capacity of a drilling fluid.

**Velocity Head**
Head (relating to pressure when divided by the density of the fluid) created by movement of a fluid, equal to an equivalent height of static fluid.

**Venturi**
Streamlining up to given pipe size following a restriction (as in a jet in a mud hopper) to minimize turbulence and pressure drop.

**Vibrating Screen**
A screen with motion induced as an aid to solids separation. *See:* Shale Shaker.

**Vibration Isolaters**
Elastomers, ranging from solid to air-pressured, or springs that allow the shale shaker screens to vibrate but do not transmit the vibratory motion to the rest of the machine.

**Vibrators**
Weights rotated about an axis that does not pass through the center of mass.

**Viscometer**
An apparatus to determine the viscosity of a fluid or suspension. Viscometers vary considerably in design and testing methods.

**Viscosifiers**
Material added to a drilling fluid to increase the low-shear-rate viscosity.

**Viscosity**
The ratio of shear stress to shear rate is defined as the viscosity of a fluid. If the shear stress is measured in dynes/sq. cm and the shear rate in reciprocal seconds, the ratio is the viscosity in poise. Viscosity may be viewed as the internal resistance offered by a fluid to flow. This phenomenon is attributable to the attractions between molecules of a liquid, and is a measure of the combined effects of adhesion and cohesion to the effects of suspended particles, and to the liquid environment. The greater this resistance, the greater the viscosity. (2) A characteristic property of a fluid, liquid, or slurry crudely defined as resistance to flow (by accurate definition the ratio of shear-stress to shear-rate). *See:* Apparent Viscosity, Plastic Viscosity, API Bulletin RP-13B.

**Viscosity-Gravity (V.G.) Meter**
The name more commonly used for a direct indicating viscometer. *See:* Viscometer.

**Viscous Flow**
*See:* Laminar Flow.

**Volatile Matter**
Normally gaseous products given off by a substance, such as gas breaking out of live crude oil that has been added to a drilling fluid. In distillation of drilling fluids, the volatile matter is the water, oil, gas, and so forth, that is vaporized, leaving behind the total solids, which can consist of both dissolved and suspended solids.

**Volt**
The unit of electrical "pressure" or electromotive force. One volt produces a current flow of one ampere through a resistance of one ohm.

**Volume Percent**
The number of volumetric parts of the total volume. Volume percent is the most common method of reporting solids, oil, and water contents of drilling fluids. *See:* Weight Percent, Ppm.

**Vortex**
A cylindrical or conical shaped core of air or vapor lying along the central axis of the rotating slurry inside a hydrocyclone.

| | |
|---|---|
| **Vortex Finder** | A cylinder extending into the upper end of a hydrocyclone, which causes drilling fluid to move in a circular spiral direction within the cone and prevents the entering fluid from short-circuiting directly to the hydroclone overflow. |
| **Wall Cake** | The solid material deposited along the wall of the hole resulting from filtration of the drilling fluid. *See:* Cake Thickness, Filter Cake. |
| **Wall Sticking** | *See:* Differential Pressure Sticking. |
| **Warp** | In a woven cloth, the direction of the wires running parallel with the loom or running the length of a roll of cloth. In wire cloth production, these are the long or longitudinal wires. *See:* Loom. |
| **Water-Based Drilling Fluid** | Common, conventional drilling fluid. Water is the suspending medium for solids and is the continuous phase, whether or not oil is present. *See:* Water Loss, Filtration. |
| **Water-Based Mud** | *See:* Water-Based Drilling Fluid. |
| **Water Block** | A reduction in the permeability of the formation caused by the invasion of water into the pores (capillaries). The decrease in permeability can result from the swelling of clays, thereby shutting off the pores, or in some cases by a capillary block of wetted pores due to surface tension phenomena. |
| **Water Feed** | Water added to a centrifugal separator for the purpose of diluting the mud feed. *See:* Dilution Water. |
| **Water Loss** | *See:* Filtration, Fluid Loss. |
| **Water Wet** | Not oil wet. A surface on which water easily spreads. If the contact angle of a water droplet on a surface is less than 90°, the surface is considered water wet. *See:* Hydrophilic, Oil Wet. |
| **Water-In-Oil Emulsion** | *See:* Invert Oil-Emulsion Drilling Fluid. |
| **Weft** | *See:* Shute. |
| **Weight** | In drilling fluid terminology, this refers to the density of a drilling fluid. This is normally expressed in either lb/gal, lb/cu ft, psi hydrostatic pressure per 1,000 ft of depth, or specific gravity related to water. *See:* Density. |
| **Weight Material** | Any of the high specific gravity materials used to increase the density of drilling fluids. This material is most commonly barite or hematite and in special applications, limestone. |
| **Weight Percent** | The number of weighted parts of the total weight. Weight percent is the most common method of reporting oil in solids discharges and mass balance calculations. *See:* Volume Percent, Ppm. |
| **Weight-Up** | To increase the weight of a drilling fluid, usually by the addition of weight material. |
| **Weighted Drilling Fluid** | A drilling fluid to which commercial solids have been added to increase the slurry weight. |
| **Weighted Mud** | *See:* Weighted Drilling Fluid. |

| | |
|---|---|
| **Well Logging** | *See:* Electric Logging, Mud Logging. |
| **Wetting** | The adhesion of a liquid to the surface of a solid. |
| **Wetting Agent** | A substance which, when added to a liquid, increases the spreading of the liquid on a surface or the penetration of the liquid into a material. |
| **Whipstock** | A device inserted in a wellbore that causes the drill bit to exit the established path of the existing wellbore. The whipstock is the tool used for the initiation of directional drilling. |
| **Wildcat** | A well in unproved territory. |
| **Windage Loss** | (1) The resisting power of air, or air friction, acting against a rapidly rotating armature or cooling fan to create a power loss. (2) The resisting power of air, or air friction, against the rotating bowl of a centrifuge. |
| **Wire Cloth** | Screen cloth of woven wire. *See:* Screen Cloth. |
| **Working Pressure (WP)** | The maximum pressure to which equipment should be exposed in order to comply with the manufacturer's warranty and to be within industry codes and safety standards. |
| **Workover Fluid** | Any type of fluid used in the work-over operation of a well. |
| **Yield** | A term used to define the quality of a clay by describing the number of barrels of a given viscosity (usually 15 cp) slurry that can be made from a ton of the clay. Based on the yield, clays are classified as bentonite, high-yield, low-yield, and so forth. Not related to yield point as described below. *See:* API Bulletin RP 13B. |
| **Yield Point** | (1) A term derived from a direct-reading viscometer (Fann V.G. or equivalent) based on subtracting the plastic viscosity from the 300 RPM reading. (2) An extrapolated shear stress at zero shear rate created by assuming a linear relationship between shear stress and shear rate and determining the intercept on the shear stress axis. The linear relationship between shear stress and shear rate which results in a yield point is called a Bingham Plastic Model. *See:* Viscosity, Plastic Viscosity, API Bulletin RP-13B. |
| **Yield Value** | *See:* Yield Point. |
| **Zero-Zero Gels** | A condition wherein the drilling fluid fails to form measurable gels during a quiescent time interval (usually 10 minutes). The measurements of gel are made with a direct-reading viscometer at intervals of 10 seconds and 10 minutes. *See:* Progressive Gel, Flat Gel. |
| **Zeta Potential** | The electro-kinetic potential of a particle as determined by its electrophoretic mobility. This electric potential causes colloidal particles to repel each other and remain suspended. |
| **Zinc Bromide** | $ZnBr_2$. A very soluble salt used to increase the density of water or brine to more than double that of water. Normally added to calcium chloride/calcium bromide mixed brines. |

# Index

## A
Acidity, drilling fluid, 217–218
Additions section equipment, 140, 226
Auto-ignition temperature, 211

## B
Back tanks, 93
Baffles, 228
Balanced elliptical motion shakers, 96, 101–102
   defined, 97–98
Barite
   calculating discard, 176–177
   development of, 3
   *See also* Weighting agents
Baskets, 93
Bonded screens, 137–138

## C
Cascade system, 111–113
Cavitation, 195
Centrifugal pumps, 179–181, 198
   corrosion resistance, 195
   equipment guidelines, 232–234
   and head, 179, 181
   and head losses, 187–194
   horsepower, 183–187
   installing, 195
   operating, 199–200
   parts of, 196–197
   performance curves, 184–194
   and suction, 181–184
   and vibration, 197–198
   and viscous liquids, 194–195
Centrifuges, 155–162
   applications, 160, 161–162
   decanting, 157–158
   and discharge dryness, 160
   operating, 162
   and pond depth, 160
   rotary mud separators, 158–160
   and Stokes' law, 155–157
   two-stage operations, 161

Chemically enhanced centrifugation (CEC), 216
   *See also* Dewatering units
Circular motion shakers, 99
   advantages, 96
   deck design, 103
   defined, 97
   vibrating system, 102
Closed loop system (CLS), 113, 161–162, 216
   *See also* Dewatering units
Coagulation, 219–220
   terms, 223
   *See also* Flocculation
Collection pans, 93
Colloids, 223
Continuous duty electric motors, 201
   *See also* Motors
Continuous-cloth screens, 131–132
Costs
   drilling fluid disposal, 216
   and drilling fluids, 89–90
   drilling solids removal, 109–110
   for pre-well checklist, 236
Cut points, 125–127
   calculating, 127, 170–172
   defined, 125
   median, 135, 146, 148–149
   *See also* Separation potential
Cuttings, defined, 92
   *See also* Drilled solids

## D
$D_{50}$ cut point, 146
   *See also* Cut points
Decanting centrifuges, 157–158
Decks, 93, 103
Degassers, 143, 162
   equipment guidelines, 229, 230
Desanders, 147–148, 152
   development of, 3
   equipment guidelines, 227
Desilters, 148, 152

development of, 3
equipment guidelines, 227
Dewatering units, 216–223
   described, 162, 216–218
   development of, 4
   and mud systems, 216–218
   and screen size, 110–111
   *See also* Flocculation; Waste management
Diaphragm pumps, 178–179
Differential sticking, 88
Dilution, 164–169
   defined, 87, 164
   overview, 164–165
Discharge lines, 141
Drilled solids, 87
   calculating concentrations of, 173–177
   density of, 173–174
   disintegration of, 88, 96
   and shale shaker capacity, 95–96
   as weighting agents, 89
Drilling fluids
   acidity, 217–218
   carrying capacity, 88–89
   conserving, 92
   development of, 1
   and shale shaker capacity, 95–96
   troubleshooting, 238
Dryers, 4, 113–114
   *See also* Dewatering units

## E
Electric motors. *See* Motors
Electrical power. *See* Power system
Elliptical motion shale shakers, 96, 97
   balanced, 101–102
   unbalanced, 98–99
Emulsion polymers, 220–223
   inversion, 221–222
   stability, 222–223
   types of, 221

Emulsion polymers *(continued)*
  See also Polymers
Environmental issues. See Waste management
Equalizers, 141–142, 162
  equipment guidelines, 227
Equipment, 139–163
  arrangement of, 226, 229
  guidelines, 226–234
  list of, 87
Explosion protection, 210

**F**

Fiber-optic cable drilling, 114
Filter cakes, 87–88
Flashpoint, 210
Flocculation, 217–220
  and coagulation, 219–220
  development of, 4
Flow rates
  calculating, 110
  and centrifugal pumps, 184–187
  estimating, 91
  and screen conductance, 130
Flow-back trays, 99
Fluid routing equipment, 141, 229

**G**

"G" factor, 103–105
Generators, 105–106
Gumbo, 95, 143
Gumbo chain, 6
Gumbo slide, 109

**H**

Hazardous locations, 210–213
Head, 181–184
  calculating, 184–187
  defined, 179, 181
  losses, 187–194
High-shear devices, 231–232
Hook strip screens, 135–136, 137
Hoppers, 231
Hydrocyclones, 143–154
  capacity, 147
  and cut points, 146, 148–149
  desanders and desilters, 3, 147–148, 152
  and discharge, 146–147
  installing, 141, 152
  operating, 151–153
  overview, 143–144, 152–154
  and particle size, 145
  and Stokes' law, 149–151
  tanks, 147
  vortex finder, 144–145

**I**

Ignitable mixture, 210–211
Ignition temperature, 211

Impeller. See Centrifugal pump
Installing, 142
  centrifugal pumps, 195
  hydrocyclones, 152
  motors, 202
  mud cleaners, 141
  pumps, 162–163, 195
  screens, 116, 128
  shale shakers, 115–116

**J**

Jet hoppers, 231
Jet pumps, 179

**K**

Kindling temperature, 210

**L**

Layered screens, 120-121, 135–138
Linear motion dryers, 113
Linear motion shakers, 97–98
  advantages, 93, 96
  deck design, 103
  vibrating system, 102
Liquid salvage. See Dewatering units; Dryers; Waste management

**M**

Median cut points, 135, 146, 148–149
Mesh count, 121–125
Metal screens, 136
Micro-tunneling, 114
Molded screens, 136
Motors, 106, 201–215
  for centrifugal pumps, 215
  for centrifuges, 214–215
  enclosures, 204–206
  frames, 206–207
  and heat, 201, 202
  installing, 202
  safety issues, 115–116, 206, 208, 210–213
  selecting, 201
  for shale shakers, 214
  speed-torque characteristics, 203
  standards and nomenclature, 204, 207–208, 212–213
  troubleshooting, 202–204, 237
  types of, 213–214
  and voltage, 106–107, 201–202, 209
Mud box, 93
Mud cleaners, 154–155
  development of, 3–6
  installing, 141
Mud conditioner, 4
Mud guns, 228, 231

Mud systems
  and centrifuges, 161
  development of, 1
  and dewatering units, 216–218
  equipment guidelines, 230, 231
Multiple-deck shale shakers, 111–113

**N**

Net positive suction head (NPSH), 182–183
  See also Head

**O**

Oil-based muds
  and centrifuges, 161
  dewatering, 218

**P**

Partially-hydrolyzed polyacrylamide (PHPA), 95
Partition curve. See Separation potential
Percent separated curve. See Separation potential
Pipes
  and fluid routing, 141
  and head loss, 187–194
Plastic viscosity (PV). See Viscosity
Platforms (screens), 136
Polyelectrolytes, 218
  See also Polymers
Polymer screens, 136
Polymers, 218–223
  coagulants, 219–220
  defined, 218
  drilling fluids, 89–90
  emulsions, 219, 220–223
  flocculants, 220
  and fluid capacity, 95
  and screening problems, 91
  types of, 218–219
Possum belly, 93
Power system, 105–108
  See also Motors
Pressure. See Head
Pressure gauges, 151, 233
Pre-well checklist, 235–236
Progressive cavity pumps, 179
Propagation of flame, 211
Pumps
  centrifugal, 179–200
  installing, 162–163
  troubleshooting, 151–152
  types of, 178–179
  See also Centrifugal pumps
PV (plastic viscosity). See Viscosity
Pycnometer, 173

## R

Reciprocating pumps, 178
Removal section, 141
    arranging, 162–163
    compartments, 226, 227
River and road crossing drilling, 114
Rope discharge, 146–147
Rotary mud separators, 158–160
Rotary pumps, 178

## S

Sand traps, 143, 226-227, 228–229
Saybolt viscometers, 194
Scalper shakers, 111
Scalping shakers, 99, 109
Screens, 120–138
    blinding, 95-96, 135–136
    and cascade systems, 113
    and circular motion shakers, 99
    classifying, 121–125, 128–129
    cloth weaves, 120–121
    construction materials, 136
    and cut points, 125–127
    development of, 1
    and dewatering, 110–111
    and elliptical motion shakers, 98–99
    and flow rate, 110, 130–131
    grades, 123–124
    installing, 116
    layered and three-dimensional, 120–121, 135–138
    and linear motion shakers, 100
    maintenance, 116–117
    mesh count, 121–125
    opening size and open area, 122–125
    panels, 131–132, 136–138
    selecting, 91, 109
    separation potential, 129–135
    tensioning, 128
    troubleshooting, 128, 238–239
Separation potential (screens)
    conductance, 130–131
    percent separated curves, 131–134
    potential separation curve, 129–130
    summary, 134–135
Shale shakers
    balanced elliptical, 97–98, 101–102
    capacity, 94–96
    circular motion, 97, 99
    defined, 87
    design, 97–108
    development of, 1–12, 96
    and "G" factor, 103–105
    installing, 115–116
    linear motion, 97–98, 99–101
    manufacturers, 240–259
    need for, 224
    operating, 92–93, 116–117, 224–225
    parts of, 93–94
    and power systems, 105–108
    selecting and evaluating, 109–114, 117–119
    types of, 91, 96, 97
    unbalanced elliptical, 97–99
    and vibrating systems, 102–103
Shear pumps, 6
Shear rates
    and flow rate, 91
    and fluid capacity, 95
Skids, 93
Slug pits, 140
Slugging tanks, 231
Slurry
    and hydrocyclones, 151
    and rotary mud separators, 159
Solids control
    development of, 1–12
    equipment, 139–163
    need for, 87–90, 139
Spray bars, 116
Spray discharge, 146
Starch, 91, 109
Stokes' law
    centrifuges, 155–157
    hydrocyclones, 149–151
Submersible pumps, 179
Suction
    and centrifugal pumps, 181–184
    lines, 141, 233
Suction and testing section, 139–140
Sumpless system. See Dewatering units
Surface tanks, 142

## T

Tanks
    equipment guidelines, 226, 232
    hydrocyclone, 147
    slugging, 231
    surface, 142
    trip, 140, 232
Three-dimensional screens, 120–121, 138
Trenchless drilling, 114
Trip tanks, 140, 232
Tromp curve. See Separation potential
Troubleshooting, 237–239
    for crew, 225
    hydrocyclones, 153
    motors, 202–204, 237
    pre-well, 235–236
    pumps, 151–152, 199–200
    screens, 238–239
Type II pumps, 179

## U

Unbalanced elliptical motion shaker, 98–99
    advantages, 96
    defined, 97
    vibrating system, 102
Underflow, 145, 155
    See also Slurry

## V

Vacuum filtration, 218
Venturi tubes, 179, 231
Vibrating systems, 93, 102–103
Viscometers, 194–195
Viscosifiers, 155
Viscosity
    and drilling fluid, 92–93
    effect of, 88
    and fluid capacity, 95
    measuring, 188, 194–195
    and screen cut points, 127
Voltage, 106–107, 201–202, 209
    safety issues, 210–213

## W

Waste management
    costs, 90, 216
    dryers, 113–114
    pre-well checklist, 236
    See also Dewatering units
Weighting materials
    and centrifuges, 160
    and desanders, 148
    development of, 1
    and dilution, 164–165
    discarded, 90, 91, 176–177
    drilled solids as, 89
    and equipment sequence, 142
    and mud cleaners, 154–155
    and screen selection, 109
*World Oil's Composite Catalog*
    advertising from, 13–86
    chronology of devices, 6–12